AF478454

Probability Theory for Quantitative Scientists

Based on the long-running Probability Theory course at the Sapienza University of Rome, this book offers a fresh and in-depth approach to probability and statistics, while remaining intuitive and accessible in style. The fundamentals of probability theory are elegantly presented, supported by numerous examples and illustrations, and modern applications are later introduced, giving readers an appreciation of current research topics. The text covers distribution functions, statistical inference and data analysis, and more advanced methods, including Markov chains and Poisson processes, widely used in dynamical systems and data science research. The concluding chapter, "Entropy, Probability, and Statistical Mechanics", unites key concepts from the text with the authors' impressive research experience, to provide a clear illustration of these powerful statistical tools in action. Ideal for students and researchers in the quantitative sciences, this book provides an authoritative account of probability theory, written by leading researchers in the field.

Luca Leuzzi is CNR Research Director at the Institute of Nanotechnology (CNR-NANOTEC) in Italy. His research focuses on disordered systems and complex photonics. He has taught undergraduate courses in probability theory, computational physics, the theory of stochastic processes, statistical physics, and machine learning at Sapienza University of Rome.

Enzo Marinari is a Professor of Theoretical Physics at the Sapienza University of Rome, renowned for his contributions to statistical mechanics and condensed matter theory. Over decades, he has taught numerous courses at Sapienza, including probability theory, computational physics, and statistical mechanics.

Giorgio Parisi is Professor Emeritus of Theoretical Physics at Sapienza University of Rome, and is widely regarded as one of the leading theoretical physicists of our time. He has made seminal contributions to both statistical physics and condensed matter physics, earning him the Nobel Prize in Physics in 2021. Parisi was also awarded the prestigious Wolf Prize in 2021 for his ground-breaking discoveries in disordered systems, particle physics, and statistical physics.

Probability Theory
for Quantitative Scientists

LUCA LEUZZI

National Research Council of Italy

ENZO MARINARI

Sapienza Università di Roma

GIORGIO PARISI

Sapienza Università di Roma

Shaftesbury Road, Cambridge CB2 8EA, United Kingdom

One Liberty Plaza, 20th Floor, New York, NY 10006, USA

477 Williamstown Road, Port Melbourne, VIC 3207, Australia

314–321, 3rd Floor, Plot 3, Splendor Forum, Jasola District Centre,
New Delhi – 110025, India

103 Penang Road, #05–06/07, Visioncrest Commercial, Singapore 238467

Cambridge University Press is part of Cambridge University Press & Assessment,
a department of the University of Cambridge.

We share the University's mission to contribute to society through the pursuit of
education, learning and research at the highest international levels of excellence.

www.cambridge.org
Information on this title: www.cambridge.org/9781009580694

DOI: 10.1017/9781009580656

When citing this work, please include a reference to the DOI 10.1017/9781009580656

First published 2025

Printed in the United Kingdom by CPI Group Ltd, Croydon CR0 4YY

A catalogue record for this publication is available from the British Library

A Cataloging-in-Publication data record for this book is available from the Library of Congress

ISBN 978-1-009-58069-4 Hardback

Cambridge University Press & Assessment has no responsibility for the persistence
or accuracy of URLs for external or third-party internet websites referred to in this
publication and does not guarantee that any content on such websites is, or will
remain, accurate or appropriate.

Contents

Preface

This book, which for many years has affectionately been for us "Il Trattatello" (The Treatise) – we thank Giovanni Gallavotti for this inspiration – was born and grew with the Probability Theory course that we taught to physics students, through different ordinations – a relay race so long that one of the students from the beginning (L.L.) made it in time to become the bearer of the baton.

Starting with a dry presentation of the foundations of modern probability theory, to which we devote the first chapter, and an overview of probability distributions and their properties, in Chapter 2, we intend to provide readers with the fundamental concepts, computational methods, and their applications to the study of probabilistic problems – not always in that order.

We first delve into the study of the law of large numbers, the central limit theorem, and the theory of large deviations (in Chapters 3 and 4). We have made this choice because we have experienced that sometimes it is didactically more incisive to start with applications, with focused approaches, and then abstract to more general theories.

Chapters 5 and 6 give numerous examples of the analysis of experimental, scalar, and vector data, alternating between the application of specific methods and an in-depth study of the theory on which they are based.

In Chapter 7 we study the first example of a correlated memory-less phenomenon: the random walk (the famous drunkard's walk). We could not do this without, of course, resisting the temptation of a foray into the continuous regime, with the Fokker–Planck diffusion equation (which, we unveil, is what physicists call a Schrödinger equation in imaginary time) and its close relative, the stochastic differential Langevin equation.

In Chapters 8 and 9 we see other *forgetful* discrete processes: chain reactions and recurrent events. We realize, then, that they are all Markov chains and, in Chapter 10, we derive a general theory, continuing, also in Chapter 11, with further applications. These include, for example, the sorting of Web pages by search engines and computer simulations of the dynamics of statistical physics models.

Chapter 12 is devoted to correlations. We take up the central limit theorem once again, first with a couple of specific examples solved with considerable – but instructive – effort, and then with the central limit theorem for correlated events, which greatly simplifies things.

We return to Markov processes in Chapter 13, this time with continuous-time processes and their master equations. We finish on a high note with Chapter 14, devoted to entropy, information theory, dynamical systems, and statistical physics, tying up some knots left untied in previous chapters.

In this passionate journey of writing, teaching, and interacting with each other and with colleagues, friends, and students, we have learned and poked around a lot, and, above all, we had a lot of fun. Our wish now is that at least some of this fun will trickle down to our readers.

Luca Leuzzi, Enzo Marinari, and Giorgio Parisi

For the sake of lecturers, in Table 1 we list possible paths that can be followed in courses on different subjects, simply pointing at the chapters where such topics are presented and developed.

In the following, furthermore, we provide a number of possible course proposals that could be carried out using the material included in our book, identifying particular sections inside the chapters when necessary.

Table 1 Possible coarse-grained list of walks through the book contents.

Chapter	Basic tools in probability	Central limit and large deviations	Bayesian inference and data analysis	Stochastic Markov processes	Probability in statistical physics	Statistical inference and information theory
1	•		•		•	•
2	•				•	
3		•			•	
4		•			•	
5			•			•
6			•			•
7				•	•	
8				•		
9				•		
10				•		
11				•		•
12		•		•	•	
13				•		
14		•			•	•

A   Probability of independent and correlated events

    – Chapter 1   Introduction to probability
    – Chapter 2   Probability distributions

Introduction to Probability

1.1 Definition of Probability

A book dedicated to probability should contain a direct definition of this concept. Mathematicians could limit themselves to provide the necessary conditions to be able to call something a "probability": for example, if p_n is the probability that the nth event occurs, it is necessary that $p_n \geq 0$ and that $\sum_n p_n = 1$. More generally, the probability can be defined as a non-negative measure, normalized to 1.

This solution may leave you unsatisfied, but it is difficult to give a purely non-mathematical definition that does not bite its tail. Traditionally, following a *frequentist* approach [1], probability is defined as the limit of the frequency with which an event is observed when the number of observations tends to infinity. The frequency f_N of a certain event is the number of times the event occurs in N observations divided by N. For example, when you flip a coin, the *head* outcome is said to have probability $1/2$ if the frequency f_N tends to $1/2$ when N tends to infinity.

How can this statement be justified? Actually, for each value of N, the sequence of all heads is always possible, albeit very, very unlikely. As we will see in Chapter 2, the probability of $|f_N - 1/2|$ being greater than $\epsilon > 0$ tends to zero when N tends to infinity. We can be more precise by saying that the probability of getting a head is $1/2$ if the probability that the frequency deviates too much from $1/2$ tends to zero for large N. In this way, we have brilliantly defined probability in terms of probability!

We can, however, get away with it using a primitive concept, intuitively defined: an event whose probability tends to zero. Such an event tends never to occur; indeed, when the probability is sufficiently low, most probably it never happened in the entire history of the universe and will not happen for many lifetimes of the universe to come. This frequentist view is not the one shared by all probabilists [2].

Let us try to think of a definition of probability that uses our knowledge of the type of event we are analyzing. For example, we can use our knowledge of geometry to determine the probability of an event falling in a certain region of space. A classic example is to take a circle of radius 1 and its inscribed equilateral triangle, and try to ask ourselves what is the probability of drawing with a pencil (without looking) a chord longer than the side of the triangle. That is, if we can blindly draw something on the figure! To increase the probability of centering the circle, we recommend (empirically, without proof) drawing a very large circle.

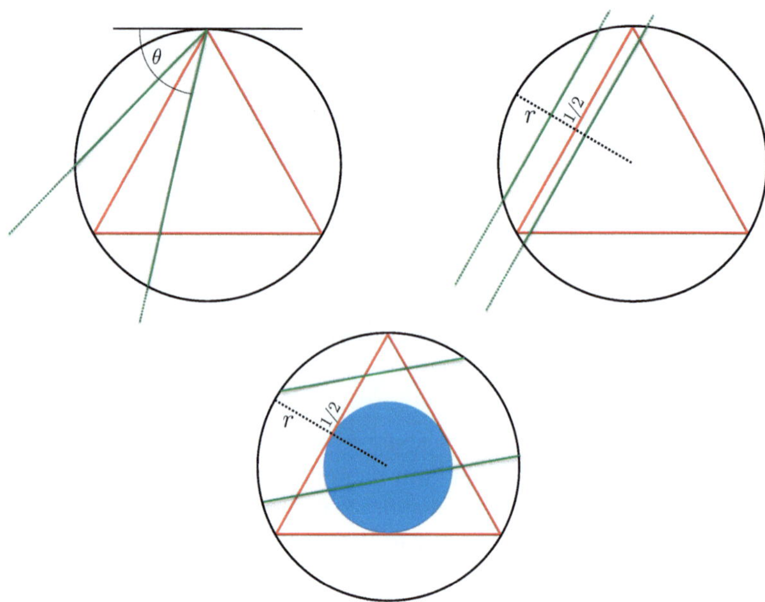

Figure 1.1 Bertrand paradox, a graphical aid for the three direct geometry-driven definitions of the probability that a chord of a circle (in green) is greater than the side of the inscribed equilateral triangle (in red).

We know everything about circles and triangles. We know that the side of the inscribed equilateral triangle is $\sqrt{3}$ (as the radius $= 1$), that the angle to the center subtended by the side is $60°$, that the arc of the circle joining the ends of one side of the triangle is $1/3$ of the entire circumference of the circle in which the triangle is inscribed, and that the radius of the circle inscribed in the triangle is $1/2$ of that of the circle in which the triangle is inscribed, etc.

We now use our vast knowledge. We refer to one end of the drawn chord and rotate the inscribed triangle so that one of its angles coincides with this end (the triangle is best cut out from a separate sheet of paper so that it can be conveniently turned), as in Figure 1.1 (top left). If the other end of the chord falls on one of the two arcs from the first end to another corner of the triangle, it means that the string is shorter than the side, whereas if it falls on the arc between the two opposite corners, it is longer. At this point it is easy. The three arcs are all equal in length and each one of them has length $1/3$ of the circumference, so the probability of having drawn a chord longer than the side of the inscribed equilateral triangle is $1/3$. Equivalently, we can say that if the angle θ between the chord and the tangent to the circle at the first chord end (the one coinciding with the corner of the triangle) is $\theta < 60°$ or $\theta > 120°$, the chord is shorter than the side of the inscribed triangle, and if, on the other hand, $\theta \in [60°, 120°]$ the chord is longer. So the probability of a long chord is always $60°/180° = 1/3$. And without drawing $N \to \infty$ strings to estimate the frequency!

We have made an implicit assumption, that the *probability* (or rather the probability density, as we shall see in a moment) that the second chord extremity falls at a point on the circle is uniform along the circumference, i.e., that the probability is uniform in the angle θ. We are again using the concept of (uniform) probability in defining a probability. Is this so bad?

To find out, we attempt the same task using different geometric information. We turn the triangle so that one of its sides is parallel to the chord and draw a radius r perpendicular to the chord, as shown in the top right panel of Figure 1.1. If the chord intersects the radius at a distance shorter than $1/2$, then it is larger than the side of the triangle. This event occurs with probability $1/2$, which is not equal to the result of $1/3$ we just derived. Here we have implicitly assumed that the probability density (whatever it is) is uniform along the radius of the circle. We then (also) have another problem: not only do we implicitly use the definition of probability to explicitly define probability, but the result also depends on the implicit choice we make!

Not yet satisfied, we continue in the paradox, proposed by Joseph Bertrand [3] in 1889, and put forward a third definition that uses the information that the radius of the circle inscribed in the equilateral triangle is half the radius of the circle in which the triangle is inscribed, as in the bottom panel of Figure 1.1. We look at whether the midpoint of the drawn chord is inside or outside the inscribed circle. If it is inside, the chord will be longer than the side of the triangle. The probability that the midpoint is inside the small circle is $\pi(1/2)^2/\pi = 1/4$. Yet a different estimate! Here we have assumed that the points in the circle are distributed with uniform probability over the surface. Again, we have used geometric information and assumed a uniformity property of a certain probability.

The three separate pieces of information used in each case are not sufficiently complete to make the definition unambiguous, and the three relative probabilistic assumptions are not equivalent to each other: the definition of probability becomes subjective.

A vast literature is devoted to the *subjectivist* approach, and various interpretations have been developed, some of which disagree with others. To mention the contributions of some of the most famous authors, we refer to the works of John Maynard Keynes [4], Émile Borel [5], Bruno de Finetti [6], and Frank Ramsey [7]. The reader interested in deepening this approach can find a fascinating guide in the recent work by Maria Carla Galavotti [8] and in the brilliant philosophical and statistical text by Persi Diaconis and Brian Skyrms [2]. According to this view, probability indicates a prediction of ours conditioned by our ignorance, expressing our degree of confidence in the realization of a certain event. For example, the meaning of the sentence "the probability of red in a rigged roulette wheel is greater than black" corresponds to "it is better to bet on red". Probability, thus, allows us to decide what to do in a world we know imperfectly.

According to de Finetti [6] (for an exciting discussion see Diaconis and Skyrms [2]), the degree of confidence is well represented by how much one is willing to bet on an event. One of the most direct representations of the subjectivist approach is the one stated in terms of betting on games, of wagers. This is not meant from the point of view of the individual player, who, for undoubtedly exquisitely subjective reasons, decides to place his or her trust in a number, a horse, or a combination of cards, but rather from the point of view of the bookmaker: the one who has to establish the probability of winning and losing in a whole system of bets, rationally setting a fair price for all of them. A fair price means that the bookie must follow certain criteria (which happen to coincide with the theorems of the theory of probability) in allocating the various probabilities to events, to avoid losing with certainty (with probability 1) against one or the other of the bettors at his/her bank. A bookmaker who manages to distribute the odds to the various bets in such a way that not a single one of his bettors is sure to win against the bank is called a *coherent person* by de Finetti. For the bookie, the computation of probability is "the mathematical theory that teaches one to be consistent".

Although the frequentist approach may seem to us much more familiar, the point of view we most frequently adopt is probably the subjectivist one, sometimes without realizing it. In this approach, probabilities are often given in the form of betting N against M. Giving one team victory 9 to 1 against another one in some game means, for example, that the bookie has reliable information that the first team is much stronger than the second, to the extent of risking paying 10 times the stake for a possible winning bet on the second team. On the contrary, it does not mean that in a series of 10 repetitions of the match (with the same line-up, in the same weather conditions, physical fitness of the players, etc.) the probability that the first team wins nine times and does not win once is close to 1.

Let us take the sentence: "The experimental data imply that the mass of the proton is within a certain range with a probability of 90%." The literal frequentist interpretation is as follows. "If we consider many possible universes in which the proton mass is different, but in which the same experimental data have been obtained, in 90% of these universes, the proton mass lies in that range." The sentence from the subjectivist point of view is very clear: "Given the results of the experiments, a bet 9 to 1 that the mass of the proton is in that range is a balanced bet."

In this book, we have been inspired, in some cases, by the subjectivist viewpoint, but the methods and theories we discuss apply regardless of this approach. A longer discussion of the different approaches would be out of place here. For the time being, we assume that we have given provisionally a reasonable idea of what probability is, and we move on to the mathematical treatment. One can, indeed, conceive a probability theory functioning independently of the direct definition of probability, i.e., of the chosen probabilistic view.

Still, in 1900, Hilbert pointed to the axiomatization of probability in his famous list of open problems to which solutions were urgently needed [9]. It was Andrey Nikolaevich Kolmogorov, in 1933 [10], who established the fundamental properties of probability satisfied in any approach. These properties are taken as axioms of the theory. The strategy is, thus, not to give a direct definition of probability, but, rather, to postulate axioms from which all other properties, valid in any approach, are deduced as theorems. Kolmogorov's was not the only axiomatization proposed. Also, Richard von Mises, one of the supporters of the frequentist conception of probability, proposed an axiomatization based on the existence of limit frequencies [11]. However, Kolmogorov's theory, which was considered more complete and coherent by the probabilist community, prevailed. A historical turning point was the conference "*Colloque consacré au calcul des probabilités*" (Colloquium Dedicated to Probability Calculus) held in Geneva in 1937 and devoted to the foundations of probability theory, the proceedings of which were published in the journal *Actualités Scientifiques et Industrielles* (1938). True fans of the debate on the choice of axiomatic theory can consult M. van Lambalgen's work [12].

In general, the crucial point is that, if none of the observers has complete information, or if each makes different assumptions, the definition of probability ends up being subjective.

1.2 Basic Properties

1.2.1 Probability of Events, Discrete Variables

Let us start with the simplest case, in which the elementary events are identified by a positive integer n. The index n can take either a finite number of values, in which case the set of events is finite, or an infinite number of values, in which case the set of events is countable.

Let us suppose that these events are self-excluding and that one (and obviously only one) of them certainly will occur. We denote by p_n the probability that the nth event occurs, in the space of all elementary events (mutually self-excluding). The following conditions are necessarily satisfied:

$$p_n \geq 0 \quad \forall n \,,$$
$$\sum_n p_n = 1 \,. \tag{1.1}$$

The latter equation is the *normalization* condition, for reasons that will be evident in the next section. To each subset A of the integers we can associate the probability $P(A)$ that an event belonging to A will occur:

$$P(A) \equiv \sum_{n \in A} p_n \,. \tag{1.2}$$

With this definition, it is easy to demonstrate the fundamental properties of set probabilities:

$$P(A \cup B) = P(A) + P(B) - P(A \cap B) \,,$$
$$P(A) \in [0, 1] \,, \tag{1.3}$$
$$P(\emptyset) = 0 \,,$$

where $\emptyset$ denotes the empty set. A function defined on sets that satisfy the previous properties is what mathematicians call a *normalized, positive measure*. In the case we are considering (in which the sets consist of a countable quantity of elements), it is possible to construct the theory from p_n or similarly from $P(A)$ quantities.

1.2.2 Axioms of Probability Theory

If we wanted to limit ourselves to probabilities defined on countable sets, finite or infinite (denumerable), the above definitions would suffice. However, we are interested, as well, in defining probability on uncountable sets, such as, for example, an interval of real numbers. In this case, things get more difficult. One possibility would be to use Lebesgue's measure theory. In Lebesgue's theory [13, 14, 15, 16], if A is a subset of the real numbers, a measure $\mu(A)$ is defined. Taking the technical definition of the Lebesgue measure for granted, we could simply say that $\mu(A)$ is a probability measure if $0 \leq \mu(A) \leq 1$, $\forall A$. This choice might seem satisfactory, but its main flaw is related to the existence of non-measurable sets,[1] i.e., sets on which $\mu(A)$ cannot be defined, such as, for example, the set of Vitali [19]. We will not use Lebesgue's measure theory in the beginning but rather present an axiomatic definition of probability in general.

We call E the generic *elementary event* or element of the *space of events*, denoted by the symbol $\mathcal{I}$. The space $\mathcal{I}$ is a set of elementary, mutually exclusive events (as we shall see, the set $\mathcal{I}$ has probability 1). We call $\mathcal{F}$ the set of subsets of $\mathcal{I}$, including $\mathcal{I}$ itself and the empty set. We call *random events* the elements of $\mathcal{F}$. We denote by E the generic event. Kolmogorov introduced five axioms in his original treatise [10]. Putting together the first two original axioms, using the above definitions, we enunciate the following four *axioms of probability theory*.

1. Algebra – The set $\mathcal{F}$ is an algebra of sets, that is,

    - the set $\mathcal{F}$ contains the null element, $\emptyset \in \mathcal{F}$,
    - if the event $E \in \mathcal{F}$ then the *complement event* $\bar{E} \equiv \mathcal{I} - E \in \mathcal{F}$ (complement normalization), and
    - if the events $E_1, E_2 \in \mathcal{F}$ then $E_1 \cup E_2 \in \mathcal{F}$ (finite union normalization).

    It follows from the previous properties that[2] $\mathcal{I} \in \mathcal{F}$ ($\mathcal{I}$ is the complement of $\emptyset$) and that, if $E_1, E_2 \in \mathcal{F}$, then $E_1 \cap E_2 \in \mathcal{F}$ (finite intersection normalization).

2. Probability – To each set $E \in \mathcal{F}$ a non-negative real number $P(E)$ is assigned, called the *probability* of the random event E.

3. Completeness – The probability of the set of all elementary events is $P(\mathcal{I}) = 1$.

4. Additivity – If any two random events E_1 and E_2 have no elements in common, i.e., if $E_1 \cap E_2 = \emptyset$, then the probability of the union event $E_1 \cup E_2$ is $P(E_1 \cup E_2) = P(E_1) + P(E_2)$. More generally, if we have a finite number N of *disjoint* events $\{E\}$, i.e., such that $E_n \cap E_k = \emptyset$, $\forall n, k = 1, 2, \ldots, N$, then the probability of their union is the sum of the probabilities of each event:

---

[1] Non-measurable sets can only be constructed using the axiom of choice [13, 17]. There is an alternative set of axioms, not including the axiom of choice but including other axioms, where it can be proved that all sets are measurable [18].

[2] This was Kolmogorov's second axiom in the original work [10]. The first axiom here includes the first and second axioms of the original presentation of the theory.

$$P\left(\bigcup_{n=1}^{N} E_n\right) = \sum_{n=1}^{N} P(E_n).$$

This additivity property can hold also for a set of infinite events, as long as it is countable.

From these four axioms, one can derive all the fundamental properties of probability, including Eq. (1.3), for discrete, countable events.

1.2.3 Probability of Events, Continuous Variables

If events are denoted by a *random variable x* taking values on real numbers, it is better to take a more general point of view and introduce the probability of a set E as

$$P(E) \equiv \int_{x \in E} dP(x) = \int_{x \in E} dx \, p(x). \tag{1.4}$$

In one dimension, the event E is a continuous interval, or a combination of intervals, on the real axis. The event E occurs if the random variable x takes values in the continuous interval $x \in [x_E^{\inf}, x_E^{\sup})$ (or in multiple, disjoint intervals). Here and in the following we will denote by "[" or "]" an endpoint included in the interval, and by "(" or ")" an endpoint excluded. By $dP(x)$ we denote the probability that x belongs to the infinitesimal interval $[x, x + dx)$.

We must be careful in carrying out the last step of Eq. (1.4), where we introduced the function $p(x)$, called the *probability density*. In general, $p(x)$ may not be a true regular function, but it can be a generalized function, i.e., a *distribution*. Indeed, $p(x)$ may contain terms proportional to Dirac delta functions. For example, a density $p(x) = \delta(x - 1)$ means that the event that x acquires exactly the value 1 along the real axis ($x = 1$) occurs certainly, i.e., with probability 1. Using the formula (1.4), this probability reads

$$P(1) = \int_{1^-}^{1^+} dx \, \delta(x - 1) = 1.$$

Another simple example is the distribution

$$p(x) = \frac{1}{2}[\delta(x - 1) + \delta(x + 1)],$$

also known as the Rademacher distribution, which is the probability density that $x = 1$ or $x = -1$ with equal weight ($1/2$). Pedantically, one can write the probabilities as

$$P(1) = \int_{1^-}^{1^+} dx \, \frac{1}{2}[\delta(x - 1) + \delta(x + 1)] = \frac{1}{2}$$

and

$$P(-1) = \int_{-1^-}^{-1^+} dx \, \frac{1}{2}[\delta(x - 1) + \delta(x + 1)] = \frac{1}{2}.$$

In the case where the number of elements in the sets is finite, there are no problems, with respect to the discrete case. In the case where a set has an infinite number of elements (countable or uncountable), it is necessary to proceed with care. To this end, mathematicians have constructed measure theory [13, 14, 15, 16]. Such a theory deals with events that are uncountable. The Kolmogorov probability axioms 1–4 of the previous section must, therefore, be modified to include these sets. In particular, a new kind of additivity (axiom 4) and, therefore, a new kind of algebra (axiom 1) have to be postulated. The finite union of events E_n is generalized into the so-called *countable union* of uncountable events E_n:

$$\bigcup_{n=1}^{N} E_n \quad \Rightarrow \quad \bigcup_{n=1}^{\infty} E_n .$$

When the events E_n are disjoint sets, the finite additivity property yielding the probability of the union for mutually exclusive events is, consequently, generalized to the so-called *countable additivity* property. In formulas, the probability of the union in axiom 4, when N is finite, is modified from

$$P\left(\bigcup_{n=1}^{N} E_n\right) = \sum_{n=1}^{N} P(E_n)$$

into a new axiom of countable additivity, as follows.

4′. Countable additivity – If we have an infinite number of disjoint events E_n:

$$P\left(\bigcup_{n=1}^{\infty} E_n\right) = \int_{x\in\bigcup_{n=1}^{\infty} E_n} dP(x) = \sum_{n=1}^{\infty} \int_{x\in E_n} dP(x) = \sum_{n=1}^{\infty} P(E_n) .$$

The algebra $\mathcal{F}$ introduced in the first axiom is, therefore, also modified. In order to deal with uncountable sets of events, such as real numbers, for instance, the algebra will be closed under countable union, i.e., the union of uncountable events is still part of $\mathcal{F}$. The subset of all uncountable events that is closed under countable additivity is termed the set of *measurable* events (see footnote 1). The algebra of measurable uncountable random events is called a σ-algebra.

Since in any experiment the number of events, however large, is always finite, one might be tempted to argue that we do not need anything more than a probability theory for finite spaces of events. On the contrary, we want to stress that the measure theory, or, equivalently, the probability theory including the countable additivity axiom, is essential to be able to deal rigorously with random variables defined on a continuous domain, like real numbers. We will explicitly see this when dealing with infinite sequences of random numbers (cf. Appendix 3.I), particularly in the framework of the law of large numbers (cf. Section 3.1.2). We will not elaborate further on Lebesgue's measure theory, instead referring interested readers to the cited texts [13, 14, 15, 16].

1.3 Probability of Set Intersections and Unions

We have begun to discuss the behavior of sets of events, using basic concepts such as the probability of the event *union* of two events A and B, $P(A \cup B)$, and the event *intersection* of two events, $P(A \cap B)$. In Figure 1.2 we graphically represent two events

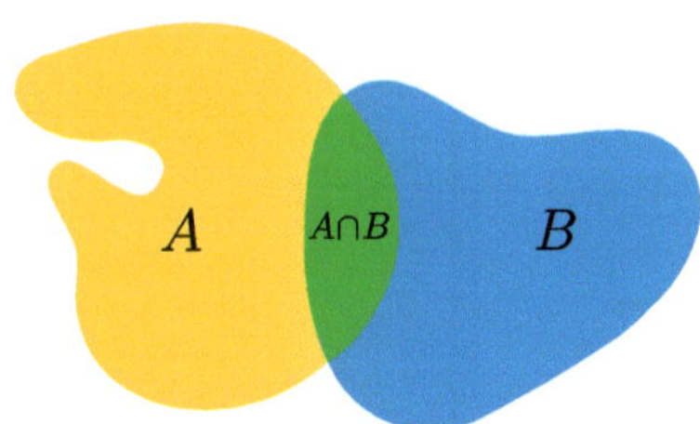

Figure 1.2   Graphical representation of the event: union of events A and B.

A (yellow) and B (blue) in terms of surfaces: their intersection (green) and their union (the total surface).

We have already seen in Eq. (1.3) that

$$P(A \cup B) = P(A) + P(B) - P(A \cap B), \tag{1.5}$$

i.e., that the probability of the union of A and B is given by the probability of A, plus the probability of B, minus the probability of the intersection of A and B. Figure 1.2 helps us visualize this relationship, once we associate the probability of an event with an area. By summing the areas of A and B, we count the intersection area twice, common to the two events. Once we subtract it, we obtain the right area representing the probability of the event $A \cup B$.

If there is no way to realize both the event A and the event B, then the intersection of A and B is null, $A \cap B = \emptyset$. Two such sets are called *mutually exclusive* (or *incompatible* or *disjoint*, as we anticipated) and, in this case, their joint probability is

$$P(A \cap B) = P(\emptyset) = 0. \tag{1.6}$$

The probability that two incompatible events occur simultaneously is the probability of the null event, i.e., zero, according to the fourth axiom stated in Section 1.2.2. This consideration, which at first sight resembles a simple and good-natured petition of common sense, can be used as one of the key points of probabilistic theory. Indeed, the entire theory can essentially be constructed based on the fact that the probabilities of simple or composed events in general have values between 0 and 1, and on the fact that a *complete* set basis of incompatible events exists. Intuitively, this indicates that we can completely characterize our space of events by means of a series of sets whose graphical counterparts never cover each other and tile the entire space.

1.3.1 Probability Completeness

Formally, we can define as a *complete mutually exclusive set* the collection of the event sets $\{H_i\}$, with $i = 1, \ldots, M$, such that

$$H_i \cap H_j = \emptyset \quad \forall i \neq j,$$

$$\bigcup_{i=1}^{M} H_i = \mathcal{I}, \tag{1.7}$$

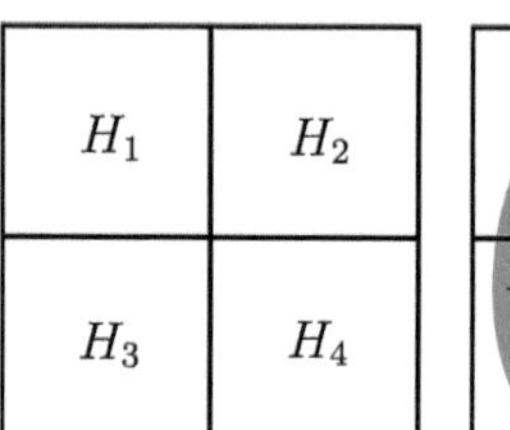
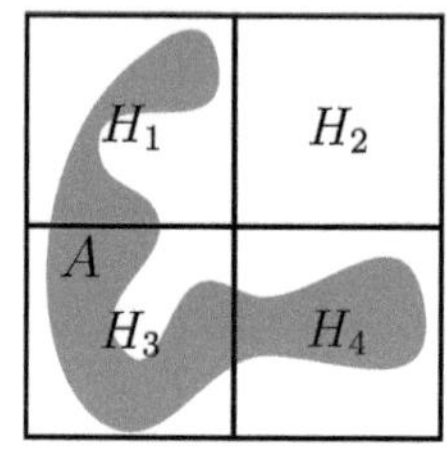

Figure 1.3 Left: Illustrative representation of the tiling of the space of the events (the big square) by means of a complete set of incompatible events (the small squares) H_k, with $k = 1, 2, 3, 4$. The set H_k fully covers the space of the events. Right: The intersection of event A (gray) with the events H_k. The union of all intersections $A \cap H_k$ yields event A.

i.e., whose union is the set of all events in the space of events. If such a set exists, using the fourth axiom we obtain the thesis of the complete probability theorem:

$$P(A) = P(\mathcal{I} \cap A) = P\left(\bigcup_{i=1}^{M} H_i \cap A\right) = \sum_{i=1}^{M} P(A \cap H_i). \tag{1.8}$$

In fact, the mutually exclusive sets $\{H_i\}$ tile the space of events without ever covering themselves. Considering all their intersections with A, then, just amounts to considering the whole set A. In Figure 1.3, we take as an example of a space of events the square divided into four parts. Summing over the four parts is equivalent to summing over the whole square (without the need for excessive precautions, since we have drawn four parts that do not overlap).

Let us now generalize the formula (1.5) to more than two sets. In Figure 1.4 we graphically display the case of three sets of events A, B, and C, overlapping in a generic way. We have used the following notation:

R_1 is the probability of the composed event represented by the surface in very light gray,consisting of the occurrence of only one of the three sets (part of set A, part of set B, and part of set C),

R_2 is the probability of the set consisting of the mid-gray surface, denoting the union of the intersection of any two (and no more than two) of the sets, and

R_3 is the probability of the darkest gray subset, shared by the three sets.

We then want to show that the quantity

$$P(A \cup B \cup C) \equiv R_1 + R_2 + R_3 \tag{1.9}$$

coincides with

$$P(A \cup B \cup C) = S_1 - S_2 + S_3, \tag{1.10}$$

where we have defined

$$\begin{aligned}
S_1 &\equiv P(A) + P(B) + P(C), \\
S_2 &\equiv P(A \cap B) + P(B \cap C) + P(C \cap A), \\
S_3 &\equiv P(A \cap B \cap C).
\end{aligned} \tag{1.11}$$

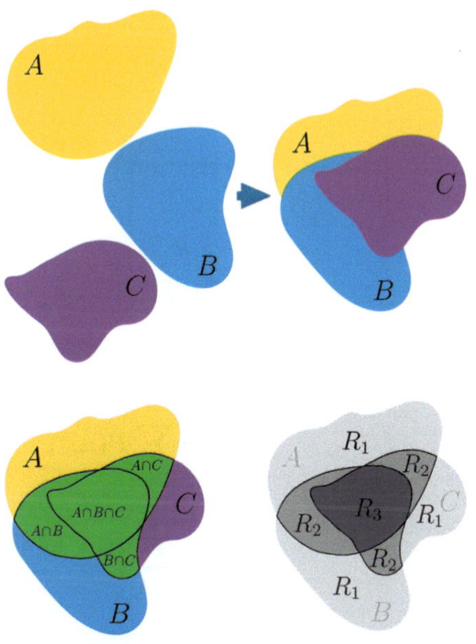

Figure 1.4 Three compatible events A, B, and C. The intersection set is shown in green in the bottom left panel. In the bottom right panel, we classify the areas on the grounds of how many events intersect. A part of the space of events is realized by only one event, A or B or C (very light gray). The probability of these regions of subsets is termed R_1. Another part (in mid-gray) is the intersection of two (and only two) compatible events ($A \cap B$ or $A \cap C$ or $B \cap C$), whose probability is denoted by R_2. The last part (darkest gray) is the region of compatibility of all three events, of probability R_3.

Basically, our way of reasoning is simple. If we add the three probabilities of A, B, and C we overestimate the probability, because we overcount the intersection areas. But by subtracting double intersections (S_2) we exaggerate a little, by subtracting three times the part where the three sets overlap (we should only subtract it twice, instead). By adding S_3 we put things back into place.

In somewhat more formal terms, the sum of the probabilities of the three sets can be written as

$$P(A) + P(B) + P(C) = R_1 + 2R_2 + 3R_3 = S_1 \,, \tag{1.12}$$

i.e., adding the light gray area of Figure 1.4 on the bottom right panel, plus twice the gray area plus three times the dark gray area. This is *not* the probability of the union of the three sets, of course, for which the relation (1.9) applies. Furthermore, for the double intersections the following holds,

$$S_2 = P(A \cap B) + P(B \cap C) + P(C \cap A) = R_2 + 3R_3 \,, \tag{1.13}$$

and for the triple intersection,

$$S_3 = P(A \cap B \cap C) = R_3 \,. \tag{1.14}$$

A suitable linear combination of the three previous relations is precisely our thesis, i.e., the fact that

$$R_1 + R_2 + R_3 = S_1 - S_2 + S_3 \,. \tag{1.15}$$

It is easy to generalize this relation to an arbitrary number of sets of N events that generically overlap. It is found that

$$P\left(\bigcup_{i=1}^{N} A_i\right) = \sum_{k=1}^{N} (-1)^{k+1} S_k \,, \tag{1.16}$$

where

$$S_1 = \sum_{i=1}^{N} P(A_i) \,, \tag{1.17}$$

$$S_2 = \sum_{i<k}^{1,N} P(A_i \cap A_k) \,, \tag{1.18}$$

$$S_3 = \sum_{i<j<k}^{1,N} P(A_i \cap A_k \cap A_j) \,, \tag{1.19}$$

$$\vdots$$

$$S_q = \sum_{i_1 < i_2 < \cdots < i_q}^{1,N} P\left(\bigcap_{j=1}^{q} A_{i_j}\right) \,, \tag{1.20}$$

$$\vdots$$

$$S_N = P\left(\bigcap_{j=1}^{N} A_j\right) \,. \tag{1.21}$$

The above formulas can be easily proved algebraically in the case of continuous or discrete variables by introducing a *characteristic function* for the set A,

$$\chi_A(x) = \begin{cases} 1 & \text{if } x \in A \,, \\ 0 & \text{if } x \in \bar{A} \,. \end{cases}$$

Thus we can write the probability of the event A as

$$P(A) \equiv \int_{x \in A} dP(x) = \int dP(x)\, \chi_A(x) \,. \tag{1.22}$$

The characteristic function of $\mathcal{I}$ is $\chi_{\mathcal{I}}(x) = 1$. One can easily see that

$$\begin{aligned} \chi_{A \cap B}(x) &= \chi_A(x)\,\chi_B(x) \,, \\ \chi_{A \cup B}(x) &= 1 - [1 - \chi_A(x)][1 - \chi_B(x)] = \chi_A(x) + \chi_B(x) - \chi_{A \cap B}(x) \,. \end{aligned} \tag{1.23}$$

The second relation follows from the set identity

$$\mathcal{I} - A \cup B = (\mathcal{I} - A) \cap (\mathcal{I} - B) \,, \tag{1.24}$$

i.e., the complement of the union is equal to the intersection of the complements (an equivalent way to write this is $\overline{A \cup B} = \bar{A} \cap \bar{B}$). We finally obtain the relationship (1.5),

$$P(A \cup B) = \int dP(x)\,\chi_{A \cup B}(x) = \int dP(x)\,[\chi_A(x) + \chi_B(x) - \chi_{A \cap B}(x)]$$
$$= P(A) + P(B) - P(A \cap B),$$

for the union of two compatible, i.e., non-incompatible, sets.

In the same way, extending the identity (1.24) to the union of three compatible sets,

$$\mathcal{I} - A \cup B \cup C = (\mathcal{I} - A) \cap (\mathcal{I} - B) \cap (\mathcal{I} - C), \tag{1.25}$$

we obtain the relation

$$\chi_{A \cup B \cup C}(x) = 1 - [1 - \chi_A(x)][1 - \chi_B(x)][1 - \chi_C(x)]$$
$$= \chi_A(x) + \chi_B(x) + \chi_C(x) - \chi_A(x)\chi_B(x) - \chi_A(x)\chi_C(x) - \chi_C(x)\chi_B(x)$$
$$+ \chi_A(x)\chi_B(x)\chi_C(x)$$
$$= \chi_A(x) + \chi_B(x) + \chi_C(x) - \chi_{A \cap B}(x) - \chi_{A \cap C}(x) - \chi_{C \cap B}(x) + \chi_{A \cap B \cap C}(x), \tag{1.26}$$

which immediately implies the inclusion–exclusion relation (1.10) previously graphically proved for three sets.

Moving on to a generic number of sets, for which

$$\overline{\bigcup_k A_k} = \bigcap_k \bar{A}_k, \tag{1.27}$$

we have (omitting the dependence of the variable x)

$$\chi_{\bigcup_k A_k} = 1 - \prod_k \chi_{\bar{A}_k}$$
$$= \sum_k \chi_{A_k} - \sum_{k_1 < k_2} \chi_{A_{k_1} \cap_{k_2}} + \sum_{k_1 < k_2 < k_3} \chi_{A_{k_1} \cap_{k_2} \cap A_{k_3}}$$
$$- \cdots + (-1)^{n+1} \sum_{k_1 < \cdots < k_n} \chi_{\bigcap_n A_{k_n}}$$
$$= \sum_n (-1)^{n+1} \sum_{k_1 < \cdots < k_n} \chi_{\bigcap_n A_{k_n}}, \tag{1.28}$$

from which, using the definition (1.22), we get the general formula (1.16) for the sum of probabilities of the union of events.

1.4 Conditional Probabilities

The probability of the intersection of A and H is also called *joint probability*. We can express it by writing

$$P(A \cap H) = P(A|H)\,P(H), \tag{1.29}$$

where the symbol | indicates the *condition* of one set A with respect to another set H. Equation (1.29) introduces the *conditional probability* $P(A|H)$ of the set A occurring if the set H occurs. The joint probability of A and H is equal to the probability of A conditioned by H multiplied by the probability of H. In other words, the conditional probability is the probability that A will occur (or occurred, or is occurring), once we know that H occurs. The reader should note that we are not making any assumptions on the relative occurrence time and causality of the events, such as H causes A or H occurs before A.

Obviously the conditional probability is only defined if $P(H) \neq 0$; in this case we can write that

$$P(A|H) \equiv \frac{P(A \cap H)}{P(H)} .\tag{1.30}$$

The condition symbol, |, is, in general, non-symmetric:

$$P(H|A) \equiv \frac{P(A \cap H)}{P(A)} = P(A|H)\frac{P(H)}{P(A)} \neq P(A|H) .\tag{1.31}$$

For example, if it is night and a cat goes by in front of us, we will most certainly see it in a very dark colour; whereas the mere fact that a cat has a very dark fur does not necessarily mean that it is night. Maybe it is a bright August morning, but the cat appears to be black anyway. One has that

$$P(\text{black cat}|\text{night}) = 1 ,$$
$$P(\text{night}|\text{black cat}) < 1 .\tag{1.32}$$

In Figure 1.5 we propose a simple illustration to help understand some relations and implications of conditional probability. We pick four events A, H, K, and L, with their probabilities $P(A) = 0.29$, $P(H) = 0.12$, and $P(K) = P(L) = 0.11$. Some events are chosen not disjoint and their joint probabilities in the example are $P(A \cap H) = 0.07$

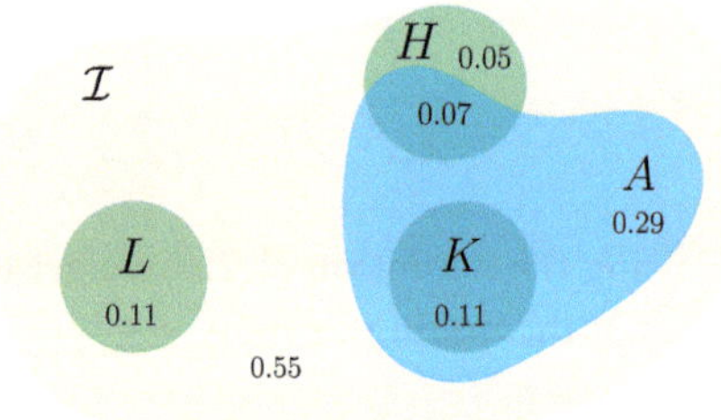

Figure 1.5 Graphical illustration of the formula (1.29) for the conditional probability. Here $P(A|H)$ is the probability of the joint occurrence of A and H divided by the probability of H: $P(A \cap H) = 0.07$, $P(H) = 0.05 + 0.07 = 0.12$, so $P(A|H) = 0.07/0.12 = 0.5833$. The set K is completely included into the set A, therefore $P(A \cap K) = P(K) = 0.11$, and $P(A|K) = 1$: if K takes place, then A occurs for sure. The events L and A are incompatible, $P(A \cap L) = 0$, yielding $P(A|L) = 0$ whatever the probability $P(L)$. As usual, we denote by $\mathcal{I}$ the complete space of events, for which we know that $P(\mathcal{I}) = 1$.

and $P(A \cap K) = 0.11$. Through these events, we are able to exemplify the following various situations:

- $A \cap H \neq 0$, A and H may occur together and may condition each other, so

$$P(A|H) = \frac{P(A \cap H)}{P(H)} = \frac{0.07}{0.12} = 0.5833,$$

$$P(H|A) = \frac{P(A \cap H)}{P(A)} = \frac{0.07}{0.29} = 0.2414;$$

- $A \supset K$, $A \cap K = K$, whenever K occurs also A takes place (but not vice versa), so $P(A|K) = 1$, while

$$P(K|A) = \frac{P(K \cap A)}{P(A)} = \frac{0.11}{0.29} = 0.3793;$$

- $L \cap K = L \cap A = L \cap H = \emptyset$, L is disjoint from A, H, and K, thus $P(L|A) = P(L|H) = P(L|K) = 0$; and
- $H \cap K = 0$, H and K are incompatible, and so $P(H|K) = P(K|H) = 0$.

We note that when conditioning an event with another, their conditional probability cannot be smaller than their joint probability. For example $P(A|H) > P(A \cap H)$, $P(H|A) > P(A \cap H)$, and $P(K|A) > P(A \cap K)$. In general, the condition of the occurrence of another non-independent (and compatible) event narrows the space of events, and in this smaller space the probabilities of the remaining possible events cannot decrease: they either grow or remain unchanged.

1.4.1 Joint Probability of Independent Events

If A and H are compatible ($A \cap B \neq \emptyset$) but independent events, the probability of A conditional on H does not depend on H: $P(A|H) = P(A)$. This implies for the joint probability (1.29) that

$$P(A \cap H) = P(H)P(A|H) = P(H)P(A). \tag{1.33}$$

The joint probability factorizes into the product of the probabilities of the individual events.

1.5 Bayes' Formula

To begin discussing Bayes' formula, let us start by recalling the importance of a complete set of mutually exclusive events: the basis of incompatible events of the space of the events $\mathcal{I}$, defined in (1.7). If the events A_k form a complete basis, then, for the probability of the event B, the complete probability theorem (1.8) applies,

$$P(B) = \sum_k P(A_k \cap B), \tag{1.34}$$

and through the definition of conditional probability (1.29) we arrive at the expression

$$P(B) = \sum_k P(A_k)P(B|A_k). \tag{1.35}$$

Formula (1.31) applied to the case of one of the events A_k and the event B tells us that

$$P(A_k|B) = \frac{P(A_k)P(B|A_k)}{P(B)}, \tag{1.36}$$

which using (1.35) gives us the *Bayes' formula*

$$P(A_k|B) = \frac{P(A_k)P(B|A_k)}{Z}, \tag{1.37}$$

where the normalization factor Z is

$$Z \equiv P(B) = \sum_k P(A_k)P(B|A_k). \tag{1.38}$$

1.5.1 Marginal Probability

The factor Z depends only on the event B and is called *marginal probability*, as it is found by summing the joint probability $P(B \cap \bigcup_j A_j)$ on all hypotheses A_j.

The term "marginal" comes from the practice of noting on the joint probability table the sums at the margins of the table, along the row (or column) corresponding to a given variable: for illustration purposes see Table 1.1. Notice that in the table we start by writing the joint probability of the events x_j and y_k with the comma instead of the intersection symbol: $P(x_j \cap y_k) = P(x_j, y_k)$.

Table 1.1 An example of marginalization. The *marginal probability* of a stochastic variable is the sum over the other stochastic variables of the joint probability. For example, $P(y) = \sum_k P(x_k, y)$. In the continuous case of a probability density, see Section 5.2.2, it will be $P(y) = \int dx\, P(x, y)$. Similarly $P(x) = \sum_k P(x, y_k)$ in the discrete case, or $P(x) = \int dy\, P(x, y)$ in the continuous case. The summed probability is called "marginal" because it was traditionally written in the margins of the joint probability table, just as shown here.

$P(x, y)$		x		marginal on x
	$P(x_1, y_1)$	$P(x_2, y_1)$	$P(x_3, y_1)$	$P(y_1)$
y	$P(x_1, y_2)$	$P(x_2, y_2)$	$P(x_3, y_2)$	$P(y_2)$
	$P(x_1, y_3)$	$P(x_2, y_3)$	$P(x_3, y_3)$	$P(y_3)$
marginal on y	$P(x_1)$	$P(x_2)$	$P(x_3)$	

The denominator Z, cf. Eq. (1.38), which does not depend on A_k, can be easily derived by imposing the normalization condition

$$\sum_k P(A_k|B) = 1. \tag{1.39}$$

For the time being, Bayes' formula (1.37) is merely a mathematical rewriting of Eq. (1.30). In Section 5.2 we will discuss its application and make it the starting point for the analysis of experimental data and, more generally, for statistical inference. However, we can already see that if we identify the A_k as the possible *causes* of the event B (all the possible causes, of which one occurs with certainty, since we have considered a complete self-excluding set of events), the formula (1.37) can be seen as a rule that allows us to establish the probability of the causes of an event.

1.5.2 Multiplication Rule for Conditional Probability

Let us now look at the conditional probability multiplication rule if we have N non-self-excluding events $\{B_k\}$. We can write their joint probability by reiterating the definition of conditional probability:

$$
\begin{aligned}
P\left(\bigcap_{j=1}^{N} B_j\right) &= P\left(B_N \middle| \bigcap_{j=1}^{N-1} B_j\right) P\left(\bigcap_{j=1}^{N-1} B_j\right) \\
&= P\left(B_N \middle| \bigcap_{j=1}^{N-1} B_j\right) P\left(B_{N-1} \middle| \bigcap_{j=1}^{N-2} B_j\right) P\left(\bigcap_{j=1}^{N-2} B_j\right) \\
&= P\left(B_N \middle| \bigcap_{j=1}^{N-1} B_j\right) P\left(B_{N-1} \middle| \bigcap_{j=1}^{N-2} B_j\right) P\left(B_{N-2} \middle| \bigcap_{j=1}^{N-3} B_j\right) P\left(\bigcap_{j=1}^{N-3} B_j\right) \\
&\;\;\vdots \\
&= P\left(B_N \middle| \bigcap_{j=1}^{N-1} B_j\right) P\left(B_{N-1} \middle| \bigcap_{j=1}^{N-2} B_j\right) \cdots P(B_3|B_2 \cap B_1)P(B_2|B_1)P(B_1).
\end{aligned}
\tag{1.40}
$$

We can try to apply this rule to a very famous example, proposed in 1939 by Richard von Mises [20]: the birthday problem.

1.5.2.1 The Birthday Problem

Let us try to determine the probability that, in a class of n pupils, at least two have the same birthday. We will assume that a year consists of 365 days and that the probability of being born on any day of the year is uniform.

With the multiplication formula (1.40) we directly solve the problem of coincident birthdays. We sort the pupils in some arbitrary order and we define the following events:

$A = \{$at least two pupils have the same birthday$\}$,

$\bar{A} = \{$no pupils were born on the same day$\}$,

$B_1 = \{$the second pupil was not born on the same day as the first one$\}$,

$B_2 = \{$the third pupil was not born on the same day as the second or as the first one$\}$,

$$\vdots$$

$B_k = \{$the $(k+1)$th pupil was not born on any of the birthdays of the preceding k pupils$\}$.

For example, the set $B_1 \cap B_2$ will be the event "the first, second, and third pupils were born on different days".

With these definitions for a class of n pupils using conditional probability multiplication, we can express the probability that no birthdays coincide as

$$P(\bar{A}) = P\left(\bigcap_{j=1}^{n-1} B_j\right) = P\left(B_{n-1}\left|\bigcap_{j=1}^{n-2} B_j\right.\right) \cdots P(B_3|B_2 \cap B_1)P(B_2|B_1)P(B_1)\,, \quad (1.41)$$

where

$$P(B_1) = \frac{364}{365}\,, \quad P(B_2|B_1) = \frac{363}{365}\,, \quad \ldots,$$

$$P\left(B_k\left|\bigcap_{j=1}^{k-1} B_j\right.\right) = \left(1 - \frac{k}{365}\right)\,, \quad k = 1,\ldots,n-1\,. \quad (1.42)$$

Here the probability of at least two pupils, among n, having the same birthday, the complement of $P(\bar{A})$, is

$$P(A) \equiv P_n = 1 - \prod_{k=1}^{n-1}\left(1 - \frac{k}{365}\right) = 1 - \frac{365!}{365^n(365 - n)!}\,, \quad (1.43)$$

as shown in Table 1.2 and Figure 1.6.

Table 1.2 Probability that, in a group of n persons, at least two have the same birthday.

n	2	3	4	5	6	7	8	9	10	15	20
P_n	0.0027	0.0082	0.016	0.027	0.040	0.056	0.074	0.095	0.117	0.253	0.411

n	21	22	23	24	25	30	40	50	60	70	100
P_n	0.444	0.476	0.507	0.538	0.569	0.706	0.891	0.970	0.994	0.999	0.9999997

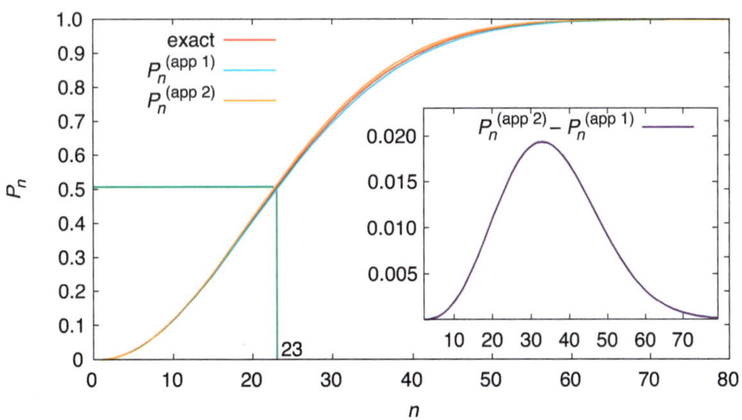

Figure 1.6   Probability that, in a group of n persons, at least two have the same birthday. The exact solution, the "app 1" solution in Eq. (1.45), and the "app 2" solution in Eq. (1.46) are displayed for comparison. In a group of $n = 23$ persons, the probability is already slightly larger than $1/2$. In the inset, we show the relative difference between Eqs. (1.46) and (1.45). We observe that it always remains small.

Obviously, in the case of a year of N days, the above formula generalizes to

$$P_n = 1 - \prod_{k=1}^{n-1} \left(1 - \frac{k}{N}\right) = 1 - \exp\left\{\sum_{k=1}^{n-1} \log\left(1 - \frac{k}{N}\right)\right\}. \tag{1.44}$$

If we assume $k \ll N$ we can approximate $\log(1 - k/N)$ with $-k/N$, thus obtaining the formula

$$P(A) \simeq P_n^{(\text{app 1})} \equiv 1 - \exp\left\{-\frac{n(n-1)}{2N}\right\}. \tag{1.45}$$

This formula produces a result approximated to a few percent for $N = 365$ and n around 25, as can be seen in Figure 1.6. By approximating $\log(1 - k/N)$ with $-k/N - \frac{1}{2}k^2/N^2$, we would obtain the more accurate approximation

$$P(A) \simeq P_n^{(\text{app 2})} \equiv 1 - \exp\left\{-\frac{n(n-1)}{2N}\left(1 - \frac{2n-1}{6N}\right)\right\}, \tag{1.46}$$

which produces a result approximating a few per thousand for $N = 365$ in the worst cases.

For all practical purposes regarding birthdays in not-too-crowded classes, the formula

$$P(A) = 1 - e^{-n^2/(2N)}$$

works very well and is easy to remember.

Table 1.3 The conditional probabilities of medical tests. The rate of true positives is called sensitivity; the rate of true negatives is called specificity.

	Ill	Healthy
positive	$P(p\|M)$ sensitivity	$P(p\|S)$ false positive rate
negative	$P(n\|M)$ false negative rate	$P(n\|S)$ specificity

1.5.3 Conditional Probability in Diagnostic Tests

A diagnostic test to detect a viral or bacterial infection must feature an accurate estimation of its *sensitivity* and its *specificity*, which are two conditional probabilities. We denote by M the event of being ill and with S that of being healthy. We use p to designate the event of a positive test result and n for a negative test result. The sensitivity $P(p|M)$, therefore, indicates how likely it is that a sick patient will test positive. The specificity $P(n|S)$ indicates how likely it is that a healthy patient will test negative. Needless to say, a diagnostic test is all the more valid the closer these indicators are to 1. Their complements are, respectively, the false negative rate $P(n|M)$ and the false positive rate $P(p|S)$, as shown in Table 1.3.

Once the test has been carried out, it is fair to ask, whatever the outcome, what is the probability that it is "wrong". That is, the probability of being ill if the test reports negative,

$$P(M|n) = \frac{P(n|M)P(M)}{P(n)},$$

or the probability of being healthy if the test turns out positive,

$$P(S|p) = \frac{P(p|S)P(S)}{P(p)}. \tag{1.47}$$

These are important questions, since testing positive does not give the certainty of being ill, especially if the disease is rare, and testing negative does not give the certainty of being healthy, especially if the disease is widespread. To estimate these probabilities, it is necessary to know the incidence of the disease in the population, $P(M)$. From this information, the marginal distribution for being positive can be computed:

$$P(p) = P(p|M)P(M) + P(p|S)P(S). \tag{1.48}$$

More precisely, however, the computation of $P(M)$ should be performed over the population that is swabbed. This is not a piece of directly accessible information, as what is actually measured in the tests is the positive or negative response of the sample taken with the diagnostic test reagents.

If we have access to the rate of positives out of the total number of tests performed, we directly have an estimate of $P(p)$, while it remains to establish the fraction of sick

$P(M)$ and healthy $P(S)$, exclusively in the tested population (not in the general population). In this case, with simple algebraic steps, $P(M)$ can be derived from $P(p)$ as

$$P(M) = \frac{P(p) - P(p|S)}{P(p|M) - P(p|S)} = \frac{P(p) + P(n|S) - 1}{P(p|M) + P(n|S) - 1} \, , \tag{1.49}$$

where we have rewritten the numerator and denominator as functions of the sum of specificity and sensitivity.

Let us consider a test and ask ourselves what is the probability of not actually being ill in the case of a positive outcome. We will see this in different cases, ranging from a scenario in which the disease is very rare to one in which there is a large spread of the disease. To fix ideas, we can think of the HIV virus, on the one hand, and the Sars-Cov-2 coronavirus at times of different spread of the COVID-19 pandemic, on the other.

The incidence of HIV in Italy in the year 2020 was about 2 per 100 000.[3] The incidence of coronavirus has varied greatly during the pandemic that began in 2020. We will consider the representative incidence figures of[4] 20 and[5] 1600 positives per 100 000 persons. The incidence is the estimated $P(M)$. However, it must be specified that the incidence of the disease in the general population does not correspond to the incidence in the test population alone. The latter is clearly higher (generally much higher), as those who take the test do not do so by chance, but for reasons related to the spread of the disease (manifestation of symptoms or contact with sick people). Sampling is, therefore, biased. This correlation is presumably all the greater, the rarer or more symptomatic the disease is, particularly if it has very specific symptoms.

Let us see the two cases in which (i) we approximate $P(M)$ with the incidence in the general population and compute the positivity rate with (1.48), and (ii) we estimate a grossly higher $P(M)$ (by a factor that is greater the less widespread the disease). For this second estimate, we can use the formula (1.49) taking estimates of the so-called test positivity rates. In any case, we are interested in the order of magnitude, so we are not going too far out on a limb. In Table 1.4, fourth column, first, third, and fifth rows, we report the estimates of $P(p)$ in case (i), where we start from the incidence $P(M)$ in the general population. In the third column, second, fourth, and sixth rows, we instead compute the incidence of the disease on the testing population from the positivity rate.

We can now answer our question about the probability of not being sick even if the test is positive for all cases considered, simply by applying the formula (1.47). As an example, we consider tests for different infections with the same specificity of 0.998, and the same sensitivity, which we assume to be 0.97. The interesting results are those on the population tested, case (ii). We obtain, therefore, for the cases under investigation, i.e., HIV, Sars-Cov-2 at "low" incidence, and Sars-Cov-2 at high incidence, the probabilities that the test is wrong, either because the patient is not sick although positive or because

---

[3] Notiziario ISS, Vol. 34, No. 11, 2021.
[4] Surveillance Bulletin ISS, National Update of 30/06/2021.
[5] Surveillance Bulletin ISS, National Update COVID19 of 12/01/2022.

Table 1.4 Illustrative table for diagnostic tests with the same sensitivity $P(p|M) = 0.998$ and specificity $P(n|S) = 0.97$. The six rows indicate different cases of disease incidence. The first two columns show, for illustrative purposes only, which disease and epidemic spread inspired the chosen values of incidence or positivity rate. The third column shows the incidence $P(M)$ and the fourth column the positivity rate $P(p)$. Marked in bold are the **epidemic data** from which the other data are derived. In the fifth column, we report the probability of being healthy even if the test came out positive, and in the sixth column the probability of being sick even if the test is negative.

Test for	Incidence population	Incidence $P(M)$	Positive rate $P(p)$	Healthy if p $P(S\|p)$	Ill if n $P(M\|n)$
HIV	general	**0.00002**	0.0020	0.99	6×10^{-7}
HIV	tested	0.001	**0.003**	0.67	0.00003
Sars-Cov-2 (202106)	general	**0.0002**	0.0022	0.91	6×10^{-6}
Sars-Cov-2 (202106)	tested	0.0083	**0.01**	0.20	0.00025
Sars-Cov-2 (202201)	general	**0.016**	0.018	0.11	0.00049
Sars-Cov-2 (202201)	tested	0.15	**0.15**	0.012	0.0053

the patient is healthy although the test is negative. The results are shown in the fifth and sixth columns of Table 1.4.

We note that, the rarer the disease, the more likely it is that a positive test is wrong. Testing positive for HIV in the population tested means that one has a 67% chance of not having contracted HIV. Whereas the test is extremely accurate if it gives a negative result: the probability of having HIV if the test turns out negative is 0.003%.

2

Probability Distributions

We will present in this chapter some fundamental probability distributions that will play a crucial role in our discussion. We will first define the concepts of expectation value, moment, cumulant, and median of the distribution. We will deal with the binomial distribution, the Poisson distribution, the Gaussian or normal distribution, and the Cauchy–Lorentz distribution: we will discuss their definitions, their most relevant properties, and their asymptotic behavior.

2.1 Properties of Probability Distributions

It is of great practical (and theoretical) utility to characterize a probability distribution by a series of values, given by integrals (or sums, in the discrete case) of functions of the random variable considered.

Let X be a random variable that can assume the discrete values x_k ($k = 1, 2, \ldots$) with probability $P_k \equiv P(x_k)$. If the series $\sum_k x_k P_k$ converges absolutely, i.e., $\sum_k |x_k| P_k < \infty$, the expectation value, also called the mean value, of X, defined by

$$\langle x \rangle \equiv \sum_k x_k P_k \,, \tag{2.1}$$

is finite. If, on the other hand, the mean of the absolute value $\sum_k |x_k| P_k$ diverges, X has no unambiguously defined expectation value unless we give a precise prescription as to the order in which we perform the sum, since sums (or integrals) that do not absolutely converge have a result that depends on the order in which they are performed. We will also denote the expectation value $\langle \cdot \rangle$ with $\mathbb{E}(\cdot)$.

2.1.1 Moments

In greater generality, we can compute the expectation value of a function $f(X)$ of the random variable X on discrete values by

$$\langle f(x) \rangle \equiv \sum_k f(x_k) P_k \,, \tag{2.2}$$

or on continuous values distributed with probability density $P(x)$ by

$$\langle f(x) \rangle \equiv \int dx \, f(x) P(x) \,. \tag{2.3}$$

Obviously, the convergence of the sum (integral) is by no means guaranteed and depends on the behavior of the function f and the probability distribution P.

For the power function, we have

$$\langle x^n \rangle \equiv \sum_k x_k^n P_k , \tag{2.4}$$

which with integer $n > 0$ gives the *nth moment* of the probability distribution. In the case of a probability that goes to zero exponentially for large $|k|$, all moments are convergent. Conversely, in the case of a probability that goes to zero as a power for $|k|$ large, the moments will be divergent for sufficiently large n.

Note that these definitions, which we have given for a discrete probability distribution, are extended to a continuous distribution by replacing the sums with integrals and the discrete probability P_k with the continuous probability $P(x)\,dx$.

2.1.2 Cumulants

Certain combinations of moments are of special importance. They were called by Fisher *cumulative moments* [21], from which the contraction *cumulants* probably originates. One of these cumulants is the *variance* of the variable X,

$$\langle (x - \langle x \rangle)^2 \rangle = \sum_k P_k(x_k^2 - 2\langle x \rangle x_k + \langle x \rangle^2) = \langle x^2 \rangle - \langle x \rangle^2 \equiv \sigma^2 . \tag{2.5}$$

The standard deviation σ is the square root of the variance and gives a first quantitative assessment of the typical distance that the values of the variable X have from its mean. It is evident that the variance is a non-negative quantity (it is the sum of squares). The variance is zero if and only if the variable X can only take on a value k with non-zero probability (i.e., $P_k = 1$, $P_{i \neq k} = 0$). Exactly the same situation occurs if X is a variable with a continuous distribution given by a single $P(x) = \delta(x - x_k)$, the Dirac delta.

The combination (2.5) is the second-order cumulant of the random variable X, and is also called the second-order *central* moment.

Another combination of moments that gives relevant information is the skewness of a distribution, which indicates how far the distribution is from a symmetric one. The numerator of the skewness is the third moment of the variable's deviation from its expected value, i.e., the third-order central moment. The skewness is then normalized to the cube of the standard deviation:

$$S = \frac{\langle (x - \langle x \rangle)^3 \rangle}{\sigma^3} . \tag{2.6}$$

Here S stands for *skewness*. It is zero for a symmetric probability distribution around $x = \langle x \rangle$. Also the combination of moments in the numerator of (2.6),

$$\langle x^3 \rangle - 3\langle x^2 \rangle \langle x \rangle + 2\langle x \rangle^3 ,$$

is a cumulant of X, the third-order cumulant.

The last combination relevant for our immediate future that characterizes a probability distribution $P(x)$ is the *kurtosis*, from the Greek word $\kappa \upsilon \rho \tau \acute{o} \varsigma$ (*kyrtos*), meaning "bulging" or "swelling" (as, for example, the swelling of a vein in the hand):

$$K \equiv \frac{\langle (x - \langle x \rangle)^4 \rangle}{\sigma^4} \, . \tag{2.7}$$

In the numerator, we have the fourth-order central moment, $K\sigma^4$. As we will discuss in Section 12.3 in Chapter 12 or, before that, in Appendix 3.C.2, $K\sigma^4$ is not, however, the fourth-order cumulant, which, instead, is equal to $(K - 3)\sigma^4$. From the fourth order onwards, the cumulants and the central moments no longer match.

Cumulants are also called *connected moments*, and we will sometimes denote them by the expected value symbol with a subscript "c". For the above cases we have, for example,

$$\langle x \rangle_c = \langle x \rangle \, , \quad \langle x^2 \rangle_c = \sigma^2 \, , \quad \langle x^3 \rangle_c = \sigma^3 S \, , \quad \langle x^4 \rangle_c = \sigma^4 (K - 3) \, . \tag{2.8}$$

The generation of cumulants, their expressions, and their usefulness will become clear when we study the connected correlation functions between different variables in Section 12.3. In fact, as we will see there, when two or more variables are independent, their *connected* correlation function is null, thus immediately signaling their independence even without knowing their probability distribution. The cumulants we now consider are, very briefly, the connected correlation functions of a variable with itself.

2.1.3 Median and Mode

In addition to the moments and, in particular, in addition to the first moment, the mean value, it is useful to introduce the median value x_m, also known as the *median*, i.e., the value for which the probability of the values to the left of x_m is equal to that of the values to the right, $X > x_m$. Or, at least, their difference is as small as possible. The median is always well defined even when the mean value of X is divergent (for an interesting discussion of the importance of the difference between the mean and median value, see ref. [22]). Some other times we hear of the modal value, or *mode*, X, but we should not be frightened: it is simply the value of the random variable X that makes the probability maximum (the one *à la mode*).

2.1.3.1 Cumulative Probability

The probability that the value of a random variable x is less than x' is the *cumulative probability* of x':

$$F(x') \equiv \int_{-\infty}^{x'} dx \, P(x) \, ,$$

and the median x_m of a probability distribution is that value such that

$$\int_{-\infty}^{x_m} dx \, P(x) = \int_{x_m}^{\infty} dx \, P(x) = \frac{1}{2} \, . \tag{2.9}$$

For the normalization of probability, we have $F(\infty) = 1$, see also the third axiom Section 1.2.2, and the median corresponds to the value for which

$$F(x_m) \equiv \int_{-\infty}^{x_m} dx\, P(x) = \frac{1}{2}\,.$$

Given a set of values, the median x_m is the number such that the probability of values less than or equal to x_m is equal to the number of values greater than or equal to x_m.

When it is easy to compare relative values of a variable, computing the median can be much faster than computing the mean. Take, for example, the computation of the median of the height of persons belonging to a group (class, team, choir, orchestra, . . .). To compute the mean, the height of each individual member of the group must be measured with a tape measure. To compute the median, on the other hand, we simply need to ask the people to sit in a row in order of height and, if they are an odd number, measure the height of the person in the middle, or, if they are even, the heights of the two people in the middle of the row.

2.1.3.2 Generalized Median

The above definition may, however, be empty or ambiguous in some cases. We look at two cases where it is undetermined.

1. If the probability distribution is $P(x) = 0.7\delta(x) + 0.3\delta(x - 1)$, no number satisfies the above definition (2.9). In fact, $\int_{x_m}^{\infty} dx\, P(x)$ jumps discontinuously from 1 to 0.3 when x_m goes from a negative value to a value greater than zero.
2. If the probability distribution is $P(x) = 0.5\delta(x) + 0.5\delta(x - 1)$, all numbers between 0 and 1 satisfy definition (2.9).

A more general definition can be put forward, evaluating the absolute minimum of the function

$$A(y) = \int dx\, P(x)|x - y|\,. \tag{2.10}$$

In fact, if the function $A(y)$ is differentiable, its derivative is given by

$$A'(y) \equiv \frac{dA}{dy} = \int_{-\infty}^{y} dx\, P(x) - \int_{y}^{\infty} dx\, P(x)\,, \tag{2.11}$$

and the condition that $A'(y) = 0$ coincides with the previous definition (2.9) of the median. Obviously, if the function $P(x)$ contains Dirac delta functions, the derivative of A may not exist, but the minimum condition is well defined. In the first critical case, the median is $x_m = 0$, as can easily be verified.

However, it is possible that the function $A(y)$ does not have an isolated global minimum: indeed, in the second case, the function $A(y)$ is constant in the interval $[0, 1]$. It is easy to show that the set of global minima of the function $A(y)$ must be an interval, and, in this case, the equivocation is resolved by defining the median as the center of gravity of the mentioned interval. In the second case, the median is, therefore, $x_m = 1/2$.

This definition of the median can also be extended to the multi-dimensional case, defining it as the minimum of the function

$$A(\vec{y}) = \int d^D x \, P(\vec{x}) |\vec{x} - \vec{y}| , \tag{2.12}$$

where $\vec{x}$ and $\vec{y}$ are D-dimensional vectors.

We also observe that the usual mean value, see (2.1), can be defined as the minimum point of the function

$$A_2(y) = \int dx \, P(x) |x - y|^2 . \tag{2.13}$$

We can then generalize the definition of mean by introducing a mean of order k as the minimum of

$$A_k(y) = \int dx \, P(x) |x - y|^k . \tag{2.14}$$

For $k = 1$ we find the median and for $k = 2$ the expectation value. We remark that while for $k \geq 1$ the function $A_k(y)$ is convex[1] and thus the minimum is unique, or a single interval (as in the second example for the median above), this property is not true for $k < 1$.

2.1.3.3 Quantiles, Deciles, and Percentiles

As we have defined the median as the value for which the cumulative probability is $F(x_m) = 1/2$, we can define a quantity x_q such that the probability that the random variable is less than x_q is equal to a number $q \in (0, 1)$:

$$F(x_q) = q .$$

The number q is called a *quantile*.

The number $x_{1/2}$ is the median. The numbers $x_{1/4}$ and $x_{3/4}$ are called quartiles, the lower and upper quartiles to be precise, while the median is called, of course, the median quartile. The probability of the values of the variable $x \in [x_{1/4}, x_{3/4}]$ is equal to $1/2$, and the difference $x_{3/4} - x_{1/4}$ between the upper and lower quartiles is called the *interquartile range* (IQR) and is used in statistics as a measure of the dispersion of the data with respect to the central value.

Similarly, the quantities $x_{1/10}$, $x_{2/10}$, ..., $x_{9/10}$ are called *deciles*. The quantities $x_{0.01}$, $x_{0.02}$, ..., $x_{0.98}$, $x_{0.99}$ are called *centiles*, sometimes also expressed as *percentile scores*, e.g., $x_{96\%}$, $x_{97\%}$, $x_{98\%}$, a well-known measure for classifying infants by weight.

---

[1] Recall that a function is convex if $f(au + b(1 - u)) < uf(a) + (1 - u)f(b)$ for u in the interval $[0, 1]$. The positivity of the second derivative implies convexity.

2.2 Bernoulli Events and Binomial Distribution

We will use the concept of Bernoulli events as the basis for the following reasoning. Let us consider a sequence of repeated random trials: that is, we perform a trial (e.g., a coin toss), and then repeat the same trial many times. These are *Bernoulli events* if the following hold:

1. One of two permitted results may be obtained with each attempt.
2. The probability of each of the two allowed outcomes does not vary during the series of experiments.

Typical examples are the toss of a coin with the two possible outcomes of head or tail, or the throw of a die, where an even value gives the event A while an odd value gives the event B. In these two examples, the two allowed events are equiprobable, i.e., they both have $p = 1/2$. In the case, for example, of a throw of a die, one can, instead, consider the value 1 as the event A and a value between 2 and 6 as the event B. Or one can consider a card drawn at random from a deck of 52 cards, and define the event A as one in which an ace was drawn and B as one in which an ace was not drawn. In the latter examples, the two events A and B have different probabilities, but they are still complementary.

The fact that the "most overdue" numbers in the lotto game have exactly the same probability as the numbers drawn in the most recent draw is an expression of their being Bernoulli events. It should be noted that this is an empirical assumption, consistent with experimental data and our understanding of the mechanisms of physical forces. For example, we can argue that the probability of pulling a certain sphere out of an urn on Thursday of this week is the same as next week: the same will be the laws governing the interactions between the elementary components of the sphere, the urn, and the hand of the blindfolded person doing the pulling (again, under the assumption that the spheres are all at the same temperature and are not recognizable by touch). The binary nature of the event, which admits only a positive or a negative answer, and the constancy of the probability of the elementary event characterizes Bernoulli events.

2.2.1 Binomial Distribution, Definition, and Properties

Let us then consider N Bernoulli events, and ask what is the probability of getting K positive answers (i.e., events of type A and not B) in an unspecified order. The problem is the same as counting the ways of putting K balls into N holes, each of which may contain one or no balls. The number of ways in which we can order N objects is $N!$. The positive K answers are indistinguishable, thus causing us to eliminate a factor $K!$, just as the negative $N - K$ answers are indistinguishable, causing us to eliminate a factor $(N - K)!$. So the count gives us a result

$$\frac{N!}{K!(N-K)!} = \binom{N}{K},\qquad(2.15)$$

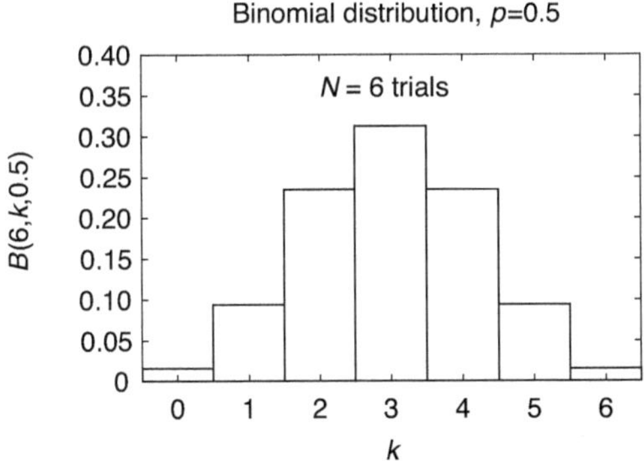

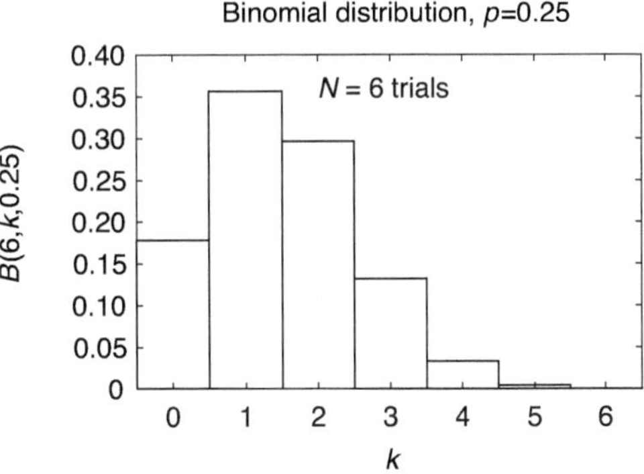

Figure 2.1 Binomial distributions per $N = 6$ trials and Bernoulli success probability of a single event $p = 1/2$ (left) and $p = 1/4$ (right).

which is, precisely, the binomial coefficient. Our final result is, therefore, that the probability that N Bernoulli events result in K positive answers and $N - K$ negative answers reads

$$B(N, K, p) = \left(\begin{array}{c} N \\ K \end{array} \right) p^K q^{N-K} \,, \tag{2.16}$$

where p is the elementary probability of a single positive response, and $q = 1 - p$. This probability distribution is called the *binomial distribution* (see Figure 2.1).

Interestingly, the coefficient $B(N, K, p)$ is precisely the Kth term of the expansion of the binomial $(q + p)^N$. It follows that the binomial distribution is well normalized, i.e., that

$$\sum_{K=0}^{N} B(N, K, p) = \sum_{K=0}^{N} \left(\begin{array}{c} N \\ K \end{array} \right) p^K q^{N-K} = (q + p)^N = 1 \,. \tag{2.17}$$

We first compute the expectation value of the number of positive responses in a series of trials governed by a binomial distribution, and the associated variance. Given the single trial probability p and the number of trials N, we must sum over all the allowed numbers of hits K, each one with its weight $B(N, K, p)$:

$$\mathbb{E}(\text{number of successes}) \equiv \langle K \rangle = \sum_{K=0}^{N} B(N, K, p)\, K \,.$$

If we use the identity

$$K\, B(N, K, p) = K \frac{N!}{K!(N - K)!}\, p^K q^{N-K} = Np\, B(N - 1, K - 1, p), \tag{2.18}$$

and we note that

$$\sum_{K=1}^{N} B(N - 1, K - 1, p) = \sum_{K'=0}^{N-1} B(N - 1, K', p) = 1 \,, \tag{2.19}$$

then

$$\mathbb{E}(\text{number of successes}) = Np\,. \tag{2.20}$$

Similarly, for the variance we see that

$$\sigma^2 \equiv \langle (K - \langle K \rangle)^2 \rangle = \sum_{K=0}^{N} K^2 B(N, K, p) - (Np)^2$$

$$= Np \left(\sum_{K=0}^{N} K\, B(N-1, K-1, p) - Np \right)$$

$$= Np((N-1)p + 1 - Np) = Np(1-p) = Npq\,. \tag{2.21}$$

Let us now look at the modal value of the binomial distribution. Let us consider a fixed number of events, N, and analyze $B(N, K, p)$ as a function of the number K of positive responses. We will show that if $(N+1)p$ is not an integer then B has a maximum in $B(N, \text{int}[(N+1)p], p)$, where by $\text{int}[\cdot]$ we denote the integer part. If, instead, $(N+1)p$ is an integer number, then we prove that the distribution has a double maximum in $B(N, (N+1)p, p) = B(N, (N+1)p - 1, p)$.

Proof   For the proof, we compute the ratio

$$\frac{B(N, K-1, p)}{B(N, K, p)} = \frac{\dbinom{N}{K-1} p^{K-1} q^{N-K+1}}{\dbinom{N}{K} p^K q^{N-K}} = \frac{K(1-p)}{(N-K+1)p}\,.$$

The function $B(N, K, p)$ of K is increasing if

$$\frac{K(1-p)}{(N-K+1)p} < 1\,,$$

i.e., if

$$K < (N+1)p\,. \tag{2.22}$$

We have shown that $B(N, K, p)$ grows with K until the largest integer $\text{int}[(N+1)p]$ satisfies this relationship, and then begins to decrease. If $(N+1)p$ is an integer number, then the maximum value is realized for

$$\frac{B(N, K-1, p)}{B(N, K, p)} = 1\,. \tag{2.23}$$

That is, we have two maxima of B, equal to $B(N, (N+1)p, p) = B(N, (N+1)p - 1, p)$, for $K = (N+1)p$ and $K = (N+1)p - 1$, respectively.   ∎

We note two interesting facts. First, $B(N, K, p)$ as a function of K grows monotonically up to a certain value of K and then decreases monotonically. Second, it should be remembered that for large N the individual terms $B(N, K, p)$ are all small. To obtain a large probability one can instead consider, for example, the cumulative probability of having *at least* K successes.

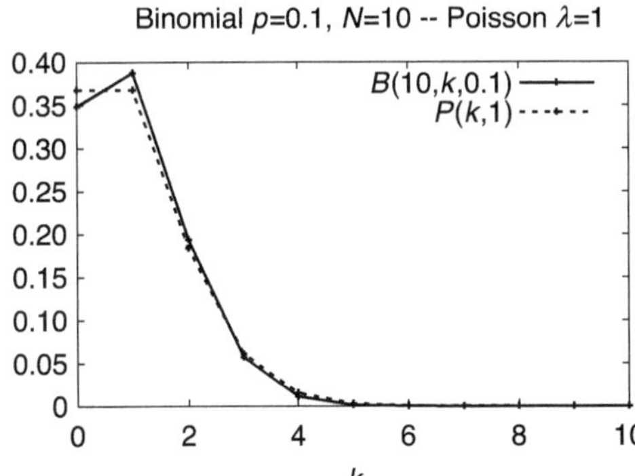

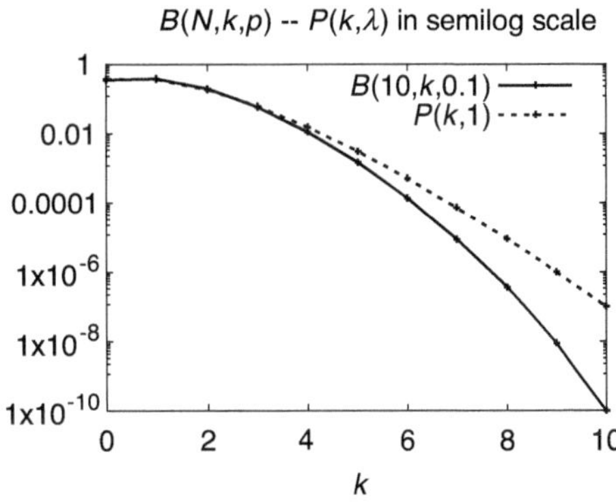

Figure 2.2 Left: Comparison between a binomial distribution of $N = 10$ variables with $p = 0.1$ and a Poisson distribution with $\lambda = 1$. Right: The same distributions in semi-log scale. Differences between the two distributions can be appreciated in the tails for large deviations from the expectation value $\langle k \rangle$.

2.2.2 Poisson Limit

It is interesting to discuss the case where the number of events N is large, the probability of a success p is small, while the expected value of successes

$$\lambda \equiv Np \tag{2.24}$$

is finite and does not increase with N.

Let us first look at the probability $P(0, \lambda)$ of not having any successes. Under these conditions,

$$B(N, 0, p) = \binom{N}{0} q^N = (1 - p)^N = \left(1 - \frac{\lambda}{N}\right)^N . \tag{2.25}$$

In the limit $N \to \infty$ we have that

$$P(0, \lambda) \equiv \lim_{N \to \infty} B(N, 0, p) = \lim_{N \to \infty} \left(1 - \frac{\lambda}{N}\right)^N = e^{-\lambda} . \tag{2.26}$$

The probability $P(1, \lambda)$ of having only one positive event is given by the term with $K = 1$. Here

$$B(N, 1, p) = Np(1 - p)^{N-1} = \lambda \left(1 - \frac{\lambda}{N}\right)^{N-1} \tag{2.27}$$

and

$$\ln B(N, 1, p) = \ln \lambda + (N - 1) \ln \left(1 - \frac{\lambda}{N}\right) \simeq \ln (\lambda) - \lambda , \tag{2.28}$$

i.e.,

$$P(1, \lambda) \equiv \lim_{N \to \infty} B(N, 1, p) = \lambda e^{-\lambda} . \tag{2.29}$$

In general we have that

$$P(K,\lambda) \equiv \lim_{N \to \infty} B\left(N, K, \frac{\lambda}{N}\right) = \frac{\lambda^K}{K!} e^{-\lambda} . \tag{2.30}$$

This is precisely the Poisson distribution that we will discuss in a moment. In Figure 2.2 we show how, already for 10 trials, the two distributions are very similar.

2.3 Poisson Distribution

The Poisson distribution, which we have already introduced in (2.30) as the natural limit of the binomial distribution, is characterized by a parameter λ. The probability of K events is given, in the case of a Poisson distribution, by

$$P(K,\lambda) = \frac{\lambda^K}{K!} e^{-\lambda} . \tag{2.31}$$

We discuss in Chapter 13 the crucial connections of this type of distribution with the theory of stochastic processes. It should also be noted that, in the limit of a large number of trials, other distributions besides the binomial distribution also tend towards the Poisson distribution. Let us cite as an example the two-deck problem, i.e., the probability distribution of the number of equal cards placed in the same positions of two well-mixed decks.

The Poisson distribution (2.31) is well normalized. Indeed,

$$\sum_{K=0}^{\infty} P(K,\lambda) = e^{-\lambda} \sum_{K=0}^{\infty} \frac{\lambda^K}{K!} = e^{-\lambda} e^{\lambda} = 1 . \tag{2.32}$$

One can also check the useful formula for the cumulative probability distribution of $P(K,\lambda)$, i.e., the probability of counting a number of events up to N,

$$\sum_{K=0}^{N} P(K,\lambda) = \frac{1}{N!} \int_{\lambda}^{\infty} dx\, x^N e^{-x} , \tag{2.33}$$

whose expression can be derived using the properties of Euler's Γ functions presented in Appendix 2.B.

To gain a deeper understanding of the meaning of a Poisson distribution, let us consider a sequence of experiments (obviously under constant time conditions). The paradigmatic example is that of the well-known radioactive emission. For simplicity, we base our reasoning on the use of discrete probabilities. Our fundamental assumption will be that disjoint time intervals are independent, i.e., the probability of the occurrence of an event in the first interval is independent of the probability in the second interval (of arbitrary length but disjoint from the first). Let us, therefore, consider a time interval of length T, and divide it into N sub-intervals of length T/N. In a sub-interval, something may happen or nothing may happen, with probabilities $p_{(T/N)}$ and $1 - p_{(T/N)}$, respectively.

Under these assumptions, therefore, the probability of K successes is, as we have seen, given by the binomial probability $B(N, K, p_{(T/N)})$. When $N \to \infty$, in order to obtain a finite probability of having $0, 1, 2, \ldots$ occurrences in a fixed macroscopic time interval T, it is necessary that the expected value $Np_{(T/N)}$ tends to a finite limit,

$$Np_{(T/N)} \to \mu T \,, \tag{2.34}$$

where μ is the average frequency of success counts. Therefore, the result already obtained on the convergence of a binomial distribution to a Poisson distribution applies. The probability of K events finally tends to

$$P(K, \mu T) = e^{-\mu T} \frac{(\mu T)^K}{K!} \,. \tag{2.35}$$

The probability of no events is $e^{-\mu T}$. If we think of the typical example in physics of the disintegration of an atom, μ^{-1} can be interpreted as the mean lifetime of an atom.

We also note that our argument for a time interval T can be extended to the analysis of a spatial distribution, where instead of T we consider a length or a volume, obtaining similar results.

We finally compute the mean and variance of a Poisson distribution with parameter λ. For the expectation value of a random variable x distributed according to Poisson, we have that

$$\langle K \rangle = \sum_{K=0}^{\infty} K P(K, \lambda) = e^{-\lambda} \sum_{K=0}^{\infty} \frac{K \lambda^K}{K!} = \lambda \sum_{K=0}^{\infty} P(K-1, \lambda) = \lambda \,. \tag{2.36}$$

Similarly, for the variance one has

$$\sigma^2 = \sum_{K=0}^{\infty} K^2 P(K, \lambda) - \lambda^2 = \lambda \left(\sum_{K=0}^{\infty} K P(K-1, \lambda) - \lambda \right) = \lambda \,. \tag{2.37}$$

2.4 Gaussian Distribution

Given a random variable X with values $x \in \mathbb{R}$, the *Gaussian* probability density of mean μ and variance σ^2 is defined by

$$p(x) = \frac{1}{\sqrt{2\pi\sigma^2}} e^{-(x-\mu)^2/(2\sigma^2)} \,. \tag{2.38}$$

In the case where the mean value μ is zero, we have a "bell-shaped" curve symmetrical around the x axis. If we take $\sigma^2 = 1$ as an example, the curve has a typical width equal to 1 and is also called the *normal distribution*.

In Appendix 2.A we derive some results on typical integrals of Gaussian functions. These show, for example, that the integral of $p(x)$ between $-\infty$ and $+\infty$ is 1, i.e., that the distribution (2.38) is well normalized.

Also, of course, for (2.38) we have that $\langle x \rangle = \mu$ and that $\langle x^2 \rangle - \mu^2 = \sigma^2$. The skewness is equal to zero while the Kurtosis is equal to $K = 3$. We can, therefore, note an important property that will be observed and used several times in this book: for a Gaussian variable, the third and fourth cumulants are zero, $S = 0$, $K = 3$; see Eq. (2.8) in Section 2.1.2. Going beyond that, as we demonstrate in Appendix 3.C.2, all cumulants of a Gaussian-distributed variable of order greater than the second are null:

$$
\begin{aligned}
\langle x \rangle_c &= \mu, \\
\langle x^2 \rangle_c &= \sigma^2, \\
\langle x^{k>2} \rangle_c &= 0.
\end{aligned}
\tag{2.39}
$$

The Gaussian distribution is the only probability distribution with this property.

Let us also consider, *en passant*, the Gaussian distribution defined on a positive support, which will come in handy, e.g., when we look at large deviations, in Section 4.3.3. In this case, the normalized probability density, setting $\mu = 0$ for simplicity and keeping σ^2 (which is no longer the variance), turns out to be

$$
p(x) = \sqrt{\frac{2}{\pi \sigma^2}}\, e^{-x^2/(2\sigma^2)},
\tag{2.40}
$$

with expectation value

$$
\langle x \rangle = \sqrt{\frac{2\sigma^2}{\pi}}
$$

and variance

$$
\langle (x - \langle x \rangle)^2 \rangle = \left(1 - \frac{2}{\pi}\right)\sigma^2.
$$

2.4.1 Relation between Binomial and Gaussian Distributions

We now want to show that, in the large-N limit, the binomial distribution is well approximated by the Gaussian distribution. This is a result that is part of a more general set of properties culminating in the central limit theorem, which we will initially address in Section 3.2. Here we show the convergence of the binomial distribution to the Gaussian distribution in the symmetric case, where $p = 1 - p = q = 1/2$. The theorem for the non-symmetric case follows a very similar line of proof, with some additional technical complications: we will simply state it, eventually.

We begin by considering the case of $2N$ binomial events with $p = 1/2$. In this case, we have shown in Section 2.2.1 that the coefficients $B(2N, K, 1/2)$ as a function of integer K have only one maximum in $K = \mathrm{int}[(2N + 1)p] = N$. We denote in the following the binomial coefficient of $N + K$ events by

$$
A_K = B(2N, N + K, 1/2)
$$

with $K \in [-N, N]$. The term in which the binomial coefficients take the maximum value is therefore $A_0 = B(2N, N, 1/2)$. Given that $A_K = A_{-K}$ we need only to consider the positive values of K. We thus compute

$$\frac{A_K}{A_0} = \frac{B(2N, N+K, 1/2)}{B(2N, N, 1/2)} = \frac{N!}{(N-K)!} \frac{N!}{(N+K)!}$$
$$= \frac{N(N-1)\cdots(N-K+1)}{(N+K)(N+K-1)\cdots(N+1)}. \tag{2.41}$$

First, let us recall that we are assuming that N is large. We also note that, for large K, $K \sim N$, the ratios we write will be very small (the denominator will be much larger than the numerator) and their probabilities can be neglected. Eventually, only the terms with small K/N will be relevant.

We then divide the numerator and denominator by N^K: each factor of the last expression is multiplied by a factor $1/N$, and has the form

$$1 + \frac{m}{N} \simeq \exp\left\{\frac{m}{N} + O\left(\left(\frac{m}{N}\right)^2\right)\right\}, \tag{2.42}$$

where m goes from $-(K-1)$ to K. In this limit, Eq. (2.41) tends to

$$\frac{A_K}{A_0} \simeq \frac{\prod_{m=1}^{K-1} e^{-m/N}}{\prod_{m=1}^{K} e^{m/N}} = e^{-K/N} \prod_{m=1}^{K-1} e^{-2m/N} = \exp\left\{-\frac{K}{N} - \frac{2}{N}\sum_{m=1}^{K-1} m\right\} = e^{-K^2/N},$$
$$\tag{2.43}$$

in which we neglect terms of order $O(K^3/N^2)$. For large values of K, where the previous formula is not valid but the probability A_k/A_0 is negligible, the normalization condition $\sum_K A_K = 1$ fixes the value of $A_0 = 1/\sqrt{\pi N}$. We finally have that the binomial distribution of K events on N trials tends to

$$A_K \simeq \frac{1}{\sqrt{\pi N}} e^{-K^2/N}, \tag{2.44}$$

a Gaussian of mean null and variance $\sigma^2 = (2N)pq = N/2$. We could have estimated A_0 using Stirling's formula derived in Appendix 2.B (see Eq. (2.73)), obtaining the same result. It would also have been possible to use directly Stirling's formula in Eq. (2.41), simplifying the proof, but perhaps making it a little too compact for our probabilistic warm-up phase.

In conclusion, we have shown that, in the limit of N tending to ∞, for values of $K < \bar{K}$, where $\bar{K}$ is such that $\bar{K}^3/N^2 \to 0$, a binomial distribution of $2N$ symmetric events (with $p = 1/2$) tends to a Gaussian distribution centered on $2Np = N = K_{\max}$ and variance $2Npq = N/2$. As we have already mentioned, this result is part of a more general set of behaviors related to the central limit theorem, which we discuss in more generality in Section 3.2.

In Figure 2.3, in the left panel we show the comparison between the binomial and the Gaussian in the symmetric case of N events and $p = 1/2$ for $N = 10$ events. The values of the binomial are already reasonably approximated by the Gaussian curve of mean $N/2$ and variance $N/4$ (note that here we are considering a binomial of N events, not $2N$).

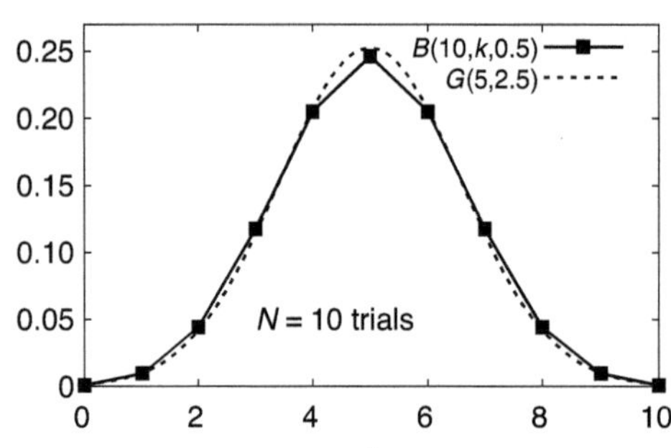

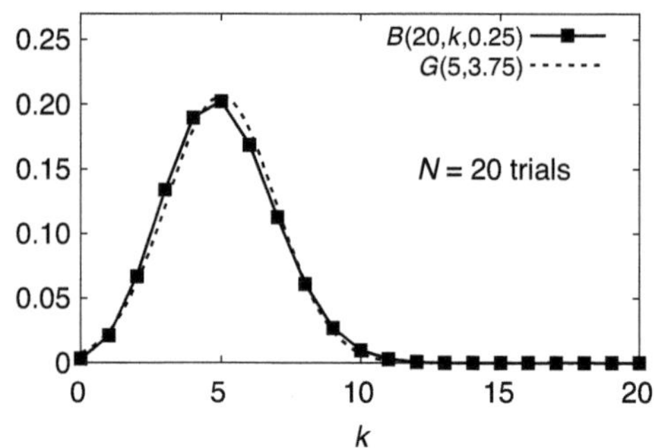

Figure 2.3 Binomial distribution $B(N, k, p)$ and Gaussian distribution $G(k, \mu, \sigma^2)$ compared. Left: $N = 10$ and $p = 0.5$. Right: $N = 20$ and $p = 0.25$.

A similar result, which we will not derive explicitly here, can be obtained for any values of p. The scheme of the proof is similar to what we have seen, with the addition of some technical complications. It is found that

$$B(N, K_{\max} + K, p) \simeq \frac{1}{\sqrt{2\pi N p(1 - p)}} \, e^{-(1/2)K^2/(Np(1-p))} , \qquad (2.45)$$

where $K_{\max}$ is the modal value of B. The Gaussian distribution we obtained has a mean value Np and a variance $Np(1 - p)$: this is obvious since these are, as we have shown above, the values that characterize the starting binomial distribution. In the right-hand panel of Figure 2.3 we show an example for $p = 1/4$ and $N = 20$.

2.4.2 Relation between Poisson and Gaussian Distributions

We have already seen that, for large N and finite values of $\lambda = Np$, the binomial distribution is well approximated by a Poisson distribution, namely

$$B(N, K, p) \simeq P(K, \lambda). \qquad (2.46)$$

It is, therefore, clear from the results of the previous section that under certain conditions a Poisson distribution will tend to a Gaussian distribution.

For small values of λ the Poisson distribution, even in the limit of N large, is appreciably different from a Gaussian. If, on the other hand, λ is not too small, the two distributions will be *similar* in the *central* part of the domain around the mean value. This result is still, in its essence, included in the central limit theorem, cf. (3.2), in whose context we will better clarify what we mean by *similar*.

If we set two suitable constants $c_1 < c_2$, we can, for example, show that, for $\lambda \to \infty$, the probability of a number of events between $K_1 = \lambda + c_1\sqrt{\lambda}$ and $K_2 = \lambda + c_2\sqrt{\lambda}$ is

$$\sum_{K=K_1}^{K_2} P(K, \lambda) \xrightarrow[\lambda \to \infty]{} \frac{1}{\sqrt{2\pi}} \int_{c_1}^{c_2} dx \, e^{-x^2/2} , \qquad (2.47)$$

where Stirling's formula (2.73) is used for the proof. In other words, the probability distribution for the variable $x \equiv (K - \lambda)/\sqrt{\lambda}$ in the limit $\lambda \to \infty$ tends to the normal distribution

$$p(x) = \frac{1}{\sqrt{2\pi}}\, e^{-x^2/2}. \tag{2.48}$$

2.5 Cauchy–Lorentz Distribution

So far we have considered probability distributions with finite mean value and variance and bound to the central limit of the Gaussian probability density. Let us now consider the probability density termed Cauchy or Cauchy–Lorentz,

$$p(x) = \frac{\gamma}{\pi}\, \frac{1}{\gamma^2 + (x - x_0)^2}\,, \tag{2.49}$$

where $p(x_0) = 1/(\pi\gamma)$ is the peak of the distribution in x_0. This is a good distribution, normalized to 1 and symmetric with respect to x_0, but none of its moments are defined. Let us take the mean value. For each finite symmetric interval, it will be

$$\int_{-a}^{a} dx\, x p(x) = 0\,, \tag{2.50}$$

but for $a \to \infty$ the integral is undetermined. The mean value of the modulus, indeed, does not converge:

$$\int_{-\infty}^{\infty} dx\, |x| p(x) = \infty\,. \tag{2.51}$$

Therefore, by defining the mean value as the limit of the integral, when the extremes tend to infinity, the limit value depends on how the extremes are chosen. For example,

$$\lim_{a\to\infty} \int_{-a}^{a} dx\, x p(x) = 0, \quad \lim_{a\to\infty} \int_{-2a}^{a} dx\, x p(x) = -\ln 2, \quad \lim_{a\to\infty} \int_{-a}^{3a} dx\, x p(x) = \ln 3, \,\ldots.$$

All higher-order moments than the expectation value are infinite.

The distribution of the sum of many Cauchy-distributed events, rather than the binomial or Poisson distribution, does not tend to the Gaussian probability density. We will see what happens in the central limit to events with distributions with divergent moments in Section 3.3.

2.5.1 Generalized Cauchy Distribution

Starting from the Cauchy distribution we can construct a family of similar probability densities of the type

$$p(x|\alpha, \gamma) = \frac{\mathcal{A}(\alpha, \gamma)}{(\gamma^2 + x^2)^{\alpha/2}}\,, \tag{2.52}$$

where the normalization coefficient

$$\mathcal{A}(\alpha, \gamma) = \frac{\gamma^{\alpha-1}}{\sqrt{\pi}} \frac{\Gamma(\alpha/2)}{\Gamma((\alpha-1)/2)}$$

can be computed using the formula (3.98), which we will introduce in the next chapter. For the distribution to be normalizable, one must have $\alpha > 1$, but as the speed at which the tails of the distribution decay varies, we will have different situations. For $\alpha \in (1, 2)$, for example, the mean value does not converge. For $\alpha = 2$ we find the Cauchy–Lorentz distribution. For $\alpha \in (2, 3]$ the mean value is well defined but the variance diverges. Finally, for $\alpha > 3$ the variance is also well defined and equal to

$$\sigma^2 = \frac{\gamma^2}{\alpha - 3} \, . \tag{2.53}$$

Mathematical Appendices

2.A Gaussian Integrals

Gaussian integrals play a crucial role throughout our book. We recall here the fundamental formula for carrying out many of our computations,

$$\int_{-\infty}^{+\infty} dx \, e^{-Ax^2/2 \pm Bx} = \sqrt{\frac{2\pi}{A}} \, e^{B^2/(2A)} \, , \tag{2.54}$$

which for $B = 0$, $A = 1$ tells us that

$$\int_{-\infty}^{+\infty} dx \, e^{-x^2/2} = \sqrt{2\pi}.$$

We first explicitly check how the latter integral is computed, or rather

$$I \equiv \frac{1}{\sqrt{2\pi}} \int_{-\infty}^{+\infty} dx \, e^{-x^2/2} \, . \tag{2.55}$$

Let us consider the two-dimensional integral

$$I^2 = \frac{1}{2\pi} \int_{-\infty}^{+\infty} dx \int_{-\infty}^{+\infty} dy \, e^{-(x^2+y^2)/2} \, , \tag{2.56}$$

and switch to polar coordinates by writing

$$I^2 = \frac{1}{2\pi} \int_0^{2\pi} d\theta \int_0^{+\infty} dr \, r e^{-r^2/2} = \int_0^{+\infty} dr \, r e^{-r^2/2} \, . \tag{2.57}$$

We now switch to the variable $y = r^2/2$, obtaining

$$I^2 = \int_0^{+\infty} dy\, e^{-y} = 1, \tag{2.58}$$

and $I = \sqrt{I^2} = 1$, which brings us to the result of our computation with the formula (2.54).

By rescaling $x \to x\sqrt{A}$ we have (2.54) with $B = 0$. The dependence on B in (2.54) can then be easily obtained by translating the variable x of $A^{-1}B$.

We will often have to deal with D-dimensional integrals. It is convenient to express multivariate integrals in matrix form. In the case of a symmetric matrix $\mathbb{A}$, of elements $A_{i,k}$, and a vector $\vec{B}$, of elements B_k, for the Gaussian integral in D dimensions we have

$$\int_{-\infty}^{+\infty} \left(\prod_{i=1}^{D} dx_i\right) \exp\left(-\frac{1}{2}\sum_{i,k} x_i A_{i,k} x_k \pm \sum_i B_i x_i\right)$$
$$= \sqrt{\frac{(2\pi)^D}{\det \mathbb{A}}}\, \exp\left(\frac{1}{2}\sum_{i,k} B_i (A^{-1})_{i,k} B_k\right). \tag{2.59}$$

When $\vec{B} = 0$, the result can easily be obtained by passing to a basis in which the matrix $\mathbb{A}$ is diagonal, if one remembers that the determinant of a matrix coincides with the product of its eigenvalues. In the same way, the dependence of $\vec{B}$ can subsequently be obtained by translating the variable x_i by the quantity $(\mathbb{A}^{-1}\vec{B}^T)_i = \sum_k (A^{-1})_{i,k} B_k$.

We note that, for $\vec{B} = 0$, the multivariate Gaussian probability density

$$p(\vec{x}|\mathbb{A}) = \sqrt{\frac{\det \mathbb{A}}{(2\pi)^D}}\, \exp\left(-\frac{1}{2}\sum_{i,k} x_i A_{i,k} x_k\right) \tag{2.60}$$

is a well-normalized probability distribution of mean null and covariance matrix $\mathbb{A}^{-1}$. By expanding both sides of (2.59) into $\vec{B}$, we obtain the following formulas:

$$\langle x_i x_k \rangle = (A^{-1})_{i,k},$$
$$\langle x_i x_k x_l x_m \rangle = (A^{-1})_{i,k}(A^{-1})_{l,m} + (A^{-1})_{i,l}(A^{-1})_{k,m} + (A^{-1})_{i,m}(A^{-1})_{l,m}. \tag{2.61}$$

In general, the formula for the average value of the product of l variables x is null for odd l and, for even l, it is given by $(l-1)!!$ terms, which are obtained from each other through permutations. For example,

$$\langle x_i x_k x_l x_m x_r x_s \rangle = (A^{-1})_{i,k}(A^{-1})_{l,m}(A^{-1})_{r,s} + 14 \text{ permutations}. \tag{2.62}$$

2.B Euler Gamma Function

The *Euler Gamma* function can be defined in many ways, for example by the Euler formula

$$\Gamma(z) = \lim_{n\to\infty} \frac{n!\,n^z}{z(z+1)(z+2)\cdots(z+n)}, \tag{2.63}$$

with z different from zero and from any negative integer number. For $\mathrm{Re}(z) > 0$ (i.e., if the real part of z is greater than zero) its integral representation is the Legendre formula

$$\Gamma(z) = \int_0^\infty dx \, x^{z-1} e^{-x} \,. \tag{2.64}$$

The function $1/\Gamma(z)$ is a function with simple zeros at points $z = -n$ (for $n = 0, 1, 2, \ldots$). In other words the function $\Gamma(z)$ has no zeros and has simple poles at zero and the negative integers. For positive integers n the function Γ coincides with the factorial

$$\Gamma(n) = (n - 1)! \,. \tag{2.65}$$

These are remarkable (and useful to know) values of the Γ function:

$$\Gamma(1/2) = \sqrt{\pi} \tag{2.66}$$

and

$$\Gamma(1) = 1 \,. \tag{2.67}$$

It should also be noted that for integer n we have

$$\Gamma(n + 1/2) = \frac{(2n - 1)!!}{2^n} \Gamma(1/2) \,. \tag{2.68}$$

The recurrence relation also applies,

$$\Gamma(z + 1) = z \, \Gamma(z) \,, \tag{2.69}$$

from which, together with (2.67), we obtain

$$\lim_{z \to 0} z\Gamma(z) = 1 \,. \tag{2.70}$$

An asymptotic formula that is often useful should be reported:

$$\lim_{z \to \infty} \frac{\Gamma(z + \alpha)}{z^\alpha \Gamma(z)} = 1 \,. \tag{2.71}$$

Finally, we highlight the formula

$$\frac{1}{\Gamma(z)\Gamma(1 - z)} = \frac{\sin(\pi z)}{\pi} \,, \tag{2.72}$$

which is extremely useful for computing the Γ function for negative values of the argument. Equation (2.72) is almost intuitive: both sides of the equation are real functions that have the same zeros (zero and all positive and negative integers), have the the same derivative at zero, and (a thing which we do not prove) have a similar behavior for $|z|$ large. There are theorems that guarantee the uniqueness of an integer function that satisfies the previous conditions [23].

2.B.1 Stirling's Formula

The asymptotic behavior of the Gamma function, for large values of the real part of its argument, is given by Stirling's formula

$$z! = \Gamma(z+1) \approx (2\pi z)^{1/2} \left(\frac{z}{e}\right)^z , \tag{2.73}$$

which can be derived using Laplace's method; see Appendix 2.C. Stirling's formula has known multiplicative corrections, which go to zero when $z \to \infty$. For example, at first order in $1/z$, the correction is $1 + 1/(12z)$.

More precisely we have that [24]

$$\Gamma(z+1) = (2\pi z)^{1/2} \left(\frac{z}{e}\right)^z \exp\left(1 + \frac{1}{12z} - \frac{1}{360z^3} + \frac{1}{1260z^5} - \frac{1}{1680z^7} + \cdots\right). \tag{2.74}$$

Although the series is not convergent, it is, however, asymptotic: for large z the remainder of the series is of the same order as the first neglected term. The form above provides an excellent and fast method of computing the function Γ with an arbitrarily small relative error. For example, if we want to compute $\Gamma(1/4)$ we can go back to the computation of $\Gamma(n + 1/4)$ by repeatedly using the recurrence relation and, already for $n = 7$, the previous formula has a relative error smaller than $O(10^{-14})$. To obtain smaller relative errors, just go to larger n values and/or use more terms of the asymptotic expansion.

Actually, although it is possible to derive Stirling's formula with Laplace's method, the fastest computation of corrections can be done using the Euler–Maclaurin formula [25]. This gives a compact formula[2] for the order-k approximant for large and non-negative z:

$$\Gamma(z+1) \simeq (2\pi z)^{1/2} \left(\frac{z}{e}\right)^z \exp\left(-\sum_{m=1}^{k} \frac{(-1)^m B_{m+1}}{m(m+1)z^m}\right), \tag{2.75}$$

where B_m denotes the mth Bernoulli number. The Bernoulli numbers are defined by the relation

$$\frac{t}{1-e^{-t}} = \sum_{m=1}^{\infty} \frac{B_m t^m}{m!} , \tag{2.76}$$

and are used in many mathematical formulas.

2.C Laplace's Method

Let us consider a regular function $f(x) \in \mathbb{R}$ that has only one maximum x_0 in the domain $[a, b]$. For large N we want to compute the integral

---

[2] The series is asymptotic [24, 26, 27]; we do not have to worry about it.

$$I = \int_a^b dx\, e^{Nf(x)}.$$ (2.77)

We expand the function in the exponent around the maximum:

$$f(x) = f(x_0) - \frac{1}{2}|f''(x_0)|(x-x_0)^2 + \frac{1}{6}f'''(x_0)(x-x_0)^3 + \frac{1}{24}f''''(x-x_0)^4 + O((x-x_0)^5).$$

We note that if $O((x-x_0)^3) < O((x-x_0)^2)$, for large N this difference is enhanced, $NO((x-x_0)^3) \ll NO((x-x_0)^2)$, and the integral can therefore be approximated as

$$\begin{aligned}
I &\simeq e^{Nf(x_0)} \int_a^b dx\, \exp\left(-\frac{N}{2}|f''(x_0)|(x-x_0)^2 \right. \\
&\qquad\qquad \left. + \frac{N}{6}f'''(x_0)(x-x_0)^3 + \frac{N}{24}f''''(x_0)(x-x_0)^4\right) \\
&\simeq e^{Nf(x_0)} \int_{-\infty}^{\infty} dx\, \exp\left(-\frac{N}{2}|f''(x_0)|(x-x_0)^2\right) \\
&\qquad\qquad \times \left[1 + \frac{N}{6}f'''(x_0)(x-x_0)^3 + \frac{N}{24}f''''(x_0)(x-x_0)^4\right] \\
&= e^{Nf(x_0)} \sqrt{\frac{2\pi}{N|f''(x_0)|}}\left[1 + \frac{1}{8N}\frac{f''''(x_0)}{(f''(x_0))^2} + O\left(\frac{1}{N^2}\right)\right].
\end{aligned}$$ (2.78)

An example of the application of Laplace's method in the computation of integrals is the computation of the Euler Gamma function, for large values of the argument:

$$\Gamma(z+1) = \int_0^\infty dx\, x^z e^{-x} = \int_0^\infty (dy\, z)(y^z z^z)e^{-yz} = z^{z+1}\int_0^\infty dy\, e^{z(\ln y - y)}.$$ (2.79)

In the exponent the real-valued function $f(y) = \ln y - y$ multiplies $z \gg 1$. Using Laplace's maximum method, the integral can then be estimated from the second-order expansion around the maximum of $f(y)$, $y_0 = 1$, where $f(y_0) = f''(y_0) = -1$, to finally obtain

$$\Gamma(z+1) \simeq z^{z+1}e^{-z}\sqrt{\frac{2\pi}{z}},$$ (2.80)

which is nothing other than Stirling's formula (2.73).

3

Law of Large Numbers and Central Limit Theorem

In this chapter, we will discuss the probability distribution of the normalized sum of N random variables

$$y_N \equiv \frac{z_N}{N}, \quad z_N \equiv \sum_{k=1}^{N} x_k, \tag{3.1}$$

for large values of N. We will assume that the random variables x_k are mutually independent and are all distributed according to the same probability distribution, i.e., that the distributions $P_k(x)$ are all identical and equal to $P(x)$. We will use the abbreviation i.i.d. which stands for independent and identically distributed. In other words, the joint probability, see (1.33), is factorized:

$$P(x_1, x_2, \ldots, x_N) = P_1(x_1)P_2(x_2) \cdots P_N(x_N) = \prod_{k=1}^{N} P_k(x_k) = \prod_{k=1}^{N} P(x_k). \tag{3.2}$$

In the last step, we expressed the identity of the distribution for all random variables. In general, two variables x and y are independent if and only if the probability of the composite event (x, y) is factorized into the product of the two separate probabilities,

$$P(x, y) = R(x)Q(y), \tag{3.3}$$

$R(x)$ being the probability distribution of the variable x and $Q(y)$ the distribution of y. Otherwise stated, the probability of x conditioned by y (see Section 1.4.1) does not depend on y, and vice versa, i.e.,

$$P(x|y) = R(x), \quad P(y|x) = Q(y). \tag{3.4}$$

The case in which the variables x_k are not independent is considerably more difficult to deal with. It will be studied in Chapter 12 dedicated to correlations between variables. On the contrary, in the case where the x_k are independent but not identically distributed, we can obtain satisfactory results with a simple generalization of the techniques introduced so far.

The concepts involved in the formulation and proof of the *law of large numbers* and the *central limit theorem* are the basis of our intuitive conception of probability, and are, therefore, to be analyzed with care.

3.1 Laws of Large Numbers

The fundamental concept expressed through the law of large numbers (in its various formulations) can be summarized by stating that, in the limit where $N \to \infty$, the empirical average of a set of variables $y_N = \sum_{i=1}^{N} x_i/N$, cf. (3.1), tends to its theoretical expectation value:

$$\mu \equiv \langle x \rangle = \int dx\, P(x) x \,. \tag{3.5}$$

We immediately note that this law is only true if the mean value of the absolute value of x is finite, $\langle |x| \rangle \equiv \int dx\, P(x)|x| < \infty$. In general, if this condition is not met, the law of large numbers does not apply. We will later discuss, in Section 3.3 dedicated to the generalized central limit theorem, examples in which this is the case.

There are two different theorems, respectively called the *weak* law and the *strong* law. Let us now state the weak one and devote the subsequent section to the strong one.

3.1.1 The Weak Law of Large Numbers

The *weak form* states that the probability that $|y_N - \mu|$ is greater than an arbitrary $\epsilon > 0$ tends to zero when N tends to infinity:

$$\lim_{N \to \infty} \mathcal{P}(|y_N - \mu| \geq \epsilon) = 0, \quad \forall \epsilon > 0. \tag{3.6}$$

In complementary terms, the probability that $|y_N - \mu|$ is less than ϵ tends to 1 when N increases. The weak form only implies that the probability that the relation $|y_N - \mu| < \epsilon$ is true tends to 1 in the limit $N \to \infty$. For every finite N there is always a probability, however small, that $|y_N - \mu|$ is arbitrarily large as N grows.

We shall merely prove the theorem in the weak form in the case with limited variance, using Chebyshev's lemma. However, we emphasize the fact that the law of large numbers does not require as a necessary condition that the probability distribution has a finite variance, as shown, for example, in ref. [28]. In the present book, we will analyze the law of large numbers in the case of unbounded variance in the framework of the generalized central limit theorem (see Section 3.3).

The weak form implies that y_N has a given probability of being arbitrarily close to μ for a given finite value of N. We note that y_N is a random sequence as N increases, and has no determined nor deterministic trend with N. In the limit of $N \to \infty$, the probability tends to 1, but the weak law gives no prescription as to how or how fast. This is called *weak* convergence or *convergence in probability*. On the contrary, we will see that the strong form implies also that the probability that experimental and theoretical averages become close increases as N increases (from a given N onwards).

It should be noted that one often erroneously attributes to the law of large numbers consequences that do not actually arise from it. An example of a wrong but widespread belief is what is known as the *player's fallacy*, i.e., the belief that if the same outcome occurs often, sooner or later in the future, it will occur more infrequently – or the other way round, if a number is never extracted in many trials, its probability in the next trial increases. If *black* comes up in a roulette 26 times in a row, the player's fallacy (or the Monte Carlo fallacy) leads to assigning a high probability that *red* will come up on the next bet, while the probability is always 18/37 (not 1/2, a detail not to be overlooked for gambling-house operators, as we will see in Chapter 10).

We now ask the reader two questions, and invite them to consider and deduce the answers.

1. Two opponents play M times, with sequences of E coin tosses each round. After a single round of E tosses, whoever has made the most heads scores a point. The game is won by who has more points out of M rounds. According to the law of large numbers, can it be expected that for large enough E and/or M there will always be a very similar number of heads and tails during the game?

2. Let us then play a second game. Consider a single very long sequence, won by the player who has chosen heads. By the law of large numbers, can we expect that, on average, in this sequence, there will be a number of heads very similar to that of tails?

That is, can we expect, in the first and in the second game, that the player who won the challenge would have been in the lead on average for half of the time of the game?

While we leave the answer to the first game question to the reader as an exercise, in Figure 3.1 we give some explanatory examples of the trend of the second game, up to 10 000 coin tosses. Let m_i be the value of the coin at the ith toss. We associate the value $m_i = +1$ with the event "head" and the value $m_i = -1$ with the event "tail" and we display 10 behaviors of the random walk

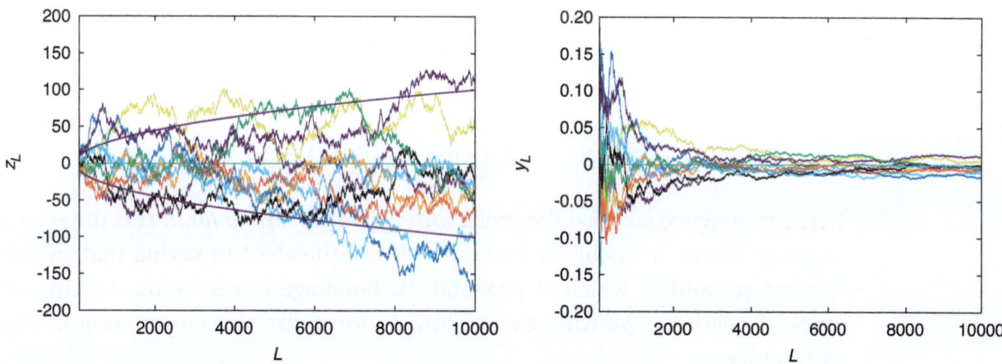

Figure 3.1 Left: Behavior of the sum z_L of heads $(+1)$ and tails (-1) versus the number of tosses L in a sequence of 10 000 total coin tosses. The continuous curves are $\pm\sqrt{L}$ and correspond to a standard deviation of the random motion of the variable z_L. Right: Behavior of the mean $y_L = z_L/L$ of the toss results versus the number of tosses L. The distance between the empirical mean and the theoretical mean $\mu = 0$ decreases as the number of throws increases. As seen in the figure, this reduction is neither smooth nor monotonic.

$$z_L = \sum_{i=1}^{L} m_i$$

as the number of tosses L increases (left panel). In the right panel of Figure 3.1 we display the empirical average

$$y_L = \frac{z_L}{L} = \frac{1}{L} \sum_{i=1}^{L} m_i$$

as L increases, for 10 different sequences.

In the left panel also the curve $\pm\sqrt{L}$ is drawn, corresponding to the mean square deviation of the dynamics of heads and tails. As some of you will have already realized, and as we will see at length in Chapter 7, this dynamics is that of a random walker who can move in one dimension by one step to the right or one step to the left.

3.1.2 Strong Stochastic Convergence and the Strong Law of Large Numbers

As we are about to see, the *strong from* of the law has more remarkable implications than the weak form. The strong formulation starts by considering *infinite sequences* of random variables x_i, $i \in [1, \infty)$, and the corresponding averages y_N. Without posing too many mathematical problems, we introduce the probability distribution on these infinite sequences (the strict mathematical construction is not straightforward). Let us give an example: In the case of Boolean random variables, $x_i = \{0, 1\}$, we can associate the infinite sequence of 0's and 1's with a real number y in the interval $[0, 1]$. Since x_i is random, y will also be random.

The strong form of the law of large numbers states that for all sequences, excluding a subset of zero probability, we have that

$$\lim_{M \to \infty} y_M = \mu . \tag{3.7}$$

In terms of probability, this is sometimes formally written as

$$\mathcal{P}\left(\lim_{M \to \infty} y_M = \mu \right) = 1 . \tag{3.8}$$

This convergence is called the *probability* 1 *limit* or *strong limit*. It is the strongest convergence found in stochastic processes. It is equivalent to saying that y_M tends to μ "almost certainly", which in probabilistic language is a locution defining an infinite series of relations. We will now see how to formalize this enunciation in slightly less abstract terms.

The strong convergence of stochastic sequences, also termed almost certain convergence or convergence with probability 1, imposes that empirical and theoretical mean values are arbitrarily close to each other for every value of N larger than some given integer number N_0, up to infinity.

Excluding a set of null probability, for every arbitrary $\epsilon > 0$ there exists a finite (sequence-dependent) $N_0 > 0$ such that for all $N > N_0$ one has $|y_N - \mu| < \epsilon$ with a probability arbitrarily close to 1. That is, for every $\delta > 0$, we can say that such probability is larger than $1 - \delta$.

The reader should note this interesting fact: if we broaden our horizon to consider infinite sequences, we can have sequences of events with strictly zero probability. This should not surprise us. For example, if we have a real random variable r taking values between 0 and 1, the probability that this variable takes a value exactly equal to 0.5, or to $\exp(-\pi)$, or to $1/\sqrt{2}$, is strictly zero. Excluding a number of sequences with null probability is like excluding a set with zero measure in Lebesgue measure theory, a so-called *null set*. In Lebesgue's theory, we can find statements of this kind: "all real numbers are rational except for a set with null measure", or "every countable set of real numbers has null measure".

The convergence of the strong law very concisely expressed in Eq. (3.8) can be formulated in more elaborate ways.

- Let us consider the probability $Q(N_0)$ of having an infinite sequence such that there exists at least one N greater than N_0 such that $|y_N - \mu| > \epsilon$. The strong law implies that the probability $Q(N_0)$ tends to 0 when $N_0 \to \infty$ for any positive ϵ. Thus, the probability $1 - Q(N_0)$ that the complementary event $|y_N - \mu| < \epsilon$ occurs for *any* $N > N_0$ tends to 1 when N_0 tends to infinity.
- The set of infinite sequences of x_i, such that the relation $|y_N - \mu| < \epsilon$ is violated an infinite number of times, has a probability equal to 0. Formally, we say that there exists an integer $N_0 < \infty$, an $\epsilon > 0$, and, for every arbitrarily small ϵ, a $\delta > 0$ such that

$$\mathcal{P}(|y_N - \mu| > \epsilon) < \delta, \quad \forall N > N_0. \tag{3.9}$$

Notice that the above formula represents an infinite list of conditions, $\forall N > N_0$, up to $N \to \infty$.

As a comparison, the weak statement only implies that the probability that the relation $|y_N - \mu| < \epsilon$ is true for $N = N_0$ tends to 1 when N_0 tends to infinity. Furthermore, the weak form, as opposed to the strong form, has no need to consider infinite sequences. The strong form, on the other hand, implies that for *all* infinite sequences (apart from a null set) the relation $|y_N - \mu| < \epsilon$ is always true for sufficiently large N.

To see that the two laws correspond to mathematically different statement, we take sequences of discrete-valued random numbers consisting of sub-sequences in which the number N of variables to be empirically averaged grows from $N = 2^k + 1$ to $N = 2^{k+1}$:

$$N \in \,]2^k, 2^{k+1}], \quad k = 1, 2, \ldots, \infty.$$

Let us assume something absurd, which we already know cannot be true. We assume that for a given ϵ the sequences are such that in each of them the relation $|y_N - \mu| > \epsilon$ is true once and only once at a randomly chosen point in each of the intervals $]2^k, 2^{k+1}]$. The strong formulation would clearly be false since for each sequence the relation $|y_N - \mu| < \epsilon$ is violated once, with frequency 2^{-k}, but the number of intervals is infinite, because $k \to \infty$, and thus $|y_N - \mu| < \epsilon$ is false an infinite number of times (one for each interval k). There is no "k_0" beyond which the probability of violation is arbitrarily small (i.e., "$< \delta$", as in (3.9)): $|y_N - \mu| > \epsilon$ once per interval, for infinite intervals. Instead, the weak formulation turns out to be true for this sequence, since the probability of violation goes to zero as 2^{-N}: the assumed violation is always one, but the interval becomes exponentially larger and larger, so the violation is not detected by the weak statement, it is not in opposition to it.

For reasons of brevity we do not present the proof of the law of large numbers in the strong form, which is valid under the same conditions as the weak form (i.e., $\langle |x| \rangle < \infty$). The interested reader may consult, for example, the books by Billingsley [13] (Chapter 1) and Feller [28] (Chapter 10).

3.1.3 Chebyshev's Lemma

Chebyshev's lemma allows us to quickly prove the weak statement of the law of large numbers with the sufficient assumption that the variance is finite, $\sigma^2 = \langle x^2 \rangle - \langle x \rangle^2 < \infty$. Before proving Chebyshev's lemma, it is convenient to prove another lemma first.

Proof   Given the probability density $P(x)$ and its expectation value μ, as in Eq. (3.5), we assume that $P(x)$ is non-zero only for non-negative x, $x \in [0, \infty)$. Given a real number $t > 0$, the pre-lemma states that the probability of the occurring event $\{x > t\}$ is less than or equal to μ/t. The proof is straightforward:

$$\mu \equiv \int_0^\infty dx \, P(x)x \geq \int_t^\infty dx \, P(x)x \geq t \int_t^\infty dx \, P(x) \quad \Longrightarrow \quad \int_t^\infty dx \, P(x) \leq \frac{\mu}{t}.$$

$$(3.10)$$

Let us take the probability density $P(x)$ of variance σ^2, assumed finite,

$$\sigma^2 \equiv \langle x^2 \rangle - \mu^2 = \int dx \, P(x)x^2 - \mu^2 = \int dx \, P(x)(x - \mu)^2. \qquad (3.11)$$

Let us consider the variable $(x - \mu)^2$, which is non-negative and has mean value σ^2. Applying the pre-lemma we find that the probability of the event $\{(x - \mu)^2 > t^2\}$ is less than or equal to σ^2/t^2, i.e., that

$$P(|x - \mu| > t) = \int_{|x-\mu|>t} dx \, P(x) < \left(\frac{\sigma}{t}\right)^2, \qquad (3.12)$$

which is the statement of Chebyshev's lemma.   ∎

As we are about to see, Chebyshev's lemma has great theoretical importance, but from a quantitative point of view the estimate (3.12) is often not an estimate of great precision. (It is easy to set tighter limits if further assumptions are made, as we shall see in Chapter 4.)

There is a final curiosity. The lemma is a special case of the Markov inequality [13] which holds for all (converging) moments of a distribution,

$$\langle x^m \rangle = \int dx \, x^m P(x) \geq t^m \int_{|x|>t} dx \, P(x) \quad \Longrightarrow \quad P(|x| > t) \leq \frac{\langle x^m \rangle}{t^m}, \quad \forall t > 0,$$

$$(3.13)$$

where either m is even or x has a non-negative support. The interested reader may observe in the example in Appendix 3.A that this upper bound is not a stringent limit and its precision for a given interval of t depends on the order m of the moment. For small t the most stringent inequality is given by small order moments and for large t by large m moments.

3.1.4 Proof of the Weak Law of Large Numbers

Proof We preliminarily observe that the variance of the sum of two or more independent variables is equal to the sum of the variances. In fact, due to independence, we have $\langle ab \rangle = \langle a \rangle \langle b \rangle$ and, therefore,

$$\langle (a+b)^2 \rangle = \langle a^2 \rangle + 2\langle a \rangle \langle b \rangle + \langle b^2 \rangle. \tag{3.14}$$

By regrouping the terms we have

$$\langle (a+b)^2 \rangle = \langle (a - \langle a \rangle)^2 \rangle + \langle (b - \langle b \rangle)^2 \rangle + \langle a + b \rangle^2. \tag{3.15}$$

Moving the last term to the left of the equals sign proves the thesis. Generalizing the reasoning to N independent variables for the sum variable $z_N = \sum_{i=1}^{N} x_i$ of i.i.d. variables, we obtain

$$
\begin{aligned}
\langle z_N^2 \rangle - \langle z_N \rangle^2 &= \left\langle \left(\sum_{i=1}^{N} x_i \right)^2 \right\rangle - \left\langle \left(\sum_{i=1}^{N} x_i \right) \right\rangle^2 \\
&= \sum_{i,j} \langle x_i x_j \rangle - \sum_{i,j} \langle x_i \rangle \langle x_j \rangle \\
&= \sum_{i=j} \langle x_i^2 \rangle + \sum_{i \neq j} \langle x_i x_j \rangle - \sum_{i=j} \langle x_i \rangle^2 - \sum_{i \neq j} \langle x_i \rangle \langle x_j \rangle \\
&= \sum_{i=1}^{N} \sigma_i^2 + \sum_{i \neq j} (\langle x_i x_j \rangle - \langle x_i \rangle \langle x_j \rangle) = N\sigma^2,
\end{aligned}
\tag{3.16}
$$

where all variances are equal because the x are identically distributed and the last term on the left-hand side of the last line is null because they are independent and their averages factorize. The variance of the quantity $z_N = N y_N$ is given by $N\sigma^2$ while its mean value is obviously $N\mu$. Chebyshev's lemma (3.12) thus implies the following inequality for the probability of the event $\{N|y_N - \mu| > t\}$:

$$P(N|y_N - \mu| > t) < \frac{\sigma^2 N}{t^2}.$$

Choosing $t = N\epsilon$ we have that

$$P(|y_N - \mu| > \epsilon) < \frac{\sigma^2}{N\epsilon^2} \tag{3.17}$$

and for any $\epsilon > 0$ (as small or large as desired) this upper bound tends to zero when N tends to ∞. ∎

3.2 Central Limit Theorem

If we consider N independent and identically distributed random variables x_i, $i \in [1, N]$, under the assumption that the mean μ and variance σ^2 are well defined, the *central limit theorem* states that the probability distribution $c_N(w)$ of the rescaled variable

$$w_N \equiv \frac{1}{\sqrt{N}} \sum_{i=1}^{N} (x_i - \mu) \tag{3.18}$$

tends to a Gaussian distribution of variance σ^2 in the limit $N \to \infty$:

$$c(w) \equiv \lim_{N \to \infty} c_N(w) = \frac{1}{\sqrt{2\pi\sigma^2}} \exp\left(-\frac{w^2}{2\sigma^2}\right). \tag{3.19}$$

Recall that the law of large numbers implies that, in the limit $N \to \infty$, the probability distribution of the variable $y_N \equiv N^{-1/2} w_N$ tends to $\delta(y_N)$.

We will assume in the following, without loss of generality, that $\mu = 0$. The mean value, in fact, can be set to zero by a simple translation of the variable x, using a new variable $x' = x - \mu$.

3.2.1 Constructive Proof of the Central Limit Theorem

Under the simplifying (but not necessary) assumption that the random elementary variables x_k are all distributed according to the same probability distribution $P(x_k)$, we denote by $\tilde{P}(q)$ the Fourier transform (see Appendix 3.C) of the probability distribution $P(x)$:

$$\tilde{P}(q) \equiv \int dx \, e^{iqx} P(x). \tag{3.20}$$

Proof   Let us proceed step by step, and first (re)define the sum variable

$$z_N \equiv \sum_{k=1}^{N} x_k = w_N \sqrt{N}. \tag{3.21}$$

Let us call $d_N(z)$ the probability distribution of the variable z. By repeatedly applying the convolution theorem reported in Appendix 3.C, we find that the Fourier transform of $d_N(z)$ is given by Eq. (3.91), which we anticipate here as

$$\tilde{d}_N(q) = [\tilde{P}(q)]^N = e^{N \ln \tilde{P}(q)}. \tag{3.22}$$

Now changing variables from z to w gives (see Appendix 3.B)

$$dw \, c_N(w) = dz \, d_N(z) = \sqrt{N} \, dw \, d_N(w\sqrt{N}) \tag{3.23}$$

and, consequently, the Fourier transforms of the distributions of w and z are linked by the relation

$$\tilde{c}_N(q) = \int_{-\infty}^{\infty} dw\, e^{iqw} c_N(w) = \sqrt{N} \int_{-\infty}^{\infty} dw\, e^{iqw}\, d_N\left(w\sqrt{N}\right)$$

$$= \int_{-\infty}^{\infty} dz\, e^{iqz/\sqrt{N}}\, d_N(z) = \tilde{d}_N\left(\frac{q}{\sqrt{N}}\right). \tag{3.24}$$

Using (3.22) we obtain

$$\tilde{c}_N(q) = \exp\left\{ N \log\left[\tilde{P}\left(\frac{q}{\sqrt{N}}\right)\right]\right\}, \tag{3.25}$$

where the factor $1/\sqrt{N}$ comes from having rescaled the variable w by a factor $\sqrt{N}$, so that $q \to q/\sqrt{N}$. We now see how the result (3.25) immediately implies that the sum of random variables of finite variance is Gaussian-distributed.

All points q such that $|\tilde{P}(q)| < 1$ yield contributions vanishing exponentially with N. In the $N \to \infty$ limit, only the points where $|\tilde{P}(q)| = 1$ and their surroundings survive. Let us consider for the moment the contribution at $q = 0$, for which $\tilde{P}(0) = 1$, because of the probability normalization, $\tilde{P}(0) = \int dx\, P(x) = 1$. In the limit of large N, we can expand around $q = 0$. By further rescaling $q \to q/\sqrt{N}$ as in Eq. (3.25), in the exponent we have

$$N \ln \tilde{P}\left(\frac{q}{\sqrt{N}}\right) = N \ln \int dx\, \exp\left(\frac{iqx}{\sqrt{N}}\right) P(x) = N \ln \left(1 - \frac{q^2\sigma^2}{2N} + O\left(\frac{q^3 \langle x^3\rangle}{N^{3/2}}\right)\right), \tag{3.26}$$

since we have set $\mu = 0$. Continuing the expansion we obtain

$$N \ln \tilde{P}\left(\frac{q}{\sqrt{N}}\right) = -\frac{q^2\sigma^2}{2} + O\left(\frac{q^3\sigma^3 S}{N^{1/2}}\right), \tag{3.27}$$

where S is the skewness defined in (2.6), which for $\mu = 0$ is

$$S = \frac{\langle x^3\rangle}{\langle x^2\rangle^{3/2}}. \tag{3.28}$$

Then, for large N,

$$\tilde{c}_N(q) = e^{-q^2\sigma^2/2} + O(N^{-1/2}), \tag{3.29}$$

and by inverse Fourier transform (see Appendix 3.C), we obtain

$$c_N(w) = \int_{-\infty}^{\infty} \frac{dq}{2\pi}\, e^{-iqw} \tilde{c}_N(q) = \frac{e^{-w^2/(2\sigma^2)}}{\sqrt{2\pi\sigma^2}} + O(N^{-1/2}), \tag{3.30}$$

and we are done. ∎

We note that for the computation of (3.30) we can use the formulas for Gaussian integrals given in Appendix 2.A, which lead to the result that the Fourier transform of a Gaussian function is still a Gaussian function.

The corrections to the asymptotic Gaussian distribution are of order $N^{-1/2}$ under the assumption that $\langle x^3\rangle$ is finite and non-zero. Recall that the skewness is null for a symmetric probability distribution around $x = \mu$. In the case of a symmetric distribution, i.e., satisfying the condition $P(x) = P(-x)$, all the odd moments $\langle x^{2k+1} \rangle$ are, actually, equal to zero. Therefore, expanding formula (3.25) to the non-trivial higher order, and assuming that the fourth moment $\langle x^4 \rangle$ is finite, we immediately have that

$$\tilde{c}_N(q) = \exp\left(-\frac{q^2 \sigma^2}{2} + \frac{q^4 \sigma^4}{24N}(K-3) + O(N^{-2}) \right), \tag{3.31}$$

where K is the *kurtosis* defined in (2.7) that characterizes the distribution of probability distribution $P(x)$. If the mean $\mu = 0$ we have

$$K \equiv \frac{\langle x^4 \rangle}{\langle x^2 \rangle^2}. \tag{3.32}$$

At this point, we can expand to the first order in $1/N$ and carry out an inverse Fourier transform. A simple computation gives

$$c_N(w) = \frac{e^{-w^2/(2\sigma^2)}}{\sqrt{2\pi\sigma^2}} \left[1 - \frac{K-3}{24N} H_4\left(\frac{w}{\sigma}\right) + O\left(\frac{1}{N^2}\right) \right], \tag{3.33}$$

where $H_4(z)$, i.e., the function governing the dominant corrections to the asymptotic Gaussian distribution, is nothing other than

$$H_4(z) \equiv z^4 - 6z^2 + 3, \tag{3.34}$$

the fourth-order *Hermite polynomial* (see Appendix 3.E for the definition and properties of Hermite polynomials).

In Figure 3.2 we show examples of how the Gaussian distribution of the central limit, with and without order corrections $1/N$, approximates the true distribution $c_N(w)$ of the variable w obtained by summing N random variables with given distribution. The exact $c_N(w)$ shown in the figure are obtained by carrying out an anti-Fourier transform of $\tilde{c}_N(q)$ to N finite, as shown in the examples in Appendix 3.F.

In the same way, when the skewness is non-zero, the dominant corrections are

$$c_N(w) = \frac{e^{-w^2/(2\sigma^2)}}{\sqrt{2\pi\sigma^2}} \left[1 - \frac{S}{6N^{1/2}} H_3\left(\frac{w}{\sigma}\right) + O\left(\frac{1}{N}\right) \right], \tag{3.35}$$

where

$$H_3(z) \equiv z^3 - 3z \tag{3.36}$$

is the third-order Hermite polynomial. Having the explicit form of the dominant corrections allows for quantitative control of how the distribution approaches its asymptotic form for large N.

It is interesting to note that the dominant corrections in the case of a symmetric distribution, proportional to $1/N$, cancels out if the kurtosis K is equal to 3. This is extremely reasonable. In fact, if the distribution $P(x)$ is Gaussian, the central limit distribution $c_N(w)$ must also be Gaussian for every N, since the sum of Gaussian variables is also a variable with a Gaussian distribution, as is easy to verify from the previous formulas: in this case all corrections in inverse powers of N are null. A probability distribution with the same kurtosis as a Gaussian, $K = 3$, therefore bears a strong

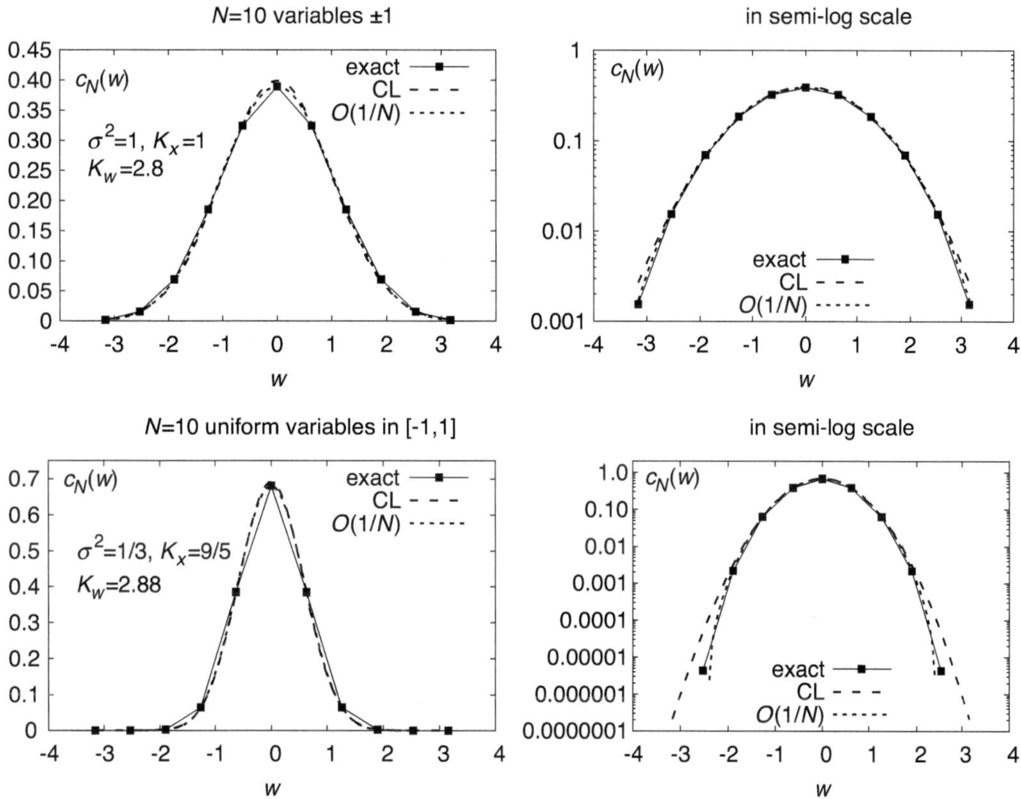

Figure 3.2 Distributions of the variable w, the rescaled sum of $N = 10$ random variables, with a bimodal, or Rademacher, distribution, i.e., $x = \pm 1$ (upper panels), and with a uniform distribution in $x \in [-1, 1]$ (lower panels). The exact distributions (computed in Appendices 3.F.1 and 3.F.2) are compared to their Gaussian central limit (CL) and to the limit distribution corrected at order $1/N$, cf. Eq. (3.33). In the left panels, the graphs are on a linear scale, and in the right panels on a semi-logarithmic scale to enhance the differences in the tails, away from the central area where they coincide.

resemblance to a Gaussian, and the dominant corrections are zero: in the latter case, for a symmetric distribution, the corrections to the central limit will be of order $1/N^2$.

We have neglected to consider the contribution to the central limit distribution of a possible other point $q^* \neq 0$ for which $|\tilde{P}(q^*)| = 1$. In general, it will be $\tilde{P}(q^*) = \exp(\iota\alpha)$. Following the procedure illustrated in the formulas (3.26) and (3.27), now expanding the Fourier transform around $q = q^*$, we can show (see Appendix 3.C.5) that this additional point gives a contribution of the type

$$c_N^*(w) \sim \exp\left(-Aw^2 + \iota N\alpha - \iota\sqrt{N}\,q^* w\right). \tag{3.37}$$

Such a contribution (irrespective of the value of α) tends to zero *in distribution* when N goes to infinity, as is better clarified in Appendix 3.I devoted to convergence in stochastic systems. The fact that this term tends to zero might seem weird at first sight, since for each value of w the probability density $c_N^*(w)$ is a non-zero quantity. However, due to the rapid oscillations of $c_N^*(w)$ when N goes to infinity, for each regular function $f(w)$, the mean value $\langle f \rangle_N$ goes like

$$\lim_{N \to \infty} \langle f \rangle_N = \lim_{N \to \infty} \int dw\, f(w) c_N^*(w) = 0, \tag{3.38}$$

as stated by the Riemann–Lebesgue theorem [29]. In the case where the function $f(w)$ is analytic in a strip around the real axis, the integral in Eq. (3.38) goes to zero exponentially with N. In other words, a limited function whose oscillations become faster and faster goes to zero with *weak convergence* (see Appendix 3.I), also called *convergence in distribution* [30].

3.2.2 Moments-Based Proof of the Central Limit Theorem

We would now like to discuss a proof of the central limit theorem that is based on more restrictive assumptions than in the previous proof. We will assume here *a priori* the existence of all moments of the function c_N. The reason that leads us to illustrate this second proof is that it is more easily generalizable to the case in which the variables x are not independent, but are, instead, correlated with each other, as we will discuss in Chapter 12.

Let us also assume $\mu = \langle x \rangle = 0$ and, for simplicity, suppose that the distribution $P(x)$ is even. Let us consider the variable w_N defined in (3.18). Our proof will be based on the theorem of the moments.

Theorem. (Theorem of the moments) *If a function $f(z)$ is positive and its moments*

$$f_j \equiv \int dz\, f(z) z^j \tag{3.39}$$

are finite and do not grow with j faster than $(j!)^2$, then $f(z)$ is univocally determined by its moments.

We will not prove this theorem here. The interested reader may consult ref. [13], for example.

To prove the central limit theorem, we will show that all even moments of $c_N(w)$ for large N tend to the moments of a Gaussian distribution. The odd moments are null by construction due to the symmetry of $P(x)$. Thanks to the theorem of moments, this is sufficient to prove that, for $N \to \infty$, $c_N(w)$ tends to a Gaussian distribution.

Proof   The moments of the Gaussian are

$$\frac{1}{\sqrt{2\pi\sigma^2}} \int_{-\infty}^{\infty} dw\, e^{-w^2/(2\sigma^2)} w^{2j} = (2j-1)!!\, \sigma^{2j}, \tag{3.40}$$

as can be computed by changing variables to eliminate the dependence on σ and switching to the variable $y = w^2/(2\sigma^2)$. In terms of y the integral can be evaluated using the properties of the Euler Gamma function Γ reported in Appendix 2.B.

Thanks to the theorem of moments, if we can prove that, as $N \to \infty$, all moments of the function $c_N(w)$ tend to those of the formula (3.40), we will have shown that the function $c_N(w)$ tends to a Gaussian distribution (assuming there are no difficulties in exchanging the order in which the limit $N \to \infty$ and the computation of the moments are taken).

The computation of the second moment is straightforward. By indicating with $\langle \cdot \rangle$ the expectation value with respect to the probability distribution $P(x)$, we have

$$\langle w^2 \rangle = \left\langle \left(\frac{1}{\sqrt{N}} \sum_{k=1}^{N} x_k \right)^2 \right\rangle = \frac{1}{N} \left\langle \sum_{k=1}^{N} x_k^2 \right\rangle = \langle x^2 \rangle = \sigma^2 , \tag{3.41}$$

since x_k are independent and the mean is null. The fourth moment has the form

$$\langle w^4 \rangle = \left\langle \left(\frac{1}{\sqrt{N}} \sum_{k=1}^{N} x_i \right)^4 \right\rangle = \frac{1}{N^2} \sum_{i=1}^{N} \sum_{k=1}^{N} \sum_{l=1}^{N} \sum_{m=1}^{N} \langle x_i x_k x_l x_m \rangle . \tag{3.42}$$

The non-zero contributions to this sum are those in which the indices are equal two by two or all four are equal (the two cases must be kept distinct): all other contributions are null due to integration over the symmetric $P(x)$. Thus, it is easily observed that

$$\langle x_i x_k x_l x_m \rangle = (\delta_{i,k}\delta_{l,m} + \delta_{i,l}\delta_{k,m} + \delta_{i,m}\delta_{k,l})\sigma^4 + \delta_{i,k}\delta_{i,l}\delta_{i,m}(\langle x^4 \rangle - 3\sigma^4), \tag{3.43}$$

where δ are Kronecker delta functions. Contributions with products of two deltas are those in which two pairs of indices are equal. They multiply $\langle x^2 \rangle^2$. Contributions with products of three deltas are those in which all indices are equal. Here we have to multiply by $m_4 = \langle x^4 \rangle$ and, in addition, for each of these contributions we subtract a factor that we have added in the terms with two deltas (where the four indices could be equal, thus adding an improper contribution). Performing the sums in (3.42) we find

$$\langle w^4 \rangle = 3\sigma^4 + \frac{(\langle x^4 \rangle - 3\sigma^4)}{N} = \left(3 + \frac{K-3}{N} \right) \sigma^4 , \tag{3.44}$$

where we note that the terms with two deltas receive a factor N^2 from the two residual sums, while the term with three deltas receives a factor N. The dominant term in the $N \to \infty$ limit is, therefore, the one in which all pairs are different, because here is the maximum number of factors N provided by sums that have not been nailed down by δ functions. It is obvious that exactly the same mechanism is valid for higher-order moments: only the terms with distinct pairs of indices do not vanish in the large N limit.

In computing the higher-order moments $\langle w^{2j} \rangle$, the problem remains of evaluating the number $M(2j)$ of non-trivial dominant contributions. Our final result will be

$$\langle w^{2j} \rangle = M(2j)\sigma^{2j} . \tag{3.45}$$

The value $M(2j)$ is given by the number of different ways in which an even number ($2j$) of objects can be grouped two by two. In the case $j = 2$, $M(4)$ is equal to 3: indeed, we can pair the first object with the second, and the third with the fourth, the first with the third, and the second with the fourth, the first with the fourth, and the second with the third. Since we have $2j - 1$ possible choices for the index to be paired with the first index, $2j - 3$ remaining choices for the index to be paired with the first index remaining (the third index), and so on, we can conclude that

$$M(2j) = (2j - 1)(2j - 3) \cdots 1 = (2j - 1)!!, \tag{3.46}$$

which is the desired result. ∎

3.3 Generalized Central Limit Theorem

We are now interested in generalizing the central limit theorem to the case of probability distributions $P(x)$ with infinite variance. Let us take, for example, the sum of random variables with Cauchy distribution; see Section 2.5. By doing the Fourier transform we obtain

$$\tilde{P}(q) = \int_{-\infty}^{\infty} dx \, e^{iqx} \frac{\gamma}{\pi} \frac{1}{\gamma^2 + x^2} = e^{-\gamma|q|}. \tag{3.47}$$

We define the rescaled sum variable as

$$w_N \equiv \frac{1}{N^\rho} \sum_{k=1}^{N} x_k, \tag{3.48}$$

where we can guess that, since there is no finite variance, the typical elements of the sum will no longer be $O(1)$ in modulus and the mean square displacement w_N of their sum will no longer scale as $\sqrt{N}$ as in the previous case of $\sigma^2 < \infty$. Therefore, we allow a stronger rescaling and we denote it by a generic (for the time being) $\rho > 1/2$. Following the building procedure of the previous section and using (3.47), we compute

$$\tilde{c}_N(q) = \tilde{d}_N \left(\frac{q}{N^\rho} \right) = \exp\left[N \ln \tilde{P} \left(\frac{q}{N^\rho} \right) \right] = \exp\left(-N\gamma \frac{|q|}{N^\rho} \right), \tag{3.49}$$

from which we see that in (3.48) we need to have $\rho = 1$ to have a central limit for $N \to \infty$ and

$$c(w) = \int_{-\infty}^{\infty} \frac{dq}{2\pi} e^{iqw} e^{-\gamma|q|} = \frac{\gamma}{\pi} \frac{1}{\gamma^2 + w^2}. \tag{3.50}$$

Eventually, in the central limit, the empirical average $w_N = (1/N) \sum_k x_k$ of Cauchy–Lorentz-distributed random variables is also distributed with the Cauchy–Lorentz probability density, with the same γ parameter.

We can treat the generic case of infinite variance by defining a probability density of the generalized Cauchy type,

$$p_\alpha(x) \propto \frac{1}{(\gamma^2 + x^2)^{\alpha/2}}, \quad 1 < \alpha \leq 3. \tag{3.51}$$

If $\alpha > 3$ the variance of $p_\alpha(x)$ is finite, equal to $\sigma^2 = \gamma^2/(\alpha - 3)$ as shown in (2.53), and we return to the Gaussian case. For $\alpha \leq 3$, however, the variance diverges. For $\alpha = 2$, in particular, we have the Cauchy–Lorentz probability density, which has infinite variance and undetermined expectation value. For $\alpha \in (1,2)$, finally, both variance and mean value diverge. In these cases, we might be inclined to say that the central limit of a suitably rescaled sum variable will reproduce itself and be of the type (3.51), i.e., for large $|w|$, we would have $c(w) \sim |w|^{-\alpha}$, $\alpha \in (1,3]$.

We will see in a moment that, indeed, the central limit exists also in these cases but it depends on the (slow) decay of the specific $p(x)$ for large $|x|$. In the case of the family of (3.51), for example, this is $p_\alpha(x) \sim |x|^{-\alpha}$. Doing the explicit computation of the central limit of distributed variables with the density (3.51) is possible, although more complicated than the simple Cauchy–Lorentz case, and involves modified Bessel functions of the second kind (see, for example, Appendix 3.F).

The discriminating behavior in determining the form of the central limit distribution is, in any case, the behavior of the tails. Let us, then, focus on the behavior for large $|x|$ of generic distributions with infinite variance and, for the sake of simplicity, consider the case of symmetric probability distributions. We will assume that for large values of x we have $p(x) \sim |x|^{-\alpha}$, which can always be written as

$$p(x) = r(x) + \frac{k}{|x|^\alpha}, \tag{3.52}$$

where k is a constant and $r(x)$ is a function that decays quickly for large $|x|$ and is such that $p(x)$ is well normalized. For $\alpha > 3$ the variance is finite. For $\alpha \leq 1$ the integral of $p(x)$ does not converge and, therefore, this cannot be a good normalizable probability distribution. The range of values of interest is that of α smaller than 3 and larger than 1. A finer analysis, which we will not carry out here, might consider cases where the function $p(x)$ does not go to zero as a simple power, but logarithmic corrections are present, for example.

As we show in detail in Appendix 3.C in the generalized case the Fourier transform $\tilde{P}(q)$ is not differentiable around $q = 0$; see also (3.47). Its second derivative is the variance, which is divergent by hypothesis. Its first derivative is the expectation value. This is not defined for $\alpha \leq 2$, and for $\alpha \in (2,3]$ we will set it to zero without loss of generality. For small q the transform eventually behaves like

$$\tilde{P}(q) = 1 + kC(\alpha)|q|^{\alpha-1} + O(q^2), \quad 1 < \alpha \leq 3, \tag{3.53}$$

where

$$C(\alpha) = \frac{\sqrt{\pi}\,\Gamma((1-\alpha)/2)}{2^{\alpha-1}\Gamma(\alpha/2)} = \frac{\pi^{3/2}}{2^{\alpha-1}}\frac{1}{\Gamma((\alpha+1)/2)\Gamma(\alpha/2)\cos(\pi\alpha/2)}. \tag{3.54}$$

In the second step, we used the relation (2.72) to explicitly show that $C(\alpha)$ is a negative function for $1 < \alpha < 3$ because of the cosine. If we define

$$\rho \equiv \frac{1}{\alpha-1}, \tag{3.55}$$

the Fourier transform of the probability density of w_N for large N is given by

$$\exp\left[N\ln\tilde{P}\left(\frac{q}{N^\rho}\right)\right] \propto \exp[-k|C(\alpha)|\,|q|^{\alpha-1}]. \tag{3.56}$$

In this case, the generalized central limit theorem implies that the fluctuations from the mean are of order $N^{-\rho}$, with $\rho \in (1/2, \infty)$. Furthermore, when $N \to \infty$ the probability density $c_N(w)$ is asymptotically given by

$$c(w) \equiv \lim_{N \to \infty} c_N(w) = \frac{1}{2\pi} \int_{-\infty}^{\infty} dq \, \exp\left[\iota q w - k|C(\alpha)| \, |q|^{\alpha-1}\right] \qquad (3.57)$$

which is not a Gaussian distribution but a distribution that for large w goes like

$$c(w) \propto |w|^{-\alpha},$$

just like $p(x)$ for large $|x|$.

In the case $\alpha = 2$, the integration is elementary and the formula (3.50) is obtained again in its full glory,

$$c(w) = \frac{k|C(\alpha)|}{\pi} \frac{1}{w^2 + k^2 C(\alpha)^2}, \qquad (3.58)$$

i.e., a Cauchy–Lorentz distribution (see Section 2.5).

The case $\alpha = 2$ is particularly interesting, as the mean of N variables for N large does not converge to the mean computed on the probability density, but continues to fluctuate, while remaining a number of order 1 when N goes to infinity. In general, for $\alpha < 2$, the expectation value over the limit probability distribution is divergent and the expectation value of N variables is of order $N^{(2-\alpha)/(\alpha-1)}$. In this case, the expectation value becomes larger and larger as N increases.

The situation is very different from that of the standard central limit theorem (with $\rho = 1/2$), in which a single generic term x_i cannot yield a contribution of order 1 to

$$w_N = \sum_{i=1}^{N} \frac{x_i}{\sqrt{N}}$$

because a single variable should be of order $x_i \sim N^{1/2}$. This is impossible in the case of a probability density on a finite support or it happens with extremely small probability in the case of a probability density with infinite support but finite variance (i.e., fast enough decay in the tails).

When the variance is infinite, instead, the music is completely different. If the x variables have a probability distribution that at large x behaves like $P(x) \sim |x|^{-\alpha}$, $\alpha \leq 3$, the maximum X of these N variables lies approximately in the region where

$$\int_{X}^{\infty} dx \, P(x) = O(1/N),$$

or

$$X \sim N^{\rho}.$$

Consequently, a *finite* number of variables gives a contribution of order 1 to the quantity w_N as defined in the formula (3.48). We can say more: the main contribution to Eq. (3.48) comes from a limited number M of terms. More precisely, we can rearrange

x_k in descending order, creating the sequence $\{\tilde{x}_1, \tilde{x}_2, \ldots, \tilde{x}_M, \tilde{x}_{M+1}, \ldots, \tilde{x}_{N-1}, \tilde{x}_N\}$ such that $|\tilde{x}_1| > |\tilde{x}_2| > \cdots > |\tilde{x}_M| > \cdots > |\tilde{x}_N|$. We will have that the rescaled partial sum

$$w_M = \frac{1}{N^\rho} \sum_{k=1}^{M} \tilde{x}_k, \tag{3.59}$$

is a good approximation for w_N and in the limit $N \to \infty$ the difference $w_N - w_M$ goes to zero.

3.4 Stable Distributions

A probability distribution $P(x)$ is called stable if, given two independent variables x and y both with distribution P, the values of the variable $z = A(x + y)$ are still distributed with P, for a suitable choice of A.

It is easy to convince oneself that all functions obtained from the central limit theorems must be stable distributions (although the inverse may not be true). Consider an infinite sequence of independent random numbers $x(i)$ with $i = [0, \infty)$, and a finite sub-sequence of $L = 2^n$ variables. Given an appropriate value of ρ, we can define the quantities

$$x^{(n)}(i) = \frac{1}{L^\rho} \sum_{k=0}^{L-1} x(Li + k). \tag{3.60}$$

Obviously, we have that

$$x^{(0)}(i) = x(i). \tag{3.61}$$

Then for $n = 1$ we have

$$x^{(1)}(i) = 2^{-\rho}[x(2i) + x(2i + 1)] = 2^{-\rho}[x^{(0)}(2i) + x^{(0)}(2i + 1)], \tag{3.62}$$

and so on. The formula (3.60) is a series in i, of rescaled averages, where each term is computed on a disjoint set of variables. Therein, $x^{(n)}(0)$ is the average, appropriately normalized, over the first L values $k \in [0, L - 1]$; $x^{(n)}(1)$ is the average over the second L values $k \in [L, 2L - 1]$; and so on. The $x^{(n)}(i)$ are, therefore, like the $x(i)$, also independent and identically distributed. It is evident that, if ρ is appropriately chosen, the central limit theorem ensures that the variables $x^{(n)}(i)$ have a distribution limit when n tends to infinity.

It is useful to note the recursion formula

$$x^{(n+1)}(i) = 2^{-\rho}[x^{(n)}(2i) + x^{(n)}(2i + 1)], \tag{3.63}$$

which, when combined with (3.61), allows us to recursively compute all $x^{(n)}$ and their probability distributions. Indeed, by the convolution property of the Fourier transform, we obtain

$$\tilde{P}_{n+1}(q) = [\tilde{P}_n(2^{-\rho}q)]^2. \tag{3.64}$$

Since in the limit $n \to \infty$ both $x^{(n+1)}$ and $x^{(n)}$ are distributed according to the probability distribution $P_\infty(x)$, the latter must be a stable probability distribution. The probability distribution $P_\infty(x)$ is then obtained by solving Eq. (3.64) in the limit $n \to \infty$:

$$\tilde{P}_\infty(q) = [\tilde{P}_\infty(2^{-\rho}q)]^2 . \tag{3.65}$$

The general solution of the previous equation is

$$\tilde{P}_\infty(q) = \exp[\, |q| f(\ln|q|) + qg(\ln|q|)]\,, \tag{3.66}$$

where f and g are periodic functions with period $\rho \ln 2$. If we define, for example, $A = 2^{-\rho}$, the period is $|\ln A|$.

The stable distributions we have encountered in the previous paragraphs are characterized by having $g = 0$ and $f = \text{constant}$. The Gaussian distributions have $\rho = \frac{1}{2}$ and then $A = 2^{-1/2}$; the other distributions considered, in the case of infinite variance, have larger ρ and, therefore, smaller values of $A = 2^{-\rho}$. However, if the functions f and g are arbitrary, it should not be taken for granted that the previously written function $\tilde{P}_\infty(q)$ is the Fourier transform of a probability (i.e., a non-negative function). Stable probability distributions have been classified by mathematicians and many results are available [28, 31, 32, 33].

Mathematical Appendices

3.A Markov Limit

Let us take a Gaussian distribution and compute its moments. From (3.40) we know that

$$\langle x^m \rangle = \int_{-\infty}^{\infty} \frac{dx}{\sqrt{2\pi\sigma^2}}\, x^m e^{-x^2/(2\sigma^2)} = (m-1)!!\,\sigma^m$$

if m is even, and 0 otherwise. By Markov's inequality (3.13) this implies, for m even, that the probability that $|x|$ is larger than t is

$$\mathcal{P}(t) \equiv P(|x| > t) \le (m-1)!!\,\frac{\sigma^m}{t^m} .$$

If we have a Gaussian distribution we can compute $\mathcal{P}(t)$ explicitly. The result is a function of t (let us put $\sigma^2 = 1$),

$$\mathcal{P}(t) = 2 \int_t^{\infty} \frac{dx}{\sqrt{2\pi}}\, e^{-x^2/2} = \text{erfc}\left(\frac{t}{\sqrt{2}}\right), \quad t > 0, \tag{3.67}$$

where the complementary error function erfc is defined as

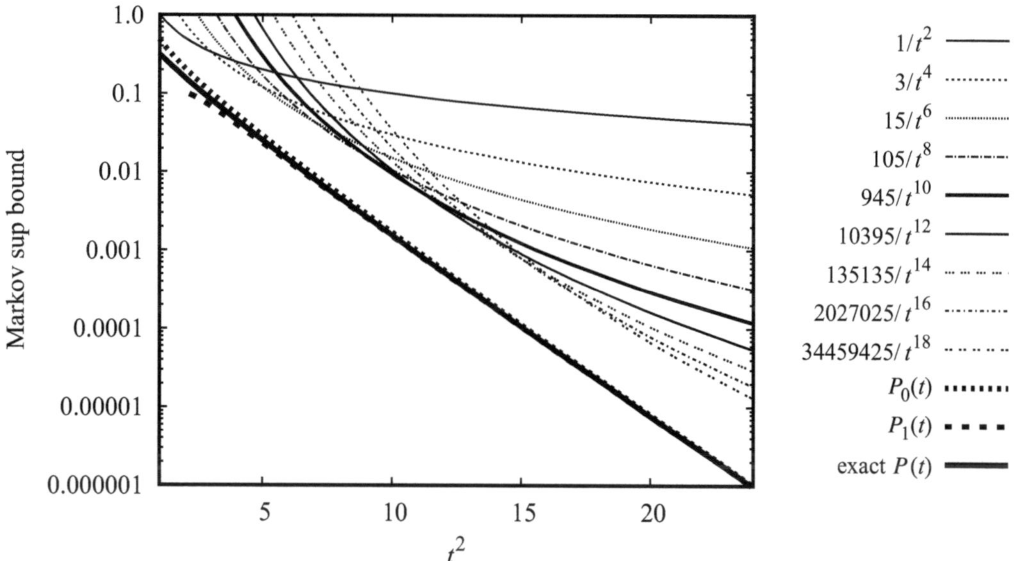

Figure 3.3   Markov upper bounds for moments of increasing order and comparison to the exact expression (3.67) of the probability $\mathcal{P}(t)$ of having a value of the Gaussian variable with modulus greater than t. Also reproduced in the graph are the first two approximants of $\mathcal{P}(t)$ for large t, $\mathcal{P}_0(t)$ and $\mathcal{P}_1(t) \equiv \mathcal{P}_0(t)(1 - 1/t^2)$, according to the formula (3.69).

$$\text{erfc}(w) \equiv \frac{2}{\sqrt{\pi}} \int_w^\infty dy\, e^{-y^2} . \tag{3.68}$$

Its expansions at small and large t are known and can be plotted.

In Figure 3.3 we show the upper Markov approximations up to the 16th moment, together with the exact $\mathcal{P}(t)$ from (3.67). For large t, $\mathcal{P}(t)$ goes like

$$\mathcal{P}(t) = \text{erfc}\left(\frac{t}{\sqrt{2}}\right) = \mathcal{P}_0(t)\left[1 - \frac{1}{t^2} + O(t^{-4})\right], \tag{3.69}$$

with

$$\mathcal{P}_0(t) = \sqrt{\frac{2}{\pi}} \frac{e^{-t^2/2}}{t} .$$

The plot is in semi-log scale and the curves are plotted as a function of t^2, so that the Gaussian is a straight line. It can be seen immediately that the differences between the exact formula of $\mathcal{P}(t)$ and the approximations (3.69) are almost invisible for $t > 4$. It can also be observed from the figure that the Markov limit of lower moments is more exact for small t, while, as t increases, the Markov inequalities at higher-order moments become more and more stringent. In any case, we observe that the upper Markov limit (of which the Chebyshev limit is a special case, for $m = 2$) for the probability $P(|x| > t)$ is rather cautious.

3.B Change of Variables

Suppose that we know the probability density of the random variable x, which we call $P(x)$. We propose to compute the probability density of the quantity

$$y = g(x), \tag{3.70}$$

which we will term $Q(y)$, from knowledge of $P(x)$.

The way to proceed is as follows. We first solve Eq. (3.70) for the variable x. Let $x_\alpha(y)$, with $\alpha = 1, \ldots, n$, be the real roots of this equation, where n can also be infinite. Then the distribution of y is equal to

$$Q(y) = \sum_{\alpha=1}^{n} \left| \frac{dg(x)}{dx} \right|_{x=x_\alpha(y)}^{-1} P(x_\alpha(y)), \tag{3.71}$$

where the derivatives are computed at the roots of Eq. (3.70). We have assumed that there exists no more than a numerable infinity of roots of the equation: this is certainly true, for example, if $g(x)$ is continuous and never constant.

Let us discuss, as an example, the transformation $g(x) = a/x$. Here the equation $y = a/x$ only admits the solution $x_0(y) = a/y$. On the solution, we have that

$$\left. \frac{dg(x)}{dx} \right|_{x=x_0(y)} = -\frac{y^2}{a}, \tag{3.72}$$

which implies that

$$Q(y) = \frac{|a|}{y^2} P\left(\frac{a}{y}\right).$$

If, for example, the starting distribution is a Cauchy–Lorentz distribution, defined by the formula (2.49),

$$P(x) = \frac{\gamma}{\pi} \frac{1}{x^2 + \gamma^2},$$

then we can immediately see that the probability density of $y = a/x$ is still a Cauchy distribution characterized by the parameter $|a|/\gamma$.

Another example is the function $g(x) = a \ln x$. In this case $x_0(y) = e^{y/a}$ and

$$\left. \frac{dg(x)}{dx} \right|_{x=x_0(y)} = \frac{a}{x_0(y)} = ae^{-y/a},$$

$$Q(y) = \frac{e^{y/a}}{|a|} P(e^{y/a}). \tag{3.73}$$

In this case, if someone tells us, as in Section 5.3.3, that the distribution of $y = a \ln x$ is uniform, we can derive the distribution of x from (3.73):

$$\text{constant} = Q(a \ln x) = \frac{x}{|a|} P(x) \quad \Longrightarrow \quad P(x) \propto \frac{1}{x}. \tag{3.74}$$

Similar formulas are true in multiple dimensions. If $\vec{x}$ and $\vec{y}$ are vectors of dimension N whose elements are functions of each other, $y_i = g_i(\vec{x})$, we have that the joint probability of $\vec{y}$ is related to the joint probability density of $\vec{x}$ as

$$Q(\vec{y}) = P(\vec{x}(\vec{y}))|J(\vec{y})|\,,\tag{3.75}$$

where the Jacobian J is the determinant of the $N \times N$ matrix with components

$$J_{i,k} = \frac{\partial x_i(\vec{y})}{\partial y_k}\,.$$

3.C Fourier Transform, Generating Function

Given a function $F(x)$, it is convenient to introduce its *Fourier transform* $\tilde{F}(q)$ defined as

$$\tilde{F}(q) \equiv \mathcal{F}[F(x)] = \int_{-\infty}^{\infty} dx\, e^{iqx} F(x)\,,\tag{3.76}$$

where we have denoted the application of the Fourier transform by the symbol $\mathcal{F}[\,\cdot\,]$. The function $F(x)$ can be obtained again by application of the *inverse Fourier transform* or antitransform, which we denote by $\mathcal{F}_I[\,\cdot\,]$:

$$F(x) = \mathcal{F}_I[\tilde{F}(q)] \equiv \frac{1}{2\pi} \int_{-\infty}^{\infty} dq\, e^{-iqx} \tilde{F}(q)\,.\tag{3.77}$$

Note how we have (legitimately) chosen to distribute the factors π between the transform and antitransform operation and that we have reversed the sign in front of the exponent i. A widely used alternative assignment is to put the factor $1/\sqrt{2\pi}$ in both the transform and the antitransform. The important thing is that the product of the factors in front of the transform and the antitransform gives $1/(2\pi)$. Even the sign on the exponent is not important as long as it is inverse between transform and antitransform. The only thing that matters is that the series combination of the two operations returns the original function. If we insert (3.76) in (3.77) we have, indeed,

$$F(x) = \frac{1}{2\pi} \int_{-\infty}^{\infty} dq\, e^{-iqx} \int_{-\infty}^{\infty} dx'\, e^{iqx'} F(x') = \int_{-\infty}^{\infty} dx'\, F(x') \int_{-\infty}^{\infty} \frac{dq}{2\pi} e^{-iq(x-x')}$$

$$= \int_{-\infty}^{\infty} dx'\, F(x')\delta(x - x') = F(x)\,.\tag{3.78}$$

Here we have introduced the generalized Dirac delta distribution as the antitransform of 1:

$$\delta(x) = \frac{1}{2\pi} \int_{-\infty}^{\infty} dq\, e^{-iqx}\,.\tag{3.79}$$

We note that reversing the sign of the exponent does not change the expression of the delta (as $\delta(x) = \delta(-x)$) and we leave it as an easy exercise to change the factors in (3.76) and (3.77) while keeping their product unchanged.

3.C.1 Generating Function of Moments

If we take a probability density function $P(x)$, we can see that the series expansion of its Fourier transform $\tilde{P}(q)$ around the point $q = 0$ is connected to the values of the moments of the function $P(x)$. If the moments of $P(x)$ are finite, we have

$$\tilde{P}(q) = \sum_k \frac{m_k(iq)^k}{k!}, \tag{3.80}$$

where

$$m_k \equiv \langle x^k \rangle \equiv \int dx\, P(x) x^k. \tag{3.81}$$

Here $\tilde{P}(q)$ is also called the *generating function* of the moments of $P(x)$, as these, instead of integrating $P(x)$, can be obtained by differentiating $\tilde{P}$:

$$m_k = (-\imath)^k \left. \frac{\partial^k \tilde{P}(q)}{\partial q^k} \right|_{q=0}. \tag{3.82}$$

If the probability distribution $P(x)$ has finite variance, the function $\tilde{P}(q)$ is analytic and differentiable at least twice around the point $q = 0$. We will return at length to the generating functions in Chapter 8, dedicated to them, and in Section 12.3.

3.C.2 Generating Function of Cumulants

To generate the cumulants, on the other hand, we can use the logarithm of the Fourier transform of the probability density:

$$C(q) = \ln \tilde{P}(q). \tag{3.83}$$

In this case we will have that the cumulant of order k, $c_k \equiv \langle x^k \rangle_c$, can be obtained by differentiation,

$$c_k = (-\imath)^k \left. \frac{\partial^k C(q)}{\partial q^k} \right|_{q=0}. \tag{3.84}$$

In particular, as we have seen in Section 2.1.2, $c_2 = \sigma^2$, $c_3 = S\sigma^3$, and $c_4 = (K-3)\sigma^4$. By expanding the generator $C(q)$ for small q, we have, indeed,

$$C(q) = -\imath q \langle x \rangle_c - q^2 \frac{1}{2} \langle x^2 \rangle_c + \imath q^3 \frac{1}{6} \langle x^3 \rangle_c + q^4 \frac{1}{24} \langle x^4 \rangle_c + O(q^5)$$

$$= -\imath q \mu - q^2 \frac{1}{2}\sigma^2 + \imath q^3 \frac{1}{6} S\sigma^3 + q^4 \frac{K-3}{24}\sigma^4 + O(q^5).$$

3.C.3 Cumulant Generator of a Gaussian Variable

For a variable with Gaussian distribution (2.38), the exact expression of the cumulant generator can be immediately computed using the formula (2.54). It is

$$C(q) = \ln \int_{-\infty}^{\infty} \frac{dx}{\sqrt{2\pi\sigma^2}} \, e^{\iota q x - (x-\mu)^2/(2\sigma^2)} = \ln \exp\left[-\frac{\sigma^2}{2}\left(q - \iota \frac{\mu}{\sigma^2} \right)^2 \right] - \frac{\mu^2}{2\sigma^2}$$

$$= -\frac{q^2\sigma^2}{2} + \iota q \mu \,, \tag{3.85}$$

which is a simple quadratic polynomial. For this reason, all derivatives of $C(q)$ of order higher than 2 will be zero, thus proving that a Gaussian has only the first two non-zero cumulants; see the formula (2.39).

We will see in Section 12.3 of Chapter 12 the multivariate case and we will elaborate on the generalization of the cumulants for many correlated variables.

3.C.4 Convolution Theorem

The convolution theorem for Fourier transforms is fundamental in many proofs. Let x and y be independent variables, and let us assume

$$z = x + y \,. \tag{3.86}$$

Let $a(x)$, $b(y)$, and $c(z)$ denote the probability distributions of x, y, and z, and $\tilde{A}(q)$, $\tilde{B}(q)$, and $\tilde{C}(q)$ the respective Fourier transforms. We find that the Fourier transform of the probability distribution of z is given by

$$\tilde{C}(q) = \tilde{A}(q)\tilde{B}(q) \,. \tag{3.87}$$

We give here the proof of the result (3.87), which is very simple to obtain.

Proof   The probability that the variable $z = x + y$ takes on a given value is obtained by multiplying the probability of a given value of x by the probability of $y = z - x$ and summing over all the ways in which x can be chosen:

$$c(z) = \int_{-\infty}^{+\infty} dx \, a(x) b(z - x) \,. \tag{3.88}$$

This operation is called the *convolution* product, and we will denote it with an asterisk $*$, i.e.,

$$c = a * b \,.$$

We then prove that

$$\mathcal{F}[c] = \mathcal{F}[a * b] = \mathcal{F}[a]\,\mathcal{F}[b] \,, \tag{3.89}$$

starting with the Fourier transform of (3.88):

$$\tilde{C}(p) = \int_{-\infty}^{\infty} dz \, e^{\iota p z} c(z) = \int_{-\infty}^{\infty} dz \, e^{\iota p z} \int_{-\infty}^{\infty} dx \, a(x) b(z - x)$$

$$= \int_{-\infty}^{\infty} dx \, a(x) \int_{-\infty}^{\infty} dz \, e^{\iota p z} b(z - x) = \int_{-\infty}^{\infty} dx \, a(x) \int_{-\infty}^{\infty} dy \, e^{\iota p (y + x)} b(y) \tag{3.90}$$

$$= \int_{-\infty}^{\infty} dx \, e^{\iota p x} a(x) \int_{-\infty}^{\infty} dy \, e^{\iota p y} b(y) = \tilde{A}(p)\tilde{B}(q) \,,$$

which completes the proof.   ■

A quick alternative proof is to write down the distribution (3.88), $c(z)$, of the sum variable (3.86) as the double integral of the product of the two distributions of the random variables x and y with the constraint that the sum of the two is equal to z, $\delta(z - (x + y))$,

$$c(z) = \int_{-\infty}^{\infty} dx \int_{-\infty}^{\infty} dy \, \delta(z - x - y) a(x) b(y).$$

The Fourier transform thus yields

$$\tilde{C}(p) = \int_{-\infty}^{\infty} dz \, e^{\iota p z} \int_{-\infty}^{\infty} dx \int_{-\infty}^{\infty} dy \, \delta(z - x - y) a(x) b(y)$$

$$= \int_{-\infty}^{\infty} dx \, e^{\iota p x} a(x) \int_{-\infty}^{\infty} dy \, e^{\iota p y} b(y) = \tilde{A}(p) \tilde{B}(p).$$

Generalizing to the convolution of N independent variables x_k, all governed by the same probability distribution $a(x)$, we trivially obtain that the Fourier transform $\tilde{C}_N(q)$ of the probability distribution $c(z)$ of the variable $z = \sum_{k=1}^{N} x_k$ is given by

$$\tilde{C}_N(q) = [\tilde{A}(q)]^N . \tag{3.91}$$

A direct way of seeing this is to write the probability density of z as the integral over the joint distribution of x_k, $p(x_1, x_2, \ldots, x_N) = \prod_{k=1}^{N} A(x_k)$,

$$c_N(z) = \int \prod_{k=1}^{N} dx_k \, p(x_1, x_2, \ldots, x_N) \delta\left(z - \sum_{k=1}^{N} x_k\right) = \int \prod_{i=1}^{N} dx_k \, A(x_k) \delta\left(z - \sum_{k=1}^{N} x_k\right),$$

for which the Fourier transform $\tilde{C}_N(q)$ becomes

$$\tilde{C}_N(q) = \int \prod_{k=1}^{N} dx_k \, A(x_k) \int_{-\infty}^{\infty} dz \, e^{\iota q z} \delta\left(z - \sum_{k=1}^{N} x_k\right)$$

$$= \prod_{k=1}^{N} \int dx_k \, A(x_k) e^{\iota q x_k} = [\tilde{A}(q)]^N .$$

3.C.5 Some Details on the Fourier Transform in the Central Limit Theorem Proof

In the proof of Section 3.2.1 we mentioned the contribution of a possible other point $q^* \neq 0$ where $|\tilde{P}(q)| = 1$. Considering the generic form $\tilde{P}(q^*) = \exp(\iota \alpha_0)$ we have stated, see formula (3.37), that this contribution to the probability distribution of the variable $w = \sum_{k=1}^{N} x_k / \sqrt{N}$ has the behavior

$$c_N^*(w) \propto \exp\left(- A w^2 + \iota N \alpha_0 - \iota \sqrt{N} q^* w\right).$$

Such a contribution, according to the Riemann–Lebesgue theorem [13], tends to zero in weak topology (see Appendix 3.I) when N goes to infinity and does not count in the central limit.

Let us see how this trend emerges. We define the generalized moments of the distribution $e^{\iota q^* x} P(x)$:

$$x_n(q^*) \equiv \int dx\, x^n e^{\iota q^* x} P(x) = A_n e^{\iota \alpha_n}, \quad \text{with} \quad \alpha_n(q^*), A_n(q^*) \in \mathbb{R},$$

$$= \int dx\, x^n \cos(q^* x) P(x) + \iota \int dx\, x^n \sin(q^* x) P(x). \tag{3.92}$$

We notice that in the case where (without loss of generality) we translate the variables so that $\mu = \int dx\, x P(x) = 0$, the generalized moment $x_1(q^*)$ can be non-zero.

If $P(x)$ is symmetric, from (3.92) we have that the imaginary part of the even generalized moments is zero, as well as the real part of the odd generalized moments:

$$\mathrm{Im}\,[x_{2k}] = \mathrm{Re}\,[x_{2k+1}] = 0. \tag{3.93}$$

Therefore, $|x_0| = x_0$, and, by definition $\tilde{P}(q^*) = e^{\iota \alpha_0}$, one must have $\alpha_0 = 0$.

The Fourier transform of the distribution of the variable $w = \sum_{k=1}^{N} x_k / \sqrt{N}$ is given by the formula (3.25) as

$$\tilde{c}_N(q) = \tilde{d}_N(q/\sqrt{N}) = e^{N \ln \tilde{P}(q/\sqrt{N})}.$$

To compute it in general, with $\alpha_0 \neq 0$ (thus $P(x)$ not symmetric), we expand around a point $q \neq q^*$ where $|\tilde{P}(q^*)| = |e^{\iota \alpha_0}| = 1$:

$$\tilde{P}(q) = \int dx\, e^{\iota q x} P(x) = e^{\iota \alpha_0} + \iota x_1(q^*)(q - q^*) - \frac{x_2(q^*)(q - q^*)^2}{2} + O((q - q^*)^3).$$

Hence

$$\tilde{d}_N(q) \simeq \exp\left(\iota N \alpha_0 + \iota N (q - q^*) \bar{x}_1(q^*) - \frac{N(q - q^*)^2}{2} \sigma^2(q^*) + O(N(q - q^*)^3) \right),$$

$$\tag{3.94}$$

where we have defined the functions

$$\bar{x}_1(q^*) \equiv \frac{x_1(q^*)}{x_0(q^*)} = A_1(q^*) e^{-\iota(\alpha_0 - \alpha_1(q^*))}, \tag{3.95}$$

$$\sigma^2(q^*) \equiv \frac{x_2(q^*)}{x_0(q^*)} - \left(\frac{x_1(q^*)}{x_0(q^*)} \right)^2 = A_2 e^{-\iota(\alpha_0 - \alpha_2(q^*))} - A_1^2(q^*) e^{-2\iota(\alpha_0 - \alpha_1(q^*))}. \tag{3.96}$$

From (3.23) we know that

$$c_N(w) = \sqrt{N}\, d_N(w\sqrt{N}),$$

which rewritten in Fourier antitransforms gives the identity

$$c_N^*(w) = \frac{1}{2\pi} \int_{-\infty}^{\infty} dq\, e^{-\iota q w} \tilde{c}_N(q) = \frac{\sqrt{N}}{2\pi} \int_{-\infty}^{\infty} dq\, e^{-\iota q \sqrt{N} w} \tilde{d}_N(q) = \sqrt{N} d_N^*(w\sqrt{N}). \tag{3.97}$$

Substituting (3.94) into this we obtain

$$c_N^*(w) = \frac{\sqrt{N} e^{\iota N \alpha_0}}{2\pi} \int_{-\infty}^{\infty} dq\, \exp\left(-\iota q \sqrt{N} w + \iota N (q - q^*) \bar{x}_1(q^*) - \frac{N(q - q^*)^2}{2} \sigma^2(q^*) \right).$$

Rescaling $k \equiv q - q^*$ and integrating in k finally gives the expression

$$c_N^*(w) = \frac{1}{\sqrt{2\pi\sigma^2(q^*)}} \exp\left(-\frac{w^2}{2\sigma^2(q_*)} + \imath N\alpha_0 - \imath\sqrt{N}wq^*\right)$$
$$\times \exp\left(\sqrt{N}w\frac{\bar{x}_1(q*)}{\sigma^2(q^*)} - \frac{N[\bar{x}_1(q^*)]^2}{2\sigma^2(q^*)}\right),$$

whence (3.37). Since $q^* \in \mathbb{R}$, for growing N the oscillating term $e^{\imath\sqrt{N}wq^*}$ makes any integral over the probability density $c_N^*(w)$ equal zero and does not contribute to the central limit. As explained in Appendix 3.I, the distribution $c_N^*(w)$ is weakly convergent to zero.

Finally, we observe that for $q^* \to 0$ ($\alpha_0 \to 0$), we have $\bar{x}_1(q^*) \to \mu = 0$ and $\sigma^2(q^*) \to \sigma^2$, and we obtain the expression of Section 3.2.1:

$$c_N^*(w) \to c_N(w) = \frac{1}{\sqrt{2\pi\sigma^2}} \exp\left(-\frac{w^2}{2\sigma^2}\right).$$

3.C.6 Fourier Transform of Power-Law Decay Functions

We want to discuss here the asymptotic behavior of Fourier transforms. Using physicists' jargon, we can say that typically we will be interested in the behavior at small moments q (where q is the conjugated variable in Fourier space), i.e., at large distances x (where x is the original variable). In particular, we are interested in dealing with those distributions that have such fat tails as $|x| \to \infty$ that their variance is infinite. This information is essential to extend the central limit theorem also to the sum of random variables of this type.

We first compute the Fourier transform of the function $|x|^{-\alpha}$. Formally, our computation is valid for $\alpha < 1$. The result we are most interested in for the generalized central limit (see Section 3.3) is, however, for $1 < \alpha \le 3$. As we shall see, it can be obtained by analytic continuation, explicitly verifying the correctness of this procedure.

3.C.6.1 Inverse Power Transform, with Exponent $\alpha \in [0, 1)$

Though we will be mainly interested in the result on the real axis $\mathbb{R}$, i.e., in $D = 1$, here we will perform the computation for a generic dimensionality D.[1]

We base our computation on the use of the integral representation

$$A^{-\gamma} = \frac{1}{\Gamma(\gamma)} \int_0^\infty dz\, z^{\gamma-1} e^{-Az}, \tag{3.98}$$

which can be immediately derived by introducing the variable $y \equiv zA$ and using the definition of the Euler Gamma function (2.64). We then compute the Fourier transform of $|\vec{x}|^{-\alpha}$, where $\vec{x}$ is a D-dimensional vector:

---

[1] For the sole case of dimensionality $D = 1$, the same result is obtained simply by dividing the integral into two integrals from $-\infty$ to 0 and from 0 to $+\infty$ and using the integral representation (3.98) of the inverse power.

$$\tilde{F}(\vec{q}) \equiv \int d^D x \, e^{i\vec{q}\cdot\vec{x}}(\vec{x}\cdot\vec{x})^{-\alpha/2} = \frac{1}{\Gamma(\alpha/2)} \int_0^\infty dz \, z^{\alpha/2-1} \int d^D x \, e^{i\vec{q}\cdot\vec{x}} e^{-z(\vec{x}\cdot\vec{x})} . \tag{3.99}$$

Here we have used the representation (3.98) with $A = |\vec{x}|^2$ and $\gamma = \alpha/2$. We can now compute the Gaussian integral in $d^D x$ using Eq. (2.59), obtaining

$$\tilde{F}(\vec{q}) = \frac{\pi^{D/2}}{\Gamma(\alpha/2)} \int_0^\infty dz \, z^{\alpha/2-D/2-1} \exp\left(-\frac{\vec{q}\cdot\vec{q}}{4z}\right) . \tag{3.100}$$

The behavior in $\vec{q}$ can be determined from dimensional analysis, as explained in Appendix 3.G. Alternatively, a direct computation can be performed: by changing the variable $y = |\vec{q}|^2/(4z)$ the integral in dz represents a Γ function, cf. Eq. (2.64). We then find

$$\tilde{F}(q) = C(\alpha, D)|\vec{q}\,|^{\alpha-D} , \tag{3.101}$$

with

$$C(\alpha, D) \equiv \frac{\pi^{D/2}\Gamma((D-\alpha)/2)}{2^{\alpha-D}\Gamma(\alpha/2)} , \tag{3.102}$$

which is the final result of our computation. We notice that, as $q \to 0$, the transform is non-differentiable.

We stress that the computation is formally correct for $\alpha < D$, otherwise, the integrals would not converge at the origin ($|q|^{\alpha-D} \to \infty$ for $q \to 0$). In the one-dimensional case, this condition becomes $\alpha < 1$, which is not a useful range of power exponents for the proof of the generalized central limit theorem (see Section 3.3). To move on to the regime we are interested in, we must make some further considerations.

First, let us consider a function $F(x)$ generically different from $|x|^{-\alpha}$, but behaving like it asymptotically, for large $|x|$. In this case the non-differentiable behavior of the Fourier transform $\tilde{F}(q)$ at small q does not change. This is true, for example, if $F(\vec{x})$ differs from $|\vec{x}|^{-\alpha}$ for a function $G(\vec{x})$ which goes to zero quickly for $x \to \infty$ and makes $F(\vec{x})$ a regular function at $\vec{x} = \vec{0}$.

In the following we set $D = 1$ for simplicity and write

$$F(x) = A|x|^{-\alpha} + G(x). \tag{3.103}$$

In this case

$$\tilde{F}(q) = c(\alpha)A|q|^{\alpha-1} + \tilde{G}(q), \tag{3.104}$$

where

$$c(\alpha) = C(\alpha, 1) = \frac{\pi^{1/2}\Gamma((1-\alpha)/2)}{2^{\alpha-1}\Gamma(\alpha/2)} , \tag{3.105}$$

and $\tilde{G}(q)$ is a regular function of q also at $q = 0$. Unfortunately, the argument fails for $\alpha > 1$, as in this case, even though $F(x)$ is regular for $x \to 0$, both $x^{-\alpha}$ and $G(x)$ have non-integrable singularities.

The first temptation would be to close one's eyes to this double problem: derive formula (3.104) in the $\alpha < 1$ range and, keeping our fingers crossed, extend it analytically to $\alpha > 1$. This way of guessing formulas, somewhat blindly, sometimes works; at other times it gives terrifying results. So some skepticism may be necessary.

3.C.6.2 Inverse Power Transform, with Exponent $\alpha \in [1, 3[$

Let us, hence, first test the validity of the approach by considering a concrete example, where the maths can be carried out explicitly. We have already seen at the beginning of Section 3.3 that for $\alpha = 2$ the generalized central limit can be constructed easily and leads to a Cauchy–Lorentz distribution. We now compute the Fourier transform of the generalized Cauchy function (see Section 2.5.1):

$$F(x) = \frac{1}{(\gamma^2 + x^2)^{\alpha/2}} \, . \tag{3.106}$$

With manipulations similar to those used in the previous computation, we obtain

$$\tilde{F}(q) = \frac{\sqrt{\pi}}{\Gamma(\alpha/2)} \int_0^\infty dz \, z^{(\alpha-1)/2-1} e^{-q^2/(4z)-\gamma^2 z} \, . \tag{3.107}$$

The singular behavior at small q is controlled by the region at small z: for small z, indeed, the term in γ^2 is small compared to the term in q^2. This region does not depend on the term in γ^2, and has the same properties as the function with $\gamma = 0$ that we analyzed earlier. In this particular case, the integration can be carried out explicitly *also for $\alpha > 1$* (and we invite the reader to do so): it leads to a modified second-order Bessel function of the type $\mathcal{K}$ (see Eqs. (3.127) and (3.128) in Appendix 3.F), and it is possible to explicitly analyze the form of the singular terms.

Instead, we now go on presenting a slightly more detailed argument helping to overcome the difficulties previously exposed for generic functions with inverse power decay and achieve the final result for $\alpha > 1$ in a clean way. The interesting case of values of α between 2 and 3 can easily be analyzed by considering the Fourier transform of the function $F_2(x) \equiv F(x)x^2$. Indeed, it holds that

$$\tilde{F}_2(q) = -\frac{d^2 \tilde{F}(q)}{dq^2}. \tag{3.108}$$

Moreover, Eq. (3.104) with an exponent $\alpha - 2$ yields

$$\tilde{F}_2(q) = Ac(\alpha - 2)|q|^{\alpha-3} + \tilde{G}_2(q), \quad 0 < \alpha - 2 < 1, \tag{3.109}$$

where we denote by $\tilde{G}_2(q)$ the regular terms in q. Integrating with respect to q twice, from (3.108) we have

$$\tilde{F}(q) = A \frac{-c(\alpha - 2)}{(\alpha - 1)(\alpha - 2)}|q|^{\alpha-1} + \tilde{G}(q), \quad 2 < \alpha < 3, \tag{3.110}$$

which is exactly the desired formula, as a simple computation shows that the following recursion relation is valid:

$$\frac{-c(\alpha - 2)}{(\alpha - 1)(\alpha - 2)} = c(\alpha), \tag{3.111}$$

as we could reasonably expect. We emphasize that for $1 < \alpha < 3$ one has $c(\alpha) < 0$ and

$$\tilde{F}(q) = -A|c(\alpha)| \, |q|^{\alpha-1} + \tilde{G}(q), \tag{3.112}$$

where $\tilde{G}(q) = g_0 + g_1 q + g_2 q^2 + \cdots$ is a regular function of q for small q. If $g_1 < \infty$, by a translation of x we can always take $g_1 = 0$.

In conclusion, for $\alpha < 3$ the singular terms we have found are dominant over the regular terms proportional to q^2, while in the region $\alpha > 3$ the opposite occurs. However, in the latter case, we do not need singular terms for applications to the central limit, because we are back to the case of finite variance. If $\alpha \in (1, 2]$ we will have that the singular term also dominates over the order q.

In the case where $F(x) = P(x)$ is a probability density, the constant g_0 is 1 because of the normalization condition $\int dx\, P(x) = \tilde{P}(0) = 1$.

3.D Laplace Transform

Given a real function $p(x)$ of real variable x, its Laplace transform between 0 and ∞, sometimes called *unilateral*, is defined as

$$\hat{P}(h) \equiv \int_0^\infty dx\, e^{-hx} p(x) \tag{3.113}$$

with $h \in \mathbb{C}$. In addition to this, there is also the *bilateral* Laplace transform, between $-\infty$ and ∞, defined as

$$\hat{P}(h) \equiv \int_{-\infty}^\infty dx\, e^{-hx} p(x). \tag{3.114}$$

The antitransform is given by the formula

$$p(x) = \lim_{L \to \infty} \int_{\gamma - \iota L}^{\gamma + \iota L} \frac{dh}{2\pi \iota} e^{hx} \hat{P}(h), \tag{3.115}$$

where the real parameter γ is a number such that the entire integration path lies in the convergence disk of $\hat{P}(h)$. Eventually, γ may tend to infinity. In the applications of interest to probability densities, as we shall see in the case of generating functions in Chapter 8, the integration path is often closed and the inverse transform can be computed with the residue theorem.

If h is a pure imaginary number in formula (3.114) and we write it as $h = -\iota q$, with $q \in \mathbb{R}$, we obtain the Fourier transform (3.76), where $\tilde{F}(q) = \hat{F}(-\iota q)$. Similarly, by making the same substitution in the integration variable of (3.115), we obtain the formula of the Fourier antitransform (3.77) again.

3.E Hermite Polynomials

If we call

$$c(x) = \frac{1}{\sqrt{2\pi}} e^{-x^2/2}$$

the Gaussian distribution with null mean and unity variance, also called the normal distribution, the *probabilistic* definition of the Hermite polynomials is

$$H_n(x) \equiv \frac{(-1)^n}{c(x)} \frac{d^n c(x)}{dx^n}, \tag{3.116}$$

where $x \in \mathbb{R}$. These polynomials have several interesting properties. One is that polynomials of different degrees are orthogonal to each other with respect to the normal measure $dC(x) = c(x)\,dx$, i.e.,

$$\int_{-\infty}^{\infty} dx\, c(x) H_n(x) H_k(x) = n!\,\delta_{n,k}. \tag{3.117}$$

Let us now take a Hilbert space of complex-valued functions $(a(x), b(x), \ldots)$ in which the inner product is defined as

$$\langle a, b \rangle \equiv \int_{-\infty}^{\infty} dx\, c(x) a(x) \overline{b(x)}$$

and for which

$$\int_{-\infty}^{\infty} dx\, c(x) |a(x)|^2 < \infty.$$

The orthogonality condition (3.117) tells us that, in the Hilbert space of functions in L^2 with normal measure, the Hermite polynomials form a complete orthonormal basis.

If we set $x = w/\sigma$ in (3.116) we obtain the formulas (3.34), for $n = 4$, and (3.36), for $n = 3$.

3.F Exact Distributions of Sums of Stochastic Variables

3.F.1 Sum of Uniformly Distributed Variables

We want to know the exact distribution $c_N(w)$ for any N of the variable

$$w \equiv \frac{1}{\sqrt{N}} \sum_{i=1}^{N} (x_i - \mu),$$

i.e., the sum of variables with uniform distribution in the domain $x \in [-1, 1]$,

$$P(x) = \tfrac{1}{2}[\Theta(x+1) - \Theta(x-1)] = \tfrac{1}{4}[\text{sign}(x+1) - \text{sign}(x-1)].$$

The result can be derived by following the constructive proof of the central limit theorem up until the large N limit is taken. In particular, we make use of the result (3.91) of the convolution. The above two expressions of $P(x)$ take into account that

$$\text{sign}(z) = 2\Theta(z) - 1, \tag{3.118}$$

where Θ is the Heaviside step function.

The Fourier transform of this distribution is

$$\tilde{P}(q) = \int_{-\infty}^{\infty} dx \, e^{\imath qx} P(x) = \frac{1}{2} \int_{-1}^{1} dx \, e^{\imath qx} = \frac{e^{\imath q} - e^{-\imath q}}{2\imath q} = \frac{\sin q}{q}.$$

The Fourier transform of the distribution of w is the same as the Fourier transform of the convolution of N variables, as in (3.91), rescaled by $1/\sqrt{N}$. Thus

$$\tilde{c}_N(q) = \left[\tilde{P}\left(\frac{q}{\sqrt{N}} \right) \right]^N = N^{N/2} \frac{\left(e^{\imath q/\sqrt{N}} - e^{-\imath q/\sqrt{N}} \right)^N}{2^N (\imath q)^N}$$

$$= \frac{N^{N/2}}{2^N (\imath q)^N} \sum_{K=0}^{N} \binom{N}{K} e^{\imath q(N-K)/\sqrt{N}} (-1)^K e^{-\imath q K/\sqrt{N}}$$

and its antitransformation leads to the required result:

$$c_N(w) = \frac{N^{N/2}}{2^N} \sum_{K=0}^{N} \binom{N}{K} (-1)^K \int_{-\infty}^{\infty} \frac{dq}{2\pi} \frac{e^{-\imath qw} e^{\imath q(N-2K)/\sqrt{N}}}{(\imath q)^N}.$$

The last thing left to do is to evaluate the integral

$$I(t) = \int_{-\infty}^{\infty} \frac{dq}{2\pi} \frac{e^{\imath qt}}{(\imath q)^N} \quad \text{where} \quad t \equiv \frac{N - 2K}{\sqrt{N}} - w.$$

At least two approaches are possible. In one, the residue theorem is applied to the function $e^{\imath qt}/(\imath q)^N$, which has a pole of order N at $z \equiv \imath q = 0$. Considering two different integration paths for $t > 0$ and $t < 0$ we obtain

$$I(t) = \int \frac{dz}{2\pi\imath} \frac{e^{zt}}{z^N} = \frac{\text{sign}(t)}{2\pi\imath} \frac{\imath\pi}{(N-1)!} \frac{d^{N-1}}{dz^{N-1}} e^{zt} \bigg|_{z=0} = \frac{\text{sign}(t) t^{N-1}}{2\Gamma(N)}. \tag{3.119}$$

In the second approach, we use the identity (3.98) and

$$I(t) = \frac{1}{\Gamma(N)} \int_0^{\infty} d\zeta \, \zeta^{N-1} \int_{-\infty}^{\infty} \frac{dq}{2\pi} e^{\imath q(t-\zeta)} = \frac{1}{\Gamma(N)} \int_0^{\infty} d\zeta \, \zeta^{N-1} \delta(t-\zeta) = \frac{\Theta(t) t^{N-1}}{\Gamma(N)}.$$

We thus obtain, respectively, the two forms

$$c_N(w) = \frac{N^{N/2}}{2^{N+1}\Gamma(N)} \sum_{K=0}^{N} \binom{N}{K} (-1)^K \left(w - \frac{2K - N}{\sqrt{N}} \right)^{N-1} \text{sign}\left(w - \frac{2K - N}{\sqrt{N}} \right),$$

$$\tag{3.120}$$

$$c_N(w) = \frac{N^{N/2}}{2^N \Gamma(N)} \sum_{K=0}^{N} \binom{N}{K} (-1)^K \left(w - \frac{2K - N}{\sqrt{N}} \right)^{N-1} \Theta\left(w - \frac{2K - N}{\sqrt{N}} \right),$$

$$\tag{3.121}$$

which are equivalent, as can be demonstrated using the relation (3.118) and unwinding the sums in K.

3.F.2 Sum of Rademacher-Distributed Variables

The exact distribution $c_N(w)$ for any N of the variable

$$w \equiv \frac{1}{\sqrt{N}} \sum_{i=1}^{N} (x_i - \mu),$$

i.e., the sum of random variables with a bimodal distribution,

$$P(x) = \tfrac{1}{2}[\delta(x+1) + \delta(x-1)],$$

also called a Rademacher distribution, can be computed using Fourier transforms, exploiting the convolution property (3.91) for the sum of N variables, obtaining

$$c_N(w) = \frac{1}{2^N} \sum_{k=0}^{N} \binom{N}{k} \delta\left(w - \frac{2k-N}{\sqrt{N}}\right). \tag{3.122}$$

3.F.3 Sum of Exponentially Distributed Variables

The exact distribution $c_N(w)$ for any N of the variable

$$w \equiv \frac{1}{\sqrt{N}} \sum_{i=1}^{N} (x_i - \mu),$$

i.e., the sum of variables with exponential distribution of the modulus

$$P(x) = \frac{1}{2\tau} e^{-|x|/\tau}, \tag{3.123}$$

can be computed from the Fourier transform of $P(x)$,

$$\tilde{P}(q) = \frac{1}{q^2 \tau^2 + 1},$$

using (3.98), i.e.,

$$c_N(w) = \frac{1}{2\pi} \int dq \, \frac{e^{-iqw}}{[1+q^2\tau^2/N]^N} = \frac{\sqrt{N}}{2\tau\sqrt{\pi}} \frac{1}{\Gamma(N)} \int_0^{\infty} dz \, z^{N-3/2} \exp\left(-z - \frac{Nw^2}{4\tau^2 z}\right). \tag{3.124}$$

To carry out the computation of $c_N(w)$ we can start from the useful result (for $N = 1$, $w = x$, and $A = x/(2\tau)$)

$$\int_0^{\infty} \frac{dz}{\sqrt{z}} \exp\left(-z - \frac{A^2}{z}\right) = \sqrt{\pi}\, e^{-2|A|}, \tag{3.125}$$

which returns the exponential distribution from the starting $c_1(x) = P(x)$ (3.123). We can also observe that (3.124) can be rewritten as

$$c_N(w) = \frac{1}{\sqrt{N\pi}\,\tau\Gamma(N)} \left(\frac{|w|}{2\tau\sqrt{N}}\right)^{N-1/2} K_{1/2-N}\left(\frac{\sqrt{N}|w|}{\tau}\right), \tag{3.126}$$

where

$$\mathcal{K}_M(|y|) = \frac{1}{2^{M+1}|y|^M} \int_0^\infty dz\, z^{-M-1} \exp\left(-z - \frac{y^2}{4z}\right) \tag{3.127}$$

is one of the possible integral expressions of the modified Bessel function of the second kind of order M, the solution of the differential equation

$$y^2 \frac{d^2\mathcal{K}}{dy^2} + y\,\frac{d\mathcal{K}}{dy} - (y^2 + M^2)\mathcal{K} = 0. \tag{3.128}$$

We have mentioned this type of function in Section 3.3 and in Appendix 3.C.6 for the case of the generalized central limit theorem. The same function can also be found expressed in other integral representations. Two of the most useful are [34, 35]:

$$\mathcal{K}_M(y) = \frac{\sqrt{\pi}}{\Gamma(M+1/2)} \left(\frac{y}{2}\right)^M \int_1^\infty dz\,(z^2-1)^{M-1/2} e^{-zy} \tag{3.129}$$

$$= \frac{\Gamma(M+1/2)(2y)^M}{\sqrt{\pi}} \int_0^\infty dz\,\frac{\cos z}{(z^2+y^2)^{M+1/2}}. \tag{3.130}$$

Alternatively, using the binomial expansion with negative powers, the expression (3.124) is equivalent to

$$c_N(w) = \frac{1}{2\pi} \int_{-\infty}^\infty dq\, \frac{e^{-\imath q w}}{[1 + q^2\tau^2/N]^N}$$

$$= \sum_{k=0}^{N-1} (-1)^k \binom{N+k-1}{k} \left(\frac{\tau^2}{N}\right)^k \frac{1}{2\pi} \int dq\, e^{-\imath q w} q^{2k},$$

where, in the sense of distributions, one can formally write

$$\frac{1}{2\pi} \int dq\, e^{-\imath q w} q^n = \delta^{(n)}(w),$$

as the nth derivative of the Dirac delta,

$$\delta^{(n)}(w) \equiv \imath^n \frac{d^n}{dw^n}\delta(w). \tag{3.131}$$

3.G Dimensional Analysis

Dimensional analysis, which we will briefly introduce here, is one of the most fundamental tools in all mathematical sciences and, all things considered, in everyday life as well. We are often told that we cannot add two apples to twenty-seven cherries, because they are quantities of different kinds: if we want to obtain a meaningful result, it is allowed to add only congruent quantities, of the same kind. (Of course, if I am only interested in counting objects of the "fruit" type, then it is reasonable and acceptable to say that I have twenty-nine fruits: but this would lead us into a discussion about the quality and depth of classifications that we are not interested in here.) On the other hand, it is possible to multiply or divide non-congruent quantities: in our example, we have 13.5 cherries per apple (useful information if we need to determine the quantities of ingredients needed to make a cake).

The main observation is that magnitudes (physical, typically) have *dimensions*, which we will discuss a little more later from a more mathematical point of view. The ones we have introduced are *appleness* and *cherryness*, but in a more standard context, we will typically be dealing with lengths, times, and masses. Only quantities with equal dimensions can be added together, and ratios of quantities of various dimensions are characterized by the appropriate combination of the quantities considered. Ratios of quantities with the same dimensions give *adimensional quantities*, i.e., pure numbers.

It is useful to consider a set of *basic quantities*, A, B, C, ..., the choice of which is reasonably arbitrary. This choice is associated with a *system of units*. A common choice in physics is the SI, *Système International*, where the complete set of basic quantities consists of length L (meter, m), time t (second, s), mass M (kilogram, kg), (absolute) temperature T (kelvin, K), current I (ampere, A), amount of substance that contains exactly $6.02214076 \times 10^{23}$ elementary entities (mole, mol), and light (luminous) intensity (candela, cd). In the SI these are the basic units, whose dimensions are fundamental, and every other quantity has dimensions composed by these basic dimensions. In the study of mechanics, we are used to the basic quantities L, M, and t, and, for example, velocity has dimensions of a length divided by a time, while acceleration has the dimension of a length divided by a time squared. Newton teaches us that the force is proportional to the acceleration through a mass, namely that $\vec{F} = m\vec{a}$ shows us that the force has the dimension of a mass times a length divided by a time squared. We also note that the temperature scale is in fact connected to the energy scale by a constant, which effectively determines the temperature scale, and necessitates the addition of an additional basic quantity.

In general, dimensional quantities will be given by monomials that are combinations of powers of basic quantities and pure numbers,

$$k, \quad A_1^a, \quad A_2^b, \quad A_3^c, \quad ..., \tag{3.132}$$

where k is a pure dimensionless number, the coefficient of our monomial, A_i are basic dimensional quantities, and exponents a, b, c are themselves pure numbers. A function such as, for example, $\tanh(x)$ is a dimensionless function whose argument must necessarily be dimensionless (we will come back to this later) and can be part of the coefficient k.

It is therefore interesting to explicitly introduce the concept of the dimension of a quantity O, in terms of the dimensions of the basis, through the symbol

$$[O] = \dim_O. \tag{3.133}$$

Thus, for example, we have $[\text{area}] = L^2$. Similarly, if x has the dimensions of a length, at this point we can immediately see that

$$\left[\int dx\, x^3\right] = L^4. \tag{3.134}$$

This last example also suggests that dimensional analysis gives indications as to the dominant behavior of an expression, but does not allow the evaluation of numerical coefficients. For example, the behavior in q obtained in (3.101) can be obtained from simple dimensional considerations: this is not the case for the coefficient $C(\alpha, D)$ that we had laboriously computed. More generally, dimensional analysis is often useful for qualifying asymptotic behavior, e.g., for large or small values of the argument. If, for example, we know that a function behaves as cx^4, where c is a dimensionless value, but are unable to compute c, it is useful to note that

$$\log(cx^4) = \tilde{c} + 4\log(x), \tag{3.135}$$

and that therefore when $x \to \infty$ the constant value c becomes irrelevant. Suggestions obtained through dimensional analysis may prove unfounded in the presence of singularities of functions (typically, for example in statistical mechanics, only in cases where the volume of the system tends to infinity). In these situations, anomalous dimensions arise, and deciphering the true asymptotic behavior of the functions of interest requires a much greater amount of work.

If we consider, for example, the function

$$a = b + \frac{c+d}{f} + ge^{-h} + m\sin(n), \tag{3.136}$$

dimensional analysis immediately tells us that a, b, c/f, d/f, g, and m must all have the same dimensions, that c and d must (obviously) have the same dimension, and that h and n must be dimensionless quantities.

Adimensional functions play an important role in dimensional analysis. A dimensionless quantity, i.e., of dimension zero, can only be a function of other dimensionless quantities. For example, the ratio of the area of a circle on a plane to the square of its radius r cannot depend on r. If, then, we are on a sphere of radius R (the Earth, for example), this ratio can only depend on the dimensionless quantity $(r/R)^2$. Let us now consider a simple one-dimensional example, where the only dimensional variable is the length. Let us consider a function f, which has dimension L^{F_1}. Let us assume that the function depends on three-dimensional variables with dimensions D_1, D_2, and D_3, respectively. We naturally always refer to length, which in this problem is the only dimensional variable, with $[x_k] = L^{D_k}$. Then we have that

$$f(\lambda^{D_1} x_1, \lambda^{D_2} x_2, \lambda^{D_3} x_3) = \lambda^{F_1} f(x_1, x_2, x_3). \tag{3.137}$$

In other words, in the one-dimensional problem, where we have only one type of dimensional quantity, the function f is necessarily a homogeneous function of degree F_1.

Let us conclude this brief analysis with a mathematical problem, which dimensional analysis solves immediately. Let us consider eight lead spheres of unit radius. Let us fuse them together, constructing a new sphere by joining the eight original spheres. What radius will this new sphere have? Dimensional analysis immediately tells us that the new radius will be 2. In fact, the new volume is equal to eight times that of the initial spheres, and the new sphere will have a radius equal to $8^{1/3} = 2$.

3.H Dirac Delta Distribution Miscellanea

Recall that the Dirac distribution δ can be expressed by its integral representation (3.79), the Fourier antitransform of 1:

$$\delta(x) = \frac{1}{2\pi} \int_{-\infty}^{\infty} dz\, e^{izx}\,.$$

It is interesting to note that

$$\delta(Ax - b) = |A|^{-1}\delta(x - bA^{-1})\,, \tag{3.138}$$

which is easy to reconstruct from the relation

$$\pi\delta(x) = \lim_{\epsilon \to 0} \frac{\epsilon}{x^2 + \epsilon^2}\,. \tag{3.139}$$

In D dimensions we have that

$$\delta^D(\mathbb{A}\vec{x}) = |\det(\mathbb{A})|^{-1}\delta^D(\vec{x})\,, \tag{3.140}$$

where $\mathbb{A}$ is a square $D \times D$ matrix. More generally we have that

$$\delta^{(D)}(\vec{f}(\vec{x})) = \sum_n c_n \delta(\vec{x} - \vec{x}(n))\,, \tag{3.141}$$

where the $\vec{x}(n)$ are all the solutions of the system of equations $\vec{f}(\vec{x}) = 0$ and

$$c_n = \det|\mathbb{A}|^{-1}\,, \tag{3.142}$$

where the components of the matrix $\mathbb{A}(n)$ are given by

$$A_{i,k}(n) \equiv \left.\frac{\partial f_i(\vec{x})}{\partial x_k}\right|_{\vec{x}=\vec{x}(n)}\,. \tag{3.143}$$

3.H.1 Multivariate Integral of the Dirac Delta of a Sum of Variables

For the multiple integral of a sum of N variables $p_i \in [0, p_{\max}]$ bound to have a fixed sum of value $Z \leq p_{\max}$, it holds that

$$\int \left(\prod_{k=1}^{N} dp_k\right) \delta\left(\sum_{k=1}^{N} p_k - Z\right) = \frac{Z^{N-1}}{(N-1)!}\,. \tag{3.144}$$

Notice that the upper bound of the support of the p_k variables needs to be taken just slightly larger than Z, i.e., in the limit $p_{\max} = Z^+$, in order to have a non-zero integral, but it can take any value larger than Z.

We propose two proofs of the formula (3.144). For the first one, we make a change of variables,

$$y_1 = p_1,$$
$$y_2 = p_1 + p_2,$$
$$\vdots$$
$$y_k = \sum_{i=1}^{k} p_i,$$
$$\vdots$$
$$y_N = \sum_{i=1}^{N} p_i,$$

where by construction $0 \le y_1 \le y_2 \le \cdots \le y_{N-1} \le y_N \le \bar{y}_N \equiv p_{\max}$. The Jacobian of the transformation is the inverse of the matrix of elements,

$$J^{-1} \equiv \frac{\partial y_m}{\partial p_i} = \begin{cases} 1 & \text{if } m \ge i, \\ 0 & \text{if } m < i. \end{cases} \tag{3.145}$$

This is a matrix which has all elements equal to 1 to the left of and along the diagonal and equal to 0 to the right of the diagonal. The determinant of the Jacobian is therefore $\det J = 1/(\det J^{-1}) = 1$. We then rewrite the formula (3.144) and obtain the desired result by iteratively integrating starting from the variable y_1 with the smallest domain, $y_1 \in [0, y_2]$:

$$\int \left(\prod_{k=1}^{N} dy_k \right) \delta(y_N - Z) = \int_0^{\bar{y}_N} dy_N\, \delta(y_N - Z) \int_0^{y_N} dy_{N-1} \cdots \int_0^{y_3} dy_2 \int_0^{y_2} dy_1$$

$$= \int_0^{\bar{y}_N} dy_N\, \delta(y_N - Z) \int_0^{y_N} dy_{N-1} \cdots \int_0^{y_k} dy_{k-1} \frac{y_{k-1}^{k-2}}{(k-2)!}$$

$$= \int_0^{\bar{y}_N} dy_N\, \delta(y_N - Z) \frac{y_N^{N-1}}{(N-1)!} = \frac{Z^{N-1}}{(N-1)!}.$$

For the second computation, we use the expression of the Dirac delta as the Fourier antitransform of 1, (3.79), and the expression with the inverse power with the Γ function, (3.98). Opening the delta on the left-hand side of Eq. (3.144) gives

$$\int \prod_{k=1}^{N} dp_k\, \frac{1}{2\pi} \int_{-\infty}^{\infty} dq\, \exp\left(-\imath q \sum_k p_k + \imath q Z \right) = \frac{1}{2\pi} \int_{-\infty}^{\infty} dq\, e^{\imath q Z} \prod_{k=1}^{N} \int_0^{p_{\max}} dp\, e^{-\imath q p}$$

$$= \frac{1}{2\pi} \int_{-\infty}^{\infty} dq\, e^{\imath q Z} \frac{(1 - e^{-\imath q p_{\max}})^N}{(\imath q)^N}. \tag{3.146}$$

Using (3.98) for the inverse power representation and using the expansion of the binomial, we get

$$\frac{1}{2\pi} \int_{-\infty}^{\infty} dq\, e^{\iota q Z} \sum_{k=0}^{N} \binom{N}{k} (-1)^k\, e^{-\iota q k p_{\max}} \frac{1}{\Gamma(N)} \int_{0}^{\infty} dz\, z^{N-1} e^{-\iota q z}$$

$$= \frac{1}{\Gamma(N)} \sum_{k=0}^{N} \binom{N}{k} (-1)^k \int_{0}^{\infty} dz\, z^{N-1} \delta(z - (Z - k p_{\max})) = \frac{Z^{N-1}}{(N-1)!},$$

where in the last step only the contribution for $k = 0$ remains non-zero since, for each $k \geq 1$, one has[2] $Z - k p_{\max} < 0$ because $Z < p_{\max}$.

3.1 Convergence of Functions

In this appendix we will outline the problem of the convergence of functions, trying to list and clarify the main assumptions that apply in our text. This is one of the moments when probability theory shows most clearly its aspect of measure theory. We will limit ourselves to providing the fundamental concepts and formulas, with a few useful examples. The interested and motivated reader is invited to delve into the subject in one of the many texts that discuss measure theory. See, for example, as a basic reference on functional analysis, the book by Reed and Simon [36], familiar to physicists, and the reference texts [13, 14, 37].

Very often we will consider sequences of probability density functions $p_n(x)$, defined in Section 1.2.3. The index n labels several probability density functions, and our primary interest is in the limit where $n \to \infty$. Is there, in some sense, a limit to the sequence $p_n(x)$ when $n \to \infty$? What is it? In general, the notion of distance between two functions has a fundamental character, that is necessary to understand a large number of concepts.

Our aim will be to mathematically define what is meant by *a probability density function $f(x)$ tends to the function $g(x)$* in a certain limit. Various notions of convergence are useful in different contexts: the convergence (or non-convergence) of a function to a certain limit in one scheme or another is related to specific properties of the system considered.[3]

In what follows, we shall merely distinguish the case in which the topological space of interest is equipped with a *strong topology* and the case in which it is equipped with a *weak topology*. For detailed definitions, see one of the cited texts [13, 14, 36, 37].[4]

---

[2] We can include the limit case $p_{\max} = Z$ if we adopt for the Dirac delta the "extremes convention" that includes the upper extreme and excludes the lower extreme: $\int_a^b f(x)\delta(x - a) = 0$ and $\int_a^b f(x)\delta(x - b) = f(b)$.

[3] We remind our physicist readers that this issue is particularly relevant to the study of the properties of renormalization group transformations: see, for example, refs. [38, 39, 40] and the numerous references contained therein.

[4] We quote from page 213 of Reed and Simon's *Functional Analysis* text [36]: "The reader may be bewildered by the many topologies we have introduced on $\mathcal{L}(\mathcal{H})$: the weak, strong, and uniform operator topologies,

Let us consider the square integrable functions $f(x)$: they satisfy the normalization property $\int dx\, f(x)^2 < \infty$, and form the so-called space of functions that are converging in L^2.

A sequence of functions $f_n(x)$ tends to zero in the *strong topology* if, for $n \to \infty$,

$$\int dx\, f_n(x)^2 \to 0. \tag{3.147}$$

Conversely, convergence of $f_n(x)$ to zero occurs in the *weak topology* if, for $n \to \infty$,

$$\int dx\, f_n(x)g(x) \to 0, \tag{3.148}$$

for any function $g(x)$ belonging to L^2.

If a sequence of functions in L^2 tends to zero strongly, it must tend to zero also weakly, since

$$\left| \int dx\, f_n(x)g(x) \right| \le \left(\int dx\, f_n(x)^2 \int dx\, g(x)^2 \right)^{1/2}, \qquad \forall\, f_n, g \in L^2. \tag{3.149}$$

On the contrary, it is common for functions to tend to zero weakly while not tending to zero strongly. Let us look at some examples.

- Let us consider functions $f(x) \in L^2$ defined on the interval $x \in (-\infty, \infty)$. The sequence of functions of type $f_n(x) \equiv f(x-n)$ tends to zero weakly, but it does not tend to zero strongly. Indeed, $\int_{-\infty}^{\infty} dx\, f(x-n)^2$ does not depend on n.

A good exercise for the reader is to discuss explicit examples of these functions. What happens, for example, when both $f_n(x)$ and $g(x)$ are Gaussian distributions?

- Let us consider the Hilbert space of functions in L^2 with inner product defined as

$$\langle f, g \rangle \equiv \int dx\, f(x)g(x). \tag{3.150}$$

The inner product of a function with itself is the squared norm

$$\| f \|^2 = \langle f, f \rangle \equiv \int dx\, f(x)^2. \tag{3.151}$$

The weak convergence in a Hilbert space to a limit function f is denoted by the limit

$$\lim_{n \to \infty} \langle f_n, g \rangle = \langle f, g \rangle,$$

cf. Eq. (3.148) when $f(x) = 0$. The strong convergence corresponds to the convergence in norm

$$\lim_{n \to \infty} \| f_n \| = \| f \|,$$

cf. Eq. (3.147) when $f(x) = 0$.

the weak Banach space topology, the ultraweak topology. Later on we will encounter the ultrastrong topology. Why is it necessary to introduce all these topologies?" We hope the reader will be intrigued enough by this list to consult ref. [36] to get the answer to the question.

It is interesting to note that the weak convergence of infinite orthogonal sequences of functions, such as, for example, the Hermite polynomials in Appendix 3.E, tends to zero. Indeed, if we take a complete basis of orthonormal functions $\hat{f}_n(x)$,

$$\langle \hat{f}_n, \hat{f}_k \rangle = \delta_{n,k},$$

in which we can express any function in the Hilbert space as

$$g(x) = \sum_{n=1}^{\infty} \langle \hat{f}_n, g \rangle \hat{f}_n(x),$$

for Bessel's inequality we have

$$\sum_{n=1}^{\infty} |\langle \hat{f}_n, g \rangle|^2 \leq \|g\|^2 < \infty.$$

Since the sum on the left has infinite terms, one must have

$$\lim_{n \to \infty} \langle \hat{f}_n, g \rangle = 0,$$

which is nothing other than the weak expression (3.148) of convergence, $\lim_{n \to \infty} \hat{f}_n \to 0$.

If the Hilbert space is finite-dimensional, weak and strong convergence are equivalent. They differ only if the space is infinite-dimensional, as we shall see explicitly in the next example.

- Let us consider the functions defined on the interval $[0, 1]$. The sequence of functions $f_n(x) \equiv \cos(2\pi n x)$ does not tend to zero strongly; in fact, $\int dx \, f_n(x)^2 = 1/2$ and does not depend on n. However, it does tend to zero weakly. Indeed, if we define

$$c_n \equiv \int dx \, \cos(2\pi n x) g(x), \quad g(x) \in L^2, \tag{3.152}$$

it is an intuitive fact that the sequence of c_n tends to zero for $n \to \infty$ (the cosine oscillates faster and faster and the oscillations eventually mediate and thus the sequence $f_n(x)$ tends to zero weakly).

This fact can best be seen by also defining the coefficient

$$s_n \equiv \int dx \, \sin(2\pi n x) g(x), \tag{3.153}$$

and noting that from the Fourier series properties of a function $g(x)$ it follows that

$$\sum_{n=0}^{\infty} (c_n^2 + s_n^2) = \int dx \, g(x)^2 < \infty. \tag{3.154}$$

The latter expression is finite by the assumption that $g(x)$ converges in L^2, from which it follows that c_n and s_n must go to zero when $n \to \infty$, because otherwise the sum would diverge. The same type of argument applied to the Fourier transform implies that the integral defined in Eq. (3.38) tends to zero when N tends to infinity.

- Let us finally see the case of the function $2\cos(2\pi nx)^2$ in the interval $x \in [0, 1]$ when $n \to \infty$. We know from Eq. (3.152) that $\cos(2\pi nx) \to 0$ weakly. Its square instead tends to $1/2$. Indeed, $2\cos(2\pi nx)^2 = 1 + \cos(4\pi nx)$ and, as in Eq. (3.152), $\cos(4\pi nx)$ weakly tends to zero when n tends to infinity. This example shows that, if a function tends weakly to zero, the square of that function may not tend to zero.

3.1.1 Convergence in Distribution

The use of a mathematical scaffold based on different topologies is relevant also in probability theory (as we mentioned earlier). In this section, we shall stick to only one of the possible definitions. We will indeed say that the probability density sequence $p_n(x)$ tends to the probability density limit $p(x)$ if the following two conditions are true:

1. For all functions $g(x)$ sufficiently regular (e.g., $\forall\, g(x) \in L^2$) we have

$$\lim_{n\to\infty} \int dx\; p_n(x)g(x) = \int dx\; p(x)g(x), \tag{3.155}$$

   or, in other words, $p_n(x)$ tends to $p(x)$ in the sense of the distributions.
2. The limit function $p(x)$ is well normalized, i.e.,

$$\int dx\; p(x) = 1. \tag{3.156}$$

It is possible to prove in a very simple and general way that the inequality $\int dx\; p(x) \le 1$ holds, i.e., that the probability limit is normalized to a value less than or equal to one.

Proof   Let us consider a system with infinite discrete states labeled by the variable k, and recall that the individual probabilities $p_n(k) \ge 0$ are well normalized for each finite value of n, i.e.,

$$\sum_k p_n(k) = 1, \quad \forall n.$$

We want to prove that if $\lim_{n\to\infty} p_n(k) = p_\infty(k)$ then $\sum_k p_\infty(k) \le 1$.

To do this we consider, at finite n, the sum over a finite number S of the infinite countable states of the system. Given that the probabilities are positive and that we are summing over a finite number of states, not on all of them, we have that $\sum_{k=1}^{S} p_n(k) \le 1$. We can take the limit as $n \to \infty$ of a sum made over a finite number of terms, obtaining $\sum_{k=1}^{S} p_\infty(k) \le 1$. This finally shows that

$$\lim_{S\to\infty} \lim_{n\to\infty} \sum_{k=1}^{S} p_n(k) = \lim_{S\to\infty} \sum_{k=1}^{S} p_\infty(k) = \sum_{k=1}^{\infty} p_\infty(k) \le 1, \tag{3.157}$$

and we are done.   ∎

This style of proof, which will serve us again later (see the case of Markov chains, in Section 10.6.1), is based on the fact that the limit $n \to \infty$ can be brought inside a finite sum. Then the finite sum to S states (with $n = \infty$) can be extended to infinite states and it remains not greater than one.

3.J Generation of Pseudo-Random Numbers

In many contexts, the numerical generation (typically on electronic computers) of *pseudo-random numbers* plays a crucial role. We therefore want to discuss this briefly here. For more precise details, we refer the reader to Knuth's text [41], and the numerical methods text [42], which contains a chapter, and several specific programs, on pseudo-random numbers. For reasons that we will clarify below, the problem of pseudo-random number generation constitutes a constantly evolving field: for this reason, we feel it is useful to cite the relevant Wikipedia entry, which follows the evolution of the field, together with a recent review [43].

In principle, we would be interested in having sets of true random numbers, which have exactly the theoretical properties postulated by us. An electronic computer will not do this, respecting its character as an eminently deterministic object in rejecting all random behavior. The sequences of numbers generated by a program of an (unbroken) electronic computer will be a purely deterministic sequence of numbers, even if generated by an algorithm so complex that it is capable of confusing us and the system we are studying and of appearing, in many senses that can be well defined mathematically, as random sequences.

In reality, the discussion on the meaning of random or pseudo-random sequences does not only concern the generation of random numbers on computers but applies more generally to the discussion of most phenomena studied by the natural sciences. Let us forget for a moment quantum mechanics, which is today our fundamental intellectual tool for understanding nature, where random behavior is hypothesized as a fundamental mechanism of physical laws, and think instead about the study of phenomena that do not occur on too large energy scales. The typical situation is where a behavior that we define as random is essentially a deterministic behavior, however, governed, in effect, by laws too complex for us. In the case of rolling a die, for example, if we were able to know with very high precision the initial conditions, the forces acting on the die, and the details of its structure, we could compute the exact outcome: the problem is that we are unable to do so. On the relevant energy scales of the process, an eminently deterministic problem such as the roll of a die ends up, due to its great complexity, representing precisely the prototype of a random problem: the roll of dice or *alea* for Latin scholars. Let us not go into the details of so-called *chaotic behaviors* here, which might open up another branch of this discussion.

Probably the most useful way of looking at this issue in the following is to assume that a sequence of numbers generated by a computer using a given algorithm (PRNG, pseudo-random number generator) is a good pseudo-random sequence if it is apparently indistinguishable from a random sequence. In other words, we will require that, given, for example, a sequence of pseudo-random numbers $\{r_k\}$ uniformly distributed between 0 and 1 and the sufficiently regular functions f and g, relations of the following type are true to a very good approximation:

$$\lim_{N\to\infty} \frac{1}{N} \sum_{k=1}^{N} f(r_k) = \int_0^1 dr\, f(r),$$
$$\lim_{N\to\infty} \frac{1}{N} \sum_{k=1}^{N} g(r_k, r_{k+1}) = \int_0^1 dr \int_0^1 ds\, g(r,s),$$

(3.158)

even for distances in the sequence greater and much greater than one.

One of our first goals in producing a good PRNG is to generate sequences of values that have a large *period*. Since we represent a number with n bits, which may be 0 or 1 (nowadays n may typically be 64, but we will see that smaller values are often still used), the maximum period we can obtain is 2^n, and we want the true period of the sequence to be as close to this maximum value as possible. It is necessary for us that the period, although finite, can be considered as infinite for all practical purposes. A generator using a 32-bit integer has a period of the order of 10^9, which is not a frighteningly large number, whereas 64-bit generators have periods of the order of 10^{18}. It is also easy to conceive generators that have extremely large periods. A generator using, for example, fifty-seven 32-bit numbers can have a period on the order of 10^{500}.

3.J.1 Linear Congruential Method

The simplest, and probably most popular, way to generate a sequence of pseudo-random numbers uniformly distributed in the interval between 0 and 1 is based on the so-called *linear congruential* method. It starts with an initial ρ_0 value (the *seed*) which initialises the iterative relation, and then the $(i + 1)$th number is obtained, depending on the ith number, as

$$\rho_{i+1} = (A\rho_i + B)\,\%\,\mu,$$

(3.159)

where $\%$ denotes the modulo operation, A is the so-called *multiplier*, B is the *increment*, and μ is called the *modulus* of the generator. The choice of "good" values for the parameters A, B, and μ is a very delicate topic and an eminently empirical subject: the field of applied mathematics that most closely resembles *haute cuisine* is probably precisely the theory of choosing suitable coefficients.

The problem is profound and has to do with the enormous number of features that should be tested before declaring a generator acceptable. The space of possible tests is of infinite size, and completely adequate generators for some problems can fail dramatically for other problems. It may happen, for example, that a generator that had performed well to simulate systems defined on lattices of volume up to 31^3 (we are in three spatial dimensions, and the linear size is 31) collapses, showing dramatic inadequacies, on a lattice of volume 32^3, which had the misfortune to catch exactly some periodicity of the generator algorithm.

Empirically, two things can be said: first, that "good" generators are the ones that pass at least the most popular tests; and second, that experience plays a crucial role in this field. If a generator has been used in many situations at least the easiest-to-detect weaknesses will probably be ruled out: typically, with a very high probability, a generator invented by us, thinking it will make things better, will end up exhibiting serious problems. In this field, it is better to trust accumulated knowledge. In general, as Knuth points out, codes for generating random numbers do not have to be written at random, quite the contrary. In fact, for an algorithm that has no deep mathematical justification, it can very often happen that the period is extremely small. And this is not to mention the possible correlations between the numbers produced, where things could be dramatic.

A further simplification of the linear congruential generators is obtained by setting the increment B to zero, defining the *multiplicative congruential* generators

$$\rho_{i+1} = (A\rho_i) \% \mu, \tag{3.160}$$

which we will now discuss.

We first note that dramatic bad choices are possible. One historical choice, which has long been used as a machine generator on many computers, is the famous and nefarious *truly horrible* RANDU, with

$$A = 65\,539, \quad \mu = 2^{31}, \quad \textbf{not to be used!} \tag{3.161}$$

Before its inadequacy was clearly established [42] this implementation plagued generations of numerical simulations.

A widely used generator in the literature, which in some sense constitutes a standard, is the simple

$$A = 7^5 = 16\,807, \quad \mu = 2^{31} - 1. \tag{3.162}$$

It is especially good that this has been used a lot without too many macroscopic pathologies being noticed.

The simple algorithm defined in (3.160) can be further simplified: for μ, choose 2 to the power of the number of bits constituting a word in our computer (typically, nowadays, 64, or at least 32). In this case, the modulo operation is performed automatically thanks to the truncation carried out in the computer hardware itself. A first, very widespread example of this type is

$$A = 69\,069, \quad \mu = 2^{32}. \tag{3.163}$$

It is frequently possible to show, for these generators, that if $\rho_0 \neq 0$ the period is close to the maximum possible. However, problems of various kinds can arise. A typical analysis consists of grouping the numbers generated in P-tuples, and trying to understand what happens for different values of P. For example, first, in the infamous RANDU, if one groups the generated numbers three by three ($P = 3$) and represents these three-dimensional vectors on a unit cube, one can see that the vectors cluster on only 17 planes (in principle, for a good generator of the type considered, they should cluster on a thousand planes). Second, it can be seen that if one looks at correlations at large distances in the series of pseudo-random numbers, correlations often appear. The general recipe, which usually guarantees, at least as far as this last aspect is concerned, is to use random numbers generated on scales much smaller than the period of the generator.

In general, congruential generators are relatively faultless when μ is a prime number. However, the choice $\mu = 2^n$, where n is the word length in bits, makes numerical computations much faster and is, therefore, often favored.

On newer computers, it is natural and economical to work with words of 64 bits (since most of the latest generation of processors are based on operations that act on words of 64 bits). In this case, a possible and probably reasonable choice is

$$A = 6364\,136\,223\,846\,793\,005\,, \quad \mu = 2^{64}\,, \tag{3.164}$$

where, for example, in the C language, the variable containing the pseudo-random number must be defined with a real 64-bit type (typically the statement unsigned long int fulfills this task), and the form is then taken automatically.

On many computers and in many languages there is a machine function that generates random numbers: it is best not to trust these functions blindfolded, but it is usually reasonable to use them for the initial phase of expanding codes that need pseudo-random numbers. For example, often in Fortran the instruction sequence

```
REAL A(1000), RAN
INTEGER I, IRR
IRR = 1234567
DO I = 1, 1000
  A(I) = RAN(IRR)
ENDDO
```

fills the vector A with pseudo-random numbers uniformly distributed over the interval $(0, 1]$: these values depend on the initial value of IRR. The thus generated sequence typically has a very long period (almost always on the order of 2^{32}).

The generation of pseudo-random numbers has several delicate points. We limit ourselves here to pointing out a few of them, leaving it to the reader's experience to accumulate more bad (and, why not, good) memories. First, once the generator has been initialized with a seed, extract a certain number of values to discard (typically a hundred is a good amount). This is because the good statistical properties of the generator are shared by typical sequences, at equilibrium, whereas at the beginning the sequence may be very atypical.

Delicate points may also be the low bits of the number (which sometimes have a non-randomness that is difficult to uncover *a priori*). Little is known about the possible correlations of the various bits of pseudo-random numbers: using a pseudo-random number in the hope to extract 32 or 64 random and uncorrelated bits often turns out to be an incorrect procedure. Another crucial aspect, which must be kept under control, is the distribution of vectors formed by D successive pseudo-random numbers: one associates the first D numbers in a d-dimensional vector, those from $D + 1$ to $2D$

in a second vector, and so on. The analysis of the correlations of these points in a D-dimensional space often holds surprises.

A final delicate point we wish to mention concerns the parallel generation of pseudo-random numbers. We are thinking here of a parallel computer consisting of several processors producing pseudo-random numbers at the same time, which must be independent. Here, attention must be paid to the problem of generating the necessary seeds, one per processor, so that the different parallel sequences are not correlated.

Very often we need random numbers in the range $[0, 1]$. For this purpose we can, for example, consider a random number generated with the congruential generator and set

$$r_i = \frac{\rho_i}{\mu}. \tag{3.165}$$

The reader must bear in mind that the extreme values of the interval, i.e., 0 and 1, may appear among the generated numbers depending on how exactly the variables are defined. If, for example, everything is done correctly, $\mu = 2^{32}$ and the variable r is defined on 64 bits and the divisions are also done with 64-bit precision, the obtained range for the real numbers goes from $1/\mu$ to $1 - 1/\mu$.

A rule of thumb for obtaining better random numbers suggests using two sequences r and s of random numbers in the interval $[0, 1]$, and taking the sum $r_i + s_i$ modulo 1 as the resulting random number. This operation often destroys possible correlations present in each of the two sequences, but it must be done by thinking carefully about what one is doing, having at the front of one's mind the defects or correlations of each of the two sequences, to avoid running the risk of magnifying the defects instead of reducing them. In the field of random generation, it is better to avoid proceeding at random.

3.J.2 Non-Uniform Distributions

Let us now consider a problem that one often needs to solve. Given pseudo-random numbers $\{r\}$ uniformly distributed between 0 and 1, we want to obtain a sequence of pseudo-random numbers $\{\rho\}$ distributed differently, according to the probability distribution $P(\rho)$. The considerations on the change of variables reported in Appendix 3.B come to our aid. Let us consider a few specific cases of great importance.

- If we want to get exponentially distributed pseudo-random numbers,

$$P(\rho) = e^{-\rho}, \quad \text{with } 0 < \rho < \infty, \tag{3.166}$$

we must set $\rho = -\log r$.
- If we want to obtain Gaussian-distributed values, we need to start by generating two independent uniform numbers r_1 and r_2 in $(0, 1)$. We will obtain two Gaussian-distributed values in return: in some cases, it is appropriate to use only one of the two. Let

$$P(\rho) = \frac{e^{-\rho^2/2}}{\sqrt{2\pi}}, \quad \text{with } -\infty < \rho < \infty. \tag{3.167}$$

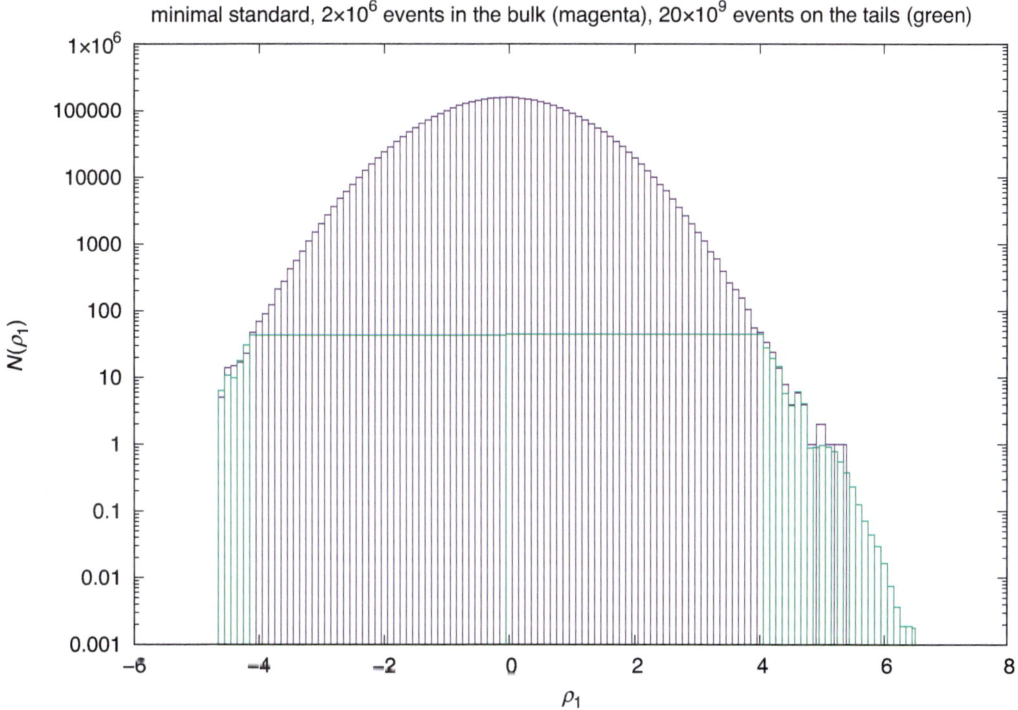

Figure 3.4   Histogram $N(\rho_1)$ of the values obtained in the bulk (two million values in total, in magenta) and in the tails (twenty billion values in total, among which only those with modulus greater than 4 are shown, in green). Bin values on the tails were renormalized by a factor of 10,000. The figure has a linear scale for the x axis and a logarithmic scale for the y axis. The values of the uniformly distributed r_1 and r_2 were produced by the standard minimal generator.

The two values ρ_1 and ρ_2 are distributed according to (3.167) by setting

$$\rho_1 = \sqrt{-2\log r_1}\,\cos\left(2\pi r_2\right) \quad \text{and} \quad \rho_2 = \sqrt{-2\log r_1}\,\sin\left(2\pi r_2\right). \tag{3.168}$$

This is the typical case where it is easy to make a mistake.

In Figure 3.4 we see the histogram of the random numbers ρ_1, where we have used for r_1 and r_2 two random numbers generated sequentially by the good minimal standard congruential generator (3.162) and we have used 2×10^6 events to represent the central part of the distribution (magenta), about the whole period. We are obviously blind to probabilities smaller than the order 10^{-8}, but it is easy to see that disaster begins at probabilities of the order 10^{-4}. The origin of the disaster is obvious. For the number ρ_1 to be large, the uniformly distributed number r_1 must be small, but as soon as $r_1 \ll 16\,807^{-1}$, r_2 will also necessarily be small. (The number r_2 is the number that follows r_1 in the pseudo-random value series, and under these conditions it is equal to r_1 multiplied by 16 807: if r_1 is very small, r_2 will be small despite this multiplicative factor.) Therefore, $\cos\left(2\pi r_2\right)$ will be very close to 1, thus causing an absolute hole for strongly negative values and an excess at positive values of the random number. The simplest solution is to use two different sequences for r_1 and r_2, obtained from the same generator but initialized with different seeds, or with two different generators (better). The 64-bit generator we have just described does not show this pathology, in the range we are interested in investigating in a typical numerical simulation.

The reader should reflect on this fact: The wrong generator works perfectly in the region where the probability of an event is large and only fails in the tails, where the probability is small. Hence, it can be promoted with flying colours by all tests that are sensitive to the bulk of the probability distribution, but it fails if it is used to generate rare events. The quality of the generator depends crucially on its use.

- Let us consider the case where we require a random number to have a Poisson distribution,

$$P(\lambda) = \frac{\lambda^k e^{-\lambda}}{k!}, \quad \text{with } 0 < \lambda < \infty. \tag{3.169}$$

We can use $k + 1$ independent uniformly distributed pseudo-random numbers and write

$$\lambda = -\sum_{i=1}^{k+1} \log(r_i). \tag{3.170}$$

- Uniformly distributed vectors on the N-dimensional unit sphere are obtained by extracting the N components with a Gaussian distribution and normalizing the result to 1. Indeed, the probability distribution is proportional to

$$\exp\left(-\sum_{i=1}^{N} x_i^2\right) = \prod_{i=1}^{N} \exp\left(-x_i^2\right), \tag{3.171}$$

which factors in the product of N independent probability distributions (one for each variable) and is invariant under rotations. As noted by Maxwell, this type of probability distribution is the only probability distribution that is simultaneously factorized and invariant by rotations.

Large Deviations

4.1 Large Deviations Theorem

We have seen in the previous chapter the law of large numbers, which assures us that the empirical average $y_N = \sum_{i=1}^{N} x_i/N$ gets closer and closer to the expected theoretical average with a probability that increases with N. Thus, the probability that $y_N - \mu$ is a quantity of order 1 goes to zero when N tends to infinity. However, the law of large numbers, via Chebyshev's lemma (3.17), imposes a very loose upper bound (see also Appendix 3.A):

$$P(|y_N - \mu| > \epsilon) < \frac{\sigma^2}{N\epsilon^2}, \quad \forall \epsilon > 0. \tag{4.1}$$

As we shall see in this section, in many cases it is possible to estimate $P(|y_N - \mu| > \epsilon)$ and to show that it is a much smaller number than $1/N$ when N goes to infinity, i.e., it turns out to be exponentially small with N.

We define the variable z as usual as the sum

$$z \equiv \sum_{i=1}^{N} x_i, \tag{4.2}$$

where the random variables x_i are independent and, for simplicity, all governed by the same probability distribution $P(x_i)$. In this chapter we will be interested in computing the probability distribution $d_N(z)$ of the random variable z in the regions of the domain of z where the probability is very small. In other words, we want to calculate the probability of having *large deviations* from the mean, significantly larger than those analyzed using the central limit approach of the previous chapter.

In dealing with the behavior of large deviations and with the theorem governing their distribution, we will consider the case in which the distribution $P(x)$ is different from zero only in a limited support, or goes to zero faster than any exponential when $|x| \to \infty$. In reality, as we shall see below, it is sufficient that the descent is not slower than an exponential, i.e., that for large values of $|x|$ we have that $P(x) \leq Ae^{-B|x|}$.

Our objective will be to estimate the probability $d_N(z)$ for large values of z in the limit where N tends to infinity. More specifically, we will be interested in the probability of the event $z = \lambda N$ in the limit where $N \to \infty$ with λ constant. In this region the probability of an event will be extremely small.

The *large deviations theorem* states that for large N the distribution of the sum of N i.i.d. variables in the domain where $z = \lambda N$ has the behavior

$$d_N(z) \sim e^{N S(\lambda)}, \tag{4.3}$$

where we will call $S(\lambda)$ the *decrease function*.[1] Formula (4.3) can be rewritten by defining the decrease function as

$$S(\lambda) \equiv \lim_{N \to \infty} \frac{\log d_N(N\lambda)}{N}. \tag{4.4}$$

We will obtain here results consistent with those implied by the central limit theorem, which will, however, allow us a more accurate control of the probability of very rare events. This control is obviously of great use in many practical situations, where the purpose is to predict, and thus avoid, an unlikely but devastating disaster.

4.2 Proof of the Large Deviations Theorem

To prove the theorem given by (4.3) we can apply the *saddle point* method, which is a generalization of Laplace's method to the case of integrals in the complex plane and which will be presented and demonstrated in detail in Appendix 4.A. The same result can be obtained in a less direct but mathematically simpler way using techniques similar to those used in statistical mechanics, as we shall see in Section 4.4.

Proof To derive our result, we start from the Fourier transform $\tilde{d}_N(q)$ of the probability distribution $d_N(z)$. As in (3.22), by the properties of the convolution product, we have that

$$\tilde{d}_N(q) = [\tilde{P}(q)^N] = e^{N \ln \tilde{P}(q)}. \tag{4.5}$$

The assumptions we have made about the function $P(x)$ imply that its Fourier transform $\tilde{P}(q)$ is a function devoid of singularities in the entire complex plane $q \in \mathbb{C}$.

By carrying out a Fourier antitransform we find that

$$d_N(z) = \frac{1}{2\pi} \int_{-\infty}^{\infty} dq \, e^{-\imath z q} \tilde{d}_N(q) = \frac{1}{2\pi} \int_{-\infty}^{\infty} dq \, e^{N[\ln \tilde{P}(q) - \imath \lambda q]}, \tag{4.6}$$

where in the last step we have explicitly inserted $z = \lambda N$ to indicate that we study the extreme values of z (and not the central values $z = O(N^{1/2})$). The factor N multiplying the entire argument of the exponential means that for large N the integrand oscillates very rapidly. The path of integration in the complex plane must be deformed from the real axis into a path passing through the saddle point of the real part of the exponent. This operation must be done in such a way that the argument of the exponential assumes its maximum along the path to the saddle point, which is identified by setting its derivative with respect to q to zero. There are, indeed, several cautions to be taken

---

[1] As we will see in Section 14.5.2 the large deviations decrease function is basically a rescaled entropy. Sometimes it is defined with a minus sign and it is called the *rate function* or Crámer function.

in this procedure that are considered in detail in Appendix 4.A dedicated to the saddle point theorem.

To determine the saddle point of the argument of our exponential we must, therefore, solve the equation

$$\frac{d \ln \tilde{P}(q)}{dq} = \imath \lambda. \tag{4.7}$$

Let us call $\bar{q}(\lambda)$ the solution of this equation. Then we have that

$$d_N(z) = d_N(N\lambda) \propto e^{NS(\lambda)}, \tag{4.8}$$

where the decrease function is the exponent function computed on the saddle point $\bar{q}(\lambda)$:

$$S(\lambda) \equiv \ln \tilde{P}(\bar{q}(\lambda)) - \imath \lambda \bar{q}(\lambda), \tag{4.9}$$

and we are done. $\blacksquare$

If the saddle point equation (4.7) does not admit a solution, the probability $d_N(N\lambda)$ decays to zero faster than an exponential with N or, as we shall see later, it is exactly zero.

We first look for a solution for small values of q. In this case, using the definitions (3.5) and (3.11), similarly to what was seen in Eq. (3.26), we have that

$$\ln \tilde{P}(q) = \log\left(1 + \imath q \mu - \frac{q^2(\sigma^2 + \mu^2)}{2} + O(q^3)\right) = \imath q \mu - \frac{q^2 \sigma^2}{2} + O(q^3). \tag{4.10}$$

In this limit, the saddle point equation (4.7) is

$$\imath \mu - q\sigma^2 = \imath \lambda, \tag{4.11}$$

which admits the solution, valid for small $(\mu - \lambda)/\sigma^2$,

$$\bar{q}(\lambda) = \imath \frac{\mu - \lambda}{\sigma^2} + O(\bar{q}(\lambda)^2). \tag{4.12}$$

On the solution of the saddle point equation, we obtain the decrease function

$$S(\lambda) = -\frac{(\mu - \lambda)^2}{2\sigma^2} + O\left(\left(\frac{\mu - \lambda}{\sigma}\right)^3\right), \tag{4.13}$$

from which the well-normalized distribution of $\lambda = z/N$ turns out to be

$$\hat{d}_N(\lambda) = N d_N(z) = \sqrt{\frac{N}{2\pi\sigma^2}} \exp\left(-\frac{N}{2\sigma^2}(\mu - \lambda)^2\right).$$

This result is obviously consistent with the predictions of the central limit theorem and the law of large numbers. Indeed, it tells us that $\lambda = z/N$ has a small probability of being very different from the expectation value μ. If we switch to the central limit variable

$$w \equiv \frac{1}{\sqrt{N}} \sum_{i=1}^{N} (x_i - \mu) = \sqrt{N}(\lambda - \mu)$$

and replace $\lambda - \mu$ with $w/\sqrt{N}$ in the distribution, this becomes the Gaussian distribution of the central limit. At this order of the expansion, the large deviations reproduce the distribution of the Gaussian central limit, barring orders $O(1/\sqrt{N})$. We will find this expression again in the case of the formula (4.20).

It is very interesting that the saddle point is located on the pure imaginary axis. In general, we will assume (and we will confirm *a posteriori*) that $\bar{q}(\lambda)$ always assumes a pure imaginary value: the results obtained under this assumption will be supported by the analysis in the next section. We will, thus, set $\iota q \equiv h$ and define a function in the real field h,

$$R(h) \equiv \ln \tilde{P}(-\iota h) = \ln \int_{-\infty}^{\infty} dx\, e^{hx} P(x) , \tag{4.14}$$

which is the logarithm of the Laplace transform (bilateral) of $P(x)$; see Appendix 3.D. As we showed in Appendix 3.C.3 and as we will see again later in Section 4.6 and in Chapter 8, $R(h)$ is a generating function of the cumulants of $P(x)$. In these new variables, Eq. (4.7) of the saddle point becomes

$$R'(\bar{h}(\lambda)) = \lambda , \tag{4.15}$$

where $R'(h) = dR(h)/dh$ and the function $S(\lambda)$ is given by[2]

$$S(\lambda) = R(\bar{h}(\lambda)) - \bar{h}(\lambda)\lambda . \tag{4.16}$$

We note that if we introduce the probability distribution

$$P_h(x) \equiv \frac{P(x)e^{hx}}{\displaystyle\int dx\, P(x)e^{hx}} , \tag{4.17}$$

Eq. (4.15) can also be rewritten as

$$R'(h(\lambda)) = \langle x \rangle_h = \lambda , \tag{4.18}$$

where $\langle \cdot \rangle_h$ is the mean value computed from the probability distribution (4.17). It is evident that, if $P(x)$ is different from zero only in the interval $[a, b]$, the function $P_h(x)$ will share the same property, and Eq. (4.18) will have one (and only one) solution for $\lambda \in [a, b]$ (if $a \neq b$), with the possible exception of the extremes of the interval. The solution is unique since $R'(h)$ is an increasing function:

$$R''(h) \equiv \frac{d^2 R}{dh^2} > 0 . \tag{4.19}$$

We leave to the reader the easy exercise of proving that for $\lambda > b$ or for $\lambda < a$ Eq. (4.18) admits no solutions. We will explicitly consider a few examples in the following sections.

---

[2] As we shall see in Section 4.6.1, this relation is a special case of Crámer's theorem.

4.3 Examples of Large Deviations

In this section, we are going to explicitly illustrate some instructive examples of the application of the large deviations theorem.

4.3.1 The Gaussian Case

The computation of large deviations in the case of a sum of Gaussian-distributed variables is trivial, and the demonstration is left as an exercise for the reader. The final result for a Gaussian probability density with mean μ and variance σ^2 is given by the decrease function

$$S(\lambda) = -\frac{(\lambda - \mu)^2}{2\sigma^2}, \tag{4.20}$$

which coincides with the dominant term in (4.13).

4.3.2 Large Deviations for the Rademacher Distribution

We now explicitly derive $S(\lambda)$ in the case where the elementary probability distribution of the variables x is the bimodal distribution

$$P(x) = \tfrac{1}{2}[\delta(x - 1) + \delta(x - 1)], \tag{4.21}$$

known as the Rademacher distribution, i.e., in the case where the variable x can take only the two values ± 1, each with probability $1/2$. This is a case of some importance in physics and in the schematization of many statistical problems, where it is assumed that the elementary variables of the problem (e.g., spins, see Chapter 11, or bits, see Chapter 14) can take on only two values, ± 1, *up* and *down*, *head* and *tail*, or *true* and *false*, each one with probability $1/2$. The Fourier transform of (4.21) is

$$\tilde{P}(q) = \frac{e^{iq} + e^{-iq}}{2} = \cos(q), \tag{4.22}$$

hence

$$R(h) = \ln(\cosh(h)). \tag{4.23}$$

From (4.15) the saddle point equation has the form

$$\tanh(h) = \lambda, \tag{4.24}$$

and the solution, for $|\lambda| < 1$, is given by

$$\bar{h}(\lambda) = \operatorname{arctanh}(\lambda) = \frac{1}{2} \ln\left(\frac{1 + \lambda}{1 - \lambda}\right). \tag{4.25}$$

For $|\lambda|$ greater than 1 the saddle point equation does not admit solutions, and this is consistent with the cancellation of the function $d_N(z)$ for $|z| > N$ (the sum of N variables that can be ± 1 remains between $-N$ and N).

The saddle point equation (4.25) may be rewritten as

$$-\bar{h}(\lambda)\lambda = -\frac{\lambda}{2}[\ln(1+\lambda) - \ln(1-\lambda)] \tag{4.26}$$

and

$$\ln\cosh(\bar{h}(\lambda)) = \ln\left(\frac{1}{\sqrt{(1+\lambda)(1-\lambda)}}\right) = -\frac{1}{2}[\ln(1+\lambda) + \ln(1-\lambda)]. \tag{4.27}$$

Thus, we obtain the decrease function

$$S(\lambda) = R(\bar{h}(\lambda)) - \bar{h}(\lambda)\lambda = -\frac{1+\lambda}{2}\ln(1+\lambda) - \frac{1-\lambda}{2}\ln(1-\lambda). \tag{4.28}$$

The function $d_N(\lambda)$ is, therefore, symmetric under the exchange $\lambda \leftrightarrow -\lambda$. For small λ we also have

$$S(\lambda) = -\frac{\lambda^2}{2} + O(\lambda^4), \tag{4.29}$$

which returns the central limit. The relationship at the extremes is also remarkable:

$$S(\pm 1) = -\ln(2), \tag{4.30}$$

expressing the fact that the probability that all N variables x_i are equal to 1 (or all equal to -1) is evidently 2^{-N}. The same result could have been obtained using the exact expression of the binomial (Rademacher) distribution and applying Stirling's formula (2.73). In general, by expanding the function $R(h)$ around the saddle point, one can systematically compute the corrections proportional to the powers of $1/N$.

The fact that, for a given value of λ, the saddle point equation does not admit any solution is a signal that, for such a value of λ, the probability function is strictly zero. This occurs in our example for $|\lambda| > 1$. More generally, as we have already noted in the formula (4.18), if the probability distribution $P(x)$ is non-zero only in the interval $[a, b]$, then $d_N(\lambda)$ will also be different from zero only in that interval and zero outside that interval.

In Figure 4.1 we show, as a function of λ, the logarithm of the probability distribution (properly normalized) for the sum of 24 and 48 binary variables. The symbols represent the true probability distribution computed with a sampling of 10^9 events extracted with a computer using a pseudo-random number generator (Appendix 3.J), whereas the lines represent the estimate obtained using the saddle point within the large deviations theorem (LD) and the Gaussian estimate obtained from the central limit theorem (CL).

When the absolute value of λ is small, the estimate provided by the central limit theorem is a good approximation, which gets worse as $|\lambda|$ increases: in regions of large values of $|\lambda|$ the distribution provided by the large deviations theorem is much better.

For larger N the approximation, even for very small orders of magnitude of the probability, improves. Note that in Figure 4.1 we are plotting the logarithm of the probability distribution, and that we are hence reasoning about very small values of the probability.

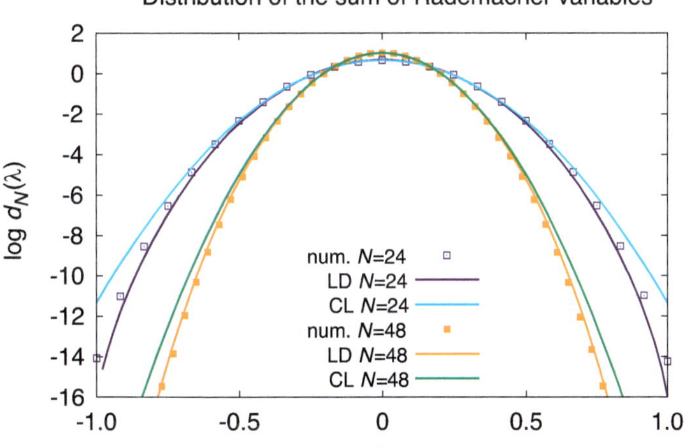

Figure 4.1   The logarithm of the probability distribution for the sum of 24 and 48 Rademacher variables. The symbols represent the true probability distribution computed with a numerical sampling of 10^9 events (num.). The lines, on the other hand, represent the estimate obtained by means of the saddle point within the large deviations theorem (LD), formula (4.28), and the Gaussian estimated with the central limit theorem (CL), see (3.19) with $\lambda = z/N = w/\sqrt{N}$.

4.3.3 Large Deviations for the Semi-Gaussian Distribution

Let us now consider the distribution (2.40),

$$P(x) = \sqrt{\frac{2}{\pi\sigma^2}}\, e^{-x^2/(2\sigma^2)}\theta(x)\,, \tag{4.31}$$

whose domain is no longer restricted as in the previous example. However, $P(x)$ decays fast enough with $x \to \infty$ to fall under the assumptions of the large deviations theorem. The generating function $R(h)$ of $P(x)$ is

$$R(h) = \frac{h^2\sigma^2}{2} + \log\left[1 + \mathrm{erf}\left(\frac{h\sigma}{\sqrt{2}}\right)\right], \tag{4.32}$$

where the error function is defined as[3]

$$\mathrm{erf}(x) \equiv \frac{2}{\sqrt{\pi}}\int_0^x dy\, e^{-y^2}\,. \tag{4.33}$$

The corresponding saddle point equation has the form

$$\frac{\sqrt{2\sigma^2/\pi}\; e^{-h^2\sigma^2/2}}{1 + \mathrm{erf}(h\sigma/\sqrt{2})} + h\sigma^2 = \lambda\,, \tag{4.34}$$

where the first term tends to 0 for $h \to -\infty$ and is ∞ for $h \to \infty$. This equation is not easily solved in analytical form to obtain $\bar{h}(\lambda)$. By expanding for small h, as in (4.12), one can, however, obtain a polynomial equation of the desired order and

---

[3] Recall that in the formula (3.68) we defined the complementary function erfc.

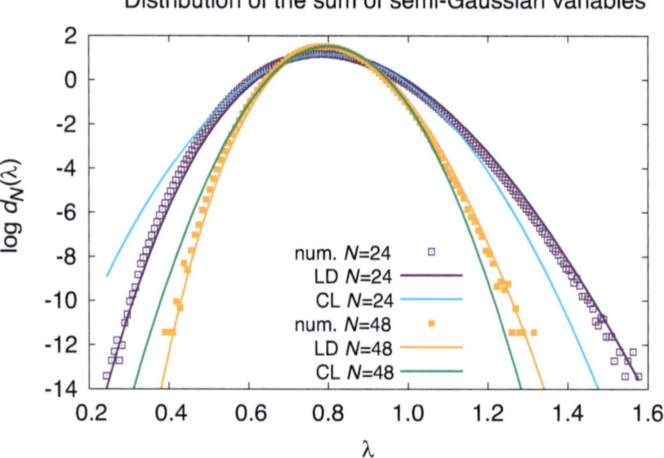

Figure 4.2 Logarithm of the probability distribution for the sum of 24 and 48 variables distributed with (4.31). The symbols represent the true probability distribution computed with a sampling of 10^8 events (num.). The lines are the estimate obtained by means of the large deviations theorem (LD) and the behavior estimated by the central limit theorem (CL).

derive a function $S(\lambda)$ defined in the domain $\lambda \in [0, \infty)$. In Figure 4.2 we show the case in which the true probability distribution was estimated empirically as for the previous figure. The saddle point estimate was computed from (4.34) with a fourth-order expansion in h and is a very good approximation already for $N = 24$.

4.3.4 Large Deviations for the Exponential Distribution

Let us finally consider the case where the quantities x_i are distributed according to an exponential law, which does not satisfy the assumptions of the large deviations theorem:

$$P(x) = e^{-x}, \quad x \geq 0, \tag{4.35}$$

while $P(x) = 0$ for $x < 0$. We ask ourselves what is the distribution of the sum of N variables distributed in this way. In this case the very restrictive conditions we used to prove the large deviations theorem (decrease of $P(x)$ to large $|x|$ faster than an exponential function and consequent analyticity of the function $R(h)$) are not satisfied. However, as we shall see, $R'(h)$ diverges at the point of non-analyticity, and the distribution of the large deviations is obtained correctly.

The decrease function $S(\lambda)$ can be estimated without using the large deviations theorem, by directly computing the integral

$$d_N(N\lambda) = \int \prod_{i=1}^{N} dx_i \, \delta\left(\sum_{i=1}^{N} x_i - N\lambda\right) \exp\left(-\sum_{i=1}^{N} x_i\right) \theta(x_i)$$

$$= \int \frac{dq}{2\pi} e^{-\iota q \lambda N} \left(\frac{1}{1 - \iota q}\right)^N, \tag{4.36}$$

where we have used the expression (3.79) of the Dirac delta as the Fourier antitransform of 1. By making a variable change $q = \imath w$ and computing the saddle point with (4.72), we find that the asymptotic behavior has the form

$$d_N(\lambda) \simeq \sqrt{\frac{N}{2\pi \lambda^2}}\, e^{N S(\lambda)}, \tag{4.37}$$

$$S(\lambda) = -\lambda + \ln(\lambda) + 1. \tag{4.38}$$

Equivalently, the integral (4.36) can be computed exactly using Eq. (3.144) of Appendix 3.H and obtaining

$$d_N(\lambda) = N\,\frac{(N\lambda)^{N-1}}{(N-1)!}\, e^{-N\lambda}. \tag{4.39}$$

Then applying Stirling's formula (2.73) we find, for N large, the expression (4.37).

Let us instead try to apply the theorem derived in the previous section. The Laplace transform $\tilde{P}(-\imath h)$ of the function $P(x)$ is given by $(1-h)^{-1}$ (which exists only for $\mathrm{Re}(h) < 1$). The function $R(h)$ is given by

$$R(h) = -\ln(1-h), \quad h < 1, \tag{4.40}$$

and the equation to solve is therefore

$$\lambda = (1-h)^{-1}.$$

By substituting in the equations in the previous section we find the result already obtained in (4.37),

$$S(\lambda) = R(\bar{h}(\lambda)) - \bar{h}(\lambda)\lambda = \ln \lambda - \lambda + 1.$$

We show in Figure 4.3 the analogue of Figure 4.1 for the exponential distribution. Here, too, the true distribution was obtained by empirically sampling 10^9 events (both

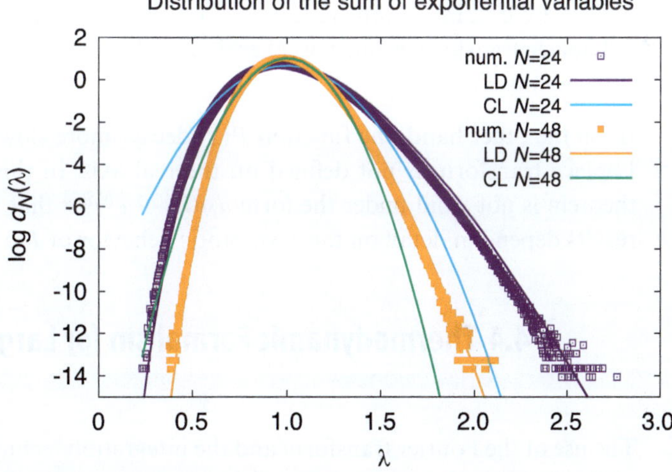

Figure 4.3   Logarithm of the probability distribution for the sum of 24 and 48 exponential variables, distributed according to the density (4.35). The symbols represent the true probability distribution computed with a sampling of 10^9 events (num.). The lines, on the other hand, represent the estimate obtained by means of the large deviations theorem (LD), see formula (4.37), and the estimate using the central limit theorem (CL), see (3.19) with $\lambda = z/N = w/\sqrt{N}$.

for the $N = 24$ case and for $N = 48$). Here, again, it can be clearly seen that the saddle point estimate is extremely reasonable for values of N already on the order of a few tens. The difference between the rough estimate given by the central limit theorem and the more sophisticated one obtained thanks to the large deviations theorem is in this case even larger than for the previous cases.

4.3.5 A Comparison of Large Deviations Distributions

In the previous three subsections we analyzed three different cases, of the sum of variables with Rademacher distribution, with semi-Gaussian distribution, and with exponential distribution.

- In the first case, exemplified by the bimodal distribution, the function $P(x)$ is defined in a compact interval. In this case also the values of λ for which the probability of an event is non-zero belong to the same interval: when λ approaches the edges of the interval, $|h|$ diverges and $d_N(\lambda) = 0$.
- In the second case, exemplified by the Gaussian and semi-Gaussian distributions, the function $P(x)$ is non-zero in an infinite domain, and for large values of $|x|$ decays faster than an exponential, meeting the assumptions of the large deviations theorem. Again, the values of λ for which the probability of an event is non-zero belong to the same domain of x: $|h|$ goes to infinity when λ tends to the extremes of the domain. For example, the sum of variables distributed with (4.31) has a non-zero probability in the interval $\lambda \in [0, \infty)$, bounded by the points where $|h|$ diverges.
- In the third case, the function $P(x)$ decreases exactly exponentially. In this case the Laplace transform has a strong singularity on the real axis, and h tends to a finite value when λ goes to infinity. More complex situations can occur due to various asymptotic behaviors of the function $P(x)$: for example, much more complicated ones are obtained in the case of $P(x) \approx x^{-2} \exp(-x)$, where the singularity of the Laplace transform is much weaker than in the case of a pure exponential decay, with diverging $R'(h)$.

If, on the other hand, the function $P(x)$ decays more slowly than an exponential, the Laplace transform is not defined on the real axis. In this case, the large deviations theorem is not valid under the form $d_N(\lambda) \sim e^{N S(\lambda)}$ that we have presented, and the results depend in detail on the asymptotic behavior of $P(x)$.

4.4 Thermodynamic Formalism for Large Deviations

The use of the Fourier transform and the integration technique in the complex field on which we based the computation of the previous section allows us to solve the problem in a technically simple, but maybe not very intuitive, way. For this reason, we will review the previous results using a method in common use in the study of physical systems, a method that, perhaps, allows a better understanding of the relevant mechanisms.

We begin by assuming, for the sake of argument, that for large values of $z = \lambda N$ the function $d_N(\lambda N)$ behaves like $e^{NS(\lambda)}$: our aim is now to determine the function $S(\lambda)$. We thus consider the following mean value for large N:

$$\langle e^{hz} \rangle = \int dz\, d_N(z) e^{hz} \propto \int d\lambda\, e^{N(S(\lambda)+h\lambda)} , \tag{4.41}$$

with real h, where the last step is, indeed, true for the hypothesis (4.3), and assume that the integral exists.

We can directly compute the same expected value without the assumption (4.3) but by integrating over all the distributions of x_i, $i \in [1, N]$, constrained to sum up at the value z. We have that

$$\langle e^{hz} \rangle = \int dz \int \prod_{i=1}^{N} dx_i\, P(x_i)\, \delta \left(\sum_{i=1}^{N} x_i - z \right) e^{hz} = e^{NR(h)} , \tag{4.42}$$

where $R(h)$ is the function defined in (4.14),

$$e^{R(h)} = \int dx\, P(x) e^{hx} .$$

We know that, for large N, the integral (4.41) that interests us can be computed using Laplace's maximum point method, i.e., by evaluating the function $S(\lambda) + \lambda h(\lambda)$ at $\lambda = \bar\lambda(h)$ determined by the equation of the maximum

$$\left. \frac{dS(\lambda)}{d\lambda} \right|_{\lambda = \bar\lambda(h)} = -h . \tag{4.43}$$

Comparing, therefore, the formulas (4.41) and (4.42), we obtain the condition

$$R(h) = S(\bar\lambda(h)) + h\bar\lambda(h) , \tag{4.44}$$

which gives us $R(h)$ as a function of $S(\lambda(h))$. We are interested, instead, in the inverse relationship, i.e., in computing the function of the large deviations $S(\lambda)$ given the function $R(\bar h(\lambda))$, which we obtain by Laplace transforming $P(x)$. The function R in (4.44) is called the *Legendre transform* of the function S, a transformation that we will consider in more detail in Section 4.5.

If we differentiate (4.44) with respect to h and use (4.43), we find that

$$\frac{dR}{dh} = \left. \frac{dS}{d\lambda} \right|_{\bar\lambda(h)} \frac{d\bar\lambda}{dh} + h\frac{d\bar\lambda}{dh} + \bar\lambda(h) = -h\frac{d\bar\lambda}{dh} + h\frac{d\bar\lambda}{dh} + \bar\lambda(h) = \bar\lambda(h) , \tag{4.45}$$

i.e., that the saddle point equation (4.15) holds,

$$\frac{dR(h)}{dh} = \lambda , \tag{4.46}$$

which allows us to determine the inverse function $\bar h(\lambda)$. The function $S(\lambda)$ is therefore given by $R(\bar h(\lambda)) - \lambda\bar h(\lambda)$, exactly as in the result of the previous derivation.[4]

---

[4] This approach can be seen as a special case of the Gärtner–Ellis theorem to which we will refer in Section 4.6.2.

The relationship between this approach to the problem of large deviations and the thermodynamic description of the state of physical systems will be discussed in more detail in Chapter 14. First, however, let us formalize and explore the properties of the Legendre transform.

4.5 The Legendre Transform

Let us consider the function $f(x)$, which, for simplicity, we assume to be twice differentiable, with $f''(x) > 0$. Functions with this property are very important in mathematics and are called *convex functions* (there is a definition of a convex function that is also applicable to non-differentiable functions, as we will see in Section 4.5.3). We then define the function of two variables

$$F(x, p) \equiv px - f(x), \tag{4.47}$$

where p is a real number. Having chosen a given value for p, we call $\bar{x}(p)$ the maximum point of the function F at p fixed, i.e., the solution of

$$\left. \frac{\partial F}{\partial x} \right|_{x=\bar{x}(p)} = 0 \quad \Longrightarrow \quad \left. \frac{df}{dx} \right|_{x=\bar{x}(p)} = p.$$

The convexity of $f(x)$ implies that the solution $\bar{x}(p)$, if it exists, is unique.

We, then, define the function of p,

$$g(p) \equiv \sup_x F(x, p) = F(\bar{x}(p), p) = p\bar{x}(p) - f(\bar{x}(p)), \tag{4.48}$$

which we term the *Legendre transform* of the function $f(x)$. The Legendre transform maps functions on a given vector space into functions on the dual space, as we will discuss in more detail below.

Let us now consider as an example the function

$$f(x) = \frac{x^\alpha}{\alpha},$$

with $\alpha > 1$, and compute its Legendre transform. The equation $dF/dx = p - x^{\alpha-1} = 0$ gives

$$\bar{x}(p) = p^{1/(\alpha-1)},$$

and the Legendre transform is

$$g(p) \equiv F(\bar{x}(p), p) = \bar{x}(p)p - \frac{\bar{x}^\alpha(p)}{\alpha} = \frac{p^{\alpha/(\alpha-1)}}{\alpha/(\alpha-1)} = \frac{p^{\tilde{\alpha}}}{\tilde{\alpha}}, \tag{4.49}$$

with $\tilde{\alpha} = \alpha/(\alpha - 1)$. That is, we have found that $g(p)$ has the same functional form as $f(x)$. It is also interesting to note that

$$\frac{1}{\alpha} + \frac{1}{\tilde{\alpha}} = 1.$$

Let us now transform the function $g(p)$ via a Legendre transformation. For this purpose we introduce the function

$$G(x, p) = xp - g(p).\tag{4.50}$$

Now, using x as a transformation parameter, we obtain the Legendre transform of $g(p)$, i.e.,

$$h(x) \equiv \sup_p G(x, p) = G(x, \bar{p}(x)) = x\bar{p}(x) - g(\bar{p}(x)),$$

where $\bar{p}(x)$ is determined by the condition

$$\left.\frac{\partial G}{\partial p}\right|_{p=\bar{p}(x)} = 0 \quad\Longrightarrow\quad \left.\frac{dg}{dp}\right|_{p=\bar{p}(x)} = x.\tag{4.51}$$

4.5.1 Properties of the Legendre Transform

4.5.1.1 Involutive Property

We will now explicitly see that $h(x) = f(x)$, i.e., we have performed an inverse Legendre transform operation, which here coincides with the direct transform. In other words, the Legendre transform of the Legendre transform returns to the original function. This is called an *involutive* operation and its square is equal to the identity transform. Let us prove it.

First, using Eqs. (4.50), (4.51), and (4.48) we have that

$$\left.\frac{\partial G}{\partial p}\right|_{p=\bar{p}(\bar{x})} = x - \left.\frac{\partial F(x, p)}{\partial p}\right|_{\substack{x=\bar{x}(p),\\p=\bar{p}(x)}} = x - \bar{x}(\bar{p}(x)) = 0,\tag{4.52}$$

which implies that

$$x = \bar{x}(\bar{p}(x)),\tag{4.53}$$

and thus $\bar{p}(x)$ is the inverse function of $\bar{x}(p)$. In other words, the first Legendre transform introduces a correspondence between the points x and p and this relationship persists when performing the second Legendre transform. At this point it is immediate to verify that the Legendre transform is involutive. Indeed, using the relation (4.53), it holds that

$$h(x) = x\bar{p}(x) - g(\bar{p}(x)) = x\bar{p}(x) - \bar{p}(x)\bar{x}(\bar{p}(x)) + f(\bar{x}(\bar{p}(x))) = f(x).\tag{4.54}$$

4.5.1.2 Geometric Interpretation

It may be useful to recall a simple geometric interpretation of the Legendre transformation that we show in Figure 4.4. The Legendre transform $g(p)$ of a function $f(x)$ is the ordinate (with the opposite sign) of the tangent to $f(x)$ at the point $\bar{x}(p)$ of slope p.

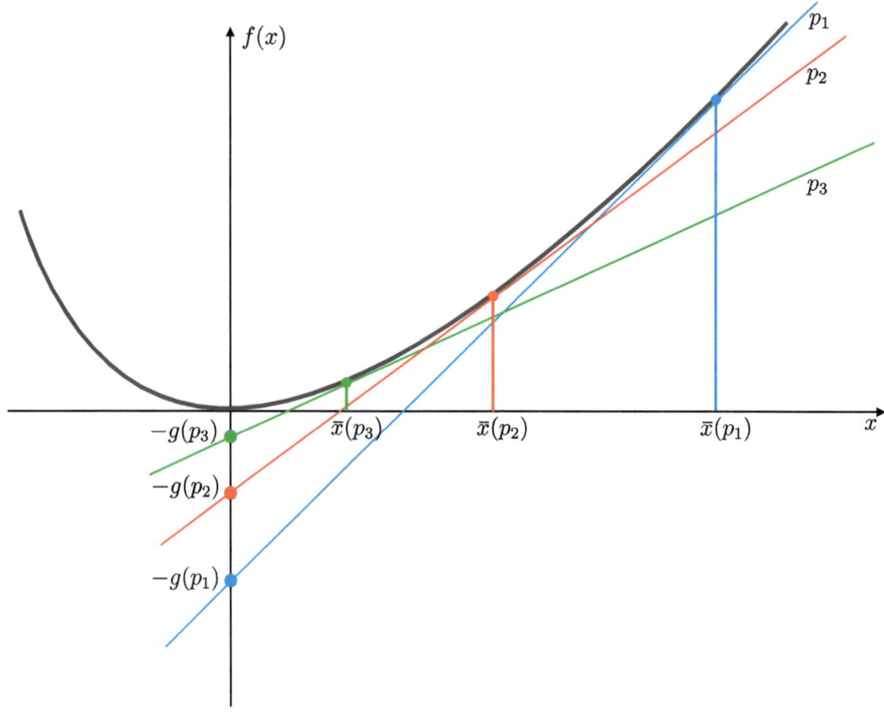

Figure 4.4 Graphical interpretation of the Legendre transform $g(p)$ of a convex function $f(x)$. The value $-g(p)$ is the intercept of the tangent of slope p to the function $f(x)$. The point on the abscissa at which $f(x)$ takes slope p is $\bar{x}(p)$. Three points are marked in the figure, with their slopes and intercepts as examples. The tangents have equation $y(x) = xp_k - g(p_k)$. At the tangency point $f(\bar{x}(p_k)) = \bar{x}(p_k)p_k - g(p_k)$, from which the Legendre transform $g(p)$ can be obtained.

4.5.1.3 Young's Inequality in the Legendre Transform

Two functions $f(x)$ and $g(p)$ which are the Legendre transforms of one another are called *dual*. They satisfy *Young's inequality*:

$$px \leq f(x) + g(p), \tag{4.55}$$

which follows trivially from the fact that the function $F(x, p) = px - f(x)$ that we defined in (4.47) is, by construction, smaller than $g(p)$ for all values of x and p.

4.5.1.4 Multi-Dimensional Legendre Transform

The generalization of the construction of the Legendre transform to the vector case is immediate:

$$g(\vec{p}) \equiv \sup_{\vec{x}} [\vec{p} \cdot \vec{x} - f(\vec{x})] = \vec{p} \cdot \vec{\bar{x}}(\vec{p}) - f(\vec{\bar{x}}(\vec{p})), \tag{4.56}$$

where

$$\vec{p} = \vec{\nabla}_{\vec{x}} f \quad \Longrightarrow \quad \vec{x} = \vec{x}(\vec{p}) \,.$$

4.5.2 Legendre Transform in Physics

The Legendre transform is a tool of great importance in many fields of theoretical physics. The first example to be mentioned is probably the relationship between the Lagrange function, or Lagrangian, and the Hamilton function, or Hamiltonian, in classical mechanics (and then, by extension, to field theory). Let us briefly recall that, having defined a Lagrangian function as $L = T - U$, i.e., the kinetic energy minus the potential energy of the system, the Euler–Lagrange equations,

$$\frac{d}{dt}\left(\frac{\partial L}{\partial \dot{q}_i}\right) - \frac{\partial L}{\partial q_i} = 0 \,, \tag{4.57}$$

are equivalent to Hamilton's equations,

$$\dot{p}_i = -\frac{\partial H}{\partial q_i} \,, \quad \dot{q}_i = \frac{\partial H}{\partial p_i} \,, \tag{4.58}$$

where the Hamiltonian function of $\vec{p}$,

$$H(\vec{p},\vec{q},t) \equiv \vec{p}\cdot\dot{\vec{q}} - L(\vec{q},\dot{\vec{q}},t) \,, \tag{4.59}$$

is the Legendre transform of the Lagrange function of the variables $\dot{\vec{q}}$ (conjugate to $\vec{p}$).

Other relevant examples concern thermodynamics, statistical mechanics, and field theory. In thermodynamics first and then in statistical mechanics, free energy and entropy can be seen as connected by a Legendre transformation. In the statistical mechanical approach, this establishes a relationship between the canonical and the microcanonical statistical ensembles, and connects the temperature variable to the energy variable. Indeed, by defining the inverse temperature $\beta = (kT)^{-1}$, where k is the Boltzmann constant and T is the absolute temperature, the Legendre transform of the free energy function (multiplied by the inverse temperature, to be precise) $\beta F(\beta)$ is the entropy $S(E)$ as a function of the internal energy E:

$$S(E) = \beta E - \beta F \,, \quad \text{with } E = \frac{\partial \beta F}{\partial \beta} \,. \tag{4.60}$$

As we will discuss in more detail in Section 14.3.1, in a dedicated chapter, entropy can be considered as a large deviations function.

Similar considerations can be made for the relationship between the canonical ensemble and the grand canonical ensemble, where pressure and free energy are linked by a Legendre transformation acting on the conjugate variables of chemical potential and mass density.

At last, we mention the relationship between magnetization and the external magnetic field, which is its conjugate parameter. Also, this type of relation is based on the Legendre transform and we will revisit such relations in Section 14.5 when we will further elaborate on the link between thermodynamics and large deviations.

4.5.3 Legendre Transform for Simply Continuous Functions

When the function $f(x)$ is continuous but not differentiable, it is still possible to define a Legendre transform, now called a Legendre–Fenchel transform, as

$$g(p) = \sup_x F(x, p).$$

This transform is always a convex function of p, which means that any two points on the curve $g(p)$ are joined by a segment that is always above (or along) $g(p)$, never below, regardless of the behavior and differentiability of $f(x)$. In contrast to convex and derivable functions, however, the transform

$$h(x) = \sup_p G(x, p),$$

is not necessarily equal to $f(x)$ for all x. Also $h(x)$, as a Legendre transform, is always convex in x, regardless of $g(p)$ and $f(x)$. In particular, $h(x)$ has the property of being the largest convex function for which $h(x) \leq f(x)$ is valid in the domain of definition of x. We denote this property by saying that $h(x)$ is the *convex envelope* of $f(x)$.

If $g(p)$ is differentiable at a point p, then both the function of which it is the transform $(f(x))$ and its transform $(h(x))$ coincide at the point $x = g'(p)$: i.e., $h(\bar{p}(x)) = f(\bar{p}(x))$. Conversely, it is also true that if $h(x) \neq f(x)$ in a certain interval x. then there exists at least one point p where $g(p)$ is non-differentiable.

If, finally, $g(p)$ is differentiable everywhere, then $h(x) = f(x)$ everywhere and the transform is completely involutive, i.e., the direct transform acts as an inverse, bringing us back to the previously examined case of the Legendre transform.

4.6 Fundamental Theorems on Large Deviations

Let us now mention some theorems that are useful in the treatment of large deviations problems and that link the decrease functions $S(\lambda)$ to functions like $R(h)$, see formula (4.14), which are the logarithm of the Laplace transform (see Appendix 3.D) of the probability distribution. The function $e^{R(h)}$ is the generating function of the moments of the distribution $P(x)$. This means that the nth derivative of $e^{R(h)}$ computed for $h = 0$ coincides with $\langle x^n \rangle$:

$$\frac{d^n}{dh^n} e^{R(h)}\Big|_{h=0} = \frac{d^n}{dh^n} \int dx\, e^{hx} P(x)\Big|_{h=0} = \int dx\, x^n P(x) = \langle x^n \rangle.$$

Once the generating function has been computed, the moments are obtained by deriving it, rather than integrating $P(x)x^n$. This can be very convenient. Similarly, $R(h)$ is the generating function of *cumulants* of $P(x)$. Generator functions are very important objects in probability theory and will be treated at length in Chapter 8 and in Section 12.3. In particular, in the somewhat broader context of correlation functions, we will demonstrate this exponential relationship between the cumulants generator and the moments generator in Section 12.3.3. For now, we will simply take the relation for granted and present (or re-present in a formally more explicit manner) some theorems concerning the function R: Crámer's theorem, the Gärtner–Ellis theorem, and Varadhan's theorem for large deviations.

4.6.1 Crámer's Theorem

We take the variable $z = \sum_{i=1}^{N} x_i$, the sum of N random variables x_i distributed with $P(x)$ and consider their behavior for large N, $z/N \sim \lambda$. Crámer's theorem summarizes what we have seen and proved in Section 4.2.

Theorem. (Crámer's theorem) *The large deviations decrease function is the Legendre–Fenchel transform of the cumulants generating function* $R(h) = \ln \int dx \, e^{hx} P(x)$:

$$S(\lambda) = \sup_{h} \left[R(h) - \lambda h \right]. \tag{4.61}$$

In particular, for convex differentiable functions, we find the formula (4.16) derived and widely used in Section 4.2.

4.6.2 The Gärtner–Ellis Theorem

An extension of Crámer's theorem is the Gärtner–Ellis theorem.

Theorem. (Gärtner–Ellis theorem) *If the generating function of the cumulants of a distribution* $d_N(\lambda)$,

$$\mathcal{R}(h) \equiv \lim_{N \to \infty} \frac{R_N(h)}{N} \tag{4.62}$$

with

$$R_N(h) \equiv \ln \int d\lambda \, d_N(\lambda) e^{h\lambda N} \tag{4.63}$$

exists and is differentiable for any h, then

1. the distribution d_N satisfies the large deviations theorem,

$$d_N(\lambda) \propto e^{N S(\lambda)},$$

and

2. the decrease function $S(\lambda)$ is the Legendre–Fenchel transform of $\mathcal{R}(h)$,

$$S(\lambda) = \sup_{h} \left[\mathcal{R}(h) - \lambda h \right]. \tag{4.64}$$

We have already seen this relation in the formalism of Section 4.4, where $\mathcal{R}(h) = (1/N) \ln \langle e^{hz} \rangle$, see Eq. (4.42). We underline that in the present case in (4.63) we are assuming the factorization of $P(x)$. This will come in handy in Section 14.5 in Chapter 14, when we will take up the link between entropy, and more generally between thermodynamic potentials in statistical mechanics, and the decrease function of large deviations.

4.6.3 Varadhan's Theorem

A further extension is given by Varadhan's theorem, which applies to the functional generalization of the generating function. Given a function $f(\lambda)$ we define the following functional of f:

$$\mathcal{R}[f] = \lim_{N \to \infty} \frac{1}{N} \ln \int d\lambda \, d_N(\lambda) e^{N f(\lambda)}. \tag{4.65}$$

With this definition, we can state the theorem.

Theorem. (Varadhan's theorem) *If the probability distribution $d_N(\lambda)$ satisfies the large deviations theorem with decrease function $S(\lambda)$, then the functional $\mathcal{R}[f]$ can be expressed as*

$$\mathcal{R}[f] = \sup_{\lambda} [S(\lambda) + f(\lambda)]. \tag{4.66}$$

In the particular case where $f(\lambda) = h\lambda$, we have $\mathcal{R}[f] = R(h)$ and the application of Varadhan's theorem leads to the Legendre–Fenchel transform,

$$R(h) = \sup_{\lambda} [S(\lambda) + h\lambda],$$

i.e., Crámer's theorem (4.61).

Mathematical Appendices

4.A The Saddle Point Method

The saddle point method, also known as the *pass method* or the *method of steepest descents*, allows one to estimate the asymptotic contribution of integrals in a complex field of the type

$$I(N) = \int_{C(z_1, z_2)} dz \, \Phi(z) e^{N F(z)}$$

for large N. Here $C(z_1, z_2)$ is a path in the complex plane from the point z_1 to the point z_2, and $\Phi(z)$ and $F(z)$ are complex analytic functions defined on a domain that includes completely C.

To compute the asymptotic expression of $I(N)$ we recall some basic notions about harmonic functions and complex functions.

4.A.1 Some Properties of Harmonic and Analytic Functions

Harmonic functions are the solutions of the Laplace equation $\nabla^2 G = 0$, and satisfy the so-called *maximum principle*: if ζ is a domain in the complex field and $G(x)$ has a local maximum x_0 within the domain ζ, i.e.,

$$\exists\, x_0 \in \zeta \text{ such that } G(x_0) \geq G(x) \;\forall\, x \in \zeta,$$

then $G(x) = G(x_0)$ is a constant function. The functions do not have maxima or minima except on the boundary of their domain of existence.

Furthermore, constant value curves of harmonic functions, $G(x) = \kappa$, always rest on the boundary of the domain and cannot be closed (otherwise they would contain a maximum or a minimum). The curve $G(x) = \kappa$ therefore divides the domain ζ into at least two sectors, one in which $G(x) > \kappa$ (positive sector) and one in which $G(x) < \kappa$ (negative sector).

Turning to the complex differentiable (or analytic) functions $F(z) = u(x, y) + \imath v(x, y)$, with $z = x + \imath y$, we recall that for the real part u and the imaginary part v of the function F, the Cauchy–Riemann relations apply,

$$\frac{\partial u(x, y)}{\partial x} = \frac{\partial v(x, y)}{\partial y}\,, \quad \frac{\partial u(x, y)}{\partial y} = -\frac{\partial v(x, y)}{\partial x}\,, \tag{4.67}$$

and from these it follows that

$$\nabla^2 u = \nabla^2 v = 0,$$

i.e., u and v are harmonic functions in the plane (x, y), and

$$\nabla u \cdot \nabla v = 0, \tag{4.68}$$

at every point in the definition domain, and, in particular, along every path of integration. This means that if $\nabla u \neq 0$ near to a point (x_0, y_0), we can find a path C passing through that point along which the gradient component of v is zero because $v(x, y) = v(x_0, y_0) = $ constant in that neighborhood. This particular path is called a *pass*, or a steepest descent (or ascent) path. The descent (ascent) obviously refers to the behavior of the real part $u(x, y)$.

For analytic functions in a simply connected domain ζ, the Cauchy integral theorem applies whereby on a closed path C completely contained in ζ it holds that

$$\int_C dz\, F(z) = 0$$

and, as a corollary, the integral on an open path from z_1 to z_2,

$$\int_{z_1}^{z_2} dz\, F(z) = I(z_1, z_2),$$

exclusively depends on the extremes and not on the path in the complex plane.

Furthermore, if the integrand function has a pole within the closed path C, as for example in the function $G(z) = F(z)/(z - \lambda)$, with $F(z)$ a pole-free analytic function for $z \in C$, then

$$\int_C dz\, G(z) = \int_C dz\, \frac{F(z)}{z - \lambda} = 2\pi \imath\, G(\lambda),$$

regardless of the contour, as proved by the residue theorem (see also formula (8.8)).

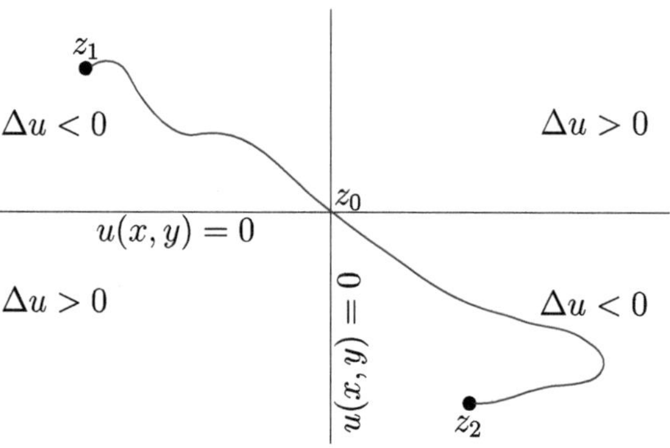

Figure 4.5　Division of the complex plane into four sectors according to the sign of $\Delta u = u(x, y) - u(x_0, y_0) < 0$ around the saddle point z_0.

4.A.2 Properties in a Saddle Point Neighborhood

At this point we make a few remarks on the properties of the real part $u(x, y)$ of $F(z)$, which we report schematically in Figure 4.5. Let us assume that u, harmonic, has a saddle point $z_0 = (x_0, y_0)$ in the complex plane. In its neighborhood the domain will be divided into sectors where $\Delta u = u(x, y) - u(x_0, y_0) > 0$ and sectors where $\Delta u = u(x, y) - u(x_0, y_0) < 0$. The curves $u(x, y) = u(x_0, y_0) = $ constant subdivide the entire plane into four sectors, because they cannot be closed within the domain of definition. We will have two $\Delta u > 0$ sectors and two $\Delta u < 0$ sectors, disjoint from each other.

Let us see the different cases for computing the integral

$$I(N) = \int_{z_1}^{z_2} dz \, \Phi(z) e^{NF(z)} . \tag{4.69}$$

If z_1 and z_2 are in the same sector and $F(z_2) > F(z_1)$ (or vice versa) one can always deform the path so that F grows monotonically between z_1 and z_2.

If, instead, z_1 and z_2 are in two disjoint negative sectors, the path can be deformed to pass through the saddle point z_0. This is the case we are interested in (but in the course of the discussion we will also deal with the previous case). In an interval around z_0, in negative sectors we have that $\Delta u < 0$. We choose, therefore, the path L passing through z_0 where the behavior of u is steepest. As mentioned above, along the direction of steepest slope one has $\nabla u \neq 0$ (at every point except the saddle point) and $\nabla v = 0$. Therefore, in that particular direction $v(x, y)$ is constant: $v(x, y) = v(x_0, y_0)$. We emphasize that the condition (4.68) is always satisfied.

Exactly at the saddle point, however, $\nabla u(x_0, y_0) = 0$. In the line variable s along the curve L of steepest slope, with origin at z_0 such that $z(s_0) = z_0$, we can then write

$$\frac{\partial u(x(s), y(s))}{\partial s}\bigg|_{s=0} = \frac{\partial v(x(s), y(s))}{\partial s}\bigg|_{s=0} = 0 ,$$

and this implies that $F'(z_0) = 0$, since the derivative in the complex plane of an analytic function does not depend on the direction considered. Taylor-expanding $F(z)$ around the saddle point of $u(x, y)$ we have

$$F(z) = F(z_0) + \tfrac{1}{2}F''(z_0)(z - z_0)^2 + O((z - z_0)^3). \tag{4.70}$$

In polar coordinates, where

$$z - z_0 \equiv \rho e^{\iota\phi}, \quad \tfrac{1}{2}F''(z_0) \equiv r_0 e^{\iota\theta_0},$$

we can rewrite

$$F(z) - F(z_0) = \rho r_0^2\, e^{\iota(2\phi+\theta_0)},$$

$$\Delta u = u(x, y) - u(x_0, y_0) = \rho r_0^2 \cos{(2\phi + \theta_0)},$$

$$\Delta v = v(x, y) - v(x_0, y_0) = \rho r_0^2 \sin{(2\phi + \theta_0)}.$$

The directions in which Δu changes sign are $\phi_k = k\pi/2 + \pi/4 - \theta_0/2$, with $k = 0, 1, 2, 3$, and mark the directions of the lines of constant u and the transitions from one sector to another. The directions of steepest descent are those for which $\Delta v = 0$ and $\Delta u < 0$ and correspond to the phases

$$\phi_0 = \frac{\pi - \theta_0}{2}, \quad \phi_1 = \frac{\pi - \theta_0}{2} + \pi = \phi_0 + \pi, \tag{4.71}$$

one the opposite of the other.

4.A.3 Proof of the Saddle Point Theorem

We then proceed to state and prove the saddle point theorem.

Theorem. (Saddle point theorem) *Given the integral*

$$I(N) \equiv \int_{z_1}^{z_2} dz\, \Phi(z) e^{NF(z)},$$

if the following conditions are satisfied:

1. *$\Phi(z)$ and $F(z)$ are two functions in ζ;*
2. *$u(x, y) = \mathrm{Re}[F(z)]$ has a single saddle point $z_0 \in \zeta$ where $F''(z_0) < 0$;[5]*
3. *there exists a $\delta > 0$ such that, along the L curve of steepest descent, for $|z - z_0| > \delta$ there exists an h such that $u(x_0, y_0) - u(x, y) = -\Delta u \geq h > 0$; and*
4. *there exists N_0 such that the line integral*

$$\int_C |\phi(z(s))| ds\, e^{N_0 u(x(s), y(s))} z'(s) \leq M < \infty,$$

where the path C joins two extremes z_1 and z_2 in distinct $\Delta u < 0$ sectors (and can, therefore, be deformed to include the steepest descent curve L by means of connections γ_1 and γ_2 along which $-\Delta u \geq h$)[6];

---

[5] These are sufficient but not necessary assumptions. The integral can be computed by the same method even if there are several saddle points, even infinite ones. If the second derivative is null, a generalization of the theorem in terms of higher-order derivatives can be written.

[6] Also the finite length of the connecting paths $\gamma_{1,2}$ is not necessary. The extremes can go to infinity and the theorem is still valid.

then for each $N > N_0$ there exists the integral $I(N)$ and it is equal to

$$I(N) \equiv \int_{z_1}^{z_2} dz\, \Phi(z) e^{NF(z)} = e^{NF(z_0)} \left\{ \sqrt{\frac{2\pi}{N|F''(z_0)|}}\, \Phi(z_0) e^{\imath \phi_k} + O\left(\frac{1}{N^{3/2}}\right) \right\}, \quad (4.72)$$

$$\phi_k = \frac{\pi - \theta_0}{2} + k\pi, \quad k = 0, 1, \quad \theta_0 = \arg(F''(z_0)).$$

The phase term $e^{\imath \phi_k}$ determines the direction of the integration path, see (4.71), and the two possible choices for ϕ_k simply correspond to a change of sign of $I(N)$.

Some of these assumptions are sufficient but not necessary. The saddle point theorem can be proved, with a final formula somewhat different from (4.72), even if: (i) there are multiple saddle points, (ii) if $F''(z_0) = 0$ (provided some some higher-order derivative is not), and (iii) if the connecting paths $\gamma_{1,2}$ each go to infinity in its sector.

Proof We will use the listed assumptions to simplify the proof. We divide the computation into

$$I(N) = I_{\gamma_1}(N) + I_L(N) + I_{\gamma_2}(N). \quad (4.73)$$

Let us see the contribution for the path γ_1, entirely included in the sector $\Delta u < 0$ where z_1 is located. Using hypothesis 3 of the theorem we can write

$$I_{\gamma_1}(N) = \int_{\gamma_1} dz\, \Phi(z) e^{NF(z)} = e^{NF(z_0)} \int_{\gamma_1} dz\, \Phi(z) e^{N(\Delta u + \imath \Delta v)}$$

$$\leq e^{NF(z_0)} e^{-hN} \int_{\gamma_1} dz\, \Phi(z) e^{\imath \Delta v N} = e^{NF(z_0)} O(e^{-hN}). \quad (4.74)$$

The same applies for γ_2 in another disjoint sector $\Delta u < 0$.

Let us now turn to the contribution around the saddle point along path L:

$$I_L(N) = \int_L dz\, \Phi(z) e^{NF(z)} = \int_{-a_L}^{b_L} ds\, \frac{dz(s)}{ds}\, \Phi(z(s)) e^{NF(z(s))}$$

$$= e^{\imath v(x_0, y_0)} \int_{-a_L}^{b_L} ds\, z'(s) \hat{\Phi}(s) e^{N\hat{u}(s)}, \quad (4.75)$$

where the integration line variable $s \in [-a_L, b_L]$ is chosen equal to zero at the saddle point and where, in the last step, we used the fact that $v(x, y)$ is constant along the steepest descent curve L. The integral of the last line is an integral of a function $\hat{u}(s)$ of a real variable that can be approximated as

$$\hat{u}(s) = \hat{u}(0) + \tfrac{1}{2}\hat{u}''(0)s^2 + O(s^3), \quad \hat{u}''(0) < 0$$

around the point of maximum on L and, hence, the asymptotic form is obtained by Laplace's method (Appendix 2.C):

$$\int_{-a_L}^{b_L} ds\, z'(s) \hat{\Phi}(s) e^{N\hat{u}(s)} = e^{N\hat{u}(0)} \left\{ \sqrt{\frac{2\pi}{N|\hat{u}''(0)|}}\, \hat{\Phi}(0) z'(0) + O\left(\frac{1}{N^{3/2}}\right) \right\} \quad (4.76)$$

and

$$I_L(N) = e^{N[\hat{u}(0)+\iota v(x_0,y_0)]} \left\{ \sqrt{\frac{2\pi}{N|\hat{u}''(0)|}}\, \hat{\Phi}(0)z'(0) + O\left(\frac{1}{N^{3/2}}\right) \right\}. \qquad (4.77)$$

We must now go back from the line variable to the complex variables. It holds that $\hat{\Phi}(0) = \Phi(z_0)$ and $\hat{u}(0) = u(x_0, y_0)$. The imaginary part $\hat{v}(s)$ is constant throughout L, so its second derivative is zero. This implies that the only contribution to $F''(z_0)$ comes from the second derivative of $\hat{u}(s)$ along L:

$$\hat{u}''(0) = \frac{d^2\hat{u}(s)}{ds^2}\bigg|_{s=0} = \frac{d^2 F(z(s))}{ds^2}\bigg|_{s=0}$$
$$= [F''(z(s))(z'(s))^2]_{s=0} + [F'(z(s))z''(s)]_{s=0} = F''(z_0)(z'(0))^2.$$

Along L, around the saddle point z_0, we can write $z-z_0 \simeq se^{\iota\phi}$, from which $z'(s) = e^{\iota\phi}$, where $\phi = \phi_0 = (\pi - \theta_0)/2$ or $\phi_1 = \phi_0 + \pi$, with $\theta_0 = \arg(F''(z_0))$. Consequently

$$\hat{u}''(0) = |F''(z_0)|e^{\iota \arg(F''(z_0))}e^{\iota(\pi-\arg(F''(z_0)))} = -|F''(z_0)|.$$

Substituting everything into (4.77) we obtain

$$I_L(N) = e^{NF(z_0)} \left\{ \sqrt{\frac{2\pi}{N|F''(z_0)|}}\, \Phi(z_0)e^{\iota\phi_k} + O\left(\frac{1}{N^{3/2}}\right) \right\}, \qquad k = 0, 1, \qquad (4.78)$$

and neglecting terms of $O(e^{-hN})$ we obtain Eq. (4.72).   ∎

Let us now make some observations.

- If z_1 and z_2 are in the same sector $\Delta u < 0$, then the integral $I(N) = I_\gamma(N) = e^{NF(z_0)}O(e^{-hN})$, with $h > 0$.
- If there are several saddle points, the method can be used to compute the asymptotic value of $I(N)$ which will contain as many contributions of the type of (4.72) as there are saddle points.
- If $F''(z_0) = 0$ we can modify the formula (4.72) using further terms in the expansion (4.70) and the polar coordinate representation of the higher-order derivatives.

4.A.4 Example of an Integral Computation Using the Saddle Point Method

As an exercise, we see how to perform an integral in the complex field with the saddle point method and to test the various hypotheses. We choose the so-called Hankel function of the first kind,

$$H_\nu(N) \equiv \frac{1}{\pi} \int_\Gamma dz\, e^{\iota(-\nu z + N \sin z)}, \qquad (4.79)$$

where Γ is any path leading from a point belonging to the region of the complex plane

$$x \in (-\pi/2, \pi/2], \quad y \in [0, \infty),$$

to a point in the domain

$$x \in [\pi/2, 3\pi/2), \quad y \in (-\infty, 0].$$

Referring to the notation in Eq. (4.72) of the saddle point theorem, we define $\Phi(z) \equiv e^{-\imath v z}$ and $F(z) = \imath \sin z$. The latter is the function whose saddle point we need to identify. We first compute its real and imaginary parts as functions of the coordinates x and y of the complex plane:

$$F(z) = u(x, y) + \imath v(x, y), \tag{4.80}$$

$$u(x, y) = -\cos x \sinh y, \tag{4.81}$$

$$v(x, y) = \sin x \cosh y. \tag{4.82}$$

We now verify the four hypotheses of the theorem.

1. $\Phi(z)$ and $F(z)$ are analytic functions. To show this, it suffices to show that they satisfy the Cauchy–Riemann relations (4.67). For $\Phi(x, y) = e^{vy} \cos(vx) - \imath e^{vy} \sin(vx)$ we have

$$\frac{\partial}{\partial x} e^{vy} \cos(vx) = -\frac{\partial}{\partial y} e^{vy} \sin(vx) = -v e^{vy} \sin(vx),$$

$$\frac{\partial}{\partial y} e^{vy} \cos(vx) = \frac{\partial}{\partial x} e^{vy} \sin(vx) = v e^{vy} \cos(vx).$$

For $F(z)$ we have

$$\frac{\partial u(x, y)}{\partial x} = \frac{\partial v(x, y)}{\partial y} = \sin x \sinh y,$$

$$\frac{\partial u(x, y)}{\partial y} = -\frac{\partial v(x, y)}{\partial x} = -\cos x \cosh y.$$

2. The real part $u(x, y)$ has at least one saddle point $z_0 = (x_0, y_0)$ where $F''(z_0) \neq 0$. The saddle points turn out to be

$$F'(z) = \imath \cos z,$$

$$F''(z) = -\imath \sin z \neq 0,$$

for all points $z = k\pi + \pi/2$, $\forall k \in \mathbb{Z}$. In the following we will look at the contribution to the integral of the saddle point $z_0 = \pi/2$, such that $F''(z_0) = -\imath$.

3. There exists a $\delta > 0$ such that, along the curve L of steepest descent, outside the interval $|z - z_0| > \delta$, we have $-\Delta u \equiv u(x_0, y_0) - u(x, y) \geq h > 0$. We derive the relationship between h and δ such that these conditions are satisfied. Recall that the curve of steepest descent is the one along which the imaginary part of F remains constant and its line derivative is zero. The conditions are, then, written as

$$-\Delta u \geq h \implies \cos x \sinh y \geq h,$$

$$|z - z_0| \geq \delta \implies (x - \pi/2)^2 + y^2 \geq \delta^2,$$

$$v(x, y) = v_0 \implies \sin x \cosh y = 1.$$

It is perhaps more convenient to work with the variable $x' = x - \pi/2$, so the conditions become

$$-\sin(-x')\sinh(y) \geq h, \quad (x')^2 + y^2 \geq \delta^2, \quad \cos(x')\cosh(y) = 1.$$

The third condition tells us that $\cosh(y) = 1/\cos(x')$, hence around the saddle point $(x', y) = (0, 0)$, y and x' grow in concord. Using this relationship and the properties of hyperbolic functions in the first inequality, we obtain

$$\sin^2(x')\left(\frac{1}{\cos^2(x')} - 1\right) = \sin^2(x')\tan^2(x') = \frac{\tan^4(x')}{1 + \tan^2(x')} \geq h^2.$$

The latter is an inequality for $\tan^2(x')$ which has a solution for

$$\tan^2(x') \geq \frac{h^2 + h\sqrt{h^2 + 4}}{2} \quad \Longrightarrow \quad x' \geq x'_{\min}(h) \equiv \arctan\sqrt{\frac{h^2 + h\sqrt{h^2 + 4}}{2}}.$$

Further, δ^2 is the minimum value of $(x')^2 + y^2$:

$$\delta^2 = (x'_{\min}(h))^2 + y_{\min}(h)^2,$$

where

$$y_{\min}(h) = \cosh^{-1}\left(\frac{1}{\cos(x'_{\min}(h))}\right).$$

From here we have the relation $\delta(h)$ for which the hypothesis of the theorem is satisfied.

4. The fourth hypothesis to be satisfied is that, given a boundary $\Gamma(z_1, z_2)$ joining two extremes belonging to distinct sectors in which $\Delta u < 0$, there must be an N_0 such that the line integral

$$\int_\Gamma ds \, |\Phi(z(s))| \, e^{N_0 u(x(s), y(s))} z'(s) \leq M < \infty.$$

We notice that $|\Phi(z)| = e^{vy}$ and we call $\Phi_{\max} \equiv \max_\Gamma e^{vy}$. As for the term $e^{N_0 u}$, its maximum is at the saddle point between the two domains $\Delta u < 0$, where $u_0 = u(x_0, y_0)$ and the maximum is $e^{N_0 u_0}$.

Having set these positions, we proceed to maximize the integral

$$\int_\Gamma ds \, |\Phi(z(s))| \, e^{N_0 u(x(s), y(s))} z'(s) \leq \Phi_{\max} e^{N_0 u_0} \int_\Gamma ds \, |z'(s)|.$$

Furthermore, the contour length $\Gamma(z_1, z_2)$ is finite. This is also expressed as

$$\mathcal{L}_\Gamma = \int_\Gamma ds \, |z'(s)|,$$

whereby the previous increment is increased by a finite number:

$$\int_\Gamma ds \, |\Phi(z(s))| \, e^{N_0 u(x(s), y(s))} z'(s) \leq \Phi_{\max} \, e^{N_0 u_0} \mathcal{L}_\Gamma < \infty.$$

Having verified the assumptions, we adopt the theorem and compute the Hankel function for large N around a saddle point z_0. We have

$$F(z_0) = \iota\,,$$
$$\Phi(z_0) = e^{-\iota \nu \pi/2}\,,$$
$$\arg F''(z_0) = 3\pi/2\,,$$
$$|F''(z_0)| = 1\,,$$

from which

$$\phi_k = \frac{\pi - \arg F''(z_0)}{2} + k\pi = -\frac{\pi}{4} + k\pi\,, \quad k = 0,1\,,$$

and finally

$$H_\nu(N) = \frac{1}{\pi} e^{\iota N} \left[\sqrt{\frac{2\pi}{N}}\, e^{-\iota\pi(\nu/2+1/4-k)} + O\left(\frac{1}{N^{3/2}}\right) \right]$$
$$= \pm\sqrt{\frac{2}{\pi N}}\, e^{\iota[N - \pi(\nu/2+1/4)]} \left(1 + O\left(\frac{1}{N}\right)\right).$$

Statistical Inference and Experimental Data Analysis

5.1 Experimental Data Analysis in the Simple Case

Analyzing the situation from a very general point of view, it can be seen that experimental data are affected by two different types of errors: *systematic errors* and *statistical errors*. A correct treatment of systematic errors (due, for example, to imperfections in the measuring apparatus or the presence of a background of spurious events that are impossible to eliminate) is extremely important but is beyond the scope of this text.

Instead, we will discuss here how to deal with statistical errors correctly: this is also a task of great importance. Without a good estimate of statistical errors, the experimental results themselves have no unambiguous meaning. We begin our remarks by noting that we must avoid falling into two opposing traps:

1. the underestimation of errors, which leads to the risk of reaching arbitrary and unfounded conclusions; and

2. the overestimation of errors, which leads to the loss of relevant information that is, instead, contained in the data set.

Let us consider a set of experimental data, and a quantity x, distributed according to an unknown probability law $D(x)$. We denote by X the mean expected value of x, defined as

$$X \equiv \int dx \, D(x)x \equiv \langle x \rangle. \tag{5.1}$$

Suppose we measured the quantity x a number of times $N > 1$, and obtained the results x_k, with $k = 1, \ldots, N$. The x_k are independent: the result of the kth measurement does not depend on the results of the other measurements. We are interested in reaching conclusions on the behavior of the mean value X from these experimental data. More precisely, we are interested in deducing what is the probability distribution of X suggested by an analysis of the experimental data. Even more generally, we would like to know what we can infer from the experimental data about the probability distribution $D(x)$ itself, asking what is the probability distribution $\mathcal{P}[D]$ of the probability distribution $D(x)$. Determining $\mathcal{P}[D]$ is technically a very complicated problem, as this is a functional of the probability D (which is a function).

A simplified approach, based on the central limit (3.19), leads to the conclusion that the probability that the mean value of x takes on the value X is proportional to

$$P(X) \propto \exp\left(-\frac{(N-1)(X-\bar{x})^2}{2\bar{\sigma}^2}\right), \tag{5.2}$$

where

$$\bar{x} \equiv \frac{1}{N}\sum_{k=1}^{N} x_k \tag{5.3}$$

is the empirical average estimated from the data and

$$\bar{\sigma}^2 \equiv \overline{(x-\bar{x})^2} = \frac{1}{N}\sum_{k=1}^{N}(x_k-\bar{x})^2 \tag{5.4}$$

is the empirical estimate of the variance.

In general, in this chapter, we will denote with a bar above the symbol the average over the experimental data and with the symbol $\langle \cdot \rangle$ the mean value with respect to the (unknown) probability law according to which the data are actually distributed.

The law of large numbers guarantees us that, if the function $D(x)$ goes to zero fast enough for large x, then, in the limit $N \to \infty$,

$$\bar{x} \to \langle x \rangle,$$
$$\bar{\sigma}^2 \to \sigma^2 \equiv \langle(x-\langle x\rangle)^2\rangle. \tag{5.5}$$

Note, however, that while $\bar{x}$ is an *unbiased* estimate of $\langle x \rangle$, in the sense that the expected value of $\bar{x}$ is precisely $\langle x \rangle$, i.e., $\langle \bar{x} \rangle = \langle x \rangle$, this is not true for $\bar{\sigma}^2$, which is a *biased* estimate of σ^2. Indeed, a simple computation, which we invite the reader to perform, shows that the expected value of the variance estimator

$$\langle \bar{\sigma}^2 \rangle = \frac{N-1}{N}\sigma^2, \tag{5.6}$$

is not equal to the variance. The best empirical estimate of the variance is, therefore, given not by $\bar{\sigma}^2$ but by the quantity

$$\frac{N}{N-1}\bar{\sigma}^2. \tag{5.7}$$

This fact is the origin of the factor $N-1$ which appears in the exponent in Eq. (5.2). For large N the difference is irrelevant and (5.5) applies. For small N the difference is serious and the bias, or distortion, is equal to

$$\sigma^2 - \langle\bar{\sigma}^2\rangle = \frac{\sigma^2}{N}.$$

In the limiting case $N=1$ the estimator $\bar{\sigma}^2$ is null and estimates nothing, and the formula (5.7) gives 0/0. The result (5.2) is commonly stated in a shortened form by writing $X = \bar{x} \pm e$, where the error e on the mean is given by

$$e = \sqrt{\frac{\bar{\sigma}^2}{(N-1)}} = \left[\frac{1}{N(N-1)} \sum_{i=1}^{N} (x_i - \bar{x})^2 \right]^{1/2}.$$

If the estimate of a variable is unbiased, this does not mean that a given function of the estimate will also be the unbiased estimate of that function. For example, $\bar{x}$ is an unbiased estimate of the expected value $X \equiv \langle x \rangle$ because $\langle \bar{x} \rangle = X$, but $\bar{x}^2$ is a biased estimate of X^2, since $\langle \bar{x}^2 \rangle = X^2 + \sigma^2/N$. In the generic case, therefore, we will have that $g(\bar{x})$ will be a distorted estimate of $g(X)$. It is therefore reasonable to guess how the distortion also enters into the error propagation from x to its function $g(x)$, whose error on the mean is given by

$$e_g^2 = \left(\frac{dg}{dx}\bigg|_{x=\bar{x}} \right)^2 \frac{\bar{\sigma}^2}{N-1}. \tag{5.8}$$

As we shall see in the rest of this chapter, the procedure leading to (5.2) is essentially correct if the following assumptions are true:

- We are in the limit of large N, i.e., many measurements have been made, and we have recorded many statistically independent events.
- The $D(x)$ distribution does not contain important tails for large values of x, and in this limit it goes to zero fast enough to have finite variance and not a huge kurtosis. We could say that often the non-Gaussian corrections are of the order of $(K-3)/N$.
- We are interested only in the region of events where the probability is not too small.[1]

These three assumptions are not always true, and it is often necessary to consider a more general approach that is also valid for small N and non-Gaussian final distributions. In these cases, a correct procedure can lead to very different results from (5.2). In our opinion, the best method to set up the study of statistical errors is to start from Bayes' rule, which we have already mentioned in Section 1.5, and which we will expose in more detail in the next section.

5.2 Use of Bayes' Rule

Bayes' rule is the basis for the correct use of the statistical inference method. In general, continuing the discussion begun in Section 1.5, we can state the problem in these terms: If at a given instant we have a certain (probabilistic) knowledge of a phenomenon, and subsequently an event is observed that has never been observed before, we ask ourselves how this new observation changes our knowledge of reality.

We would first like to note that an attempt to frame *all* knowledge within the framework of Bayesian inference encounters considerable difficulties, as the use of Bayesian inference always presupposes the existence of prior knowledge. The origin of prior

---

[1] Tails where the probability of an event is very low must always be treated separately: see Chapter 4 on large deviations.

knowledge could serve as the starting point for a lengthy discussion with philosophical and epistemological implications. Here, we merely mention the existence of at least three different points of view.

- A first point of view is that the *a priori* of one experiment is the *a posteriori* of another experiment. For example, we can trace the entire history of humankind backwards through a long chain of addition of knowledge that serves as an *a priori* framework for new advances, without posing the question of how primitive humans acquired the first knowledge.
- A second point of view is based on the assumption that human beings do not naturally reason according to the principles of Bayesian inference. In this perspective, it is assumed that Bayesian inference is used by persons who have already acquired a certain world view and possess *a priori* expectations.
- A third possible perspective postulates the existence of *a priori* knowledge that has accumulated during the phylogeny of the species through natural selection. This is, perhaps, a crueler mechanism than Bayesian inference and it is certainly very different.

To apply the Bayesian inference procedure, let us assume that M different hypotheses are plausible to explain experimental data. One and only one of these hypotheses is certainly true. Suppose we have an *a priori* estimate of the probabilities of the different hypotheses H_j, $j = 1, \ldots, M$. This is an elaborate estimate carried out *before* having done the experiment whose results we are considering. We denote by the number $A(H_j)$ the estimate of the probability that the jth hypothesis will occur, without looking at the experimental data. We want to know how these probabilities change after performing an experiment that results in the set $\mathbf{x}$ of experimental data values. That is, what are the probabilities of the hypotheses H_j conditional on the observed experimental data?

We have already seen with Eq. (1.37) that

$$P(H_j|\mathbf{x}) = \frac{P(\mathbf{x}|H_j)A(H_j)}{Z(\mathbf{x})}, \tag{5.9}$$

where $A(H_j)$ is the *a priori* probability, unconditional on the hypothesis H_j, $P(H_j|\mathbf{x})$ is the *a posteriori* probability of the jth hypothesis conditioned by the experimental data, while $P(\mathbf{x}|H_j)$ is the probability (or likelihood) of obtaining as a result of a measurement the experimental data actually observed given the hypothesis H_j. The $P(H_j|\mathbf{x})$ is the probability we seek to know. We assume that $P(\mathbf{x}|H_j)$ is known, i.e., we assume we know how the probability distribution of the experimental data depends on the various hypotheses. As we saw in Section 1.5, Z is a normalization factor that guarantees that

$$\sum_{j=1}^{M} P(H_j|\mathbf{x}) = 1, \tag{5.10}$$

and is given by

$$Z(\mathbf{x}) = \sum_{j=1}^{M} P(\mathbf{x}|H_j)A(H_j). \tag{5.11}$$

The factor $Z(\mathbf{x})$ depends only on the experimental data $\mathbf{x}$, and it is called the marginal probability because it is found by summing over all hypotheses the joint probability of hypotheses and data. Since it is a probability it is also written with the notation $P(\mathbf{x}) = Z(\mathbf{x})$. We explained the origin of the term "marginal" when we introduced Bayes' formula in Section 1.5.1; see also Table 1.1.

In general, Bayes' formula can be used to estimate probabilities from empirical observations. Often the *a priori* knowledge is quite obvious. For example, if I consider the length of a rigid object standing inside a room, it is extremely reasonable to expect the length of this object to be less than that of the room itself. In other cases, as we shall see later, the *a priori* knowledge is much less trivial and more difficult to quantify.

5.2.1 Inference of Urns

To fix ideas let us discuss a very simple application of the method. Suppose we have two urns: each one of the two contains a large number of white or black balls, and we know with certainty that the first urn contains a known fraction q of white balls (and thus a fraction $1 - q$ of black balls) and that the second urn contains instead a fraction $1 - q$ of white balls (and a fraction q of black balls). The first urn is conventionally referred to as the white urn (H_W), and the second is referred to as the black urn (H_B). If making N draws from one of the two urns (unknown and chosen at random) yields $N_W = M$ white balls and $N_B = N - M$ black balls, Bayes' rule tells us that the probability that the urn is the white one is given by

$$P(H_W | N_W = M, N_B = N - M) = \frac{P(N_W = M, N_B = N - M | H_W)A(H_W)}{Z}$$

$$= \frac{q^M(1-q)^{N-M}}{q^M(1-q)^{N-M} + (1-q)^M q^{N-M}}. \tag{5.12}$$

Let us explain the reasoning in detail. Bayes' rule tells us that the probability that, given the results of the draw, a white urn was chosen is equal to the probability of obtaining, once the white urn is selected, precisely the results of the draw, multiplied by the probability of having chosen the white urn and divided by a normalization factor Z.

The number Z is the sum of two terms: the probability of obtaining, having chosen the white urn, the results of the draw, multiplied by the probability of having chosen the white urn, plus the probability of obtaining, having chosen the black urn, the results of the draw, multiplied by the probability of having chosen the black urn:

$$Z = P(N_W = M, N_B = N - M | H_W)A(H_W) + P(N_W = M, N_B = N - M | H_B)A(H_B).$$

The *a priori* probability of choosing either the white or the black urn is $A(H_W) = A(H_B) = 1/2$, the probability of obtaining M white balls and $N - M$ black balls is either

$$P(N_W = M, N_B = N - M | H_W) = \binom{N}{M} q^M (1 - q)^{N-M},$$

if the balls are taken from the white urn H_W, or

$$P(N_W = M, N_B = N - M | H_B) = \binom{N}{M} q^{N-M} (1 - q)^M$$

if the balls are drawn from the black urn, H_B. From this follows the expression (5.12) for the posterior probability of the white urn.

We can generalize the reasoning to a more complex situation. If we have R urns, one of which is white and the other $R - 1$ black, and one of the R urns (chosen at random) yields M white balls and $N - M$ black balls, the probability that the urn is the white one is given by

$$P(H_W | N_W = M, N_B = N - M) = \frac{q^M (1 - q)^{N-M}}{q^M (1 - q)^{N-M} + (R - 1)(1 - q)^M q^{N-M}},$$

since the *a priori* probability that the urn is white is $P(H_W) = R^{-1}$, while the *a priori* probability that the urn is black is $P(H_B) = (R - 1)/R$.

If, on the other hand, the choice of the urn is made by a friend of ours who usually has a preference for the color white, the estimation of the *a priori* probability of the color of the urn becomes much more difficult.

This example is particularly instructive in that it shows that the principle of statistical inference is not limited to the case where the number of data or observations N is large, but can also be used in the case of N small, even $N = 1$. Bayes' formula allows us to quantify how our expectations of the external world are changed as a result of one or more further observations.

5.2.2 Bayesian Inference for Continuous Events

Bayes' formula can be easily generalized from the case of a discrete set of E hypotheses $H_j, j = 1, \ldots, E$, to that of a continuous distribution of variables $h \in [h_{\min}, h_{\max}]$. For the *a posteriori* probability density we find that

$$P(h|\mathbf{x}) = \frac{P(\mathbf{x}|h)A(h)}{Z}, \tag{5.13}$$

where here

$$Z \equiv \int dh \, P(\mathbf{x}|h)A(h). \tag{5.14}$$

We can discuss at this point how to use Bayes' rule, for example in the case of analyzing experimental data, in the situation where the hypothesis h quantifies the true value of a quantity we measure.

Suppose we have already made N independent measurements of the quantity x, each one written as x_k, with $k = 1, \ldots, N$, and which we collectively denote by $\mathbf{x}$. Let us further assume that we know (*a priori*, i.e., before doing the experiment) that the probability distribution $D(x)$ is not a completely arbitrary function, but depends on a set of parameters $\vec{h}$, whose values we wish to estimate:

$$D(x) = P(x|\vec{h}). \tag{5.15}$$

We also know the *a priori* probability distribution $A(\vec{h})$ of the variables $\vec{h}$, i.e., before doing the experiment. In this case, therefore, the *a posteriori* probability of the parameters $\vec{h}$ conditioned by the experimental data $\mathbf{x}$ is

$$P(\vec{h}|\mathbf{x}) = \frac{P(\mathbf{x}|\vec{h})A(\vec{h})}{Z}, \tag{5.16}$$

where normalization is given by the multivariate integral

$$Z \equiv \int d\vec{h}\, P(\mathbf{x}|\vec{h})A(\vec{h}). \tag{5.17}$$

Since we assume that the measurements of the observable x are not correlated, we have that

$$P(\mathbf{x}|\vec{h}) = \prod_{k=1}^{N} P(x_k|\vec{h}). \tag{5.18}$$

We denote by $X(\vec{h})$ the mean value of the data set $\mathbf{x}$ for a certain fixed set of parameter values $\vec{h}$:

$$X(\vec{h}) \equiv \int dx\, P(x|\vec{h})x. \tag{5.19}$$

By varying the parameter values $\vec{h}$, distributed according to the posterior distribution $P(\vec{h}|\mathbf{x})$, we can, therefore, express the probability distribution of the mean value X of the experimental data in the form

$$P(X) = \int d\vec{h}\, P(\vec{h}|\mathbf{x})\delta(X - X(\vec{h})) = \frac{\int d\vec{h}\, P(\mathbf{x}|\vec{h})A(\vec{h})\delta(X - X(\vec{h}))}{\int d\vec{h}\, P(\mathbf{x}|\vec{h})A(\vec{h})}, \tag{5.20}$$

where (5.16) was used in the second step. It is important to note that if the *a priori* distribution $A(\vec{h})$ is almost constant in the region where the function $P(\mathbf{x}|\vec{h})$ is significantly different from zero, the above formula can be successfully approximated by the expression

$$P(X) \simeq \frac{\int d\vec{h}\, P(\mathbf{x}|\vec{h})\delta(X - X(\vec{h}))}{\int d\vec{h}\, P(\mathbf{x}|\vec{h})}. \tag{5.21}$$

The two formulas (5.20) and (5.21) give essentially the same result in the case where the function $P(\mathbf{x}|\vec{h})$ has a very narrow peak around its most probable value. This is the case that is normally realized for sufficiently large values of N, where the explicit form of $A(\vec{h})$ is irrelevant. As we shall soon see, however, the prior is not at all irrelevant when the number of measurements is small.

In the case where no *a priori* knowledge is given about the possible form of the function $D(x)$, so that we do not have a form such as (5.15), the problem is mathematically ill-posed, and, strictly speaking, we would not be able to derive any expression for $P(X)$. We will later see how to deal with this case as well, and how to develop reasonable working hypotheses.

5.3 Choice of the *A Priori* Distribution

When the likelihood does not have a narrow enough peak around the mean value and there are few measurements available or the events are rare, the estimate of errors from the standard deviation is not correct because the final distribution is not a Gaussian. This is also the case where an overly naive choice of *prior choice* leads to biased estimates of the parameters of the posterior distributions in Bayesian inference.

The problem of characterizing an "objective" *a priori* distribution that represents ignorance or vague knowledge of the phenomenon without biasing the data and letting them speak for themselves is a *vexata questio*, a very complicated issue that has been debated for almost a century without arriving at an agreed solution [44]. The concept of objective *a priori* knowledge (or ignorance) is extremely delicate to define. To be intellectually honest, even if only for operational purposes, in order to obtain reliable estimates it is important to realize that data can never speak for themselves, not completely. "Let the data speak" is not something feasible, because any formalization of *a priori* knowledge has predictive implications and influences the results of the inference. Even when we want to represent the absence of knowledge, the moment we formalize it into an *a priori* distribution we are imposing a choice that in some way, however tiny, influences the *a posteriori* knowledge we obtain from the experimental data. There is no objective *a priori* that represents complete ignorance.

By renouncing vain pretensions to objectivity and arming ourselves with an operational spirit, however, we will see that it is possible to enunciate reasonable criteria adherent to experimental reality, in order to identify *a priori* distributions that are a valid reference for a correct statistical analysis of the data, producing statistically significant errors in the measurements and allowing us to infer unbiased parameters.

A complete and thorough treatment of Bayesian inference is beyond the limits of this book, but interested readers may consult the recommended texts in the bibliography and, in particular, the text by José Bernardo and Adrian Smith [45].

5.3.1 Binomial-Distributed Data and a Paradox

Let us start with something we are now familiar with: binomial-distributed data. Let us take a Bernoulli process in which an event occurs or does not occur on each experimental measure. We call the number of event counts K and the number of trials N.

The frequency of the event is K/N, and the expected value of the frequency was called the probability p (it was $\langle K \rangle = Np$), in Section 2.2.1). Given the binomial likelihood (2.16)

$$P(K, N|p) = B(N, K, p) = \binom{N}{K} p^K (1 - p)^{N-K},$$

after N measurements and K counts, we compute the *a posteriori* distribution of p using Bayes' formula (5.13):

$$P(p|K, N) = \frac{P(K, N|p)A(p)}{\int_0^1 dp\, P(K, N|p)A(p)} = \frac{p^K(1 - p)^{N-K} A(p)}{\int_0^1 dp\, p^K(1 - p)^{N-K} A(p)}.$$

As was done in the formula (5.21) we can consider constant the prior distribution in the domain $p \in [0, 1]$: $A(p) = 1$. This way the posterior distribution turns out to be

$$P(p|K, N) = \frac{p^K(1 - p)^{N-K}}{\int_0^1 dp\, p^K(1 - p)^{N-K}} = \frac{(N + 1)!}{K!(N - K)!} p^K(1 - p)^{N-K}, \tag{5.22}$$

where we used the integral

$$\beta(K_1 + 1, K_2 + 1) \equiv \int_0^1 dx\, x^{K_1}(1 - x)^{K_2} = \frac{\Gamma(K_1 + 1)\Gamma(K_2 + 1)}{\Gamma(K_1 + K_2 + 2)}, \tag{5.23}$$

which is also called the *beta function*.

If we performed an experiment with N measurements and counted K events, we expect the mean value of the probability conditional on the experimental measurements to be equal to the experimental frequency: $\langle p \rangle = K/N$. We try to do the explicit computation with the distribution (5.22), with the help of the formula (5.23),

$$\langle p \rangle = \int_0^1 dp\, p P(p|K, N) = \frac{(N + 1)!}{K!(N - K)!} \int_0^1 dp\, p^{K+1}(1 - p)^{N-K}$$

$$= \frac{(N + 1)!}{K!} \frac{(K + 1)!}{(N + 2)!} = \frac{K + 1}{N + 2}. \tag{5.24}$$

This is not the empirical frequency! In particular, it gives us an estimate $\langle p \rangle > 0$ of the mean value of the frequency even if the event never occurred: for $K = 0$ we have $\langle p \rangle = 1/(N + 2)$. This seems to be a real paradox.

As pointed out by probabilists such as Haldane [46] and Jeffreys [47], a uniform choice of *prior* creates a distortion of the *a posteriori* distribution. The issue lies in the fact that maximum ignorance about the *a priori* choice of probability is not given by considering all values as equiprobable. We know that if we consider rare events, such as, for example, genetic mutations, the probability will be small. But how small: 10^{-2} or 10^{-8}? One would think that, instead of considering the values of p as equiprobable, our ignorance would manifest itself more clearly in considering their order of magnitude as equiprobable:

$$\mathrm{Prob}(p \simeq 10^{-2}) = \mathrm{Prob}(p \simeq 10^{-8}),$$

which corresponds to saying

$$\mathrm{Prob}(\log p \simeq -2) = \mathrm{Prob}(\log p \simeq -8).$$

Perhaps maximum prior ignorance can be exemplified by a uniform distribution of the logarithm of p.

We are considering rare events, and we know that in this limit the binomial tends to the Poisson distribution. Therefore, to better understand how to get out of this paradox, we focus on Poisson-distributed experimental data. For interested readers, in Appendix 5.A we report the computation of a sensible *a posteriori* distribution even in the binomial case.

5.3.2 Poisson-Distributed Data

We have already introduced the Poisson distribution in Section 2.3. Let us reconsider here the example we described in that section, i.e., that of an elementary event that, with a time interval, has a certain probability of occurring or of not occurring. Let us assume that in a time interval T, K elemental events have occurred, i.e., the event has been experimentally observed with an empirical frequency $f = K/T$. We propose to compute the probability distribution of the expected value of the frequency $\mu = \langle K \rangle / T$ of the phenomenon, $P(\mu)$.

Under the simple assumption, already discussed in Section 2.3, that the individual elementary events are uncorrelated with each other, the probability distribution of the number of events is a Poisson distribution. In this case, the probability $P(K|\lambda)$ depends on only one parameter, $\lambda = T\mu = \langle K \rangle$. The formulas derived in Section 2.3 for the Poisson distribution tell us that

$$P(K|\lambda) = \frac{\lambda^K}{K!} e^{-\lambda}, \tag{5.25}$$

where λ denotes here the average number of events expected in the time period T.

Let $A(\lambda)$ be the *a priori* probability of the parameter λ (the *a priori* probability of our hypothesis) that characterizes the probability distribution of the events. At the moment we do not really know what is a reasonable form for $A(\lambda)$, a prior distribution, to be chosen independently of the experimental data. Obviously, if we already have some other information about the phenomenon, we can use it to determine the function $A(\lambda)$; the serious problem is what form to choose for $A(\lambda)$ if we have no information about the possible values of λ.

Applying Bayes' rule in the form (5.16), we find, for the *a posteriori* probability,

$$P(\lambda|K) = \frac{\lambda^K e^{-\lambda} A(\lambda)}{\displaystyle\int_0^\infty d\lambda\, \lambda^K e^{-\lambda} A(\lambda)}. \tag{5.26}$$

It may seem reasonable to assume that $A(\lambda)$ has a constant value: this is equivalent to assuming that $A(\lambda)$ varies little in the area where $P(K|\lambda)$ is concentrated. The exact value of the constant will be simplified between numerator and denominator, it will not enter into our computations, and we need not discuss it.

Hence, under the assumption

$$A(\lambda) = \text{constant}, \tag{5.27}$$

Bayes' rule implies that

$$P(\lambda|K) = \frac{\lambda^K e^{-\lambda}}{\displaystyle\int_0^\infty d\lambda\, \lambda^K e^{-\lambda}} = \frac{\lambda^K e^{-\lambda}}{K!}. \tag{5.28}$$

For large values of K (i.e., in the case of a large number of events), the above formula is reasonable since $P(\lambda|K)$ is centered around the value $\lambda = K$: the mean value of events expected in the period T has a probability centered around the number of observed events.

For small values of K, especially for $K = 0$ or $K = 1$, the choice of $A(\lambda)$ becomes crucial. To develop a quantitative understanding of the phenomenon, we thus compute the expected value $\langle \lambda \rangle$ of the number of events conditional on the data. We use the results about Euler Gamma functions given in Appendix 2.B. It is found that

$$\langle \lambda \rangle = \int_0^\infty d\lambda\, \lambda\, P(\lambda|K) = \frac{1}{K!} \int_0^\infty d\lambda\, \lambda^{K+1} e^{-\lambda} = K + 1. \tag{5.29}$$

As in the binomial case in the previous section, see the formula (5.24), the estimate is biased. We expect that, in the absence of other information, the correct relationship is $\langle \lambda \rangle = K$. The above formula does not make a lot of sense for small K and constitutes the paradox anticipated in the previous section. The formula (5.29) might appear particularly unreasonable for $K = 0$ (no observed events), since, in the absence of any other information, it gives an estimate of the mean value of expected events equal to 1, i.e., something non-zero. The posterior information inferred, after having observed no events, is that one event should be expected.

5.3.3 *A Priori* Probability for Maximum Ignorance: Paradox Resolution

The solution to the apparent paradox (paradoxes are apparent by definition) of the previous section consists in noting that when we stated that the *a priori* probability of λ is constant we unintentionally added arbitrary information to the data of the problem. On the other hand, it is clear that writing the *a priori* probability of λ necessarily implies the addition of information. The problem we would like to address is how to determine an *a priori* probability $A(\lambda)$ that best matches the abstract statement that *we have no prior information about the system.*

From this perspective, the choice of constant $A(\lambda)$ does not seem reasonable. With a small variation in the procedure, we can, actually, obtain different results from those obtained before, though equally unreasonable. For example, let us assume *a priori* that the mean time distance between two events, $1/\mu = T/\lambda$, is uniformly distributed. In this case, the function $A(\lambda)$ is proportional to $1/\lambda^2$ (see Appendix 3.B, with $y = 1/x$). We, therefore, have the posterior

$$P(\lambda|K) \propto \lambda^{K-2}e^{-\lambda}. \tag{5.30}$$

For $K = 0$ (where there were no events) this distribution cannot be normalized, which is not entirely unreasonable since, in the absence of other information, we cannot reasonably make predictions about the average number of events of which we have not observed a single one. Unfortunately, $P(\lambda|K)$ remains non-normalizable even for $K = 1$, which is not reasonable (having observed an event should allow some form of prediction). Furthermore, for $K > 1$ the above formula implies that

$$\langle \lambda \rangle = K - 1, \tag{5.31}$$

which clashes with common sense for small K.

The question *"What is the most reasonable choice of function $A(\lambda)$ in the total absence of information about the system?"* is not particularly well-posed. However, we can try to give an answer using the following criteria for a constructive procedure.

(a) We can make explicit the predictions implied by the different possible choices of $A(\lambda)$, and compare them with the predictions obtained by applying the most elementary common sense.
(b) The results obtained must be as stable as possible with respect to changes in the observable chosen to describe the events. For example, the results should not change if the frequency or average waiting time is taken as an unknown quantity.
(c) Although, as we shall see, it is not possible to find a reasonable probability distribution, in the interval $[0, \infty)$, that avoids setting a scale for the values of λ and is normalizable at the same time, it is nevertheless reasonable to require that the integral $\int_0^\infty d\lambda\, A(\lambda)$ diverges as slowly as possible (logarithmically).
(d) The result should be independent of the way the events are classified. If we sum two Poisson processes of mean values λ_1 and λ_2 whose value distributions are unknown *a priori*, the *a priori* distribution of the sum of the processes must also reflect the same ignorance. In other words, the probability of the mean value of the sum of two Poisson events will be homogeneous with mean value $\lambda_1 + \lambda_2$, independent of the classification of the events into event 1 or event 2.

In agreement with these criteria, we can assume that *the logarithm of λ is uniformly distributed*. Having no *a priori* information about the phenomenon, we assume that a number of characteristic events of the order 10^{10} is as likely as a number of events of $O(1)$ and that characteristic times of the order of 10^{-6} seconds are as likely as times of the order of 10^6 seconds. The statement *"I do not know the order of magnitude of the phenomenon"* or *"In our phenomenon various orders of magnitude are equally probable"* expresses well our maximum ignorance of the features of the phenomenon. Under this assumption, we have that (see Appendix 3.B):

$$A(\log \lambda) = \text{constant} \quad \Longrightarrow \quad A(\lambda) \propto \frac{1}{\lambda}, \tag{5.32}$$

and by computing the normalization we obtain the *a posteriori* probability density

$$P(\lambda|K) = \frac{\lambda^{K-1}}{\Gamma(K)} e^{-\lambda}. \tag{5.33}$$

Looking at the criteria we stated earlier, let us formulate some considerations on this result:

(a) This formula is in good agreement with common sense: for $K = 0$ the probability is not normalizable, because we have no information about the phenomenon. Indeed, if we have seen zero unicorns, we cannot use this result to argue that the probability of seeing unicorns is greater than zero. For $K = 1$ the mean value of λ is 1, exactly as it is intuitive, and, in general, for K counts we have that

$$\langle \lambda \rangle = K, \tag{5.34}$$

as it is reasonable to expect.

(b) A logarithmic distribution for λ coincides with a logarithmic distribution for the mean waiting time T/λ: in this approach, as is reasonable, the results do not change if we consider frequencies or waiting times.

(c) The normalization of a uniform distribution in the logarithm only diverges logarithmically, i.e., very slowly, and is not very far from being normalizable. Other simple probability distributions (e.g., probability distributions that behave like powers) diverge much faster.

(d) In the case of the sum of events of different types, we will see in Section 5.3.5 that this distribution (5.33) also reproduces itself by changing the classification scheme of the events. And it is the only distribution that allows such stability.

In many cases, a uniform choice on a logarithmic scale is most appropriate. As we see by applying these criteria, the choice of maximum ignorance turns out to be the one we have discussed here (a uniform measure on a logarithmic scale).

5.3.4 *A Posteriori* Probability of No Counts

A simple and interesting application of the Bayesian method is the following. Suppose we measured the variable k (which can take non-negative integer values by hypothesis) once, in a single experiment, and got a value $K \neq 0$ as the answer. Let us further assume that we know that the probability distribution of k is a Poisson distribution determined by an unknown characteristic parameter λ. We propose to estimate the probability $P_0(K) = P(k = 0|K)$ of finding, in a subsequent measure, the result $k = 0$.

Naively, we can assume that λ is close to K. For λ exactly equal to K the probability of having zero events is given by

$$P_0(K) = e^{-K}. \tag{5.35}$$

This is not, actually, a correct estimate, but only an approximation. The exact result, which we will now derive, is infact

$$P_0(K) = 2^{-K}. \tag{5.36}$$

Proof   The computation is very simple. The *a posteriori* probability of finding a given value of λ after measuring $k = K$ is given, under the assumption of the equiprobability of logarithms, by Eq. (5.33). From (5.25) we know that the probability of $k = 0$ events for fixed λ is $P_0(\lambda) = P(0|\lambda) = e^{-\lambda}$. Averaging $P_0(\lambda)$ over all possible values of λ, we then have

$$P_0(K) = \int d\lambda\, P(\lambda|K)P_0(\lambda) = \frac{1}{\Gamma(K)} \int d\lambda\, \lambda^{K-1} e^{-2\lambda} = 2^{-K}, \tag{5.37}$$

the exact result.   ∎

The naive estimate obtained in (5.35) appears to be correct in a "central" sense, since it goes to zero when K goes to infinity. However, if we are interested, as in the case of large deviations, in an estimate with a small relative error (i.e., estimating $\lim_{K\to\infty} K^{-1} \log[P_0(K)]$), the naive result is completely wrong.

5.3.5 Several Poisson Categories

A check of the correctness of the previous choice can be made by applying it to the case of several categories. Let us analyze the case in which the experiment considered consists of checking whether each of two events of type (1) and of type (2) occurred or not. Let the probability distribution of each of the two events be a Poisson distribution, with parameters λ_1 and λ_2, respectively. Suppose we have observed K_1 times the event of type (1) and K_2 times the event of type (2).

If we assume *a priori* that the values λ_1 and λ_2 are uniformly distributed in the logarithm, i.e., that

$$A(\lambda_1, \lambda_2) \propto \frac{1}{\lambda_1 \lambda_2}, \tag{5.38}$$

we obtain that the *a posteriori* probability of the two parameters λ_1 and λ_2 conditional on the experimental measurement made is

$$P(\lambda_1, \lambda_2|K_1, K_2) = P(\lambda_1|K_1)P(\lambda_2|K_2) \propto \lambda_1^{K_1-1}\lambda_2^{K_2-1}e^{-\lambda_1-\lambda_2}. \tag{5.39}$$

The average number of events is $\lambda_1 + \lambda_2$. Of these events, λ_1 are of type (1) and λ_2 are of type (2).

What happens if we change the classification and no longer distinguish between the two categories but are only interested in the sum of the variables? As we saw in Appendix 3.C, the *a priori* probability distribution of the sum λ is given by the convolution of the distributions of the summed variables,

$$A(\lambda) = \int_0^\lambda d\lambda_1\, A(\lambda - \lambda_1)A(\lambda_1) = \int_0^\lambda d\lambda_1\, \frac{1}{\lambda_1(\lambda - \lambda_1)} = \frac{1}{\lambda} \int_0^1 dy\, \frac{1}{y(1 - y)}, \tag{5.40}$$

from which $A(\lambda) \propto \lambda^{-1}$, if we can hold off the integral that is divergent, even if only logarithmically. The introduction of a cutoff $\epsilon > 0$ makes the integral finite,

$$A(\lambda) \simeq \frac{1}{\lambda} \int_{\epsilon}^{1-\epsilon} dy \, \frac{1}{y\,(1-y)} = \frac{2}{\lambda} \ln\left(\frac{1-\epsilon}{\epsilon}\right).$$

In practice, with the cutoff, we are reasonably assuming that the average frequencies of the two event types (occurs/does not occur) are not strictly zero. The cut ϵ may also be very small, but it guarantees the reproducibility of the *a priori* in the sum of Poisson processes. In Appendix 5.A we discuss the same approximation for Bernoulli events. Regarding the *a posteriori* probability, from the convolution we find that

$$P(\lambda|K_1, K_2) = \int_0^{\lambda} d\lambda_1 \, P(\lambda_1|K_1)P(\lambda - \lambda_1|K_2) = \int_0^{\lambda} d\lambda_1 \, e^{-\lambda}\lambda_1^{K_1-1}(\lambda - \lambda_1)^{K_2-1}$$

$$= \frac{\Gamma(K_2)\Gamma(K_1)}{\Gamma(K_1 + K_2)} e^{-\lambda} \lambda^{K_1+K_2-1}, \tag{5.41}$$

where we have used the relation

$$\int_0^{y} dx \, x^{K_1-1}(y - x)^{K_2-1} = \frac{\Gamma(K_1)\Gamma(K_2)}{\Gamma(K_1 + K_2)} y^{K_1+K_2-1}, \tag{5.42}$$

which generalizes the formula (5.23) for beta functions and will be useful in Section 5.4.3. We invite the reader to derive it. From (5.41) we have

$$P(\lambda|K_1, K_2) \propto e^{-\lambda}\lambda^{K_1+K_2-1},$$

from which we can see that the average number of expected events is the same as if we had not initially introduced a division into two categories. This fact is of great importance: if we do not discriminate between events of type (1) and events of type (2), i.e., if $K = K_1+K_2$, we obtain once again the distribution (5.33) with $\lambda = \lambda_1+\lambda_2$. We emphasize that (if we restrict ourselves to considering simple probability distributions) only in the case of an *a priori* distribution that is uniform in the logarithm does the result have stability properties of this type.

Let us now generalize this result to the case in which events can belong to one of M different categories (the previous case corresponds to $M = 2$). Let us assume that we have observed K_i events in the ith category, of mean value λ_i (for simplicity we consider, without loss of generality, only the case in which all K_i are non-zero, i.e., we consider only those categories in which at least one event has occurred). The total number of events will be given by

$$K = \sum_{i=1}^{M} K_i . \tag{5.43}$$

Proceeding as in the previous case, defining $\lambda = \sum_{i=1}^{M} \lambda_i$ and making $M - 1$ convolutions, we obtain that

$$P(\lambda|\mathbf{K}) = \int \prod_{i=1}^{M} d\lambda_i \, P(\lambda, \lambda|\{K_i\}) = \int \prod_{i=1}^{M} d\lambda_i \, \delta\left(\lambda - \sum_{i=1}^{M} \lambda_i\right) \prod_{i=1}^{M} P(\lambda_i|K_i) \tag{5.44}$$

$$\propto \int \prod_{i=1}^{M} d\lambda_i \, \delta\left(\lambda - \sum_{i=1}^{M} \lambda_i\right) \prod_{i=1}^{M} e^{-\lambda_i} \lambda_i^{K_i-1} \propto e^{-\lambda}\lambda^{K-1} \,,$$

where $\mathbf{K} \equiv \{K_i\}$ and $\boldsymbol{\lambda} \equiv \{\lambda_i\}$.

As a byproduct, we notice that this formula also allows us to write a general formula in terms of relative frequencies,

$$p_i = \lambda_i/\lambda \,, \tag{5.45}$$

i.e., the probability that an event belongs to the ith category. Equation (5.45) can also be written as $\lambda_i = p_i\lambda$, where p_i satisfies the constraint

$$\sum_{i=i}^{M} p_i = 1 \,.$$

If we switch from the probability density for λ_i in the integral of (5.44) to that for p_i, we obtain, therefore,

$$P(1, \mathbf{p}|\mathbf{K}) \equiv P(\mathbf{p}|\mathbf{K}) \propto \delta\left(\sum_{i=1}^{M} p_i - 1\right) \prod_{i=1}^{M} p_i^{K_i-1} \,. \tag{5.46}$$

Sometimes we are interested in computing the expectation value of a quantity $g(\mathbf{p})$, or, more generally, in evaluating its *a posteriori* probability distribution, given by

$$P(g|\mathbf{K}) = \int d\mathbf{p} \, P(\mathbf{p}|\mathbf{K})\delta(g(\mathbf{p}) - g) \,. \tag{5.47}$$

The simplest way to numerically compute the probability $P(g|\mathbf{K})$, in the case where an exact analytical computation cannot be carried out, is to randomly extract the relative frequencies $\mathbf{p}$, according to the probability law (5.46), and compute the resulting histogram of the values $g(\mathbf{p})$. We will elaborate on this procedure in a short while in Section 5.4.1 where we will address the reweighting method in the case of unknown distributions. For the time being, let us remain in the domain of inference from distributions whose form is known, and analyze the case of Gaussian data.

5.3.6 Gaussian-Distributed Data

Let us consider the situation where, by hypothesis, the probability distribution of events is a Gaussian. That is, suppose that we have observed N events, each characterized by the measurement result, which is a real number x, and that we have the *a priori* information that the probability distribution is a Gaussian centered on a and of mean square deviation b, i.e.,

$$P(x|a,b) \propto \frac{1}{b} \exp\left(-\frac{(x-a)^2}{2b^2}\right) \,. \tag{5.48}$$

Typically, in this situation the quantity a has the meaning of the expected value of x, i.e., it is the value that would have been measured if the statistical errors had been reduced to zero. Suppose that the *a priori* probability distribution $A(a,b)$ that characterizes the coefficients a and b is given approximately by

$$A(a,b) \propto \frac{1}{b}.$$
(5.49)

The term $1/b$ is due to the *a priori* assumption that all orders of magnitude of the error are equiprobable. The absence of a dependence on a lies in the assumption that $A(a,b)$ is constant in a in the domain where $P(x|a,b)$ is substantially different from zero. Indeed, as we shall see, the probability distribution we obtain will be a function that goes to zero very quickly when a becomes quite different from the empirically measured average.

Proceeding similarly to the previous section, we obtain that the *a posteriori* probability of the parameters a and b conditional on the results of the experiment is

$$P(a,b|\mathbf{x}) = \frac{P(\mathbf{x}|a,b)A(a,b)}{Z} \propto \frac{1}{b}\prod_{k=1}^{N}\left[\frac{1}{b}\exp\left(-\frac{(x_k - a)^2}{2b^2}\right)\right]$$

$$= \frac{1}{b^{N+1}}\exp\left(-N\frac{(a - \bar{x})^2 + \bar{\sigma}^2}{2b^2}\right),$$
(5.50)

where the empirical estimates of the average $\bar{x}$ and of the variance $\bar{\sigma}$ are defined in (5.3) and (5.4). In deriving the above formula, we used the relation

$$\frac{1}{N}\sum_{i=1}^{N}(x_k - a)^2 = (a - \bar{x})^2 + \bar{\sigma}^2.$$
(5.51)

The probability distribution of the quantity a is obtained by marginalizing with respect to the parameter b:

$$P(a|\mathbf{x}) = \int db\, P(a,b|\mathbf{x}) \propto \frac{1}{[(a - \bar{x})^2 + \bar{\sigma}^2]^{N/2}} \propto \left[1 + \frac{(a - \bar{x})^2}{\bar{\sigma}^2}\right]^{-N/2},$$
(5.52)

obtained, for example, by rescaling the integration variable and using the formula (3.98) for Euler's Γ functions. In the case of a single event, $N = 1$, the $P(a|\mathbf{x})$ in (5.52) cannot be normalized, in agreement with common sense, which suggests that, in the presence of a single event, we cannot reasonably determine a statistical error. Starting from $N = 2$ the probability $P(a|\mathbf{x})$ is well defined (we obtain a Cauchy–Lorentz distribution, see Section 2.5), but we have to go up to $N = 4$ to have a $P(a|\mathbf{x})$ whose variance is finite.

In Figure 5.1 we show, on the left, several realizations of the *a posteriori* distributions (5.52) obtained with $N = 4$ data whose true distribution is a normal one. When varying $\bar{x}$ and $\bar{\sigma}$, the distributions based on only four measurements change a lot. In Figure 5.1 on the right we show again the distribution (5.52) but conditioned on a different number of Gaussian-distributed data: $N = 4$, 10, and 20. We can observe that,

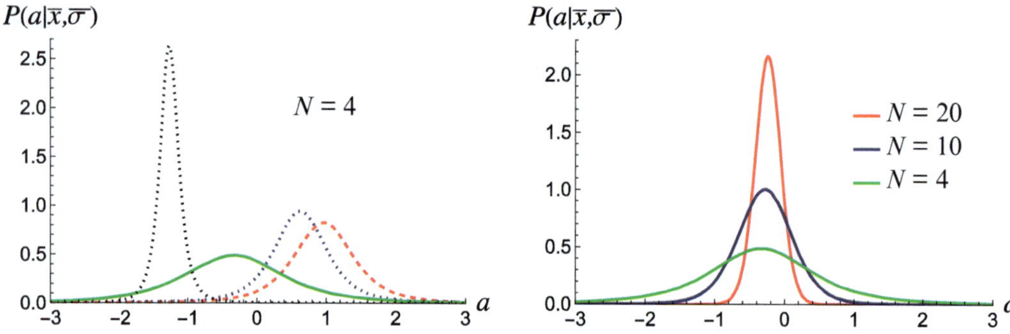

Figure 5.1 Examples of expected value distributions of Gaussian-distributed x data with normal distribution. Left: Distributions (5.52) of the mean value a conditional on $N = 4$ experimental measurements for four different samples. Right: Distributions (5.52) of the mean a value conditioned on three different experimental samples of, respectively, $N = 4, 10$, and 20 measurements.

as N increases, the function $P(a|\mathbf{x})$ concentrates around its mean value, in the area where $(a - \bar{x})^2 \ll \bar{\sigma}^2$. In this central limit, for $N \gg 1$, we have

$$P(a|\mathbf{x}) \propto \exp\left\{-\frac{N}{2}\log\left[1 + \frac{(a - \bar{x})^2}{\bar{\sigma}^2}\right]\right\} \simeq \exp\left\{-N\left[\frac{(a - \bar{x})^2}{2\bar{\sigma}^2} + O\left(\frac{(a - \bar{x})^4}{\bar{\sigma}^4}\right)\right]\right\}. \tag{5.53}$$

Neglecting the correction terms, we obtain a Gaussian distribution of width $\bar{\sigma}/\sqrt{N}$, in agreement with the simplified analysis we mentioned in the introduction to Section 5.1 in this chapter.

The average *a posteriori* value of $(a - \bar{x})^2$, computed for any value of N using (5.52), is

$$\int da\, P(a|\mathbf{x})(a - \bar{x})^2 = \frac{\bar{\sigma}^2}{N - 3} = \frac{(N - 1)\sigma^2}{N(N - 3)} \approx \frac{\sigma^2}{N}\left(1 + \frac{2}{N}\right), \tag{5.54}$$

where the relations

$$\int_{-\infty}^{+\infty} dy\,(1 + y^2)^{-N/2} = \sqrt{\pi}\,\frac{\Gamma((N - 1)/2)}{\Gamma(N/2)} \tag{5.55}$$

and

$$\int_{-\infty}^{+\infty} dy\,(1 + y^2)^{-N/2}y^2 = \sqrt{\pi}\,\frac{\Gamma((N - 3)/2)}{2\Gamma(N/2)} \tag{5.56}$$

have been used. They can be derived using the formula (3.98) for the Euler Gamma function that we introduced in Appendix 2.B. In (5.54) we used the relation (5.6) between the empirical variance estimate and the variance of the true distribution. We compare the result with the variance of the generalized Cauchy distribution (2.53) with $\alpha > 3$.

Note that, in the tails, far from the most probable value, the *a posteriori* distribution of a, (5.52), is not a Gaussian distribution. To this end, it is interesting to rewrite (5.52) in a form useful for computing large deviations:

$$P(a|\mathbf{x}) \propto e^{NS(a)},$$

with

$$S(a) \equiv -\frac{1}{2}\ln\left[1 + \frac{(a - \bar{x})^2}{\bar{\sigma}^2}\right]. \tag{5.57}$$

The results obtained using the formula (5.57) may be very different from those obtained from the formula (5.2), already when trying to compute the probability of events at four or five standard deviations from the mean value for a moderate number N of data.

5.3.7 Statistical Inference from Cauchy-Distributed Data

Suppose that we apply the formulas we have derived to the case where the probability distribution is a Cauchy–Lorentz distribution:

$$P(x|a,b) = \frac{1}{\pi}\frac{b}{b^2 + (x - a)^2}. \tag{5.58}$$

It is easy to see that, in N draws of x, there is a finite probability that a single $x = O(N)$. Indeed, the probability that $|x|$ is greater than N, for $N \gg 1$, is[2]

$$\begin{aligned}
P(|x| > N) &= \frac{2}{\pi}\int_N^\infty dx\,\frac{1}{b}\frac{1}{1 + ((x - a)/b)^2} \\
&= \frac{2}{\pi}\int_{(N-a)/b}^\infty dy\,\frac{1}{1 + y^2} = \frac{2}{\pi}\arctan y\,\Big|_{(N-a)/b}^\infty \\
&= 1 - \frac{2}{\pi}\arctan\left(\frac{N - a}{b}\right) \underset{N \gg 1}{\simeq} 1 - \frac{2}{\pi}\left(\frac{\pi}{2} - \frac{b}{N}\right) = \frac{2b}{\pi N},
\end{aligned}$$

and the probability of having no events of order N in N extractions is therefore

$$\left(1 - \frac{2b}{\pi N}\right)^N \underset{N \to \infty}{\longrightarrow} e^{-2b/\pi} > 0,$$

which does not depend on N and is strictly positive. Thus $\bar{x}$ does not converge to a and the typical value of $\bar{\sigma}$ is of order N. The formalism of Section 5.1 consistently produces an error $\bar{\sigma}/\sqrt{N}$ that in the limit $N \to \infty$ does not tend to zero. Interestingly, these considerations are the basis for the validity of a remarkable property: the empirical mean variable of N Cauchy–Lorentz-distributed variables is distributed according to the Cauchy–Lorentz distribution (as with Gaussian distributions!). We had seen this property in Section 3.3 as a special case of the generalized central limit of variables with infinite variance.

How, then, is it possible to obtain more information in the case of the analysis of experimental data distributed according to Cauchy? A very simple piece of information, e.g., the fact that there exists a point a of symmetry of the function P, such that

---

[2] It is perhaps useful to remember that, for $x \gg 1$, $(1/\pi)\arctan(x) \simeq 1/2 - x/\pi$.

$P(x) = P(2a - x)$, can help a lot. Indeed, it is sufficient to consider the median $m(N)$ of the experimental data (for the definition of the median, see Section 2.1.3): in the limit where $N \to \infty$ the median tends very quickly to a, with an error of the order of N^{-2}.

Using information of this kind we can then proceed, in the case where it is known *a priori* that the data are Cauchy-distributed, in the same way as in the Gaussian case. The *a posteriori* distribution will be

$$P(a, b | \mathbf{x}) \propto A(a, b) \prod_{i=1}^{N} \frac{b}{b^2 + (x_i - a)^2} .$$

In contrast to the Gaussian case, however, we will not be able to use any simplifications to reduce to a simple final expression, i.e., one depending on a small number of quantities, such as the empirical mean and variance of the experimental data.

5.4 General Case of Unknown Probability Distribution

So far, we have seen how to estimate the distribution of parameters of a distribution of values of an experimental observable by knowing the shape of the distribution. We now turn to the question of what happens if the form of the distribution itself is not accessible from our *a priori* knowledge.

5.4.1 Reweighting Method

The case where we have no *a priori* knowledge of the functional form of $D(x)$ is clearly much more complicated than those we have dealt with so far, and at first glance obtaining useful results might seem difficult. Since this is a case of great importance, we will try in this subsection to proceed anyway, using a heuristic approach that avoids arriving at conclusions at odds with common sense.

Suppose we have done N experiments, measuring N times a quantity x, exactly as in the previous section (we will denote by x_k the value measured in the kth experiment). Now let us assume, however, that we have no information about the probability distribution according to which the data are actually distributed, and that we cannot even assume that it is a continuous function. In such a situation, there are two obvious alternatives: the first is to give up studying the problem, while the second, more ambitious, is based on making reasonable assumptions that lead to unbiased results. In what follows, we will try to choose the second alternative.

A commonly used procedure, which is often reasonably efficient, consists of analyzing a histogram representing the experimental data. Guided by this histogram, we can hypothesize an explicit form (dependent on certain parameters) for the probability distribution, and then determine these parameters by means of the procedures we have already discussed. That is, we can proceed in analogy to what we did in the Gaussian case, with the disadvantage that, if the hypothesized probability distribution is not particularly simple, we will probably have to give up on carrying out an exact computation.

Basically, in this approach, we try to guess what the correct probability distribution is by analyzing the histogram of the experimental data. The result, of course, will not be completely reliable, and doubts will remain as to its validity.

We want to discuss here, as an alternative, how it is possible to proceed from an arbitrary number of hypotheses that is as small as possible [48]. We begin by noting that in an *extreme* situation, such as the one we are discussing, it is not possible for us to make reasonable assumptions about the probability of obtaining a value of x other than one of those observed. We will, therefore, assume that the probability distribution of x is zero for all values of x not experimentally measured.

We can summarize our procedure as follows. We have made N measurements, and we decided to consider a discrete probability distribution. The probability of the kth measured value, x_k (we will assume for now that the measured values are all different), will be called p_k, while the probability of the values that have never been measured is zero. This assumption is due to the fact that we have no information about these values. The probabilistic weights p_k will be precisely the parameters that our reasoning must lead us to determine. They are free, non-negative variables, with the only constraint being

$$\sum_{k=1}^{N} p_k = 1 .\tag{5.59}$$

By appropriately reweighting the probabilities of obtaining the measured values of x, we will now construct the appropriate probability distribution. We assume that the probability distribution $D(x)$ has the form

$$D(x) = \sum_{k=1}^{N} p_k\, \delta(x - x_k) .\tag{5.60}$$

The choice of the probability function is *data driven*: the more the number of measured data increases, the denser the delta functions become, so that in the limit $N \to \infty$ the function defined in (5.60) can tend to a continuous function.

Since we ignore the distribution of orders of magnitude of the probabilities of the values of x, reasoning as described in the previous sections (and in Appendix 5.3.1), it becomes clear that it is reasonable to assume that the *a priori* probability distribution p_k is constant on a logarithmic scale:

$$A(\mathbf{p}) \propto \prod_{k=1}^{N} \frac{1}{p_k} ,\tag{5.61}$$

and Bayes' rule tells us that the *a posteriori* distribution holds:

$$P(\mathbf{p}|\mathbf{x}) = \frac{P(\mathbf{x}|\mathbf{p})A(\mathbf{p})}{Z} .\tag{5.62}$$

What does the likelihood look like? We assume that the individual events are independent. We have N possible events, labeled by the index k, and we are interested in knowing the probability that the very events we observed occurred. The probability of each of the N events occurring once is proportional to

$$\prod_{k=1}^{N} p_k \, .$$

(5.63)

Thus, under the assumption that all events are distinct, the likelihood will be

$$P(\mathbf{x}|\mathbf{p}) \propto \prod_{k=1}^{N} p_k \, \delta \left(\sum_{k=1}^{N} p_k - 1 \right),$$

(5.64)

which obviously includes the normalization constraint, (5.59). Since the *a priori* probability is (5.61), the *a posteriori* probability of the coefficients p_k conditional on the results of our experiment turns out to be

$$P(\mathbf{p}|\mathbf{x}) = (N-1)! \, \delta \left(\sum_{k=1}^{N} p_k - 1 \right),$$

(5.65)

where the normalization factor can be computed using the formula (3.144).

Generalizing to the case of measurements yielding the same values more times, we observe that if, in N measurements, only M distinct values of x occur, we will consequently have M weights p_k and things change a little. In Eqs. (5.59), (5.60), and (5.61) the index k varies from 1 to M. The formula (5.63) must also take into account the ν_k multiplicity with which each x_k value occurs. For example, the probability of N occurrences of the event 1 is p_1^N, the probability of $(N-1)$ occurrences of the event 1 and one occurrence of the event 2 is $p_1^{N-1} p_2$, and so on. The probability that each of the events k occurred ν_k times is therefore proportional to

$$P(\mathbf{x}|\mathbf{p}) = \prod_{k=1}^{M} p_k^{\nu_k} \, ,$$

(5.66)

and the *a posteriori* distribution reads

$$P(\mathbf{p}|\mathbf{x}) \propto \prod_{k=1}^{M} p_k^{\nu_k - 1} \, \delta \left(\sum_{k=1}^{M} p_k - 1 \right).$$

(5.67)

We note that the trend is the same as in the formula (5.46) for the *a posteriori* distribution of the probability weights of the categories into which the events are divided, where the multiplicity ν_k was the number of counts of the event k, i.e., K_k.

To compute the probability distribution of the mean value of x, which we denote by X, we note that, in the case of N distinct measurements,

$$X \equiv \langle x \rangle = \sum_{k=1}^{N} p_k x_k \, .$$

(5.68)

Thus, under our assumptions, its probability distribution is proportional to

$$P(X) \propto \int \left(\prod_{k=1}^{N} dp_k \right) \delta \left(\sum_{k=1}^{N} p_k - 1 \right) \delta \left(\sum_{k=1}^{N} p_k x_k - X \right). \tag{5.69}$$

The relation (5.69) is a special case of the general method known as the *reweighting* method, based on weighting the experimental data with random weights and determining the probability distribution of the quantity of interest (in this case X) as a function of them.

Let us explain operationally what we do in the case of all different measured values.

1. We measure N times the observable x by acquiring the values $\mathbf{x} = \{x_1, x_2, \ldots, x_N\}$.
2. (a) With the distribution (5.65) we generate a set of weights $\mathbf{p} = \{p_1, p_2, \ldots, p_N\}$ for the values $\mathbf{x}$. In practice, we extract N pseudo-random numbers between 0 and 1 with uniform distribution (see Appendix 3.J), then divide each number by the sum of all drawn numbers, so that the rescaled numbers sum to 1.
   (b) With the extracted sequence $\mathbf{p}$ we compute a value of the expected value (5.68).
3. We repeat steps 2(a) of extracting the weights $\mathbf{p}$ and 2(b) of computing the average value $X[\mathbf{p}]$ for S times, so that we collect S values of X with which we construct the normalized histogram $P(X)$, i.e., the distribution of the expectation value at fixed $\mathbf{x}$ experimental data.

In Figure 5.2 we show the $P(X)$ obtained from just $N = 4$ measurements by extracting many times $\mathbf{p}$ sequences of N probabilistic weights satisfying the *a posteriori* distribution of $P(\mathbf{p}|\mathbf{x})$ weights. In this case, the formula is (5.65) because the "experimental" x are extracted on the computer as pseudo-random numbers by means of a normal distribution and, therefore, they are all distinct in double-precision floating-point numbers. In the upper panel of Figure 5.2, we show the distribution obtained by making the histogram with $S = 100$ reweighted values of X, and in the lower panel that with $S = 1000$ generations of X. We note that, reasonably, the curves become smoother and relatively more peaked as the number of reweights S of X values increases. With only four measurements, however, the fluctuations between the various $P(X)$, corresponding to different data sets $\mathbf{x}$, are very large.

In Figure 5.3 we show the behavior of $P(X)$ with number of repetitions S fixed but increasing the number of experimental measurements N. Although with consistent fluctuations, as N increases from 4 to 10 and, then, to 20, the expected value distributions tend to concentrate around the true mean value (which in this example case we know to be zero because the experimental data are "synthetic" and we generated them).

If we measure variables at integer values, the probability of measuring exactly the same value more than once is not zero any more and the correct *a posteriori* distribution of $\mathbf{p}$ is given by (5.67). In Figure 5.4 we show the posterior distribution of the expected values obtained by the same reweighting procedure as in the previous example, but, in this case, the true distribution of the data x is a bimodal Rademacher distribution; x can take values ± 1 with equal probability.

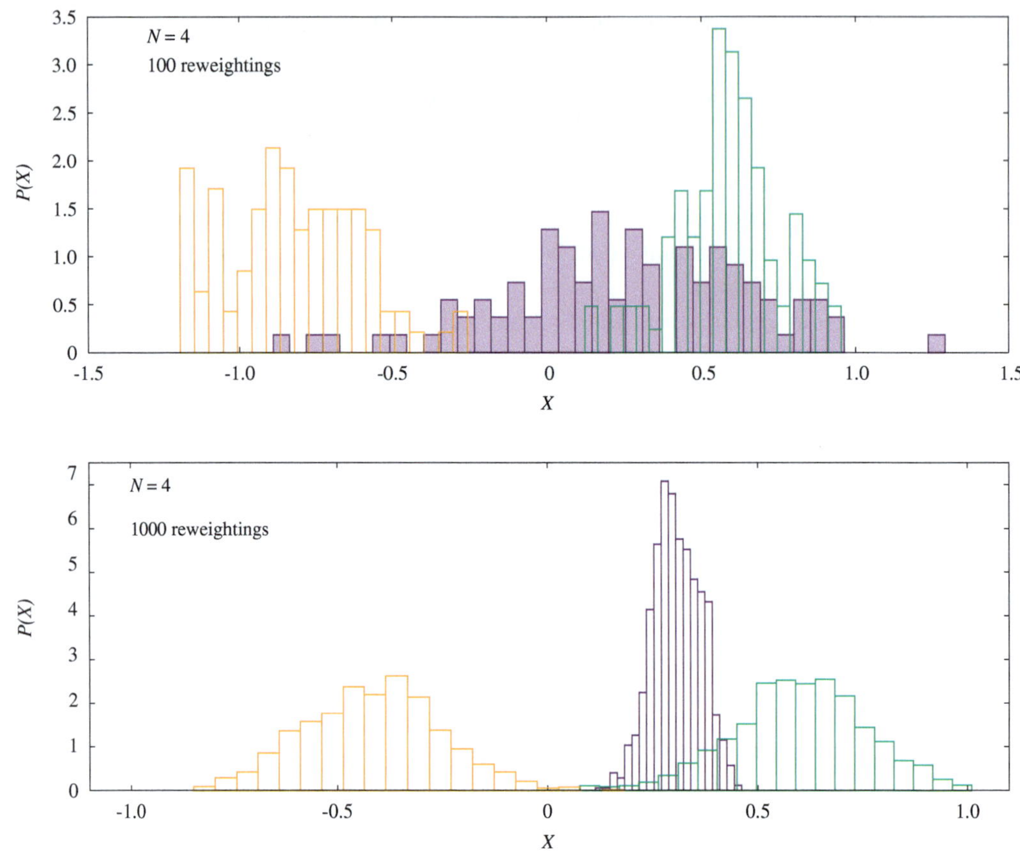

Figure 5.2 Three examples of *a posteriori* distributions $P(X)$ of the expected value X of variables x based on $N = 4$ measurements. In all cases shown, the values of x are generated with a normal distribution of pseudo-random numbers. Top: Histograms are constructed with 100 reweighting sequences p extracted with the distribution (5.65). Bottom: Histograms take X values generated by the extraction of 1000 sequences p.

The reweighting method is used very frequently, and we will discuss its application to the case of vectorial data later in Section 6.3. There are, in addition, other very popular methods of resampling data, such as *binning*, *jackknife*, and *bootstrap*, which will be discussed in Section 5.5. First, however, let us spend some time verifying the goodness of the reweighting method.

In our discussion, the relation (5.69) has been derived by making for the *a priori* probability an assumption about which one could argue at length, but that finds support in obtaining common-sense, stable, and unbiased results *a posteriori*, as we have seen in Section 5.3.3. Regardless of its derivation, however, Eq. (5.69) can be used as a starting point for estimating the probability $P(X)$ in cases where we have no precise information on the functional form of the distribution of the data x_k. We must, therefore, check that the results obtained are reasonable. We must also check how much

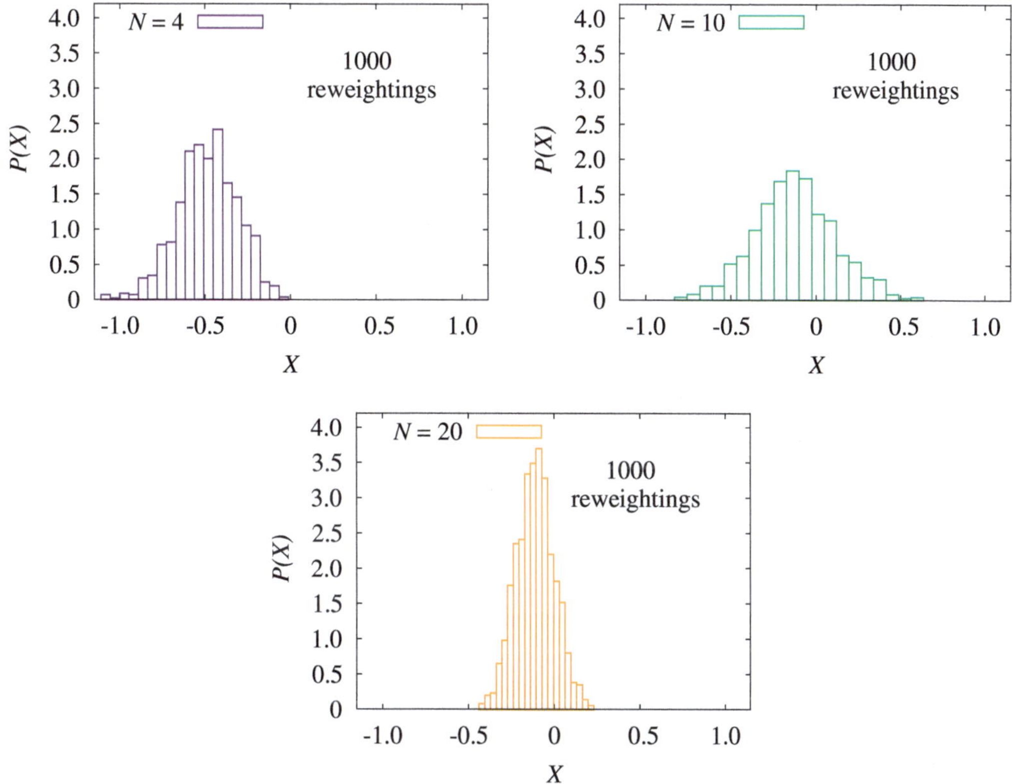

Figure 5.3 Distributions $P(X)$ from the formula (5.69) for $N = 4$ measurements (left), for 10 measurements (right), and for 20 measurements (bottom). In all reported cases, the values of x are generated with a normal distribution of pseudo-random numbers.

information is lost if we use this method in cases where we have more knowledge and we can use more efficient methods, such as in the prototypical case of Gaussian-distributed variables.

We shall see that, as the underlying distribution of experimental data varies, the relation (5.69) is capable of leading us to quite distinct forms of probability distributions. In almost all situations, the results obtained turn out to be very reasonable, not inferior in quality to those obtained by methods that explicitly use knowledge of the form of the probability distribution, and are thus less versatile. We will now discuss two concrete examples that will help to clarify this aspect.

5.4.2 A First Check of Reweighting: Poisson Distribution

We will now try to understand how the ideas of the reweighting method set out in the previous section apply to a first, simple case where we know what is going on. Suppose that the variable x has been measured N times, and has been equal to L in R cases

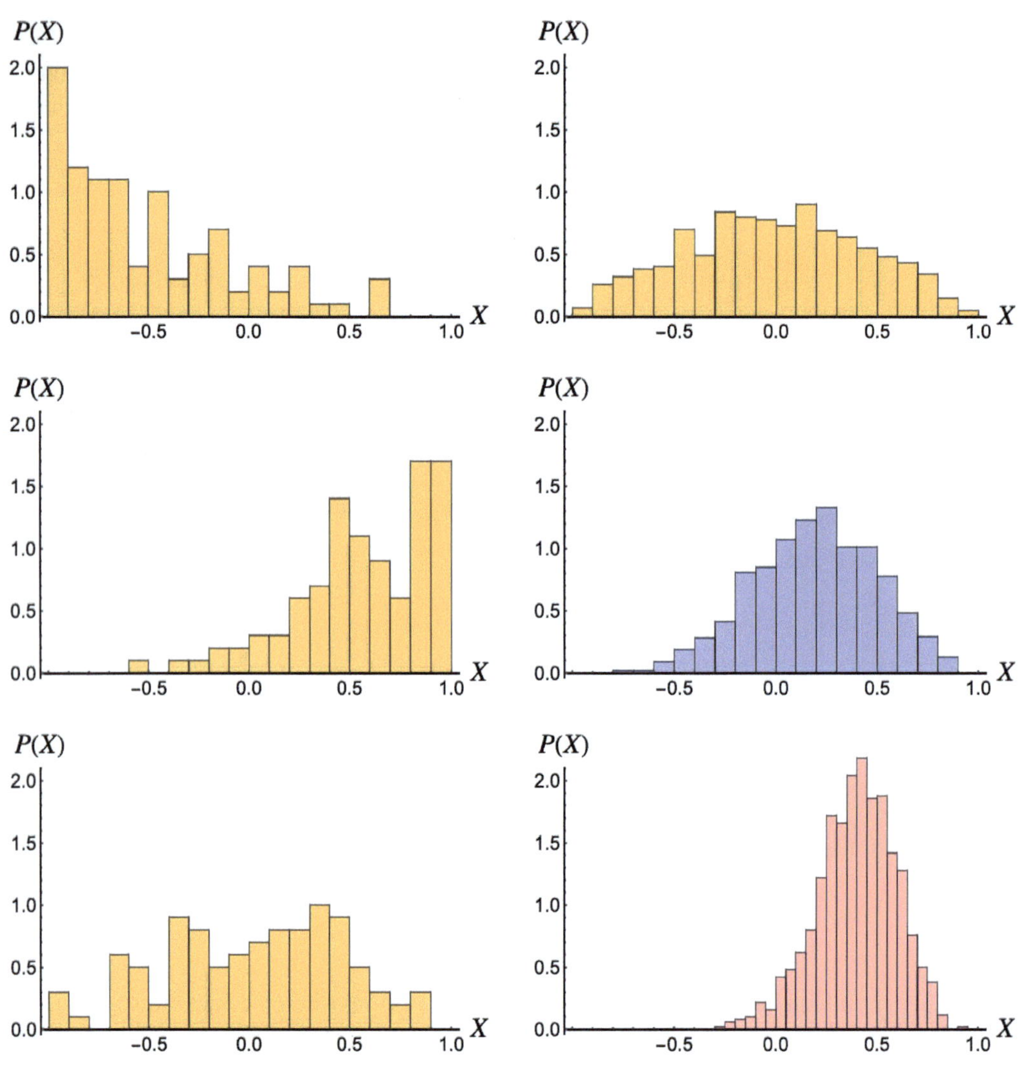

Figure 5.4 Reweighting examples obtained with distribution (5.67). The "experimental" x values are generated with a Rademacher distribution of pseudo-random numbers, $x = \pm 1$. Left column: Three examples of $P(X)$ distributions, from formula (5.69) of the posterior built with the reweighting method, with 100 reweightings and $N = 4$ (yellow histograms). Right column: Examples of $P(X)$ distributions built over 1000 reweightings for data sets of $N = 4$ (yellow), 10 (blue), and 20 (red) measurements.

and equal to 0 in the remaining $N - R$ cases. Let N be much larger than R, and R not necessarily large. In other words, the event $x = L$ has a small probability, of the order of R/N, i.e., it is a rare event. The kurtosis (2.7) of the bimodal probability distribution of x, $D(x)$, is very high, i.e., $D(x)$ is very different from a Gaussian distribution.

Following the reasoning of the previous section, we must consider here only two probabilities p_1 and p_2, subject to the constraint $p_1 + p_2 = 1$.

We can apply the reweighting procedure to the case at hand, where the two possible types of events have been recorded R and $N-R$ times, respectively. Using the constraint to replace p_2 with the quantity $(1-p_1)$, we will have that the *a priori* density is

$$A(p_1) \propto \frac{1}{p_1(1-p_1)}.$$

The likelihood, in this case, is constructed taking into account that the values of x are repeated, with $\nu_1 = R$ and $\nu_2 = N - R$:

$$P(\mathbf{x}|p_1) \propto p_1^R(1-p_1)^{N-R}.$$

We then use (5.67) for the posterior probability density and obtain

$$P(p_1|\mathbf{x}) = \frac{P(\mathbf{x}|p_1)A(p_1)}{Z} = \frac{\Gamma(N)}{\Gamma(N-R)\Gamma(R)} \frac{1}{p_1(1-p_1)} p_1^R(1-p_1)^{N-R}$$

$$= \frac{\Gamma(N)}{\Gamma(N-R)\Gamma(R)} p_1^{R-1}(1-p_1)^{N-R-1}. \tag{5.70}$$

In the limit where N is much larger than R, the distribution of p_1 will be concentrated in the region where $p_1 = \lambda/N$. The *a posteriori* probability distribution of λ, for $N \gg 1$, will therefore be proportional to

$$P(\lambda|\mathbf{x}) \propto P\left(p_1 = \frac{\lambda}{N}\middle|\mathbf{x}\right) \propto \lambda^{R-1}\left(1 - \frac{\lambda}{N}\right)^{N-R-1} \xrightarrow[N\to\infty]{} \frac{\lambda^{R-1}}{\Gamma(R)} \exp(-\lambda), \tag{5.71}$$

which coincides with (5.33). Given that $X \equiv \langle x \rangle = p_1 L + p_2 0 = \lambda L/N$, we have that $\lambda = XN/L$, and the probability distribution of the mean value X conditional on the data turns out to be

$$P(X) = \frac{N}{L\Gamma(R)}\left(\frac{XN}{L}\right)^{R-1}e^{-XN/L} = \left(\frac{N}{L}\right)^R \frac{X^{R-1}}{\Gamma(R)}e^{-XN/L}. \tag{5.72}$$

The result is exactly the same as if we had (correctly) taken into account a Poisson distribution as the probability of the occurrence of the event $x = L$. In fact, in the limit $N \to \infty$ the binomial distribution collapses onto the Poisson distribution, and, as we have already seen in (5.33), the probability distribution of the parameter λ that characterizes the Poisson probability distribution in the case of R events is the same as in $\lambda = p_1 N$, i.e., it is proportional to $\lambda^{R-1}e^{-\lambda}$. Our method, without *a priori* assumptions, successfully led us to the case of a binomial distribution which, for large values of N, collapses to a Poisson distribution.

5.4.3 A Second Check of Reweighting: Gaussian Case

Let us now assume that the true distribution of the data x is a Gaussian distribution. We want to show that if we apply the reweighting method, making no *a priori* assumptions about the type of probability distribution, for large values of N we find that the distribution of the mean value X, $P(X)$ (5.69), is a Gaussian distribution. It is crucial that this method gives the same result as the standard method in the Gaussian case and, in particular, with the same variance. If there were even a factor of 2 in the variance, it would be dramatic.

We could start from the expression (5.69) and use the Fourier antitransform representation of the $\delta(x)$ function, see (3.79). In this way (ignoring as usual constant multiplicative factors that merely enter in the normalization constant) we find that

$$P(X) \propto \int \left(\prod_{k=1}^{N} dp_k\right) \int_{-\infty}^{\infty} dz \, \exp\left[\iota z \left(\sum_{k=1}^{N} p_k - 1\right)\right]$$
$$\times \int_{-\infty}^{\infty} dw \, \exp\left[\iota w \left(\sum_{k=1}^{N} p_k x_k - X\right)\right]. \tag{5.73}$$

In principle, to compute this integral, one can first proceed by explicitly integrating over all variables p_k, and then evaluating the integral over z and w using the saddle point method (see Appendix 4.A), but this is rather laborious.

An alternative procedure is to compute, in the limit of N large, the moments of the quantity X, and check that they tend to the moments of a Gaussian distribution, somewhat as we did in the proof of the moments of the central limit theorem (see Section 3.2.2).

We begin by considering the first two moments. Using the expression of the expectation value,

$$X[\mathbf{p}] = \sum_{k=1}^{N} p_k x_k, \tag{5.74}$$

we obtain

$$\langle X \rangle \equiv \int dX \, P(X) X = \frac{1}{N} \int \left(\prod_{k=1}^{N} dp_k\right) P(\mathbf{p}|\mathbf{x}) \sum_{k=1}^{N} p_k x_k \,,$$
$$\langle X^2 \rangle \equiv \int dX \, P(X) X^2 = \frac{1}{N} \int \left(\prod_{k=1}^{N} dp_k\right) P(\mathbf{p}|\mathbf{x}) \left(\sum_{k=1}^{N} p_k x_k\right)^2. \tag{5.75}$$

We can rewrite the previous formulas as

$$\langle X \rangle = \sum_{k=1}^{N} \langle p_k \rangle \, x_k \,,$$
$$\langle X^2 \rangle = \sum_{k_1=1}^{N} \sum_{k_2=1}^{N} \langle p_{k_1} p_{k_2} \rangle \, x_{k_1} x_{k_2} \,. \tag{5.76}$$

To compute these two quantities we will use Eq. (3.144) for the integral of the Dirac delta of a sum of variables. We first note that the formula (3.144) allows us to compute the normalization factor of the *a posteriori* probability distribution $P(\mathbf{p}|\mathbf{x})$. By multiplying (5.65) by $(N - 1)!$ its integral over all p_k is equal to 1. The probability distribution of a single p_j, for example of p_1, is then obtained by marginalizing over all p_k except p_1:

$$P(p_1) = (N-1)! \int \left(\prod_{k=2}^{N} dp_k \right) \delta \left(\sum_{k=2}^{N} p_k - (1 - p_1) \right) = (N-1)(1 - p_1)^{N-2} , \quad (5.77)$$

where we again used the relation (3.144), applied this time to $N-1$ integration variables. For the sake of brevity, we have omitted writing the condition of the experimental data into the argument: to be pedantic it should read $P(p_1) = P(p_1|\mathbf{x})$. We finally find

$$\langle p_j \rangle = \frac{1}{N} , \quad \forall \, j = 1, \dots, N . \tag{5.78}$$

This relationship is quite obvious. In fact, in the absence of any other information, the mean values of the variables p_k must all be equal, and as a consequence of the normalization condition

$$\sum_{i=j}^{N} p_j = 1 , \tag{5.79}$$

they are all equal to $1/N$. In general, the moments of the distribution can be easily computed using the beta function (5.23):

$$\langle p^m \rangle = \int_0^1 dp \, P(p) p^m = \frac{m!(N-1)!}{(m+N-1)!} . \tag{5.80}$$

The joint distribution of two weights p_1 and p_2 is written, similarly to (5.77), as

$$P(p_1, p_2) = (N-1)! \int \left(\prod_{k=3}^{N} dp_k \right) \delta \left(\sum_{k=3}^{N} p_k - (1 - p_1 - p_2) \right)$$
$$= (N-1)(N-2)(1 - p_1 - p_2)^{N-3} . \tag{5.81}$$

Computing the correlation $\langle p_j p_k \rangle$ from this can be a bit complicated (we can use the formula (5.42)), but knowing the formula (5.80) we can use the trick of writing

$$\frac{1}{N} = \langle p_j \rangle = \left\langle p_j \sum_k p_k \right\rangle = \langle p_j^2 \rangle + (N-1)\langle p_j p_k \rangle,$$

from which

$$\langle p_j p_k \rangle = \frac{1}{N(N+1)} . \tag{5.82}$$

Using these results we find that

$$\langle X \rangle = \sum_{k=1}^{N} \langle p_k \rangle x_k = \bar{x} , \tag{5.83}$$

where the bar indicates the empirical average on the experimental data. This is the first moment and also the first *cumulant*. The second cumulant is the variance

$$\langle X^2 \rangle_c = \langle (X - \langle X \rangle)^2 \rangle = \langle X^2 \rangle - \langle X \rangle^2 = \frac{\overline{x^2} - \bar{x}^2}{N+1} \simeq \frac{\bar{\sigma}^2}{N} . \tag{5.84}$$

If we use the same technique to compute the expectation value of the moments $\langle X^n \rangle$ and their combinations in higher-order cumulants, see Appendix 3.C.3, after a somewhat lengthy (and tedious) computation we find that all cumulants of X of order higher than the second tend to zero *faster* than the variance (5.84).

For example, it is found that the kurtosis of X, $K(X)$ (3.32), is given by

$$K(X) = \frac{\langle (X - \langle X \rangle)^4 \rangle}{\langle (X - \langle X \rangle)^2 \rangle^2} \simeq 3 + \frac{6}{N} \frac{\overline{(x - \bar{x})^4}}{\overline{(x - \bar{x})^2}^2} , \tag{5.85}$$

and, therefore, the fourth-order cumulant tends to zero as N^{-3} :

$$\langle X^4 \rangle_c = (K(X) - 3) \left(\frac{\bar{\sigma}^2}{N} \right)^2 = O \left(\frac{1}{N^3} \right).$$

The same argument can be made for cumulants of order higher than the fourth, whose convergence to the Gaussian cumulants for large N becomes faster as the order increases. If we consider the analogue of the rescaled central limit variable defined in (3.18), $W = (X - \langle X \rangle)\sqrt{N}$, we observe that only the second cumulant $\langle W^2 \rangle_c$ has a finite limit, $\bar{\sigma}^2$, whereas all higher-order cumulants of W tend to zero for large N. According to our knowledge about cumulants and Gaussian distributions, cf. Appendix 3.C.3, the distribution obtained for the mean value X for large N then turns out to be precisely a Gaussian distribution.[3] This is a second success: our method, when applied to Gaussian experimental data, yields a Gaussian distribution, as it did in the previous section for a Poisson distribution.

In general, we can say that for distributions of the experimental data with empirical values of variance and kurtosis that are not too large, the reweighting method produces Gaussian distributions for the mean values. On the contrary, if the kurtosis is of order N (as in the example of Section 5.4.2 in the limit of N large and R finite), the distribution $P(X)$ is very different from a Gaussian one and its final form strongly depends on the features of the experimental data.

5.5 Resampling Methods

If we have a nonlinear function of the data, computing the statistical uncertainty with propagation, as shown in Section 5.1, can be tedious and induces distortion. The quality of the propagation procedure is greatly influenced by the type of function, by the variables we are observing, and it depends on the bias present in the estimation of its expected value. See, for example, the bias of the empirical variance estimate with respect to its theoretical expected value, Eqs. (5.4)–(5.6).

In addition to the reweighting method discussed in Section 5.4.1, there are also other methods of uncertainty estimation not depending on the functional dependence,

---

[3] See Section 12.3 for a comprehensive discussion of cumulants and related connected correlation functions in the case of several random variables.

which, at the same time, reduce the corrections due to bias or take into account possible correlations in the data acquisition over time or, furthermore, allow for the identification of possible anomalies in the data. We will introduce here four resampling methods that allow us to determine statistical errors and possible anomalies and distortions automatically. For further insights, we refer to some basic reviews [49, 50, 51, 52] and for examples of pragmatic applications to data analysis and interpolation we refer to Young's brilliant text [53].

5.5.1 Binary Resampling: How to Spot Anomalous Data

A method that closely resembles that of Section 5.4.1, but is slightly simpler to implement, is to resample the data as in the formula (5.69), but this time assigning a weight π_k which can be 0 or 1 with equal probability:

$$P(\pi) = \tfrac{1}{2}\delta(\pi) + \tfrac{1}{2}\delta(\pi - 1).$$

In other words, by tossing the notorious coin, we randomly choose a subsample of data, whose number fluctuates around 50% of the data, and compute the average value X of the quantity x on this subset. Then, using the same criterion, a new subsample of the data is randomly recreated and the mean value is recomputed: this operation is repeated many times. The distribution $P(X)$ of the mean values X obtained with this random sampling coincides, in the limit of a large number of experimental data, with the probability distribution of the mean X conditional on the experimental data: exactly as in the previous case, it is easy to see that in the case of a Gaussian distribution the correct results are reproduced.

This method is not very different from the previous one: almost always the results obtained in the two cases are similar, but not always. We can see the resampling method as a particular reweighting with weights of

$$p_k = \frac{\pi_k}{\sum_{j=1}^{N} \pi_j},$$

where $\sum_{j=1}^{N} \pi_j$ is the number of data resampled each time, whose average is $\sum_{j=1}^{N} \langle \pi \rangle = N/2$. By approximating the number of data from a given resampling with its mean, we can consider binary resampling as a reweighting method whose probability weights p_k are equal to either 0 or $2/N$ with probability $1/2$. The mean of the individual weights is always given by (5.78), $\langle p \rangle = 1/N$, inasmuch as the second moment is

$$\langle p^2 \rangle = \frac{1}{2}\left(\frac{2}{N}\right)^2 = \frac{2}{N^2},$$

which coincides with (5.80) for $m = 2$. Higher-order moments, however, differ from (5.80). For example,

$$\langle p^3 \rangle = \frac{1}{2}\left(\frac{2}{N}\right)^3 = \frac{4}{N^3} \neq \frac{6}{N^3}.$$

If the data are Gaussian, only the first two cumulants matter, and, therefore, the two methods give the same results. In contrast, data with different distributions, asymmetric and/or with bulges other than Gaussian, lead to different $P(X)$ when treated with the reweighting method or the binary resampling method.

One of the advantages of this resampling method is that the presence of an anomalous event (e.g., a data entry whose value should be 10^{-3}, but has become 10^3 due to a transcription error) is very well revealed. The average over half of the data gives very different results depending on whether the extracted subsample contains the anomalous event or not. If even for large values of N the histogram of the mean X values obtained by this method presents two distinct peaks of approximately equal weight, there are very strong reasons to believe that the mean is dominated by an outlier. It is, of course, possible that the outlier is not a transcription error, but a real phenomenon (and perhaps the key to a revolutionary scientific discovery). In this case, the method correctly highlights the phenomenon, and shows that it does not make sense to write the estimated error in the usual form $X \pm e$.

Let us look at a concrete example. Suppose we measured N times the variable x and found the result $x = L$ once and the result $x = 0$ for $N - 1$ times. The reweighting method discussed in the previous sections in this case generates a probability distribution of the mean X given by (5.72): $P(X) \sim \exp(-XN/L)$. The binary resampling method gives instead

$$P(X) = \frac{1}{2}\delta(X) + \frac{1}{2}\delta\left(X - \frac{2L}{N}\right). \tag{5.86}$$

Both trends (5.72) and (5.86) produce a probability distribution $P(X)$ concentrated in the region where X is of order L/N, but if an outlier exists the second method highlights it by producing a spurious peak. The tendency of the second method to produce spurious peaks quickly disappears as the number of significant data increases.

In Figure 5.5 we show another example for the study of the *a posteriori* distribution of the expected value of a variable x after $N = 20$ measurements, one of which, however, is anomalous. The variable x is, in fact, a normal variable (we have generated its pseudo-random values numerically with the formula (3.168)), but in the 20 measurements on which $P(X)$ is computed, a value $x = -20$ was presented. In the graph, we show $P(X)$ constructed by the reweighting method of Section 5.4.1 on 1000 extractions of the sequences of probabilistic weights **p** with the distribution (5.65). In the same graph we show $P(X)$ constructed with the binary resampling method described in this section, again for 1000 resamplings $p_k = 0, 1$, where the anomalous event appears in about half of the resamplings extracted.

The difference is glaring. The standard resampling method produces a fairly wide distribution of the mean X value, but fails to identify the spurious occurrence, even though this affects the mean value of the distribution (i.e., the empirical average $\langle X \rangle = \bar{x}$). In contrast, the resampling method produces a $P(X)$ with two peaks of the same order of magnitude, of which the peak on the left is due solely to the presence of the very negative spurious datum. In the figure, for comparison, we also show $P(X)$

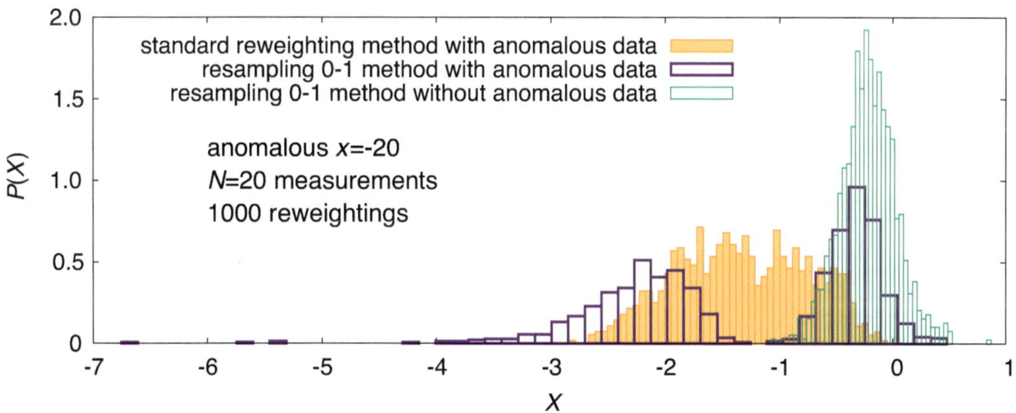

Figure 5.5 *A posteriori* distributions of the expected value of a variable measured $N = 20$ times in the presence of a spurious datum $x = -20$. In yellow, $P(X)$ computed by the reweighting method, formulas (5.65) and (5.69). In magenta, $P(X)$ computed by the binary resampling method shown in this section: each X is computed using only about half of the measurements. In green, $P(X)$ using the resampling method in the absence of outlier measurements.

in the absence of the anomalous datum (replaced by a random number of normal distribution) constructed again with the same resampling, which, therefore, appears to work fine even in the absence of anomalies.

5.5.2 Binning Method and Time Correlations

Let us now introduce a very simple method called *binning* [54] or *blocking* [52], which is useful for computing the expectation value of a function g of the data and estimating its associated error e_g without resorting to error propagation. The method, which is effective in the case of a sufficiently large number N of events, is set up by writing N as $N = K \times L$, where both K and L are sufficiently large. The data is grouped into L blocks of K elements each.

For example, if we have a set of N measurements of the observable x,

$$x_1 \quad x_2 \quad x_3 \quad x_4 \quad x_5 \quad x_6 \quad x_7 \quad x_8 \quad \ldots \quad x_i \quad \ldots \quad x_{N-3} \quad x_{N-2} \quad x_{N-1} \quad x_N \,,$$

we can split them into L blocks of K values each:

$$\left| \, x_1^{(1)} \quad x_2^{(1)} \quad \ldots \quad x_K^{(1)} \, \right| \, x_1^{(2)} \quad x_2^{(2)} \quad \ldots \quad x_K^{(2)} \, \left| \ldots \, x_i^{(\ell)} \quad \ldots \, \right| \, x_1^{(L)} \quad x_2^{(L)} \quad \ldots \quad x_K^{(L)} \, \right| .$$

We now introduce the empirical averages of a single block (*bin*), denoted by ℓ,

$$\bar{x}_\ell \equiv \frac{1}{K} \sum_{i \in \ell}^{1,K} x_i^{(\ell)} , \quad \ell = 1, \ldots, L . \tag{5.87}$$

As usual, we call $\bar{x}$ the average over all N given without grouping. Averaging over all bins, the empirical mean value, being linear, is unbiased and equal to

$$\langle x \rangle_L = \frac{1}{L} \sum_{\ell=1}^{L} \bar{x}_\ell = \frac{1}{N} \sum_{j=1}^{N} x_j = \bar{x}.$$

Having defined a fairly regular function $g(x)$ of the data, we introduce the value of the function computed on the bin average,

$$g_\ell \equiv g(\bar{x}_\ell), \tag{5.88}$$

and the average over all blocks,

$$\langle g \rangle_L \equiv \frac{1}{L} \sum_{\ell=1}^{L} g_\ell. \tag{5.89}$$

As we will review in more detail in the next chapter, when we recall error propagation also for vector quantities (Section 6.1.1), for reasonable functions g the deviation between the empirical average of a function performed on a certain number of data and the function computed on the empirical mean performed on the same data differ by an amount that decreases as the inverse of the number of data if these are numerous enough. If we take the $\bar{x}_\ell$ averages over the blocks as data, the $\langle g \rangle_L$ average is performed on L elements, so, for $L \gg 1$, this property translates into

$$|\langle g \rangle_L - g(\langle x \rangle_L)| = |\langle g \rangle_L - g(\bar{x})| = O(1/L). \tag{5.90}$$

Furthermore, we know from the central limit theorem that the deviation of the function computed in the single measure $g_i = g(x_i)$ (with $i = 1, \ldots, N$) from the empirical average is of order $N^{-1/2}$. If we place ourselves in the ℓth block of K data, the latter relationship corresponds to

$$|g_i - g(\bar{x}_\ell)| = |g_i - g_\ell| = O(1/\sqrt{K}), \quad i = 1, \ldots, K. \tag{5.91}$$

Using Eqs. (5.90) and (5.91), we can express the deviation between the function computed on the single original datum $g_i = g(x_i)$ and the empirical average of the function on all N data as

$$|g_i - \bar{g}| = |g_i - g(\bar{x})| + O(1/N) = |g_\ell - \langle g \rangle_L| + O(1/L) + O(1/\sqrt{K}) + O(1/LK). \tag{5.92}$$

The error obtained by the binning method is defined as

$$e_g^2 = \langle (g_\ell - \langle g \rangle_L)^2 \rangle_L = \frac{1}{L} \sum_{l=1}^{L} (g_l - \langle g \rangle_L)^2. \tag{5.93}$$

From (5.92) we see that this coincides with the error computed by the standard method

$$e_g^2 = \frac{1}{N} \sum_{i=1}^{N} (g(x_i) - \bar{g})^2, \tag{5.94}$$

disregarding terms going to zero, as L and K increase, as $O(1/L\sqrt{K})$, $O(1/K)$, or $O(1/L^2)$ or higher orders.

Given the need to have both K and L large, typically the binning method is applicable to cases where N is large enough (e.g., $N > 100$), taking care to check that one has

$$|\langle g \rangle_L - g(\bar{x})| \ll e_g. \tag{5.95}$$

5.5.3 Error Estimate on Correlated Measures

An additional advantage of this method is that it is very useful when the experimental data measurements are not independent. This occurs when data are taken sequentially in the same experiment and there are correlations between two measurements of the same observable taken at nearby times. It is a property of many time series, e.g., of the dynamics of a random walker (Chapter 7), of the variation of a population from generation to generation (Chapter 8) or in continuous time (Chapter 13), of price trends or of weather forecasts. We will call it *time* correlation. We will label the time by $j, k, \ldots$, i.e., the number of data taken in a series of acquisitions, and the time distance between the measurements x_k and x_j by $|k - j|$. If the time correlation is non-zero only for a distance between measurements $|k - j|$ that is not too great, e.g., if the correlation goes to zero faster than $|k - j|^{-1}$, the correlations between quantities g_l go to zero as K^{-1} decreases. Due to this fact, starting from non-independent data, the iterative application of the binning method sooner or later leads back to the case of independent data.

In practice, when no precise *a priori* estimate of the correlation time between the data is available, it is convenient to use the binning method for different values of K (e.g., with $K = 1, 2, 4, 8, 16, 32, \ldots$, satisfying the condition $K \ll N$). Typically, in the case of corelated data, the error estimates (5.94) will be underestimated for small K values and will increase as K increases, until they become the correct ones at a certain K_{corr} value (again assuming $L \gg 1$). This K value plays the role of a discrete *time* and is comparable to the *correlation time* among the measurements: numerical simulations performed with the Monte Carlo methods described in Section 11.3 are an ideal field of application for these data analysis techniques.

In the sad case where the error continues to increase as K grows (until a maximum value is reached for the maximum value used for K, compatible with the condition $K \ll N$), to extrapolate for $K \rightarrow \infty$ is impossible. The data obtained from the measurements are so correlated (or, in other words, the correlation time between the measurements is so long) that a reliable estimate of the statistical error turns out to be impossible. We cannot learn much from measurements made under these conditions, other than the fact that we are dealing with a phenomenon that develops over extremely long time scales.

In this regard, we finally notice that obviously the error estimated as K varies is itself a random variable, subject to fluctuations and statistical uncertainty. One must keep this in mind in order not to be surprised at the possible non-monotonic behavior of the estimated error as a function of the bin size. The statistical uncertainty afflicting the estimated error can in turn be estimated from experimental data.

5.5.4 Jackknife Resampling

The jackknife method, literally a multi-purpose Swiss army knife, synonymous with ready and versatile, aims to compute the statistical error on N measurements, creating from these a number N of subsets of data consisting of $N - 1$ measurements each. We denote by j the subset containing all but the jth measurements of the observable x. The "jackknife" average over the subsample j is defined as

$$\bar{x}_j^{\text{JK}} \equiv \frac{1}{N-1} \sum_{k \neq j} x_k \, . \tag{5.96}$$

We observe that this can also be rewritten as

$$\bar{x}_j^{\text{JK}} = \frac{1}{N-1} \left(\sum_{k=1}^{N} x_k - x_j \right) = \frac{N\bar{x} - x_j}{N-1} = \bar{x} + \frac{\bar{x} - x_j}{N-1} \, . \tag{5.97}$$

Let us now consider the usual generic function $g(x)$ of the observable x measured N times. We define the jackknife average on the subsample j of the function g as g computed on the jackknife average of x:

$$\bar{g}_j^{\text{JK}} \equiv g(\bar{x}_j^{\text{JK}}), \tag{5.98}$$

a bit like we did in (5.88) for the average on a bin in the binning resampling method.

As a total estimate of g we take the average of the N jackknife averages:

$$\overline{g^{\text{JK}}} = \frac{1}{N} \sum_{j=1}^{N} \bar{g}_j^{\text{JK}} \, . \tag{5.99}$$

If g is a linear function of x, the average with the jackknife method matches the usual one. If, on the other hand, g is not linear, and there is, hence, a distortion, the two averages differ.

We also define the estimation of the generic nth moment of the jackknife averages:

$$\overline{(g^{\text{JK}})^n} \equiv \frac{1}{N} \sum_{j=1}^{N} (\bar{g}_j^{\text{JK}})^n, \tag{5.100}$$

which will be useful in the following.

5.5.4.1 Jackknife Distortion Reduction

In the case of error propagation, in the presence of distortion, we can generically write a function computed on the mean value X as

$$g(X) = \langle g(\bar{x}) \rangle + O\left(\frac{1}{N} \right) = \langle g(\bar{x}) \rangle + \frac{c_1}{N} + \frac{c_2}{N^2} + O\left(\frac{1}{N^3} \right), \tag{5.101}$$

where, in the last expression, we have made explicit the first two orders of the distortion in $1/N$. Each jackknife subsample contains $N-1$ of the original N measurements. The distribution of the measured data is the same, so that, in principle, the distortion of the jackknife average will have the same form, with the same coefficients (but with one less data item). Furthermore, from the definition (5.99), the theoretical expected value of Eq. (5.98) will be equal to that of the total mean over all subsamples. Hence, we have

$$g(X) = \langle g(\bar{x}_j^{\text{JK}})\rangle + \frac{c_1}{N-1} + \frac{c_2}{(N-1)^2} + O\left(\frac{1}{(N-1)^3}\right)$$
$$= \langle \overline{g^{\text{JK}}}\rangle + \frac{c_1}{N-1} + \frac{c_2}{(N-1)^2} + O\left(\frac{1}{(N-1)^3}\right). \tag{5.102}$$

If we now multiply (5.101) by N, (5.102) by $N-1$, and subtract the latter from the former, we are able to cancel the contributions to the distortion of order $1/N$, obtaining

$$g(X) = N\langle g(\bar{x})\rangle - (N-1)\langle \overline{g^{\text{JK}}}\rangle + \frac{c_2}{N(N-1)} + O\left(\frac{1}{N^3}\right). \tag{5.103}$$

If we have few measurements and we want to reduce the bias, the jackknife method, through this formula, gives a very unbiased estimate of the function $g(X)$ computed in the expected value of x. One can go even further in reducing the bias by eliminating the order N^{-2} with what is called the second-order jackknife [49].

If N is large, however, we know that with respect to the $1/\sqrt{N}$ order of the statistical error, the dominant order of the bias, $1/N$, is also, in fact, negligible. In this case, both $g(\bar{x})$ and $\overline{g^{\text{JK}}}$ are good estimates of the mean value of g, as they differ from each other significantly less than their error bars.

5.5.4.2 Statistical Error Estimate with Jackknife

The jackknife resampling method also has the advantage of being able to compute a statistical error automatically, without having to do derivatives and propagate variances as in (5.8). To demonstrate this, we first define the squared error with the jackknife sampling:

$$e_{g^{\text{JK}}}^2 \equiv \overline{(g^{\text{JK}})^2} - \left(\overline{g^{\text{JK}}}\right)^2, \tag{5.104}$$

where we have used (5.100). To derive the relationship between this error and the error e_g^2 of the error propagation (see Section 5.1), we can expand the jackknife averages around $g(X)$:

$$g_j^{\text{JK}} = g(\bar{x}_j^{\text{JK}}) \simeq g(X) + \left.\frac{dg}{dx}\right|_{x=X}(\bar{x}_j^{\text{JK}} - X) = g(X) + g'\left[(\bar{x} - X) + \frac{\bar{x} - x_j}{N-1}\right],$$

where in the last step we shortened

$$g' \equiv \left.\frac{dg}{dx}\right|_{x=\bar{x}}$$

and used formula (5.97). With this expansion, we obtain the relations

$$\overline{g^{\mathrm{JK}}} \simeq g(X) + g'(\bar{x} - X),$$

$$\overline{(g^{\mathrm{JK}})^2} = \frac{1}{N} \sum_{j=1}^{N} [g(X) + g'(\overline{x_j}^{\mathrm{JK}} - X)]^2$$

$$= g^2(X) + 2g(X)g'(\bar{x} - X) + (g')^2 \frac{1}{N} \sum_{j=1}^{N} (\overline{x_j}^{\mathrm{JK}} - X)^2 .$$

The last term requires a bit of attention. If we use the relation (5.97) we obtain

$$\frac{1}{N} \sum_{j=1}^{N} (\overline{x_j}^{\mathrm{JK}} - X)^2 = \frac{1}{N} \sum_{j=1}^{N} \left(\bar{x} - X + \frac{\bar{x} - x_j}{N - 1} \right)^2$$

$$= (\bar{x} - X)^2 + \frac{1}{N(N - 1)^2} \sum_{j=1}^{N} (\bar{x} - x_j)^2$$

$$= (\bar{x} - X)^2 + \frac{\bar{\sigma}^2}{(N - 1)^2} , \tag{5.105}$$

where we have used the definition (5.4) of the empirical estimate of the variance. Putting the various pieces together, we can derive the expression of the jackknife error (5.104)

$$e_{g\,\mathrm{JK}}^2 = g^2(X) + 2g(X)g'(\bar{x} - X) + (g')^2 \left[(\bar{x} - X)^2 + \frac{\bar{\sigma}^2}{(N - 1)^2} \right] - (g(X) + g'(\bar{x} - X))^2$$

$$= \frac{1}{N - 1} (g')^2 \frac{\bar{\sigma}^2}{N - 1} . \tag{5.106}$$

Comparing with the error (5.8) of the propagation, we observe that the error with the jackknife is

$$e_{g\,\mathrm{JK}} = \frac{e_g}{\sqrt{N - 1}} \tag{5.107}$$

and the statistical error on g can then be estimated from the jackknife error as

$$e_g = \sqrt{N - 1}\, e_{g\,\mathrm{JK}} . \tag{5.108}$$

We note that here we multiply (not divide) by $\sqrt{N - 1}$. At first glance this may seem a little strange, but on thinking about it for a moment we realize that the fluctuations of the jackknife averages on the $N - 1$ subsamples around the overall empirical average are very small, even compared to the fluctuations between theoretical and standard empirical mean values.

For immediate application, we report the estimate (5.108) rewritten in terms of the averages over the j-knife samples:

$$e_g = \sqrt{\frac{N - 1}{N} \sum_{j=1}^{N} [g(x_j^{\mathrm{JK}}) - \overline{g^{\mathrm{JK}}}]^2}, \tag{5.109}$$

where we have used (5.99).

5.5.5 Bootstrap Resampling

The term *bootstrap* originally denotes a strap or webbing sewn behind boots to help fit them (it is still found today in well-known brands of boots). More abstractly, it indicates a tool that is simple, or at least not too sophisticated, but useful for steering things in the right direction. In short, we are using it in a sense not so different from that in which we used the term *jackknife*. In fact, the two methods follow the same philosophy of resampling the measured N data in some clever way.

In the case of the bootstrap method, N_B samples are created from the N measurements, each one composed by N data randomly chosen with uniform probability among the N measurements. This means that in a given bootstrap sample some measures will not appear, while others will appear repeated. This random choice plays the role of the bootstrap. Now let us see if the boot fits: can we automatically compute a reasonable statistical error?

Since we are reading a text on probability, we might ask ourselves what is the probability of a given measure x_k to appear repeated n_k times in a given bootstrap resampling. We start with the constraint that each of the N_B resamples contains N data:

$$\sum_{k=1}^{N} n_k = N \, . \tag{5.110}$$

If in each randomly composed sample (which we call a bootstrap sample) the event of having x_k has probability $p = 1/N$, the probability that, on N data, groups of n_k elements of value x_k are formed is binomial (see Section 2.2.1),

$$P(n_k) = \binom{N}{n_k} \left(\frac{1}{N}\right)^{n_k} \left(1 - \frac{1}{N}\right)^{N-n_k} , \tag{5.111}$$

of mean and variance

$$\langle n_k \rangle = 1 \, , \tag{5.112}$$

$$\langle n_k^2 \rangle - \langle n_k \rangle^2 = 1 - 1/N \, . \tag{5.113}$$

For N large the constraint (5.110) becomes less and less important, the variance approaches the mean, and the distribution (5.111) tends to a Poisson one. For N small, however, the constraint is important, and the number of times two different measures appear is correlated. The non-diagonal part of the covariance, indeed, is

$$\langle n_j n_k \rangle - \langle n_j \rangle \langle n_k \rangle = -\frac{1}{N} \, ,$$

as found by taking the mean of the square of Eq. (5.110) and using the expressions of $\langle n_k \rangle = 1$ and $\langle n_k^2 \rangle = 2 - 1/N$:

$$N^2 = \left\langle \sum_{j,k}^{1,N} n_j n_k \right\rangle = \sum_{j,k}^{1,N} \langle n_j n_k \rangle = \sum_{j \neq k}^{1,N} \langle n_j n_k \rangle + \sum_{k=1}^{N} \langle n_k^2 \rangle$$

$$= N(N-1)\langle n_j n_k \rangle + N\left(2 - \frac{1}{N}\right) = N(N-1)\langle n_j n_k \rangle + 2N - 1 , \tag{5.114}$$

whence

$$\langle n_j n_k \rangle = \frac{(N-1)^2}{N(N-1)} = 1 - \frac{1}{N}. \tag{5.115}$$

To fix the notation and show the procedure for computing the average and uncertainty, we start with the case where we use the bootstrap directly on the measurements (or on any linear function of the measurements). This is a case, in other words, in which traditional analysis is sufficient because there is no bias and error propagation is trivial. We define the average on a given bootstrap data set as

$$\bar{x}_b^{\text{BO}} \equiv \frac{1}{N} \sum_{k=1}^{N} n_k^{(b)} x_k, \quad b = 1, \dots, N_B. \tag{5.116}$$

Averaging over all N_B bootstrap samples we then compute the bootstrap average and bootstrap variance. For a sufficiently large number of samples N_B, in the derivation we can substitute for the mean on the bootstrap samples the theoretical values (5.112), (5.113), and (5.115). Consequently, for the bootstrap average and the variance we obtain

$$\overline{x^{\text{BO}}} = \frac{1}{N_B} \sum_{b=1}^{N_B} \bar{x}_b^{\text{BO}} = \frac{1}{N} \sum_{k=1}^{N} \frac{1}{N_B} \sum_{b=1}^{N_B} n_k^{(b)} x_k \xrightarrow[N_B \to \infty]{} \frac{1}{N} \sum_{k=1}^{N} \langle n_k \rangle x_k = \bar{x}, \tag{5.117}$$

$$\overline{(x^{\text{BO}})^2} = \frac{1}{N_B} \sum_{b=1}^{N_B} (\bar{x}_b^{\text{BO}})^2 = \frac{1}{N^2} \sum_{j,k}^{1,N} \frac{1}{N_B} \sum_{b=1}^{N_B} n_j^{(b)} n_k^{(b)} x_j x_k \xrightarrow[N_B \to \infty]{} \frac{1}{N^2} \sum_{j,k}^{1,N} \langle n_j n_k \rangle x_j x_k$$

$$= \frac{1}{N^2} \sum_{k=1}^{N} \langle n_k^2 \rangle x_k^2 + \frac{1}{N^2} \sum_{j \neq k}^{1,N} \langle n_j n_k \rangle x_j x_k = \left(2 - \frac{1}{N}\right) \frac{\overline{x^2}}{N} + \left(1 - \frac{1}{N}\right) \left[\bar{x}^2 - \frac{\overline{x^2}}{N}\right]$$

$$= \frac{\overline{x^2} - \bar{x}^2}{N} + \bar{x}^2 \tag{5.118}$$

$$e_{x^{\text{BO}}}^2 = \overline{(x^{\text{BO}})^2} - \left(\overline{x^{\text{BO}}}\right)^2 = \frac{\overline{x^2} - \bar{x}^2}{N} = \frac{\bar{\sigma}^2}{N}. \tag{5.119}$$

From (5.117) it is clear that the bootstrap average, tending to the empirical usual mean, is unbiased. Similarly, it is clear from (5.119) that the bootstrap variance, which tends to be proportional to the empirical variance, is a biased estimator, see (5.6), i.e.,

$$\langle e_{x^{\text{BO}}}^2 \rangle = \frac{\langle \bar{\sigma}^2 \rangle}{N} = \frac{N-1}{N^2} \sigma^2 = e_{\bar{x}}^2 \frac{N-1}{N}, \tag{5.120}$$

from which we have that the error of the mean can be estimated from the bootstrap error with the somewhat obvious formula

$$e_{\bar{x}} = \sqrt{\frac{N}{N-1}} \, e_{x^{\text{BO}}}. \tag{5.121}$$

The advantage of the bootstrap resampling method is for estimating the statistical error on a nonlinear function of the measured data. In analogy to the notation of the jackknife method, we first define the estimation of the function computed on the bootstrap average value of the measurable observables,

$$g_b^{\text{BO}} \equiv g(\bar{x}_b^{\text{BO}}). \tag{5.122}$$

With this we define the bootstrap moment as

$$\overline{(g^{\text{BO}})^n} \equiv \frac{1}{N_B} \sum_{b=1}^{N_B} (\bar{g}_b^{\text{BO}})^n. \tag{5.123}$$

5.5.5.1 Bootstrap Distortion Reduction

When the number of measurements is not very large, even a bias of order $1/N$ can lead to incorrect results. Similarly to what was done in the case of the jackknife in the steps (5.101)–(5.103), one can reduce the distortion on the estimate of the mean value of an observable g by an order of magnitude. One simply adopts the right combination of the function computed on the standard empirical average of the observable and the bootstrap-averaged function, which comes out to be

$$g(X) = 2g(\bar{x}) - \overline{g^{\text{BO}}} + O(1/N^2). \tag{5.124}$$

Obviously, for large N, the bias is much smaller than the uncertainty anyway, and one can estimate the mean with the bootstrap average $\overline{g^{\text{BO}}}$.

5.5.5.2 Bootstrap Uncertainty Estimation

Starting from the definitions (5.116), (5.122), and (5.123) and carrying out all the mathematical computations, along the lines of Section 5.5.4, one can compute the bootstrap error squared,

$$e_{g^{\text{BO}}}^2 = \overline{(g^{\text{BO}})^2} - \left(\overline{g^{\text{BO}}}\right)^2, \tag{5.125}$$

from which we obtain that, for large N, the propagated error (5.8) tends to the bootstrap error as

$$e_g = \sqrt{\frac{N}{N-1}}\, e_{g^{\text{BO}}}. \tag{5.126}$$

Proof Let us demonstrate this from the linear expansion

$$\bar{g}_b^{\text{BO}} = g(\bar{x}_b^{\text{BO}}) \simeq g(X) + g'(\overline{x_b}^{\text{BO}} - X).$$

Using (5.112), (5.113), and (5.115) we derive the averages of the first two moments on the bootstrap subsamples,

$$\overline{g^{\text{BO}}} = \frac{1}{N_B} \sum_{b=1}^{N_B} g(\bar{x}_b^{\text{BO}}) \simeq g(X) + \frac{g'}{N_B} \sum_{b=1}^{N_B} (\overline{x_b}^{\text{BO}} - X) = g(X) + g'(\bar{x} - X),$$

$$\overline{(g^{\mathrm{BO}})^2} = \frac{1}{N_B} \sum_{b=1}^{N_B} [g(X) + g'(\overline{x_b}^{\mathrm{BO}} - X)]^2$$

$$= g^2(X) + 2g(X)g'(\bar{x} - X) + (g')^2 \left(\frac{1}{N^2} \sum_{j,k} \langle n_j n_k \rangle x_j x_k - 2X\bar{x} + X^2 \right)$$

$$= g^2(X) + 2g(X)g'(\bar{x} - X) + (g')^2 \left(\frac{\bar{\sigma}^2}{N} + (\bar{x} - X)^2 \right),$$

where in the last line we used the definition (5.6) of the empirical variance. Putting the two previous expressions together, we obtain the relation between the squared error on the mean obtained by bootstrapping and that obtained by error propagation, see (5.8):

$$e_{g\,\mathrm{BO}}^2 = \overline{(g^{\mathrm{BO}})^2} - \left(\overline{g^{\mathrm{BO}}}\right)^2 = \frac{N-1}{N}(g')^2 \frac{\bar{\sigma}^2}{N-1} = \frac{N-1}{N} e_g^2, \tag{5.127}$$

and we are done. ∎

5.5.6 A Comparison of Resampling Methods

After analyzing the methods of binary binning, jackknife, and bootstrap resampling, the question arises as to which of these approaches is best to use. The answer depends on the type of data we have and on the statistical analysis we are interested in.

The binary resampling method, as we have seen, is very useful for detecting anomalies in the data. Using the binning method, we can deal with measurements of data that are correlated in time and determine both the correlation time and the correct estimate of the statistical uncertainty. The jackknife method, like the bootstrap method, is unable to take into account possible correlations between different measurements. These are not included in the estimation of the statistical error, which is therefore underestimated in their presence. The two methods are extremely useful, on the other hand, when it comes to estimating the errors of complicated variable functions without resorting to error propagation and reducing bias.

The bootstrap is more time-consuming to implement than the jackknife, because one has to extract all the data of the N_B samples at random each time. It is necessary to perform a fairly large number of series of extractions because there is also a need to test the goodness of the estimate of the mean (5.125) as N_B changes. If N_B is too small, in fact, the substitutions made in the estimation of the average (5.117) and in the estimation of the variance (5.119) are no longer justified and distortions may occur. Why use the bootstrap method then?

The bootstrap comes in handy when, in addition to estimating the error on the mean, we are interested in estimating the whole distribution of the data, over the entire domain of the original measurements. Indeed, the dispersion of the averages over the bootstrap samples is equal to the dispersion of the final estimates, as seen in (5.126), and the bootstrap averages are meaningful representations of the values generated by the (unknown) distribution of the observed quantities.

In contrast, for the jackknife, the deviation of the resampled averages is $1/\sqrt{N-1}$ of that of the original measurements and, although very good and fast for estimating the averages and their uncertainties, the jackknife method does not generate data distributed as the original ones.

Mathematical Appendices

5.A *A Posteriori* Distribution of Bernoulli Events

We have seen in Section 5.3.1, and analyzed in more detail in Section 5.3.3, that the best estimate of the *a priori* distribution of a parameter in Bayesian inference is that its logarithm is equiprobable: $A(\ln p) = \text{constant}$, whence $A(p) \propto 1/p$. The same applies to the probability of the complement of the occurrence of the event, $1 - p$, whence we have

$$A(p) \propto \frac{1}{p(1-p)}.$$

We see immediately, however, that this cannot be true exactly at $p=0$ or at $p=1$, because otherwise the probability density would diverge so fast as to violate the normalization condition. We can then modify the distribution in $p = 0$ and $p = 1$ [46] by assigning these two points a finite probability and putting a cut-off $\epsilon > 0$, even a very small one, in the density expression:

$$A(p) = \begin{cases} \pi_0 & \text{if } p = 0, \\ \kappa/[(p+\epsilon)(1-p+\epsilon)] & \text{if } p \in (0,1), \\ \pi_1 & \text{if } p = 1. \end{cases} \tag{5.128}$$

If we integrate over the whole domain of $p \in [0, 1]$ we obtain the normalization

$$1 = \int_0^1 dp\, A(p) = \pi_0+\pi_1+\int_0^1 dp\, \frac{\kappa}{(p+\epsilon)(1-p+\epsilon)} = \pi_0+\pi_1+\frac{2\kappa}{1+2\epsilon}\ln{(1+\epsilon^{-1})},$$

thanks to which we find the relation between κ and the probabilities π_0 and π_1 of (5.128), indicating, respectively, the *a priori* probability that the event is impossible ($p = 0$) and the *a priori* probability that the event is certain ($p = 1$):

$$\kappa = \frac{(1 - \pi_0 - \pi_1)(1 + 2\epsilon)}{2\ln(1 + \epsilon^{-1})}.$$

If one wants to be picky, one should notice that the correct *a priori* distribution is this. In any case, however, ϵ is very small and the probability that the event never occurs or always occurs is not to be taken into account. If we consider an event, it is because, however very rarely, at least once in the life of the universe, it may happen, there will be at least a 10^{-100} probability of it happening. On the other hand, if we think of rare events, as we are doing, as being far from the central limit ($K^2/N \ll 1$), the probability of N events occurring in the $K = N$ measurements is something that in the real world we can neglect altogether. Consequently, there is nothing wrong (statistically speaking) with taking

$$A(p) = \frac{\kappa}{p(1 - p)},$$

as long as we are not shocked if the normalization turns out to be a little divergent afterwards. We discussed this at length in Section 5.3.5.

Suppose we are measuring N times M simultaneous observables, or M degrees of freedom of a system, each represented by a random variable (for $M = 1$ we are back to the case of the previous chapter). This is a very common situation. Think of the simultaneous measurements of pressure, temperature, humidity, and wind speed that are made for a weather forecast. The kth of the N measurements will be represented by the vector $\vec{x}_k = \{x_k^{(\alpha)}\}$, where $\alpha = 1, \ldots, M$ denotes the αth experimentally acquired degree of freedom in the $k = 1, \ldots, N$ measurement. The methods introduced in the previous chapter can be applied without difficulty to this situation, except for certain algebraic expansions that become much more complex in the multivariate case.

6.1 Multivariate Gaussian Data

Let us begin by discussing the simplest case, in which we can assume that the data are Gaussian-distributed. The distribution will be Gaussian in shape and we will have to estimate the optimal values of the parameters characterizing it. In other words, we introduce the vector with M components $\vec{a} = \{a_\alpha\}$, representing the expected values of the M degrees of freedom, and the $M \times M$ matrix $\mathbb{B} = \{B_{\alpha\beta}\}$ which will be equal to the inverse of the covariance matrix. Let us assume, by analogy with the distribution (5.48), that the distribution of the values of the vector of variables $\vec{x}$ is

$$P(\vec{x}|\vec{a}, \mathbb{B}) = \frac{[\det(\mathbb{B})]^{1/2}}{(2\pi)^{M/2}} \exp\left(-\frac{1}{2}\sum_{\alpha,\beta}^{1,M}(x^{(\alpha)} - a_\alpha)B_{\alpha\beta}(x^{(\beta)} - a_\beta)\right). \qquad (6.1)$$

We aim to reconstruct the parameters of the probability distribution from the experimental data. If we wanted to proceed exactly as in the scalar case, where only one quantity was measured, we could try to write the *a posteriori* probability of these parameters as

$$P(\vec{a}, \mathbb{B}|\vec{\mathbf{x}}) \propto P(\vec{\mathbf{x}}|\vec{a}, \mathbb{B})A(\vec{a}, \mathbb{B})$$

$$\propto A(\vec{a}, \mathbb{B})[\det(\mathbb{B})]^{1/2} \prod_{k=1}^{N} \exp\left(-\frac{1}{2}\sum_{\alpha,\beta}(x_k^{(\alpha)} - a_\alpha)B_{\alpha\beta}(x_k^{(\beta)} - a_\beta)\right),$$

where $\vec{\mathbf{x}} \equiv \{\vec{x}_k\}$ is the set of N measurements of all M degrees of freedom and $A(\vec{a}, \mathbb{B})$ is the *a priori* distribution of the parameters. This approach would lead us to several non-trivial problems. First, we would have problems with the choice of the *a priori* probabilities for the variables $B_{\alpha\beta}$, forming an $M \times M$ matrix, and, then, it would not be easy to handle the marginalization problem in the elements $B_{\alpha\beta}$. Except for the case where the matrix $\mathbb{B}$ is diagonal and the degrees of freedom are uncorrelated, it is, indeed, extremely difficult to analytically (or even numerically) integrate the probability distribution of the elements $\mathbb{B}$ in order to obtain the *a posteriori* distribution of the only vector $\vec{a}$.

Instead, we will follow an alternative route here, proceeding simply, with the aim of highlighting some of the novel features that arise when measuring several degrees of freedom simultaneously. We first note that

$$\langle x^{(\alpha)} \rangle = a_\alpha \,, \tag{6.2}$$

$$\langle (x^{(\alpha)} - \langle x^{(\alpha)} \rangle)(x^{(\beta)} - \langle x^{(\beta)} \rangle) \rangle = \langle x^{(\alpha)} x^{(\beta)} \rangle - \langle x^{(\alpha)} \rangle \langle x^{(\beta)} \rangle = C_{\alpha\beta} \,, \tag{6.3}$$

where $C_{\alpha\beta}$ are the elements of the *covariance matrix* $\mathbb{C}$, which, as we have shown in Appendix 2.A, is the inverse matrix of the matrix $\mathbb{B}$:

$$C_{\alpha\beta} = (B^{-1})_{\alpha\beta} \,. \tag{6.4}$$

The theoretical parameters a_α and $C_{\alpha\beta}$, similarly to what we saw in the scalar case (5.5), can be estimated in an unbiased manner from the empirical average

$$\overline{x^{(\alpha)}} = \frac{1}{N} \sum_{k=1}^{N} x_k^{(\alpha)} \,, \tag{6.5}$$

and by the empirical covariance

$$\bar{C}_{\alpha\beta} = \frac{1}{N-1} \sum_{k=1}^{N} \left(x_k^{(\alpha)} - \overline{x^{(\alpha)}} \right) \left(x_k^{(\beta)} - \overline{x^{(\beta)}} \right) = \frac{N}{N-1} \overline{\left(x^{(\alpha)} - \overline{x^{(\alpha)}} \right) \left(x^{(\beta)} - \overline{x^{(\beta)}} \right)} \,. \tag{6.6}$$

We notice something almost irrelevant for large N but very important for small N, when the bias of the covariance estimate has a significant weight. The quantity

$$\overline{\left(x^{(\alpha)} - \overline{x^{(\alpha)}} \right) \left(x^{(\beta)} - \overline{x^{(\beta)}} \right)}$$

in (6.6) is the generalization to the case of several correlated variables of the empirical variance defined by (5.4). As we have seen in (5.6) for the variance, this multi-dimensional quantity also has a bias, and its mean value is not perfectly equal to the theoretical covariance, but is a fraction $(N-1)/N$ of it. The prefactor $N/(N-1)$ in the definition (6.6) of the empirical covariance is such that it exactly reproduces

$$\langle \bar{C}_{\alpha\beta} \rangle = C_{\alpha\beta},$$

thus eliminating the bias. It will be useful to recall this fact in Section 6.4, when we study resampling methods of vector data, such as jackknife and bootstrap, to reduce bias and to compute unbiased and reliable estimates of statistical error.

As we said, we can try to follow an extremely simplified procedure, based on a strong approximation. For sufficiently large N we can try to avoid all the complications associated with the fact that we do not know the parameters $B_{\alpha\beta}$, assuming that the value obtained from the estimate (6.6) for the covariance matrix is the correct one. We will then fix the matrix $\mathbb{B} \simeq \bar{\mathbb{C}}^{-1}$ using equations (6.4) and (6.6).

This extremely drastic simplification for small N is a harbinger of distortion. As we noted in 5.1, indeed, the function of an unbiased estimator of a variable is not an unbiased estimate of the same function. Thus, even if the empirical covariance, as we defined it, is an unbiased estimator, the elements of its inverse matrix (which are functions of the elements of $\mathbb{C}$) will give a biased estimate of the inverse matrix. On the other hand, the simplification becomes reasonable for large N and, assuming a uniform prior distribution of the expected values $\vec{a}$ (as we did in Section 5.3.6 for the scalar case), we can write the *a posteriori* probability distribution of the parameters $\vec{a}$ as

$$
P(\vec{a}|\vec{\mathbf{x}}) \propto \prod_{k=1}^{N} \exp\left(-\frac{1}{2}\sum_{\alpha,\beta}^{1,M}\left(x_k^{(\alpha)} - a_\alpha\right)\left(\bar{C}^{-1}\right)_{\alpha\beta}\left(x_k^{(\beta)} - a_\beta\right)\right)
$$

$$
= \mathcal{N}(\bar{\mathbb{C}})\exp\left(-\frac{N}{2}\sum_{\alpha,\beta}^{1,M}\left(\overline{x^{(\alpha)}} - a_\alpha\right)\left(\bar{C}^{-1}\right)_{\alpha\beta}\left(\overline{x^{(\beta)}} - a_\beta\right)\right), \qquad (6.7)
$$

where the proportionality factor $\mathcal{N}$ contains multiplicative factors that do not depend on $\vec{a}$, such as

$$
\mathcal{N}(\bar{\mathbb{C}}) \propto \exp\left(-\frac{N}{2}\sum_{\alpha,\beta}^{1,M}\left(\overline{x^{(\alpha)}x^{(\beta)}} - \overline{x^{(\alpha)}}\ \overline{x^{(\beta)}}\right)\left(\bar{C}^{-1}\right)_{\alpha\beta}\right)
$$

$$
= \exp\left(-\frac{N-1}{2}\sum_{\alpha,\beta}^{1,M}\bar{C}_{\alpha\beta}\left(\bar{C}^{-1}\right)_{\alpha\beta}\right) = e^{-(N-1)M/2}.
$$

In general, the covariance matrix $\mathbb{C}$ and its inverse $\mathbb{B}$ will not be diagonal, and the probability distribution (6.7) of the vector $\vec{a}$ cannot be written as a factorized function. In other words, both the experimental data and our estimates show correlations between the various components, labeled by the indices $\alpha = 1, \ldots, M$. The mean value estimated for a_α coincides precisely with the empirical averages of the components, $\langle a^{(\alpha)}\rangle = \overline{x^{(\alpha)}}$. To compute an error estimate, however, it will be useful to introduce the covariance matrix of the *a posteriori* distribution (6.7):

$$
G_{\alpha\beta} \equiv \langle\left(a^{(\alpha)} - \overline{x^{(\alpha)}}\right)\left(a^{(\beta)} - \overline{x^{(\beta)}}\right)\rangle = \frac{\bar{C}_{\alpha\beta}}{N}. \qquad (6.8)
$$

6.1.1 Error Propagation

Let us now try, given a generic observable function $g(\vec{a})$ of parameters a_α, to estimate its mean value on the *a posteriori* probability of $\vec{a}$,

$$\langle g \rangle \equiv \int \prod_{\alpha=1}^{M} da_\alpha \, P(\vec{a}|\vec{\mathbf{x}})g(\vec{a}), \tag{6.9}$$

and the associated error e_g. In general, since $P(\vec{a}|\vec{\mathbf{x}})$ for $N \gg 1$ is very peaked around $\vec{\bar{x}}$, given an arbitrary functional F of g and using Laplace's maximum method described in Appendix 2.C, we can write that

$$\langle F(g) \rangle \equiv \int \prod_{\alpha=1}^{M} da_\alpha \, P(\vec{a}|\vec{\mathbf{x}})F(g(\vec{a})) = F(g(\vec{\bar{x}})) + O(1/N), \tag{6.10}$$

where we have written explicitly the order N^{-1} of the corrections to the maximum of the integrand for finite N. In the particular case $F(g) = g$ we will then have that the *a posteriori* average value of the true (unknown) expected value can be approximated as

$$\langle g \rangle = g(\vec{\bar{x}}) + O(1/N). \tag{6.11}$$

This result is self-evident.

The somewhat more delicate problem, however, is the computation of the statistical error associated with the expected value of g. We note preliminarily that, since the quantities $a_\alpha - \overline{x^{(\alpha)}}$ are small, i.e., according to (6.8) are of order $O(1/\sqrt{N})$, we can expand around $\vec{\bar{x}}$ and write

$$g(\vec{a}) = g(\vec{\bar{x}}) + \sum_{\alpha=1}^{M} \frac{\partial g}{\partial a_\alpha}\bigg|_{\vec{a}=\vec{\bar{x}}} \left(a_\alpha - \overline{x^{(\alpha)}}\right)$$

$$+ \sum_{\alpha,\beta}^{1,M} \frac{\partial^2 g}{\partial a_\alpha \partial a_\beta}\bigg|_{\vec{a}=\vec{\bar{x}}} \left(a_\alpha - \overline{x^{(\alpha)}}\right)\left(a_\beta - \overline{x^{(\beta)}}\right) + O\left(\frac{1}{N^{3/2}}\right)$$

$$= g(\vec{\bar{x}}) + \sum_{\alpha=1}^{M} g'_\alpha \left(a_\alpha - \overline{x^{(\alpha)}}\right) + O\left(\frac{1}{N}\right), \tag{6.12}$$

where we have defined

$$g'_\alpha \equiv \frac{\partial g}{\partial a_\alpha}\bigg|_{\vec{x}=\vec{\bar{x}}} \tag{6.13}$$

and we have neglected higher-order terms. In this approximation, $g(\vec{a})$ is the linear combination of M Gaussian variables $a_\alpha - x^{(\alpha)}$ and is therefore Gaussian-distributed. The error associated with $g(\vec{a})$ with respect to $\langle g \rangle$ can be estimated by estimating its variance e_g^2. By averaging over the *a posteriori* distribution of the parameters $\vec{a}$, the square fluctuations of $g(\vec{a})$, using Eqs. (6.8), (6.11), and (6.12), we obtain

$$e_g^2 = \int \left(\prod_{\alpha=1}^{M} da_\alpha\right) P(\vec{a}|\vec{\mathbf{x}})(g(\vec{a}) - \langle g \rangle)^2$$

$$\simeq \sum_{\alpha,\beta}^{1,M} g'_\alpha g'_\beta \int \left(\prod_{\alpha=1}^{M} da_\alpha\right) P(\vec{a}|\vec{\mathbf{x}})\left(a_\alpha - \overline{x^{(\alpha)}}\right)\left(a_\beta - \overline{x^{(\beta)}}\right) = \sum_{\alpha,\beta}^{1,M} g'_\alpha G_{\alpha\beta} g'_\beta. \tag{6.14}$$

We have thus obtained the *error propagation formula for correlated variables*. Using it operationally may be impractical, especially if $g(\bar{x})$ is not a simple function, and its derivatives cannot be computed easily. The method is not foolproof, as the derivatives must be computed each time for each observable in the problem. Moreover, there is always the danger that, for a large number M of components, the $M(M-1)/2$ terms proportional to $\partial^2 g/\partial x_\alpha \partial x_\beta$, neglected in (6.12) because of order N^{-1}, will eventually yield a contribution that cannot be ignored.

6.2 Subsampling

Let us now look at a perhaps extreme but sometimes realistic case of the difficulties of estimating the inverse covariance matrix. Let us take, for simplicity, but without loss of generality, a set of M variables with null mean $\vec{x} \equiv \{x^{(\alpha)}\}, \alpha = 1, \ldots, M$, correlated with each other. Let us take N measurements of this set. Let $\bar{\mathbb{C}}$ be the $M \times M$ experimental covariance matrix whose elements

$$\bar{C}_{\alpha\beta} = \frac{1}{N} \sum_{k=1}^{N} x_k^{(\alpha)} x_k^{(\beta)} \tag{6.15}$$

are constructed by averaging the correlations between the $x^{(\alpha)}$ variable and the $x^{(\beta)}$ variable over the N measurements. By defining $M \times M$ matrices of rank one,

$$\bar{\mathbb{C}}_k \equiv \vec{x}_k \vec{x}_k^T , \tag{6.16}$$

where $\vec{x}$ are column vectors and $\vec{x}^T$ are row vectors, the covariance matrix can be rewritten as

$$\bar{\mathbb{C}} = \frac{1}{N} \sum_{k=1}^{N} \bar{\mathbb{C}}_k . \tag{6.17}$$

Each $\bar{\mathbb{C}}_k$ matrix has a unique non-zero eigenvalue. If all measures are independent, then, by construction, there will be at most N non-zero eigenvalues for the matrix $\bar{\mathbb{C}}$. Consequently, the rank of $\bar{\mathbb{C}}$ is the minimum between N and M. In the case of a limited number of measurements, less than the number of variables, $N < M$, this implies that there will be at least $M - N$ null eigenvalues and the covariance matrix $\bar{\mathbb{C}}$ is not invertible.

In such cases of subsampling, progress can be made by making *a priori* assumptions about the structure of the covariance matrix, but this is a level of complication we do not even want to think about.

6.3 Multivariate Reweighting Method

We are now also interested in applying the general method we introduced in Section 5.4 to the vector case, where it is not necessary to assume *a priori* a certain functional form of the probability distribution of the data, e.g., Gaussian. This approach also overcomes some of the algebraic difficulties present in the method described in Section 6.1.

As in Section 6.1.1, we set out again to compute the expectation value of a function $g(\vec{X})$, where $\vec{X}$ is an M-dimensional vector, the mean value of a vector of observables $\vec{x}$ with respect to an unknown $D(\vec{x})$ probability law:

$$\vec{X} = \int d\vec{x}\, D(\vec{x})\vec{x}. \tag{6.18}$$

Our only knowledge of the probability $D(\vec{x})$ is given by the results obtained in N measurements of the vectors $\vec{x}_k$, $k = 1, \ldots, N$, the components of which will be denoted x_k^α, with $\alpha = 1, \ldots, M$.

In the relatively simple case where we know that the probability distribution $D(\vec{x})$ can be parameterized as $P(\vec{x}|\mathbf{H})$, where $\mathbf{H}$ is a set of control parameters, we can write that the *a posteriori* probability of $\mathbf{H}$, conditional on the experimental data, assuming no time correlation between successive measurements, is given by

$$P(\mathbf{H}|\vec{x}) \propto A(\mathbf{H}) \prod_{k=1}^{N} P(\vec{x}_k|\mathbf{H}), \tag{6.19}$$

where $A(\mathbf{H})$ is the *a priori* distribution of the parameters $\mathbf{H}$. For each value of the set $\mathbf{H}$, we can compute the corresponding value of

$$\vec{X}[\mathbf{H}] = \int d\vec{x}\, P(\vec{x}|\mathbf{H})\vec{x}$$

and thus trace the probability distribution of $\vec{X}$. The Gaussian case we considered above is a special case of this procedure.

Here, instead, we want to consider the case in which we do not have this information. We do not know the form $D(\vec{x})$ and, hence, we are forced to use the reweighting method that we have already described in the case of the experimental measurement of a single quantity. We will denote by p_k the probability that the components of the vector of degrees of freedom $\vec{x}$ acquires values $x_k^{(1)}, \ldots, x_k^{(M)}$. The weight of each contribution will be represented by an M-dimensional Dirac delta distribution

$$D(\vec{x}) = \sum_{k=1}^{N} p_k \delta^{(M)}(\vec{x} - \vec{x}_k) = \sum_{k=1}^{N} p_k \prod_{\alpha=1}^{M} \delta(x^{(\alpha)} - x_k^{(\alpha)}). \tag{6.20}$$

We can also assign the weight p_k to the (uniform) probability that the value of the variable falls in a hypercubic cell of linear dimension Δ,

$$x^{(\alpha)} \in [x_k^{(\alpha)}, x_k^{(\alpha)} + \Delta].$$

When $\Delta \to 0$, the limit of the probability in the cell leads to the Dirac delta of Eq. (6.20).

The sequence $\{p\}$ thus constructed from the measured data is a normalized M-dimensional histogram and corresponds to the multivariate version of the *a posteriori* distribution (5.62), which we have written as (5.65) and (5.67) in the one-degree-of-freedom scalar case.

In analogy to the scalar case, we note that in the vector case an estimate of the probability distribution of each component of $\vec{X}$, conditional on the experimental data, can be obtained by writing for each degree of freedom α,

$$X^{(\alpha)}(\mathbf{p}) = \sum_{k=1}^{N} p_k x_k^{(\alpha)}, \quad \alpha = 1, \ldots, M, \tag{6.21}$$

and varying p_k appropriately, according to the *a posteriori* probability distribution $P(\mathbf{p}|\vec{\mathbf{x}})$ introduced in Section 5.4.1.

Let us now consider the generic function $g(\vec{X})$. We can write the squared error associated with the expected value as

$$e_g^2 = \langle g(\vec{X})^2 \rangle - \langle g(\vec{X}) \rangle^2, \tag{6.22}$$

where the first two moments on the posterior distribution are

$$\langle g(\vec{X})^n \rangle = \int \prod_{k=1}^{N} dp_k \, P(\mathbf{p}|\vec{\mathbf{x}}) \big[g(\vec{X}(\mathbf{p})) \big]^n, \quad n = 1, 2.$$

For large values of N, if the kurtosis of the true data distribution is not excessively large, the variables $\vec{X}$ will be Gaussian-distributed. In this case, the same considerations as in the previous section show that

$$\langle g(\vec{X}) \rangle = g(\langle \vec{X} \rangle) + O(1/N)$$

and that e_g is the Gaussian error associated with the quantity g.

In the general case that we are analyzing here, however, where the probability distribution is not assumed to be Gaussian, the histogram analysis of the values of g,

$$\mathcal{P}(g|\vec{\mathbf{x}}) = \int \prod_{k=1}^{N} dp_k \, P(\mathbf{p}|\vec{\mathbf{x}}) \delta(g - g(\vec{X}(\mathbf{p}))), \tag{6.23}$$

provides an estimate of the probability of g that is faithful even in the tails of the distribution. The reweighting method has the advantage of being easy to use even in the case of a large number M of variables and very complicated g functions.

We would like to stress that the approximate equality

$$\langle g(\vec{X}) \rangle \simeq g(\langle \vec{X} \rangle) \tag{6.24}$$

is only true under the assumption that the number N of measures is very large and that the errors are very small, as may be the case, for example, for the error on the mean of a quantity with finite variance distribution.

We would like to point out, however, that the formula (6.24) is not correct in general. In the one-dimensional case ($M = 1$), for example, if the function g is smooth and the variance is finite, we can write

$$\langle g(x) \rangle \simeq g(\langle x \rangle) + \frac{1}{2} \frac{d^2 g}{dx^2}\bigg|_{x=\langle x \rangle} (e_x)^2, \tag{6.25}$$

so that the mean of $g(x)$ is not equal to g computed for the average x unless e_x is very small and negligible compared to $\langle g(x) \rangle - g(\langle x \rangle)$ or $g(x)$ is linear. In general, one must be careful not to confuse the mean of the function $g(x)$ with the function evaluated on the mean, $g(\langle x \rangle)$. Indeed, in general, it holds that

$$\langle g(x) \rangle \neq g(\langle x \rangle).$$

This applies to both expected theoretical values and to experimental averages,

$$\bar{g}(x) \neq g(\bar{x}).$$

Let us consider as a simple example the case where x has only one component that can take only one of the two values ± 1, with equal probability $1/2$, and consider the function $g(x) = x^2$. In this case $g(\langle x \rangle) = \langle x \rangle^2 = 0$, while $\langle g(x) \rangle = \langle x^2 \rangle = 1$. If we consider empirical averages over N experimental data, we obtain that $|\bar{x}| = O(N^{-1/2})$ and thus $g(\bar{x}) = O(N^{-1})$, while we have $\overline{g(x)} = \overline{x^2} = 1$, exactly, for each value of N.

6.4 Multivariate Resampling Methods

In addition to the reweighting method discussed in Section 6.3, we also want to consider other methods to analyze experimental data. In the case of scalar data in Section 5.5, we analyzed several methods of resampling that allow one to cure biases in the estimates of the observables, to compute the errors in the presence of time correlation of the data, or to recognize anomalous data. Computing statistical uncertainty with the error propagation formula, indeed, as shown in Section 6.1.1, is a procedure whose effectiveness is greatly influenced by the type of function, by the kind and quantity of observed degrees of freedom, and by the biases in the estimation of their expected values. Of the methods studied in Section 5.5, the jackknife and bootstrap methods are immediately extendable from the scalar to the vector case, and we will, therefore, review them in the following.

6.4.1 Jackknife Resampling for Correlated Variables

In the multivariate case we consider a generic $g(\vec{x})$ function of M variables $\vec{x} = \{x^{(1)}, x^{(2)}, \ldots, x^{(M)}\}$ measured N times. The jackknife average of the g function of all $\vec{x}$ variables over the jth subsample is the generalization of (5.98), that is,

$$\bar{g}_j^{\text{JK}} \equiv g(\vec{x}_j^{\text{JK}}), \tag{6.26}$$

in terms of which the expression for the average of the N jackknife averages, $\overline{g^{\text{JK}}}$, remains formula (5.99). Only if g is a linear combination of $\vec{x}$ does the average with the jackknife method match the standard empirical one. Otherwise, due to the distortion induced by nonlinearity, the two averages differ.

In the case of error propagation, by explicitly considering the first two orders of the distortion from the formula (6.12), we obtain

$$g(\vec{a}) = \langle g(\vec{x}) \rangle + O\left(\frac{1}{N}\right) = \langle g(\vec{x}) \rangle + \frac{c_1}{N} + \frac{c_2}{N^2} + O\left(\frac{1}{N^3}\right). \tag{6.27}$$

The distortion of the jackknife average will have the same form, with the same coefficients, only with one less datum. Furthermore, as in the scalar case, see (5.99), the theoretical expected value of (6.26) is equal to that of the total mean overall subsamples. Hence we obtain the expression

$$g(\vec{a}) = \langle g(\vec{x}_j^{\text{JK}}) \rangle + \frac{c_1}{N-1} + \frac{c_2}{(N-1)^2} + O\left(\frac{1}{(N-1)^3}\right)$$

$$= \langle \overline{g^{\text{JK}}} \rangle + \frac{c_1}{N-1} + \frac{c_2}{(N-1)^2} + O\left(\frac{1}{(N-1)^3}\right), \tag{6.28}$$

which, when multiplied by $N-1$ and subtracted from N times Eq. (6.27), yields the formula

$$g(\vec{a}) = N\langle g(\vec{x}) \rangle - (N-1)\langle \overline{g^{\text{JK}}} \rangle + \frac{c_2}{N(N-1)} + O\left(\frac{1}{N^3}\right), \tag{6.29}$$

in which the contributions of order $1/N$ to the bias vanish.

Jackknife resampling is, moreover, an automatic and fast method to compute a statistical error, consistent with (6.14), without having to propagate covariances. We have demonstrated this in the scalar case in Section 5.5.4. The only things that change in the multivariate case are the jackknife average expressions for estimating the error on the mean (5.104), which now have to take into account the correlation between different degrees of freedom in each individual measurement.

To derive the relationship between the jackknife error for correlated data and the error e_g^2 obtained in (6.14) with error propagation, we consider the expansion around $g(\vec{a})$ of the jackknife averages. In the multivariate case, it will be

$$g_j^{\text{JK}} = g(\vec{x}_j^{\text{JK}}) \simeq g(\vec{a}) + \sum_{\alpha=1}^{M} g_\alpha'(\vec{a})(\overline{x_j^\alpha}^{\text{JK}} - a_\alpha)$$

$$= g(\vec{a}) + \sum_{\alpha=1}^{M} g_\alpha'(\vec{a})\left[(\bar{x}^\alpha - a_\alpha) + \frac{\bar{x}^\alpha - x_j^\alpha}{N-1}\right],$$

where in the last step we used (5.97). From here we obtain the relations

$$\overline{g^{\text{JK}}} \simeq g(\vec{a}) + \sum_{\alpha=1}^{M} g_\alpha'(\vec{a})(\bar{x}^\alpha - a_\alpha)$$

and

$$\overline{(g^{\mathrm{JK}})^2} = \frac{1}{N}\sum_{j=1}^{N}\left[g(\vec{a}) + \sum_{\alpha=1}^{M} g'_\alpha(\vec{a})\left(\overline{x_j^\alpha}^{\mathrm{JK}} - a_\alpha\right)\right]^2$$

$$= g^2(\vec{a}) + 2g(\vec{a})\sum_{\alpha=1}^{M} g'_\alpha(\vec{a})(\overline{x^\alpha} - a_\alpha)$$

$$+ \sum_{\alpha,\gamma}^{1,M} g'_\alpha(\vec{a})g'_\gamma(\vec{a})\frac{1}{N}\sum_{j=1}^{N}\left(\overline{x_j^\alpha}^{\mathrm{JK}} - a_\alpha\right)\left(\overline{x_j^\gamma}^{\mathrm{JK}} - a_\gamma\right).$$

Using the formula (5.97) for each degree of freedom x^α, the last term of the equation can be written in terms of the empirical covariance (6.6):

$$\frac{1}{N}\sum_{j=1}^{N}\left(\overline{x_j^\alpha}^{\mathrm{JK}} - a_\alpha\right)\left(\overline{x_j^\gamma}^{\mathrm{JK}} - a_\gamma\right)$$

$$= \frac{1}{N}\sum_{j=1}^{N}\left(\overline{x^\alpha} - a_\alpha + \frac{\overline{x^\alpha} - x_j^\alpha}{N-1}\right)\left(\overline{x^\gamma} - a_\gamma + \frac{\overline{x^\gamma} - x_j^\gamma}{N-1}\right)$$

$$= (\overline{x^\alpha} - a_\alpha)(\overline{x^\gamma} - a_\gamma) + \frac{1}{N(N-1)^2}\sum_{j=1}^{N}(\overline{x^\alpha} - x_j^\alpha)(\overline{x^\gamma} - x_j^\gamma)$$

$$= (\overline{x^\alpha} - a_\alpha)(\overline{x^\gamma} - a_\gamma) + \frac{\bar{C}_{\alpha\gamma}}{N(N-1)}. \tag{6.30}$$

Putting the various pieces together, we can derive the expression of the jackknife error, see (5.104):

$$e_{g\mathrm{JK}}^2 = g^2(\vec{a}) + 2g(\vec{a})\sum_{\alpha=1}^{M} g'_\alpha(\vec{a})(\overline{x^\alpha} - a_\alpha)$$

$$+ \sum_{\alpha,\gamma}^{1,M} g'_\alpha(\vec{a})g'_\gamma(\vec{a})\left[(\overline{x^\alpha} - a_\alpha)(\overline{x^\gamma} - a_\gamma) + \frac{\bar{C}_{\alpha\gamma}}{N(N-1)}\right]$$

$$- \left(g(\vec{a}) + \sum_{\alpha=1}^{M} g'_\alpha(\vec{a})(\overline{x^\alpha} - a_\alpha)\right)^2$$

$$= \frac{1}{N-1}\sum_{\alpha,\gamma}^{1,M} g'_\alpha(\vec{a})\frac{\bar{C}_{\alpha\gamma}}{N}g'_\gamma(\vec{a}). \tag{6.31}$$

Comparing with the error (6.14) derived with the propagation method, we observe that even in the vector case the statistical error on g can be estimated from the jackknife error as

$$e_g = \sqrt{N-1}\,e_{g\mathrm{JK}}.$$

We have shown that the jackknife error, computed by means of formula (6.31), contains the possible complete covariance matrix. Thus, if the measured observables are correlated with each other and the covariance matrix is not diagonal, the contribution of the correlations automatically enters the error estimate (5.104) using N subsamples composed of $N-1$ data each, by means of the formulas (5.98), (5.99), and (5.100). The method is not able to take into account, however, possible time correlations between different data intervals of the measurement series, unlike, for example, the binning resampling method, described in Section 5.5.2.

6.4.2 Bootstrap Resampling for Correlated Variables

Also for the bootstrap resampling method the estimate of the statistical error, as well as the bias reduction, are derived as in the scalar case, analyzed in Section 5.5.5. The only things that change are that the function g whose mean is to be estimated is a multivariate function and that the performed expansions must take into account the dependence on several variables and their correlations.

The estimate of the function computed on the empirical average value, with which one computes the bootstrap moments $\overline{(g^{\mathrm{BO}})^n}$, see (5.123), will be

$$g_b^{\mathrm{BO}} \equiv g(\vec{x}_b^{\mathrm{BO}}). \tag{6.32}$$

For the reduction of distortion to the order $1/N^2$, the combination of the function g computed on the empirical average and of the bootstrap-averaged g turns out to be

$$g(\vec{a}) = 2g(\vec{x}) - \overline{g^{\mathrm{BO}}} + O(1/N^2), \tag{6.33}$$

which, for large N, when the bias is much smaller than the uncertainty and $g(\vec{x}) \simeq \overline{g^{\mathrm{BO}}}$, reduces to the bootstrap average $\overline{g^{\mathrm{BO}}}$.

Starting from the definitions (5.116), (5.122), and (5.123) and carrying out all the mathematical computations, along the lines of Section 5.5.4, the bootstrap error formula (5.125) is obtained:

$$e^2_{g^{\mathrm{BO}}} = \overline{(g^{\mathrm{BO}})^2} - \left(\overline{g^{\mathrm{BO}}}\right)^2.$$

For the derivation in the multivariate case, we begin with the expansion

$$\bar{g}_b^{\mathrm{BO}} = g(\vec{x}_b^{\mathrm{BO}}) \simeq g(\vec{a}) + \sum_{\alpha=1}^{M} g'_\alpha(\vec{a}) \left(\overline{x_b^{(\alpha)}}^{\mathrm{BO}} - a_\alpha\right) \tag{6.34}$$

and, using formulas (5.112), (5.113), and (5.115), we compute the first two moments on the bootstrap subsamples:

$$\overline{g^{\mathrm{BO}}} = \frac{1}{N_B} \sum_{b=1}^{N_B} g(\vec{x}_b^{\mathrm{BO}}) \simeq g(\vec{a}) + \frac{1}{N_B} \sum_{b=1}^{N_B} \sum_{\alpha=1}^{M} g'_\alpha(\vec{a}) \left(\overline{x_b^{(\alpha)}}^{\mathrm{BO}} - a_\alpha\right)$$

$$= g(\vec{a}) + \sum_{\alpha=1}^{M} g'_\alpha(\vec{a}) \left(\overline{x^{(\alpha)}} - a_\alpha\right)$$

and

$$
\overline{\left(g^{\mathrm{BO}}\right)^2} = \frac{1}{N_B} \sum_{b=1}^{N_B} \left[g(\vec{a}) + \sum_{\alpha=1}^{M} g'_\alpha(\vec{a}) \left(\overline{x_b^{(\alpha)}}^{\mathrm{BO}} - a_\alpha \right) \right]^2
$$

$$
= g^2(\vec{a}) + 2g(\vec{a}) \sum_{\alpha=1}^{M} g'_\alpha(\vec{a}) \left(\overline{x^{(\alpha)}} - a_\alpha \right)
$$

$$
+ \sum_{\alpha,\gamma}^{1,M} g'_\alpha(\vec{a}) g'_\gamma(\vec{a}) \left(\frac{1}{N^2} \sum_{j,k} \langle n_j n_k \rangle x_j^{(\alpha)} x_k^{(\gamma)} - a_\alpha \overline{x^{(\gamma)}} - a_\gamma \overline{x^{(\alpha)}} + a_\alpha a_\gamma \right)
$$

$$
= g^2(\vec{a}) + 2g(\vec{a}) \sum_{\alpha=1}^{M} g'_\alpha(\vec{a}) \left(\overline{x^{(\alpha)}} - a_\alpha \right)
$$

$$
+ \sum_{\alpha,\gamma}^{1,M} g'_\alpha(\vec{a}) g'_\gamma(\vec{a}) \left(\frac{\overline{x^{(\alpha)} x^{(\gamma)}} - \overline{x^{(\alpha)}}\ \overline{x^{(\gamma)}}}{N} + \left(\overline{x^{(\alpha)}} - a_\alpha \right) \left(\overline{x^{(\gamma)}} - a_\gamma \right) \right)
$$

$$
= g^2(\vec{a}) + 2g(\vec{a}) \sum_{\alpha=1}^{M} g'_\alpha(\vec{a}) \left(\overline{x^{(\alpha)}} - a_\alpha \right)
$$

$$
+ \sum_{\alpha,\gamma}^{1,M} g'_\alpha(\vec{a}) g'_\gamma(\vec{a}) \left(\frac{N-1}{N^2} \bar{C}_{\alpha\gamma} + \left(\overline{x^{(\alpha)}} - a_\alpha \right) \left(\overline{x^{(\gamma)}} - a_\gamma \right) \right).
$$

Here, in the last line, we used the definition (6.6) of the empirical covariance. Putting the two previous expressions together, we obtain the relation between the square error on the mean obtained by bootstrap and by error propagation, see (6.14):

$$
e_{g^{\mathrm{BO}}}^2 = \overline{(g^{\mathrm{BO}})^2} - \left(\overline{g^{\mathrm{BO}}} \right)^2 = \frac{N-1}{N} \sum_{\alpha,\gamma}^{1,M} g'_\alpha(\vec{a}) \frac{\bar{C}_{\alpha\gamma}}{N} g'_\gamma(\vec{a}) = \frac{N-1}{N} e_g^2. \tag{6.35}
$$

From here, for large N, we see that the propagated error (6.14) tends to the bootstrap error exactly as in the scalar case (5.126):

$$
e_g = \sqrt{N/(N-1)}\, e_{g^{\mathrm{BO}}}.
$$

As with the jackknife, also with the bootstrap method, the expression (6.35) shows that the complete covariance of several observables measured at the same time is automatically included in the resampling procedure, provided that the number of bootstrap samples N_B is large enough.

6.5 Least-Squares Method

6.5.1 The Case of Uncorrelated Data

Let us consider a measurement process involving the determination of M variables. Fortunately, it is often the case that theoretical arguments allow us to fix the form of the dependence of the mean values of variables on certain parameters that determine the probability distribution. Hence, let us consider the quantities $x^{(\alpha)}$ and their expected values

$$a_\alpha \equiv \int \left(\prod_{\alpha=1}^{M} dx^{(\alpha)} \right) D(\vec{x}) x^{(\alpha)} . \tag{6.36}$$

We will assume in all generality that the theoretical dependence of the mean values on a set of L parameters $\{\lambda\}$ has the form

$$a_\alpha = f_\alpha(\{\lambda\}), \tag{6.37}$$

where the $\{\lambda\}$ have to be estimated. We denote by Λ the true value of the single parameter λ, which we obviously do not know (but whose existence we assume), and by λ_m and σ_λ the best estimate – *best fit* – of λ and its statistical error. In other words $\{\lambda\}$ is the set of free parameters that characterize the procedure of *best fit*.

Let us treat, for example, the case in which the distribution of the experimental data is Gaussian and assume for the moment that the variables labeled by the index $\alpha = 1, \ldots, M$ are uncorrelated. In (6.7) we thus set

$$N B_{\alpha\beta} = \delta_{\alpha\beta} \, \frac{1}{e_\alpha^2} , \tag{6.38}$$

where $\delta_{\alpha\beta}$ is the Kronecker delta, equal to 1 if $\alpha = \beta$ and equal to 0 if $\alpha \neq \beta$. The error associated with the mean value a_α is denoted by e_α, whose square is the variance of the theoretical mean value a_α of the degree of freedom α:

$$e_\alpha^2 \equiv \langle (a_\alpha - \langle a_\alpha \rangle)^2 \rangle = \left\langle \left(a_\alpha - \overline{x^{(\alpha)}} \right)^2 \right\rangle. \tag{6.39}$$

It is equal to the variance of the empirical average, the diagonal part of Eq. (6.8):

$$e_\alpha^2 = \frac{\bar{C}_{\alpha\alpha}}{N} = \frac{1}{N-1} \overline{\left(x^{(\alpha)} - \overline{x^{(\alpha)}} \right)^2} = \frac{1}{N(N-1)} \sum_{i=1}^{N} (x_i^\alpha - \overline{x^\alpha})^2 . \tag{6.40}$$

Please note that with the bars above we are indicating averages over the experimental data, see (6.5) and (6.6).

In the case of a large number of measurements N of uncorrelated vector data sets, with a finite variance distribution, the *a posteriori* probability density of the mean values a_α, conditional on the experimental data, will be given by

$$P(\vec{a}|\vec{\mathbf{x}}) \propto \exp \left(-\frac{1}{2} \sum_{\alpha=1}^{M} \frac{\left(a_\alpha - \overline{x^{(\alpha)}} \right)^2}{e_\alpha^2} \right) , \tag{6.41}$$

and the probability density of our parameters λ will then be given by

$$P(\lambda|\vec{\mathbf{x}}) \propto \exp \left[-\tfrac{1}{2} E(\{\lambda\}, \{\vec{\mathbf{x}}\}) \right] , \tag{6.42}$$

where we have defined the squares function

$$E(\{\lambda\}, \{\vec{\mathbf{x}}\}) \equiv \sum_{\alpha=1}^{M} \frac{\left[f_\alpha(\{\lambda\}) - \overline{x^{(\alpha)}}\right]^2}{e_\alpha^2}. \tag{6.43}$$

We now ask ourselves two questions.

1. What is the best fit for the parameters $\{\lambda\}$ and what is its statistical error?
2. How consistent is the hypothesis $\vec{a} = \vec{f}(\{\lambda\})$ with the observed phenomenon?

Let us see the answer to the first question. If the errors are small, $e_\alpha^2 \sim 1/N$, see (6.40), the function $E(\{\lambda\}, \{\vec{\mathbf{x}}\})$ grows, in absolute value, with the number of measurements N, and we can use Laplace's maximum method to compute any integral over the probability density (6.42), in particular, the moments of the distribution. In this case the large parameter of order N is played by $1/e_\alpha^2$.

To derive the best fit of λ we will, therefore, look for the minimum of the sum of squares $E(\{\lambda\}, \{\vec{\mathbf{x}}\})$ as a function of the parameters $\{\lambda\}$. Before we begin the computation, let us give a couple of definitions.

- Let us denote by $\{\lambda_m\}$ the set of parameter values $\{\lambda\}$ that minimize E to $\{\vec{\mathbf{x}}\}$, at fixed experimental data. These are the best-fit values.
- Let us call χ^2 the minimum of the function E, computed on the best-fit values $\{\lambda_m\}$:

$$\chi^2 = E(\{\lambda_m\}, \{\vec{\mathbf{x}}\}).$$

A parameter determination that allows for a good reconstruction of the experimental data and validates the theoretical hypothesis typically has a low χ^2 value. A χ^2 that is too low, however, signals that there may be problems in processing the data. We will make these statements more quantitative in a moment.

Let us consider for simplicity the case of a single parameter λ. Let us first expand E around the minimum λ_m, in powers of $(\lambda - \lambda_m)$. In this case, if our theoretical assumption (6.37) is not unreasonable, the quantities $f_\alpha(\lambda) - \overline{x^{(\alpha)}}$ are small:

$$a_\alpha - \overline{x^{(\alpha)}} = f_\alpha(\lambda) - \overline{x^{(\alpha)}} = O(1/\sqrt{N}). \tag{6.44}$$

We recall that, by construction, the best fit λ_m that minimizes E is the value around which $P(\lambda|\{\vec{x}\})$ is peaked. For large N the *a posteriori* distribution will be highly peaked. The mean value over this *a posteriori* distribution is the empirical average:

$$\langle f_\alpha(\lambda)\rangle = \langle a_\alpha\rangle = \int d\vec{a}\, a_\alpha P(\vec{a}|\vec{\mathbf{x}}) = \overline{x^{(\alpha)}}, \tag{6.45}$$

and, because of the behavior $e_\alpha \sim 1/\sqrt{N}$, one has $f_\alpha(\lambda_m) = \overline{x^{(\alpha)}} + O(1/N)$. Inasmuch as the difference between a function of the mean and the mean of the function is of order $1/N$ if the errors are small, see (6.11) or (6.24).

To determine the *a posteriori* distribution of the parameter λ, we can expand $f_\alpha(\lambda)$ around the minimum point of E as

$$f_\alpha(\lambda) = f_\alpha(\lambda_m) + \frac{\partial f_\alpha}{\partial \lambda}\bigg|_{\lambda=\lambda_m} (\lambda - \lambda_m) + O((\lambda - \lambda_m)^2)$$

$$= f_\alpha(\lambda_m) + (\lambda - \lambda_m)F_\alpha + O((\lambda - \lambda_m)^2)$$

$$= \overline{x^{(\alpha)}} + (\lambda - \lambda_m)e_\alpha V_\alpha + O((\lambda - \lambda_m)^2), \tag{6.46}$$

where we have defined the quantities F_α and V_α as

$$F_\alpha \equiv \left.\frac{\partial f_\alpha}{\partial \lambda}\right|_{\lambda = \lambda_m} \equiv e_\alpha V_\alpha. \tag{6.47}$$

For large N the probability distribution (6.43) of λ is, therefore, proportional to

$$P(\lambda|\vec{\mathbf{x}}) \propto \exp\left(-\frac{(\lambda - \lambda_m)^2}{2}\sum_{\alpha=1}^{M} V_\alpha^2\right). \tag{6.48}$$

The variable λ has a Gaussian distribution, the mean of which is the best fit λ_m and the variance is

$$\sigma_\lambda^2 = \left(\sum_{\alpha=1}^{M} V_\alpha^2\right)^{-1}. \tag{6.49}$$

Given a theoretical assumption of functional dependence on a given parameter, $\vec{f}(\lambda)$, we can then summarize the situation by writing the best estimate as

$$\lambda = \lambda_m \pm \sigma_\lambda. \tag{6.50}$$

We now move on to answer the second question. How can we tell if our hypothesis is a valid representation of the phenomenon we are measuring? We can determine the typical value of χ^2 and see if our data produces a value of χ^2 close to this typical value, or not. As the set of N experimental measurements changes, the λ_m value changes, and the entire function of squares will therefore vary stochastically. What is the distribution of the function E computed at its minimum? That is, what is the distribution of the χ^2 of the fitting hypothesis?

To answer this question, we denote by Λ the true value of λ (unknown!), of which we want to determine a meaningful estimate (the best fit λ_m). The true value of $a^{(\alpha)}$ (which, incidentally, is not the *a posteriori* average value of (6.45)), will be denoted by $A_\alpha = f_\alpha(\Lambda)$. For large enough N, we expect the value of Λ to be similar to the value of the best fit λ_m. We can assume that their difference will be of the order of $f_\alpha(\{\lambda\}) - \overline{x^{(\alpha)}}$ in the formula (6.43), if the assumptions on f_α make sense, i.e.,

$$\lambda_m - \Lambda \sim 1/\sqrt{N}.$$

To study the behavior of the distribution of the values of χ^2 as the set of experimental measurements varies, we look, therefore, at the fluctuations around the true value Λ of the best-fit parameter and rewrite the function E of the squares expanding around this value. In this case, in place of the expansion (6.46), we have

$$f_\alpha(\lambda) = f_\alpha(\Lambda) + \left.\frac{\partial f_\alpha}{\partial \lambda}\right|_{\lambda = \Lambda} (\lambda - \Lambda) + O((\lambda - \Lambda)^2) = A_\alpha + (\lambda - \Lambda)e_\alpha V_\alpha + O\left(\frac{1}{N}\right), \tag{6.51}$$

where V_α, see (6.47), must now be computed on the value Λ. Adding this expansion to the exponent of (6.43) gives the expression

$$E(\lambda, \{\vec{x}\}) = \sum_{\alpha=1}^{M} \left(\frac{\delta x^{(\alpha)}}{e_\alpha} - V_\alpha \delta\lambda \right)^2 , \tag{6.52}$$

where $\delta x^{(\alpha)} \equiv \bar{x}^{(\alpha)} - A_\alpha$ is the deviation of $\overline{x^{(\alpha)}}$ from its theoretical true value A_α and $\delta\lambda \equiv \lambda - \Lambda$ is the deviation of λ from its true value Λ. To simplify notation, we introduce the quantity

$$\delta y^{(\alpha)} \equiv \delta x^{(\alpha)} / e_\alpha .$$

The value of χ^2 is then written as

$$\chi^2 = \min_\lambda \sum_{\alpha=1}^{M} (\delta y^{(\alpha)} - V_\alpha \delta\lambda)^2 = \min_\lambda |\vec{\delta y} - \vec{V}\delta\lambda|^2 . \tag{6.53}$$

The quantity $\delta\lambda$ can be determined by the minimization implemented in the formula above, and is given by

$$\delta\lambda_m \equiv \lambda_m - \Lambda = \frac{\sum_{\alpha=1}^{M} V_\alpha \delta y^{(\alpha)}}{\sum_{\alpha=1}^{M} V_\alpha^2} = \frac{\vec{V} \cdot \vec{\delta y}}{|\vec{V}|^2} . \tag{6.54}$$

Observing that, given the independence of the different measured degrees of freedom,

$$\langle \delta y^{(\alpha)} \delta y^{(\beta)} \rangle = \delta_{\alpha\beta} , \tag{6.55}$$

it is possible to verify that the mean value of $(\delta\lambda_m)^2$ is just given by σ_λ^2, see (6.49), thus confirming the previous statement on the fluctuations of the best-fit variable.

Equation (6.55) results from the fact that, as the data set changes, the empirical averages will fluctuate around the true value $A^{(\alpha)}$. The error on the empirical mean $\overline{x^\alpha}$ is the e_α estimator, which tends to zero for large N. So the distribution of $\overline{x^\alpha}$ is very peaked around A^α and ideally averaging over all possible sets of experimental data we have

$$\langle \delta y^{(\alpha)} \rangle = 0 .$$

Since e_α^2 is the variance estimate of the mean, we have that

$$\langle (\delta y^{(\alpha)})^2 \rangle = \left\langle \frac{\left(\overline{x^{(\alpha)}} - A^{(\alpha)} \right)^2}{e_\alpha^2} \right\rangle = 1 .$$

In the sense of the probability distribution of the experimental measurements for large N, the values of the variables $\delta x^{(\alpha)} = \overline{x^{(\alpha)}} - A^{(\alpha)}$ are very concentrated around zero, with a Gaussian distribution. The distribution of the variables rescaled with the error, $\vec{\delta y}$, is, thus, a normal Gaussian distribution.

We resume our reformulation of χ^2 (6.53). Substituting the value of $\delta\lambda = \delta\lambda_m$ (6.54) into the expression (6.53) we find

$$\chi^2 = \sum_{\alpha=1}^{M} (\delta y^{(\alpha)})^2 - \frac{\left(\sum_{\alpha=1}^{M} V_\alpha \delta y^{(\alpha)}\right)^2}{\sum_{\alpha=1}^{M} V_\alpha^2} = |\vec{\delta y}|^2 - \frac{|\vec{V} \cdot \vec{\delta y}|^2}{|\vec{V}|^2}. \tag{6.56}$$

A more detailed computation, which we illustrate below, shows that χ^2 is precisely the sum of the squares of $M - 1$ independent variables. We, therefore, introduce the versor **n** of components

$$\hat{n}^\alpha \equiv \frac{V_\alpha}{\left(\sum_\beta V_\beta^2\right)^{1/2}} = \frac{V_\alpha}{|\vec{V}|}, \tag{6.57}$$

in terms of which χ^2 can be expressed as

$$\chi^2 = |\vec{\delta y}|^2 - |\hat{n} \cdot \vec{\delta y}|^2. \tag{6.58}$$

We now rotate our reference system so that the M axis coincides with the versor $\hat{n}$. We denote by r_μ ($\mu = 1, \ldots, M$) the coordinates of the vector $\vec{\delta y}$ in the rotated basis and with $\mathbb{R}$ the rotation matrix:

$$r_\mu = \sum_{\alpha=1}^{M} R_{\mu\alpha} \delta y^{(\alpha)}. \tag{6.59}$$

In the new reference system, the Mth component is $r_M = \hat{n} \cdot \vec{\delta y}$. Furthermore, the rotation leaves the modulus unchanged: $|\delta y|^2 = |r|^2$. In the rotated basis, therefore, χ^2 is constructed by subtracting the square modulus of $\vec{r}$ from the squared modulus of its Mth component, obtaining

$$\chi^2 = \sum_{\mu=1}^{M-1} r_\mu^2, \tag{6.60}$$

implying that χ^2 is the sum of $M - 1$ variables.

To recapitulate, having acquired a certain set of N measures and given theoretical assumptions $f_\alpha(\lambda)$, minimizing the function (6.52) of the squares yields a value of χ^2. Making another series of N measurements, under the same theoretical assumptions, another estimate of the value of χ^2 is obtained. In the space of the experimental values acquired for M degrees of freedom in N measurements, the χ^2 is, therefore, a random variable. According to (6.60) it can be expressed as the sum of $M - 1$ random variables r_μ squared. How are r_μ distributed as the sets of measures vary?

From (6.59) we have that the rotated variables r_μ are Gaussian-distributed, as linear combinations of Gaussian variables, and, as we are about to show, are independent of each other. Starting from the rotation (6.59), indeed, we have

$$\langle r_\mu r_\nu \rangle = \sum_{\alpha=1}^{M} \sum_{\beta=1}^{M} R_{\mu\alpha} R_{\nu\beta} \langle \delta y^{(\alpha)} \delta y^{(\beta)} \rangle = \sum_{\alpha=1}^{M} R_{\mu\alpha} R_{\nu\alpha} = \sum_{\alpha=1}^{M} R_{\mu\alpha} (R^T)_{\alpha\nu} = \delta_{\mu\nu}, \tag{6.61}$$

where in the last step we used the fact that $\mathbb{R}$ is a unitary matrix, i.e., $\mathbb{R}\mathbb{R}^T = \mathbb{I}$. The r_μ are independent and normally distributed. In the end, the mean value of (6.60) is

$$\langle \chi^2 \rangle = M - 1 . \tag{6.62}$$

In general, for L parameters $\{\lambda\}$, we must follow the same procedure as expressed in the relations (6.51)–(6.61), rewriting χ^2 in terms of L derivatives $\vec{V}^{(\ell)}$ and L deviations $\delta\lambda_\ell$, with $\ell = 1, \ldots, L$. Eventually, the mean value of χ^2 is the sum of $M - L$ independent variables and its expectation value is

$$\langle \chi^2 \rangle = M - L . \tag{6.63}$$

6.5.2 Distribution of χ^2

One can directly compute the probability distribution of the sum S of n Gaussian variables squared with unitary variance as

$$P(S) = \int \prod_{\mu=1}^{n} \frac{dx_\mu}{\sqrt{2\pi}} \, \delta\left(S - \sum_{\mu=1}^{n} (x_\mu)^2\right) \exp\left(-\frac{1}{2}\sum_{\mu=1}^{n} (x_\mu)^2\right) = \frac{S^{n/2-1}e^{-S/2}}{2^{n/2}\Gamma(n/2)} , \tag{6.64}$$

with mean value

$$\langle S \rangle = n .$$

Proof   This result can be obtained, for example, by first using the Fourier antitransform expression of the Dirac delta

$$P(S) = \int \prod_{\mu=1}^{n} \frac{dx_\mu}{\sqrt{2\pi}} \int_{-\infty}^{\infty} \frac{dq}{2\pi} \exp\left(iq\sum_{\mu=1}^{n} r_i^2 - iqS\right) \prod_{\mu=1}^{n} e^{-r_i^2/2}$$

$$= \int_{-\infty}^{\infty} \frac{dq}{2\pi} e^{-iqS} \prod_{\mu=1}^{n} \int_{-\infty}^{\infty} \frac{dx_\mu}{\sqrt{2\pi}} \exp\left(-\frac{1}{2}r_i^2(1 - 2iq)\right)$$

$$= \int_{-\infty}^{\infty} \frac{dq}{2\pi} e^{-iqS} \left(\frac{1}{1 - 2iq}\right)^{n/2}$$

and, then, using the inverse power integral expression (3.98) for $1/(1 - 2iq)^{n/2}$, from which we obtain

$$P(S) = \int_{-\infty}^{\infty} \frac{dq}{2\pi} e^{-iqS} \frac{1}{\Gamma(n/2)} \int_0^{\infty} dy \, y^{n/2-1} \exp\{-y(1 - 2iq)\}$$

$$= \frac{1}{\Gamma(n/2)} \int_0^{\infty} dy \, y^{n/2-1} e^{-y} \int_{-\infty}^{\infty} \frac{dq}{2\pi} e^{-iq(S-2y)}$$

$$= \frac{1}{\Gamma(n/2)} \int_0^{\infty} dy \, y^{n/2-1} e^{-y} \frac{1}{2}\delta\left(y - \frac{S}{2}\right) ,$$

i.e., the χ^2 probability density (6.64).   ■

6.5.2.1 General Considerations on the χ^2 in a Single Experiment

In the limit where the number of independent degrees of freedom $n = M - L$ is large, we expect the χ^2 per degree of freedom (dof) to be strongly centered around its mean value,

$$\frac{\chi^2}{M - L} \simeq 1. \tag{6.65}$$

Significant deviations from 1 very often signal serious problems.

- A value of χ^2 per degree of freedom much larger than 1 can suggest that the assumed parameter dependence $f_\alpha(\lambda)$ is incorrect, or statistical errors overestimated.
- A value of χ^2 per degree of freedom much smaller than 1 suggests that statistical fluctuations are underestimated, e.g., because the experimental data are correlated, so one must resort to the formulas we will derive in the next section.

A widely used empirical method to correct for inaccuracies that lead to a χ^2 value being too high (or low), and to obtain a reasonable error estimate, is as follows. As we have seen, the probability distribution of $\{\lambda\}$ is given by (6.42):

$$P(\{\lambda\}) \propto \exp\left[- \tfrac{1}{2} E(\{\lambda\}, \{\vec{x}\}) \right].$$

The method consists of correcting this probability by rescaling the least-squares function with the χ^2 per degree of freedom actually computed on the data:

$$P(\{\lambda\}) \propto \exp\left[-\frac{1}{2} E(\{\lambda\}, \{\vec{x}\}) \frac{M - L}{\chi^2} \right]. \tag{6.66}$$

Obviously, if the quantity (6.65) is very close to 1 (as it should be), the corrections introduced are negligible. On the other hand, large corrections are obtained if (6.65) differs greatly from 1. This method is only correct if the anomalous value of χ^2 is due to a generalized and uniform overestimation (or underestimation) of the errors. The procedure we have described corresponds to rescaling all errors by multiplying them by a constant χ^2/dof, so that the value of χ^2 is correct. In general, the method does not produce correct results, and it should be applied, needless to say, with a grain of salt, but it can provide, in the absence of anything better, a reasonable first estimate.

6.5.2.2 The χ^2 Distribution and Poisson Mean Value

Before moving on to the case of correlated degrees of freedom, let us look at a curious thing. If we place the simple change of variables $y = S/2$ in (6.64) and transform the χ^2 probability density, as described in Appendix 3.B, we formally obtain

$$P\left(y \,\Big|\, \frac{n}{2} - 1\right) = \frac{y^{n/2-1} e^{-y}}{(n/2 - 1)!},$$

which is nothing other than the *a posteriori* distribution of the mean value of Poisson variables (5.33) after $n/2$ counts. Statisticians call these types of distributions γ distributions or Pearson distributions of type III.

6.5.3 Correlated Data

The most difficult case to deal with, but which is also the most interesting and frequent, is that of correlated degrees of freedom (at each measurement). We are formally in the same situation as in the previous case, with the difference that the data are characterized by a covariance matrix $\mathbb{C}$ defined by (6.3). Typically, if we measure and analyze M quantities at the same time, we are dealing not with quantities that have nothing to do with each other, but with correlated degrees of freedom. The same arguments as in the previous section lead to the conclusion that we must minimize, with respect to the parameters $\{\lambda\}$, the least-squares function

$$\chi^2 \equiv \min_{\{\lambda\}} N \sum_{\alpha,\beta}^{1,M} \left[f_\alpha(\{\lambda\}) - \overline{x^{(\alpha)}} \right] B_{\alpha\beta} \left[f_\beta(\{\lambda\}) - \overline{x^{(\beta)}} \right], \tag{6.67}$$

where $B_{\alpha\beta}$ are the elements of the inverse matrix of the covariance matrix $\mathbb{C}$. The covariance matrix, divided by N, see (6.8), generalizes the diagonal factor e_α^2, which appeared in the uncorrelated variables approach in formula (6.43), to the case of non-zero off-diagonal correlations. If the empirical covariance matrix of the data is $\bar{\mathbb{C}}$, we will once again approximate the theoretical covariance with the empirical one, $\mathbb{B}^{-1} = \mathbb{C} \simeq \bar{\mathbb{C}}$. The expectation value of χ^2 is still given by

$$\langle \chi^2 \rangle = M - L, \tag{6.68}$$

where L is the number of independent parameters $\{\lambda\}$ on which the hypotheses $\vec{f}$ depend.

Let us now derive the result (6.68) in the case of M hypotheses $f^{(\alpha)}$ which depend on a single parameter λ. Let us consider M correlated variables $x^{(\alpha)}$, $\alpha = 1, \ldots, M$, of covariance $C_{\alpha\beta} = (B^{-1})_{\alpha\beta}$. The squares function reads now

$$E(\lambda, \vec{x}) \equiv N \sum_{\alpha\beta}^{1,M} \left[f_\alpha(\lambda) - \overline{x^{(\alpha)}} \right] B_{\alpha\beta} \left[f_\beta(\lambda) - \overline{x^{(\beta)}} \right]. \tag{6.69}$$

The least-squares law tells us that the best estimate of λ is the one that maximizes the *a posteriori* probability or minimizes the squares function E,

$$\chi^2 \equiv \min_\lambda E(\lambda, \vec{x}).$$

To test the validity of the theoretical assumptions $\vec{f}(\lambda)$, we compare the value of χ^2 computed on the data of a series of N measurements with its expectation value $\langle \chi^2 \rangle$. We have seen in Section 6.5.1 that one way to derive the expression of the mean value of χ^2 is to manipulate the function $E(\lambda|\vec{x})$ so as to write χ^2 as the sum of independent normal variables. We will proceed in the same way as for the case of independent degrees of freedom, but adding an extra step: diagonalization of the biquadratic form. In summary, the computation consists of performing the following steps:

1. expansion of $E(\lambda|\vec{x})$ around the true value Λ linearizing $\vec{f}(\lambda)$;
2. identification of best fit λ_m and computation of $\chi^2 = E(\lambda_m|\vec{x})$;

3. diagonalization of χ^2; and
4. rotation of χ^2.

Let us proceed step by step to analyze all the details of the procedure.

1. We linearize the function f_α around Λ, the true (and unknown) value of λ, such that $f_\alpha(\Lambda) = A_\alpha$:

$$f_\alpha(\lambda) \simeq f_\alpha(\Lambda) + (\lambda - \Lambda)\left.\frac{\partial f_\alpha}{\partial \lambda}\right|_{\lambda=\Lambda} = A_\alpha + \delta\lambda\, F_\alpha\,, \tag{6.70}$$

$$\delta\lambda \equiv \lambda - \Lambda\,, \quad F_\alpha \equiv \left.\frac{\partial f_\alpha}{\partial \lambda}\right|_{\lambda=\Lambda}\,.$$

We then define the difference between the empirical mean and the true value, $\delta x^{(\alpha)} \equiv \overline{x^{(\alpha)}} - A_\alpha$, such that the function $E(\lambda, \vec{x})$ can be rewritten as

$$E(\lambda, \vec{x}) \simeq \sum_{\alpha,\beta}^{1,M} (-\delta x^{(\alpha)} + \delta\lambda\, F_\alpha) N\, B_{\alpha\beta} (-\delta x^{(\beta)} + \delta\lambda\, F_\beta)$$

$$= (-\vec{\delta x} + \delta\lambda\, \vec{F})^T \mathbb{B}_N (-\vec{\delta x} + \delta\lambda\, \vec{F}) \tag{6.71}$$

where, for convenience, we have absorbed the factor N in the definition of $\mathbb{B}_N$, the matrix whose elements are $N B_{\alpha\beta}$.

2. To find the best fit of λ we set the derivative of E with respect to λ to zero and obtain the relation

$$\delta\lambda_m \equiv \lambda_m - \Lambda = \frac{\vec{F}^T\,\mathbb{B}\,\vec{\delta x}}{\vec{F}^T\,\mathbb{B}\,\vec{F}} = \vec{\psi} \cdot \vec{\delta x}\,, \tag{6.72}$$

where we have defined the vector $\vec{\psi}$ of coordinates,

$$\psi_\beta \equiv \frac{\sum_{\alpha=1}^{M} F_\alpha B_{\alpha\beta}}{\vec{F}^T\,\mathbb{B}\,\vec{F}}\,.$$

Substituting in (6.71) we obtain

$$\chi^2 = E(\lambda_m, \vec{x}) = \left(\vec{\delta x} - (\vec{\psi} \cdot \vec{\delta x})\vec{F}\right)^T \mathbb{B}_N \left(\vec{\delta x} - (\vec{\psi} \cdot \vec{\delta x})\vec{F}\right)$$

$$= \left\{\vec{\delta x}^T\,\mathbb{B}_N\,\vec{\delta x} - |\vec{\psi} \cdot \vec{\delta x}|^2 \vec{F}^T\,\mathbb{B}_N\,\vec{F}\right\}. \tag{6.73}$$

3. We now diagonalize $\mathbb{B}_N$. We first introduce the following notation:

- the eigenvectors $\vec{u}$ of the symmetric matrix $\mathbb{B}_N$, such that $\mathbb{B}_N \vec{u} = \mu_\alpha \vec{u}$, $\alpha = 1, \ldots, M$;
- the diagonal matrix $\mathbb{M}$ of the eigenvalues of $\mathbb{B}_N$, $M_{\alpha\beta} = \delta_{\alpha\beta}\mu_\alpha$; and
- the matrix $\mathbb{U}$ whose columns are the eigenvectors of $\mathbb{B}_N$ and such that $\mathbb{U}^{-1}\mathbb{B}_N\,\mathbb{U} = \mathbb{M}$; since $\mathbb{B}$ is symmetric, the eigenvalues are all real and the matrix $\mathbb{U}$ is orthogonal: $\mathbb{U}^{-1} = \mathbb{U}^T$.

With this notation, we introduce the orthogonal transformations

$$\vec{\delta x} = \mathbb{U}\,\vec{\delta w}\,, \quad \vec{\delta x}^T = \vec{\delta w}^T\,\mathbb{U}^T\,, \tag{6.74}$$

$$\vec{F} = \mathbb{U}\,\vec{\Upsilon}, \qquad \vec{F}^T = \vec{\Upsilon}^T\,\mathbb{U}^T, \tag{6.75}$$

into the expression (6.73) of χ^2 and after a few steps we obtain

$$\chi^2 = \delta\vec{w}^T\,\mathbb{M}\,\delta\vec{w} - \frac{|\vec{\Upsilon}^T\,\mathbb{M}\,\delta\vec{w}|^2}{\vec{\Upsilon}^T\,\mathbb{M}\,\vec{\Upsilon}}. \tag{6.76}$$

Next, we rescale the elements of the vectors $\delta\vec{w}$ and $\vec{\Upsilon}$ with the eigenvalues, defining the new vectors $\delta\vec{y}$ and $\vec{V}$ as

$$\delta y^{(\alpha)} \equiv \delta w^{(\alpha)}\sqrt{\mu_\alpha}, \qquad V_\alpha \equiv \Upsilon_\alpha\sqrt{\mu_\alpha}. \tag{6.77}$$

We arrive in this way, even in the case of correlated variables, formally at the same expression (6.56), which we quote here for convenience:

$$\chi^2 = |\delta\vec{y}|^2 - \frac{|\vec{V}\cdot\delta\vec{y}|^2}{|\vec{V}|^2} = |\delta\vec{y}|^2 - |\hat{n}\cdot\delta\vec{y}|^2,$$

with

$$\hat{n} \equiv \vec{V}/|\vec{V}|.$$

4. Finally, we perform a further basic rotation,

$$\vec{r} = \mathbb{R}\,\delta\vec{y}, \tag{6.78}$$

so that the Mth axis of the basis $\vec{r}$ coincides with the direction $\hat{n}$, as in Section 6.5.1. Eventually, we can write

$$\chi^2 = \sum_{\eta=1}^{M-1} r_\eta^2. \tag{6.79}$$

We draw the attention of the reader to the fact that the direction $\hat{n}$ depends on the eigenvalues of $\mathbb{B}$, which is taken *a priori* equal to the empirical covariance, and thus depends on the empirical data. It also depends on the derivatives of the hypotheses $\vec{f}(\lambda)$ that we want to test. Putting together rotation, rescaling, and diagonalization, the variables r_η are written as

$$r_\eta = \sum_{\alpha,\gamma}^{1,M} R_{\eta\alpha}\sqrt{\mu_\alpha}\,(U^{-1})_{\alpha\gamma}\,\delta x^{(\gamma)}. \tag{6.80}$$

Thus, as the experimental measurement data set varies, $\vec{r}$ are Gaussian, as linear combinations of δx^α variables distributed with Gaussian distribution. In the case of uncorrelated data, the r_η were independent, see (6.61). Will they be correlated now? We compute their covariance:

$$\langle r_\eta r_\xi \rangle_c = \langle r_\eta r_\xi \rangle - \langle r_\eta \rangle \langle r_\xi \rangle$$

$$= \sum_{\alpha,\beta,\gamma,\delta} \sqrt{\mu_\alpha \mu_\beta}\, R_{\eta\alpha}(U^{-1})_{\alpha\gamma}\, R_{\xi\beta}(U^{-1})_{\beta\delta}\{\langle \delta x^{(\gamma)}\delta x^{(\delta)}\rangle - \langle \delta x^{(\gamma)}\rangle\langle \delta x^{(\delta)}\rangle\}$$

$$\simeq \sum_{\alpha,\beta,\gamma,\delta} \sqrt{\mu_\alpha \mu_\beta}\, R_{\eta\alpha}(U^{-1})_{\alpha\gamma}\, R_{\xi\beta}(U^{-1})_{\beta\delta}(B_N^{-1})_{\gamma\delta}, \tag{6.81}$$

where we have used that $\langle \vec{\delta x} \rangle = 0$ and replaced $(B_N^{-1})_{\gamma\delta}$ with

$$\langle \delta x^{(\gamma)} \delta x^{(\delta)} \rangle = \left\langle \frac{\bar{C}_{\gamma\delta}}{N} \right\rangle = \langle (B_N^{-1})_{\gamma\delta} \rangle .$$

Given that $\mathbb{U}^{-1} = \mathbb{U}^T$ and that $\mathbb{U}^T \mathbb{B}_N^{-1} \mathbb{U} = \mathbb{M}^{-1}$, a diagonal matrix of elements $\delta_{\gamma\delta}\,\mu_\gamma^{-1}$, we have

$$\langle r_\eta r_\xi \rangle_c \simeq \sum_{\alpha,\beta} \sqrt{\mu_\alpha \mu_\beta}\, R_{\eta\alpha} R_{\xi\beta} (\mathcal{M}^{-1})_{\alpha\beta} = \sum_\alpha R_{\eta\alpha} R_{\alpha\xi}^T = \delta_{\eta\xi} . \tag{6.82}$$

We have shown that, just as in the case of correlated data, χ^2 can be written as a sum of normal-distributed independent random variables. And, thus, also in the correlated case,

$$\langle \chi^2 \rangle = M - 1 . \tag{6.83}$$

If, then, we have M laws $\vec{f}(\{\lambda\})$ that depend on L independent parameters, we will have

$$\langle \chi^2 \rangle = M - L . \tag{6.84}$$

The difference with respect to the case of correlated degrees of freedom is in the definition of χ^2, which in the case of correlated random variables will not simply be the sum of squares of fluctuations, but will have non-negligible off-diagonal components.

6.5.4 Least-Squares Method with Reweighting

In both the case of correlated variables and the simpler case of independent variables, we can estimate the parameters of a theoretical law and their statistical errors using the more general reweighting method presented in Section 6.3. The reweighting method is based on the introduction of probabilities p_k that a certain measurement of M degrees of freedom takes the set of values $\vec{x}_k$. We will see how the reweighting method allows us to construct a simple procedure for evaluating the errors that plague the estimation of the parameters $\{\lambda\}$. For simplicity of presentation, we will consider in the following the case of a single free parameter λ, i.e., $L = 1$.

We reweight the data with random weights p_k, chosen with the same *a posteriori* distribution as in Section 5.4.1 for the case of scalar data. The set of values recorded at the kth measurement of M variables is then given a p_k weight. The expectation values obtained from the experimental data with any distribution will be called $\vec{X}$ and are defined as in (6.21),

$$X^{(\alpha)}(\mathbf{p}) = \sum_{k=1}^{N} p_k x_k^{(\alpha)}, \quad \alpha = 1, \dots, M .$$

The probability distribution of the expected values $\vec{X}$ can be obtained from the *a posteriori* distribution of $\mathbf{p} = \{p\}$. For a small number of measurements and/or for variables that behave questionably (e.g., have infinite variance), this can be very complicated. When we have many measurements, Eq. (6.21) is the sum of many variables, so if their moments are well defined, $X^{(\alpha)}$ will tend to have a multivariate Gaussian distribution, with mean, covariance, and kurtosis that generalize to the M-dimensional case the expressions (5.83)–(5.85) derived in the scalar case.

For example, in the correlated case the variance (5.84) is generalized by the experimental covariance divided by N,

$$\langle (X_\alpha - \langle X_\alpha \rangle)(X_\beta - \langle X_\beta \rangle)\rangle = \frac{\bar{C}_{\alpha\beta}}{N}, \tag{6.85}$$

so that, eventually, the distribution will be

$$P(\vec{X}) \propto \exp\left(-\frac{N}{2}\sum_{\alpha,\beta}^{1,M}\left(X^{(\alpha)} - \overline{x^{(\alpha)}}\right)(\bar{\mathbb{C}}^{-1})_{\alpha\beta}\left(X^{(\beta)} - \overline{x^{(\beta)}}\right)\right). \tag{6.86}$$

Moving on to the least-squares method, we recall that for each different set of N measurements we compute an empirical average and an empirical covariance. When we compute the best fit of the parameter λ by minimizing the least-squares function (depending on empirical average and covariance), we find that λ_m varies as the experimental data set varies.

The best fit will therefore have a mean value and a statistical error. Similarly, the empirical average $\bar{x}$ is affected by statistical error. Starting from this simple observation, we can now write a slightly different least-squares function that allows us to use the reweighting method.

We introduce, as before, the laws for the theoretical expected value of the observables $\vec{X} = \vec{f}(\lambda)$ and look for the best estimate of λ, its mean value, and its uncertainty.

Once the experimental data are given, the empirical average is fixed and represents the expected value $\vec{X}(\mathbf{p})$ on the *a posteriori* distribution $P(\mathbf{p}|\mathbf{x})$, cf. (5.65): $\overline{x^\alpha} = \langle X^{(\alpha)}(\mathbf{p})\rangle$. We can reproduce the uncertainty on the empirical mean by writing the latter as in (6.21), $\vec{X}(\mathbf{p}) = \sum_k p_k \vec{x}_k$, and varying the weights $\mathbf{p}$, instead of the data. In many universes the same experimental data belong to different realizations of their weights. They are all possible (otherwise we would not measure them), but with different probabilities from universe to universe.

So, in summary, on the one hand, we have the parametric laws in λ, $\vec{f}(\lambda)$, and on the other, the empirical mean values $\langle \vec{X}(\mathbf{p})\rangle$, and we want to analyze the goodness of the data and of the theoretical hypothesis by looking at their deviations.

In the context of the reweighting method, and under the reasonable assumption that for large N the deviations of $\vec{X}(\mathbf{p})$ from $\bar{x}$ are small, of order $1/\sqrt{N}$ as in (5.84), we can construct a least-squares function by substituting $\vec{X}(\mathbf{p})$ in the distribution (6.86) in place of its expected value $\bar{x}$. This will give us a least-squares function

$$E(\lambda, \mathbf{p}, \bar{\vec{x}}) = \frac{1}{2}\sum_{\alpha,\beta}^{1,M}\left[f_\alpha(\lambda) - X^{(\alpha)}(\mathbf{p})\right] B_{\alpha\beta}\left[f_\beta(\lambda) - X^{(\beta)}(\mathbf{p})\right], \tag{6.87}$$

which depends on the $\mathbf{p}$ sequences of probabilistic weights as well as on the $\{\lambda\}$ parameters of the theoretical hypotheses. Each $\mathbf{p}$ sequence of extracted weights will yield a different value of the expected value $\vec{X}(\mathbf{p})$. We can, therefore, define a weight-dependent χ^2 as

$$\chi^2(\mathbf{p}) \equiv \min_{\lambda} E(\lambda, \mathbf{p}, \vec{\bar{x}}) . \tag{6.88}$$

When using (6.88) to determine the best-fit value by minimizing with respect to λ, we establish a relationship between λ and $\mathbf{p}$, $\lambda = \lambda_m(\mathbf{p})$. And, as usual for the reweighting method, the histogram of $\lambda_m(\mathbf{p})$ is constructed by performing various random extractions of the weights $\mathbf{p}$, eventually yielding the probability distribution of λ.

To be pedantically crystalline: having a given set of experimental measurements we can make a large number of extractions of the $\mathbf{p}$ weights and to each extraction associate a value of λ computed by minimizing $\chi^2(\mathbf{p})$. If we make the histogram of λ_m resulting from all extractions, this gives the probability distribution of λ_m *a posteriori* of the experimental data.

In formulas, given the *a posteriori* distribution $P(\mathbf{p}|\mathbf{x})$, (5.65), with the reweighting method of Sections 5.4.1 and 6.3) we find that the mean value of χ^2 is realized for

$$\bar{\lambda} \equiv \langle \lambda \rangle = \int \prod_{k=1}^{M} dp_k \, P(\mathbf{p}|\mathbf{x})\lambda_m(\mathbf{p}) , \tag{6.89}$$

and that its variance is

$$\sigma_\lambda^2 \equiv \langle (\lambda - \langle \lambda \rangle)^2 \rangle = \int \prod_{k=1}^{M} dp_k \, P(\mathbf{p}|\mathbf{x})[\lambda_m(\mathbf{p}) - \langle \lambda \rangle]^2 . \tag{6.90}$$

6.5.5 Choosing the Least-Squares Function Matrix

Let us now see what would happen if, in the expression for χ^2 in (6.67), instead of $\mathbb{B}$, which is the inverse covariance matrix $\mathbb{C}^{-1}$, we considered an arbitrary matrix $\mathbb{D}$:

$$E_D(\lambda|\vec{x}) = N \sum_{\alpha,\beta}^{1,M} \left(f_\alpha(\{\lambda\}) - \overline{x^{(\alpha)}} \right) D_{\alpha\beta} \left(f_\beta(\{\lambda\}) - \overline{x^{(\beta)}} \right) , \tag{6.91}$$

$$\chi^2 \equiv \min_{\{\lambda\}} E_D(\lambda|\vec{x}) , \tag{6.92}$$

$$\delta\lambda_m = \lambda_m - \Lambda = \frac{\vec{F}^T \mathbb{D} \, \vec{\delta x}}{\vec{F}^T \mathbb{D} \, \vec{F}} . \tag{6.93}$$

This case is interesting in that it is very often difficult to know the exact matrix $\mathbb{C}^{-1}$, as we discussed earlier in Section 6.1. Varying the set of N measures will vary the best fit λ_m and the value of χ^2. Ideally, we can estimate the error of the best fit from the variance of $\delta\lambda_m$, which is expressed as

$$\langle (\delta\lambda_m)^2 \rangle_D^c = \langle (\delta\lambda_m - \langle \delta\lambda_m \rangle_D)^2 \rangle_D = \langle (\lambda_m - \langle \lambda_m \rangle_D)^2 \rangle_D$$

$$= \frac{\sum_{\alpha,\beta,\gamma,\epsilon}^{1,M} F_\alpha D_{\alpha\beta} F_\epsilon D_{\gamma\epsilon} \langle \delta x^{(\beta)} \delta x^{(\epsilon)} \rangle_c}{\left(\sum_{\alpha,\beta}^{M} F_\alpha D_{\alpha\beta} F_\beta \right)^2}$$

$$= \frac{1}{N} \frac{\sum_{\alpha,\beta,\gamma,\epsilon}^{1,M} F_\alpha D_{\alpha\beta} C_{\beta\epsilon} D_{\gamma\epsilon} F_\epsilon}{(\vec{F}^T \, \mathbb{D} \, \vec{F})^2} . \tag{6.94}$$

We thus see that the estimated error on λ_m depends on the choice of $\mathbb{D}$. We can optimize the result by requiring the variance to be minimal as the elements of the arbitrary matrix $\mathbb{D}$ vary:

$$\frac{\partial \langle (\delta\lambda_m)^2 \rangle_D^{(c)}}{\partial D_{\alpha\beta}} = 0, \quad \forall \, \alpha, \beta . \tag{6.95}$$

By doing the derivatives, it is possible to show that the choice $\mathbb{D} = \mathbb{C}^{-1}$ is a solution of the minimum error equation, therefore coinciding with the expression

$$\langle (\delta\lambda_m)^2 \rangle_{C^{-1}}^{(c)} = \frac{1}{N} \left(\sum_{\alpha,\beta}^{1,M} F_\alpha C_{\alpha\beta}^{-1} F_\beta \right)^{-1} . \tag{6.96}$$

This result also justifies the choice we have made in the formula (6.7) of Section 6.1 to brutally replace the matrix $\mathbb{B}$ with the inverse of the empirical covariance matrix. In other words, we can say that if we do not use the inverse of the covariance matrix exactly, we only increase the estimated error, making it larger than the true error. Typically, a reasonable estimate of the covariance matrix is sufficient to obtain an error that is not too large.

In the case where the data are not terribly correlated, decent results are given by simply choosing to disregard their correlations, assuming that

$$D_{\alpha\beta} = \delta_{\alpha\beta} \frac{1}{N e_\alpha^2} , \tag{6.97}$$

where e_α is the error associated with the variable $x^{(\alpha)}$.

However, this solution has two disadvantages: the estimated errors are generically larger than the true errors, and the value of χ^2 per degree of freedom is no longer 1, so we can no longer assess the goodness of the assumptions made in characterizing the probability distribution of the experimental data. As far as possible, one should always try to start from a good estimate of the covariance matrix of the experimental data, and then compute the true χ^2 (which is different from χ_D^2 with $\mathbb{D}$ dissimilar from $\mathbb{C}^{-1}$). Furthermore, by using a $\mathbb{D}$ very different from $\mathbb{C}^{-1}$, we waste a lot of the information contained in the data.

We, nevertheless, invite the reader to pay attention to the computation of the inverse of the covariance matrix with a small number of data. In fact, if the number N of measurements is smaller than the number M of observables measured, for unavoidable mathematical reasons that have been discussed in Section 6.2, the covariance matrix has at least $M - N$ null eigenvalues and, thus, its inverse is ill-defined. In this very sensitive case, special care is required in proceeding (e.g., fingers crossed before doing anything).

Random Walkers

7.1 Random Walkers in Homogeneous Space

7.1.1 Random Walk on a Lattice

Let us consider a particle that can move on a hypercubic lattice of unitary step size in D dimensions.[1] Suppose that, at each instant of time (which we consider discrete, for the time being), the particle randomly moves by a step of unit length in one of the $2D$ possible directions. Two choices are allowed, one in the positive and one in the negative direction, for each one of the D directions. We begin by assuming that each of the $2D$ possible choices is equiprobable. We denote the position of the particle with the vector $\vec{m}$, a D-tuple of integers $\{m_\alpha\}$, $\alpha = 1, \ldots, D$. We want to compute the probability $P(\vec{m}, N)$ that after N passes the particle is located at $\vec{m}$ knowing that, at the initial time, the particle is located with certainty at the origin $\vec{m} = \vec{0}$. It is easy to convince oneself that $P(\vec{m}, N)$ is given by the ratio

$$P(\vec{m}, N) = \frac{\mathcal{C}(\vec{m}, N)}{(2D)^N} \tag{7.1}$$

of the number $\mathcal{C}(\vec{m}, N)$ of paths starting from the origin and arriving at the point $\vec{m}$ in N time steps to the number $(2D)^N$ of different paths that can be swept in N random steps.

To explicitly derive the function $P(\vec{m}, N)$ we need to do a few more computations. We note, to begin with, that the probability $P(\vec{m}, N)$ of being at the point $\vec{m}$ after N steps satisfies the recursion relation:

$$P(\vec{m}, N + 1) = \sum_{\vec{l}} J(|\vec{m} - \vec{l}|) P(\vec{l}, N), \tag{7.2}$$

where the probability $J(|\vec{m} - \vec{l}|)$ of moving in one (time) step from $\vec{l}$ to $\vec{m}$ is non-zero only if the (space) step length on the lattice is 1:

$$J(|\vec{m} - \vec{l}|) = \begin{cases} 1/(2D) & \text{if } |\vec{m} - \vec{l}| = 1, \\ 0 & \text{otherwise}. \end{cases} \tag{7.3}$$

---

[1] The points of a hypercubic lattice in a D-dimensional space are identified by a vector formed by D integer-valued components. In $D = 1$ we have a linear lattice, in $D = 2$ we have a square lattice, and in $D = 3$ a cubic lattice (simple cubic, to be precise).

We note that we are considering a random walk that is *uniform*, in which the probability of the displacement depends only on the difference of the coordinates, and not individually on the points of departure and arrival, and *isotropic*, since J depends only on the modulus of the displacement.

At step zero the particle is at the origin, i.e.,

$$P(\vec{m}, 0) = \delta_{\vec{m}, \vec{0}}, \tag{7.4}$$

where the multi-dimensional Kronecker delta denotes the product of the Kronecker deltas of each of the components m_k:

$$\delta_{\vec{m}, \vec{0}} = \prod_{k=1}^{D} \delta_{m_\alpha, 0}.$$

The Kronecker delta was introduced in Section 3.2.2; see the formula (3.43).

It is convenient at this point to introduce the Fourier transform $\tilde{P}(\vec{q}, N)$ of the discrete distribution defined on the lattice, some details of which are given in Appendix 7.A:

$$\tilde{P}(\vec{q}, N) = \sum_{\vec{m}} e^{\iota \vec{q} \cdot \vec{m}} P(\vec{m}, N). \tag{7.5}$$

The particle in the coordinate space lives on a discrete lattice, of step 1: the Fourier transform turns out, therefore, to be a periodic function with period 2π, and is thus uniquely fixed by its values in the first zone of the reciprocal lattice. Borrowing the nomenclature of crystal physics, the direct lattice is called the Bravais lattice and the reciprocal lattice is called the Brillouin lattice. Since $\tilde{P}(\vec{q})$ is periodic we can decide to evaluate it in any period domain 2π.

For a hypercubic lattice in dimension D we conventionally use the so-called first Brillouin zone, defined as

$$B: -\pi < q_\alpha \leq \pi \quad \text{for } \alpha = 1, \ldots, D.$$

The inverse Fourier transform, cf. (7.88), is then given by the discrete distribution

$$P(\vec{m}, N) = \frac{1}{(2\pi)^D} \int_B d^D q \, e^{-\iota \vec{q} \cdot \vec{m}} \tilde{P}(\vec{q}, N), \tag{7.6}$$

where the integral is taken only on the first Brillouin zone. In particular, we have the following representation for the D-dimensional Kronecker δ:

$$\delta_{\vec{m}, \vec{\ell}} = \frac{1}{(2\pi)^D} \int_B d^D q \, \exp[-\iota \vec{q} \cdot (\vec{m} - \vec{\ell})]. \tag{7.7}$$

We can now take the recursive relation we wrote earlier, involving a convolution procedure, into Fourier space (see the formula (3.87)). We Fourier transform (7.2), finding

$$\sum_{\vec{m}} e^{i\vec{q}\cdot\vec{m}} P(\vec{m}, N+1) = \sum_{\vec{m}} e^{i\vec{q}\cdot\vec{m}} \sum_{\vec{l}} J(|\vec{m}-\vec{l}|) P(\vec{l}, N)$$

$$= \sum_{\vec{n}} e^{i\vec{q}\cdot\vec{n}} J(|\vec{n}|) \sum_{\vec{l}} e^{i\vec{q}\cdot\vec{l}} P(\vec{l}, N), \qquad (7.8)$$

where we have defined $\vec{n} \equiv \vec{m} - \vec{l}$. As expected, since (7.8) is the transform of a convolution product, we obtain

$$\tilde{P}(\vec{q}, N+1) = \tilde{J}(\vec{q})\tilde{P}(\vec{q}, N).$$

Iterating we have

$$\tilde{P}(\vec{q}, N+1) = [\tilde{J}(\vec{q})]^2 \tilde{P}(\vec{q}, N-1) = \cdots = [\tilde{J}(\vec{q})]^{N+1} \tilde{P}(\vec{q}, 0).$$

Using the condition (7.4), whereby $\tilde{P}(q, 0) = 1$, we have that after N passes the transform is

$$\tilde{P}(\vec{q}, N) = [\tilde{J}(\vec{q})]^N. \qquad (7.9)$$

To compute the Fourier transform of the transition probability per unit time, we observe that the sum over $\vec{n}$ gives the following:

$$\tilde{J}(\vec{q}) = \sum_{\vec{n}} e^{i\vec{q}\cdot\vec{n}} J(|\vec{n}|) = \frac{1}{2D} \sum_{\vec{n}\,|\,|\vec{n}|=1} e^{i\vec{q}\cdot\vec{n}}$$

$$= \frac{1}{2D} \sum_{\alpha=1}^{D} \left(\sum_{\substack{n_\alpha=\pm 1, \\ n_{\beta\neq\alpha}=0}} e^{i q_\alpha n_\alpha} \right) = \frac{1}{2D} \sum_{\alpha=1}^{D} (e^{i q_\alpha} + e^{-i q_\alpha}), \qquad (7.10)$$

from which

$$\tilde{J}(\vec{q}) = \frac{1}{D} \sum_{\alpha=1}^{D} \cos(q_\alpha). \qquad (7.11)$$

This relation allows us to derive the probability (7.1) and may make its meaning more explicit. Each step corresponds to an application of $\tilde{J}(\vec{q})$ giving $2D$ contributions, and constructs a phase characterizing each path. Closed paths, which return to the initial point, have a phase 0. The interested reader can explicitly verify that using the multinomial formula to compute $[J(\vec{q})]^N$ yields the relation (7.1) written at the beginning of the section between $P(\vec{m}, N)$ and $C(\vec{m}, N)$ (hint: use (7.7)).

We are now interested in estimating the behavior of the function $P(\vec{m}, N)$ for large values of N, i.e., in the case where many steps have already been walked. The function $\tilde{J}(\vec{q})$ is a sum of cosines, and is therefore equal to 1 in modulus only at the points $\vec{q}_0 = (0, 0, \ldots, 0)$ and $\vec{q}_\pi = (\pi, \pi, \ldots, \pi)$. For $N \to \infty$ these two points will give the dominant contribution (the product of many factors smaller than 1 will give an infinitesimal contribution). For this reason we can also approximate the Brillouin zone of the integrand expansion around $\vec{q}_0$ or $\vec{q}_\pi$ with the entire $\mathbb{R}^N$ space. We begin by evaluating the contribution to $P(\vec{m}, N)$ by the integration region with all q_α close to zero. Expanding the cosine in powers of q we have that in this region

$$
P(\vec{m}, N) = \frac{1}{(2\pi)^D} \int_B d^D q\, e^{-i\vec{q}\cdot\vec{m}} \left(\frac{1}{D} \sum_{\alpha=1}^{D} \cos q_\alpha \right)^N
$$

$$
\simeq \frac{1}{(2\pi)^D} \int_B d^D q\, e^{-i\vec{q}\cdot\vec{m}} \exp\left\{ N \ln\left[1 - \frac{1}{2D}|\vec{q}\,|^2 + O(|q|^4 N) \right] \right\}, \qquad (7.12)
$$

and that

$$
P(\vec{m}, N) = \frac{1}{(2\pi)^D} \int_B d^D q\, \exp\left\{ -i\vec{q}\cdot\vec{m} - \frac{N}{2D}|\vec{q}\,|^2 + N O(q^4) \right\}
$$

$$
= \left(\frac{D}{2\pi N} \right)^{D/2} \exp\left\{ -\frac{|\vec{m}|^2 D}{2N} + O\left(\frac{m^4}{N^3} \right) \right\}. \qquad (7.13)
$$

We underline that we are using exactly the same technique that we introduced to prove the central limit theorem in Section 3.2.

The other asymptotically non-zero contribution is the one obtained by expanding around the point $\vec{q}_\pi = (\pi, \pi, \ldots, \pi)$: here we obtain a term equal to (7.13) multiplied, however, by a site-dependent phase (for an argument close to π the cosine is close to -1). Ultimately, summing the two contributions gives

$$
P(\vec{m}, N) \propto \left(\frac{D}{2\pi N} \right)^{D/2} \exp\left\{ -\frac{m^2 D}{2N} \right\} \left[1 + (-1)^{N + \sum_{\alpha=1}^{D} m_\alpha} \right]. \qquad (7.14)
$$

The additional term simply guarantees that if N is even the probability of being in $\vec{m}$ is different from zero only for even values of $\sum_\alpha m_\alpha$, while if N is odd the probability is different from zero only for odd values of $\sum_\alpha m_\alpha$. We note that this is a characteristic of the specific type of lattice we are considering, made of square/cubic cells. In a triangular lattice, for example, or in a body-centered or face-centered cubic lattice, it is possible to reach the same site in both odd or even numbers of steps.

The additional term becomes essentially negligible (and is often eliminated *a priori*) as soon as we are concerned with knowing the value of the probability that the random walker is in a given region of space (e.g., a hypercubic cell of 2^D lattice sites or even just a pair of sites next to each other), instead of at a specific site. If, indeed, we average $P(\vec{m}, N)$ over a spatial region of linear dimension L (and volume L^D), for L large we have that

$$
\sum_{\vec{m} \in L^D} (-1)^{\sum_\alpha m_\alpha} f(m/L) = O(e^{-\lambda L})
$$

goes to zero exponentially with L, where λ is an appropriate constant, if f is an analytic function.

The distribution (7.14), valid in the central limit, for $N \gg 1$, is isotropic. It is interesting to note that for small values of N the probability distribution is not isotropic. For example in two dimensions the probability of going in five steps from the origin to the point $(0, 5)$ is 4^{-5} and it is different from the probability of going in five steps to the point $(4, 3)$, which is zero. However, for large values of N, in the region where the probability is concentrated, i.e., for $|m| = O(N^{1/2})$, the non-rotationally-invariant terms go to zero as N^{-1}. Similar results can also be obtained for random paths defined on other types of lattices, or on the continuum, as we shall see in the next section.

Let us now see what happens for large N but in the region of large deviations $\vec{m} = \vec{\lambda} N$, rather than in the central region ($|\vec{m}|^2 \propto N$). We ask: What is the probability of staying in the site of coordinates $\lambda_\alpha N$ for each of the D directions, after N steps? Proceeding as in the one-dimensional case of Section 4.2 and using the complex rotation $q_\alpha = -\imath h_\alpha$, we arrive at solving a system of saddle point equations,

$$\frac{\sinh{(h_\alpha)}}{\sum_{\gamma=1}^{D} \cosh{(h_\gamma)}} = \lambda_\alpha, \quad \forall \alpha = 1, \ldots, D, \tag{7.15}$$

which has solution $\vec{h}(\vec{\lambda})$ only when

$$\sum_{\alpha=1}^{D} |\lambda_\alpha| \le 1, \tag{7.16}$$

in agreement with the fact that after N steps we must have

$$\sum_{\alpha=1}^{D} |m_\alpha| \le N. \tag{7.17}$$

The probability of $\vec{\lambda}$ after N steps is given by

$$P(\vec{\lambda}, N) \propto \exp{[N S(\vec{\lambda})]}, \tag{7.18}$$

where

$$S(\vec{\lambda}) = \log{\left[\frac{1}{D} \sum_{\alpha=1}^{D} \cosh{\bar{h}_\alpha(\vec{\lambda})} \right]} - \sum_{\alpha=1}^{D} \bar{h}_\alpha(\vec{\lambda})\lambda_\alpha. \tag{7.19}$$

The interesting observation is that the function $S(\vec{\lambda})$, in general, is not rotationally invariant, and only becomes invariant for small λ, i.e., in the central limit.

7.1.2 Random Walk in a Continuous Space

As we will see better in a moment, we can define the process of the random walker in an isotropic medium also in a continuous space, keeping, for now, the times at discrete steps. In the case of walks defined on a continuous space $\vec{x} \in \mathbb{R}^D$, we call $J(\vec{x})$ the probability density of traveling the distance $|\vec{x}|$ in one time step. Assuming that at time zero the particle is located at the origin, arguments similar to those above lead us to conclude that

$$P(\vec{x}, N) = \frac{1}{(2\pi)^D} \int d^D q \, e^{-\imath \vec{q} \cdot \vec{x}} [\tilde{J}(\vec{q})]^N,$$
$$\tilde{J}(\vec{q}) = \int d^D x \, e^{\imath \vec{q} \cdot \vec{x}} J(\vec{x}). \tag{7.20}$$

If the distribution $J(\vec{x})$ is, at least approximately, isotropic, i.e., if

$$\int d^D x\, J(\vec{x}) x_\alpha = 0\,,$$

$$\int d^D x\, J(\vec{x}) x_\alpha x_\beta = \nu \delta_{\alpha,\beta}\,, \tag{7.21}$$

where ν is the variance of J, for small q the function $\tilde{J}(q)$ turns out to be

$$\tilde{J}(\vec{q}) = 1 - \frac{\nu}{2}|\vec{q}\,|^2 + O(q^3)\,. \tag{7.22}$$

Exactly as in the previous case, we obtain that for large N the probability density tends to a Gaussian

$$\begin{aligned}
P(\vec{x}, N) &= \frac{1}{(2\pi)^D} \int d^D q\, \exp\left\{-\imath \vec{q} \cdot \vec{x} - \frac{N\nu}{2}|q|^2 + O(q^3)\right\} \\
&= \left(\frac{1}{2\pi N\nu}\right)^{D/2} \exp\left\{-\frac{|x|^2}{2N\nu} + O\left(\frac{|x|^3}{N^2}\right)\right\}.
\end{aligned} \tag{7.23}$$

7.2 Random Walkers in Non-Homogeneous Spaces

Let us now consider a more complicated problem, in which the probability of going from the point $\vec{k}$ to the point $\vec{m}$ depends on both $\vec{k}$ and $\vec{m}$. Or, equivalently, it depends on the starting point $\vec{k}$ and on the displacement $\vec{m} - \vec{k}$. We will write the transition probability from $\vec{k}$ to $\vec{m}$ as $J(\vec{k}, \vec{m} - \vec{k})$. If the medium is homogeneous, we lose the dependence on the initial point of the transition; if it is isotropic, we lose the dependence on the direction and $J(\vec{k}, \vec{m} - \vec{k}) = J(|\vec{m} - \vec{k}|)$.

Obviously one must have that

$$\sum_{\vec{m}} J(\vec{k}, \vec{m} - \vec{k}) = 1\,. \tag{7.24}$$

Equivalently, instead of summing over all arrival sites $\vec{m}$, we can sum over all displacements $\vec{\ell} = \vec{m} - \vec{k}$ from $\vec{k}$ to all possible $\vec{m}$:

$$\sum_{\vec{m}} J(\vec{k}, \vec{m} - \vec{k}) = \sum_{\vec{m}-\vec{k}} J(\vec{k}, \vec{m} - \vec{k}) = \sum_{\vec{\ell}} J(\vec{m} - \vec{\ell}, \vec{\ell}) = 1\,.$$

In this notation the recursive relation (7.2) becomes

$$P(\vec{m}, N + 1) = \sum_{\vec{\ell}} J(\vec{m} - \vec{\ell}, \vec{\ell}) P(\vec{m} - \vec{\ell}, N)\,, \tag{7.25}$$

i.e., we add up all terms in which we arrived in N steps at site $\vec{m} - \vec{\ell}$ and, then, in the last step, we go from $\vec{m} - \vec{\ell}$ to $\vec{m}$, moving by $\vec{\ell}$.

We now define two quantities that will be important in the following, when we study the structure of the transition probability for a large number N of steps: the average velocity of lattice displacement in the discrete time N,

$$c_\alpha(\vec{m}) \equiv \sum_{\vec{\ell}} \ell_\alpha J(\vec{m} - \vec{\ell}, \vec{\ell}), \tag{7.26}$$

and the covariance of the displacements,

$$\begin{aligned} v_{\alpha\beta}(\vec{m}) &\equiv \sum_{\vec{\ell}} [\ell_\alpha - c_\alpha(\vec{m})][\ell_\beta - c_\beta(\vec{m})] J(\vec{m} - \vec{\ell}, \vec{\ell}) \\ &= \sum_{\vec{\ell}} \ell_\alpha \ell_\beta J(\vec{m} - \vec{\ell}, \vec{\ell}) - c_\alpha(\vec{m}) c_\beta(\vec{m}). \end{aligned} \tag{7.27}$$

The quantities $c_\alpha(\vec{m})$ and $v_{\alpha,\beta}(\vec{m})$ have the physical meaning of *average velocity* and *diffusion constant tensor*. We note that since the time step is discrete and equal to 1, then $\vec{c}(\vec{m})$ coincides with the mean displacement. We will see that it will not be the same by rescaling the lattice and taking the limit to continuous space, with infinitesimal lattice and time steps.

7.3 Continuum Limit and the Fokker–Planck Equation

We are interested in studying the behavior of the function $P(\vec{m}, N)$ for large values of the number of time steps N and for large distance $|\vec{m}|$. The results of Section 7.1.1, see for example the relation (7.14), show that the region of interest is the one in which m^2 is of order $O(N)$. We will, therefore, analyze the behavior of $P(\vec{m}, N)$ in this central region in the continuum limit. In order to formalize this limit, we will first introduce a lattice step parameter a, rescale the position $\vec{m} \to a\vec{m}$ and likewise all the other quantities involved, and finally send the scale a to zero so that the ratio m^2/N remains finite in the central limit. To this end, we can define the scaling

$$\vec{x} \equiv a\vec{m} \quad \text{and} \quad t \equiv a^2 N, \tag{7.28}$$

and study the limit for $a \to 0$ of the probability per unit volume hypercubic cell a^D defined as

$$P_{(a)}(\vec{x}, t) \equiv \frac{1}{a^D} P(\vec{m}, N) = \frac{1}{a^D} P\left(\frac{\vec{x}}{a}, \frac{t}{a^2}\right). \tag{7.29}$$

Interestingly, the factor a^{-D} preserves the correct normalization of the probability in the continuum limit,

$$\lim_{a \to 0} \int d^D x \, P_{(a)}(x, t) = 1,$$

since $d^D x \sim a^D$ and

$$a^D \sum_{\vec{m}} \sim \int d^D x.$$

We also rescale the transition probability $J(\vec{m} - \vec{\ell}, \vec{\ell})$ at the starting point, define the distribution

$$J_{(a)}(\vec{x} - \vec{\ell}a, \vec{\ell}) = J\left(\frac{\vec{x}}{a} - \vec{\ell}, \vec{\ell}\right), \tag{7.30}$$

and, furthermore, assume that the limit

$$\lim_{a \to 0} J_{(a)}(\vec{x} - \vec{\ell}a, \vec{\ell}) \equiv J(\vec{x}, \vec{\ell}) \tag{7.31}$$

is well defined. This is the probability of a displacement $\vec{\ell}a$ in a time interval a^2, according to the rescaling (7.28). In the rescaled grid the displacement in a time interval a^2 is $\vec{\ell}a$. This means that the average displacement, i.e., the lattice velocity defined in (7.26), will also scale as a. The displacement velocity in the continuum will, instead, be defined as the limit for $a \to 0$ of

$$v_\alpha^{(a)}(\vec{x}) \equiv \frac{\sum_{\vec{\ell}} \ell_\alpha a J_{(a)}(\vec{x} - \vec{\ell}a, \vec{\ell})}{a^2}. \tag{7.32}$$

We emphasize that we have explicitly used the fact that the time interval is no longer 1, but a^2. The rescaled expression of the lattice velocity (7.26), on the other hand, using the prescription (7.30) for the probability of displacement of a lattice step a in a time step a^2, is

$$c_\alpha^{(a)}(\vec{x}) \equiv \sum_{\vec{\ell}} \ell_\alpha J_{(a)}(\vec{x} - \vec{\ell}a, \vec{\ell}) = a v_\alpha^{(a)}(\vec{x}), \tag{7.33}$$

and, therefore, of order a. The limit

$$\vec{v}(\vec{x}) = \lim_{a \to 0} \vec{v}^{\,(a)}(\vec{x})$$

represents – with the right dimensions – the average velocity in continuous space and time. The $\vec{c}(\vec{m})$ will, on the other hand, tend to zero in the continuum limit.

Finally, the rescaled diffusion matrix (7.27) is

$$v_{\alpha\beta}^{(a)} \equiv \sum_{\vec{\ell}} \ell_\alpha \ell_\beta J_{(a)}(\vec{x} - \vec{\ell}a, \vec{\ell}) - c_\alpha^{(a)}(\vec{x}) c_\beta^{(a)}(\vec{x}) = \sum_{\vec{\ell}} \ell_\alpha \ell_\beta J_{(a)}(\vec{x} - \vec{\ell}a, \vec{\ell}) + O(a^2). \tag{7.34}$$

To ease the presentation, we will also assume that the diffusion is isotropic (independent of the spatial direction considered) and spatially uniform (independent of $\vec{x}$):

$$v_{\alpha\beta}(\vec{x}) \equiv v \delta_{\alpha\beta}. \tag{7.35}$$

7.3.1 The Fokker–Planck Equation

To determine the equations of dynamics of the random walker in the limit of $a \to 0$, we transform the recursive equations (7.2) from the finite difference form to a differential form. With our new notation we have that

$$P_{(a)}(\vec{x}, t + a^2) = \sum_{\vec{\ell}} J_{(a)}(\vec{x} - \vec{\ell}a, \vec{\ell}) P_{(a)}(\vec{x} - \vec{\ell}a, t). \tag{7.36}$$

Expanding in power series in a up to second order, in Appendix 7.B.1 we find the *Fokker–Planck equation* for diffusive processes,

$$\frac{\partial P(x,t)}{\partial t} = -\sum_{\alpha=1}^{D} v_\alpha \frac{\partial P(\vec{x},t)}{\partial x_\alpha} + \frac{v}{2}\sum_{\alpha=1}^{D}\frac{\partial^2 P(\vec{x},t)}{\partial x_\alpha^2}$$
$$= -\vec{v}\cdot\vec{\nabla}P(\vec{x},t) + \frac{v}{2}\Delta P(\vec{x},t), \tag{7.37}$$

where in the last step we made explicit the choice (7.35) to isotropic diffusion. For more general cases, we refer the reader to Appendix 7.B.1.

In the limit where the diffusion is negligible ($v = 0$), we obtain the usual equation for the deterministic motion of particles in a uniform velocity field, i.e., the so-called Liouville, equation

$$\frac{\partial P(\vec{x},t)}{\partial t} = -\sum_{\alpha} v_\alpha \frac{\partial}{\partial x_\alpha}P(\vec{x},t). \tag{7.38}$$

The diffusion term $(v/2)\Delta P(x,t)$ is, however, connected to the random character of the random path. Interestingly, the Fokker–Planck equation can be rewritten as a continuity equation of the probability,

$$\frac{\partial P(\vec{x},t)}{\partial t} = -\vec{\nabla}\cdot J(\vec{x},t), \tag{7.39}$$

where we have defined the *probability current*

$$\vec{J}(\vec{x},t) \equiv \vec{v}P(\vec{x},t) - \frac{v}{2}\vec{\nabla}P(\vec{x},t). \tag{7.40}$$

We can further reformulate the latter as

$$\vec{J}(\vec{x},t) = \vec{w}(\vec{x})P(\vec{x},t), \tag{7.41}$$

with the effective velocity field $\vec{w}$ defined as

$$\vec{w}(x,t) \equiv \vec{v} - \frac{v}{2}\vec{\nabla}\log\left(P(\vec{x},t)\right). \tag{7.42}$$

In the proper context, this last term can be interpreted as the contribution of an *osmotic pressure* in liquids, but to clarify this interpretation we should introduce concepts such as the chemical potential that would take us too far.

7.4 Random Walks with Traps

Before tackling the resolution of the equation in the continuum, let us consider a variation on the theme. Our walker moves randomly in an isotropic medium but at each point, $\vec{m}$, a probability $w(\vec{m})$ is defined of interrupting the journey (e.g., a hyperspace door to return home). In the simplest (homogeneous) case we have that

$$P(\vec{m}, N+1) = [1 - w(\vec{m})]\sum_{\vec{\ell}} J(|\vec{\ell}|)P(\vec{m}-\vec{\ell}, N). \tag{7.43}$$

Obviously, in this case, the total probability of being somewhere after N steps,

$$\mathcal{P}(N) = \sum_{\vec{m}} P(\vec{m}, N), \tag{7.44}$$

is not equal to 1 any more. There is, indeed, a non-zero probability $\mathcal{A}(N)$ that the walker is *absorbed* and goes off the grid, such that $\mathcal{A}(N) + \mathcal{P}(N) = 1$. The probability $\mathcal{A}(N)$ of being absorbed can only increase with time N and $\mathcal{P}(N)$ can only decrease. If we, then, place strictly absorbing conditions at some points in the lattice (i.e., sites $\vec{m}$ of the lattice where $w(\vec{m}) = 1$), the total probability $\mathcal{P}(N)$ represents the probability of avoiding those points in N steps.

Interestingly, if we rescale the lattice with a and take the limit for a tending to zero, taking care to rescale, as well, the probability of absorption as $w(\vec{n}) = a^2 V(\vec{x})$, we obtain the partial differential equation

$$\frac{\partial P(\vec{x}, t)}{\partial t} = -V(\vec{x})P(\vec{x}, t) + \frac{\nu}{2}\Delta P(\vec{x}, t). \tag{7.45}$$

In this equation, we have placed the drift velocity term $\vec{v} = 0$. For the isotropic and homogeneous form of the displacement probability $J(|\vec{\ell}\,|)$, indeed, the continuum limit $\vec{v}$ of the drift is constant (see Appendix 7.B.1). This constant can always be set to zero for simplicity, e.g., working in a reference system in uniform motion. This leaves the diffusion-only term $(\nu/2)\Delta P(\vec{x}, t)$ of the Fokker–Planck equation plus an additional term $-V(\vec{x})P(\vec{x}, t)$.

If we now define a mass $m_S = \hbar/\nu$, rescale the trap term as $V(\vec{x}) \to V(\vec{x})/\hbar$, and perform a rotation of the time variable to a purely imaginary time $t = \iota\tau$ in the complex plane, we can recognize that (7.45) is equivalent to the Schrödinger equation at imaginary time τ for the wave function $\Psi(\vec{x}, \tau) \equiv P(\vec{x}, \iota\tau)$:

$$\iota\hbar \frac{\partial \Psi(\vec{x}, \tau)}{\partial \tau} = \left[-\frac{\hbar^2}{2m_S}\Delta + V(\vec{x}) \right] \Psi(\vec{x}, \tau) = \hat{H}\Psi(\vec{x}, \tau). \tag{7.46}$$

In the last step, we have defined the self-adjoint operator[2] $\hat{H}$. Its eigenvalue equation is

$$\hat{H}\psi_k(\vec{x}) = E_k\psi_k(\vec{x}),$$

with real eigenvalues E_k. By convention, we sort them starting with the lowest, $E_0 < E_1 < \cdots$, denoting by E_0 the energy of the ground state of the time-independent

---

[2] The precise definition of a self-adjoint operator is complicated. There are *symmetric* operators, in both finite-dimensional and infinite-dimensional spaces, which are easier to define. Given two functions f and g of a Hilbert space (see Appendices 3.E and 3.I) with inner product $\langle f, g \rangle$, an operator H is defined to be symmetric if the relation $\langle Hf, g \rangle = \langle f, Hg \rangle$ holds. A symmetric operator can move from right to left. If it is written in matrix form as $H_{i,k}$, the matrix is symmetric, $H_{k,i} = H_{i,k}$. Moreover, the concept of a *self-adjoint operator* exists, which we will not define here, which is more restrictive than symmetric. Their difference lies in the domain of the operator. If the domain is limited, the definitions of symmetric and self-adjoint coincide. Fundamental theorems such as the spectral theorem are only valid for strictly self-adjoint operators. Given, however, that non-self-adjoint symmetric operators are "rare", we follow the long tradition in physics of assuming that all symmetric operators are also self-adjoint.

Schrödinger equation and by $\psi_0(\vec{x})$ the relative eigenfunction. In terms of the eigen-functions ψ_k we can write the solution of the complete Schrödinger equation (7.46) for large times as

$$\Psi(\vec{x}, -\imath t) = P(\vec{x}, t) = \sum_k \psi_k(\vec{x}) e^{-E_k t} \simeq \psi_0(\vec{x}) e^{-E_0 t}.$$

For very large times the survival probability, i.e., the total probability – see (7.44) – is

$$\mathcal{P}(t) = \int d^D x \, P(\vec{x}, t) \propto \exp(-E_0 t).$$

We only need to know the ground state of the equivalent Schrödinger equation for the wave function of a non-relativistic particle of mass m_S subject to a potential $V(\vec{x})$ to determine the time behavior of the probability that a diffusive process with traps survives absorption.

By exploiting this correspondence in the one-dimensional case, for instance, we can compute the probability that a walker starting from the origin does not leave the interval $[-L/2, L/2]$ from the ground state of the Schrödinger equation in an infinite potential hole of width L with absorbing boundary conditions: $P(\pm L/2, t) = 0, \forall t$. The energy of the ground state is

$$E_0 = \frac{\pi^2 \hbar^2}{2 m_S L^2} = \frac{\pi^2 v}{2 L^2},$$

where in the last step we translated it into the stochastic equivalent. Thus, the probability of remaining in the interval L at time t, for long times, decays as

$$\mathcal{P}(t) \propto e^{-E_0 t} = \exp\left(-\frac{v\pi^2}{2L^2} t\right), \tag{7.47}$$

and the characteristic survival time of the random walker in the presence of traps at distance L is $\tau_L = 1/E_0 = 2L^2/(v\pi^2)$.

We conclude this analysis with an observation. The probability of the particle being in a given region tends to zero with time. Thus, there is no well-normalized stationary distribution for infinite times ($P(\vec{x}) = 0, \forall \vec{x}$ is not well normalized). At the same time, we have seen that the smallest eigenvalue, the ground-state one, is strictly positive. As can be well understood from the formulas above, the fact that $E_0 > 0$ implies that the probability vanishes. We shall see below how the existence (and uniqueness) of a stationary solution of the Fokker–Planck equation is linked to the existence of a ground state of some associated operator with eigenvalue $E_0 = 0$.

7.5 Brownian Motion Stationary Solution

In general, the study of the solutions of the Fokker–Planck equation is very compli-cated. In difficult cases where the random walker moves in a continuous space, the velocity field is generalized to a drift term related to an external force exerted on the system. To better understand the role of the different quantities involved and the properties of some very important solutions in physics (and beyond), we can consider here a paradigmatic case. Let us consider a particle, or grain, at position $\vec{x}$ and with velocity $\vec{v}$ in a liquid, which is not large but is massive enough to ignore continuous collisions with liquid molecules. This particle has a motion that is very difficult to describe in deterministic terms. The molecules of the liquid have a kinetic energy, as we know from classical physics courses, related to the temperature of the thermal bath in which the system is at equilibrium. Their collisions with the grain cause its thermal agitation and it consequently performs a diffusive motion like that of the random walker, called *Brownian motion*.[3]

The liquid exerts viscous friction on the particle, in the form of a force, the Stokes force, proportional to the latter's velocity: $F_{\text{Stokes}} = -\gamma \vec{v}$, where the friction coefficient γ is proportional to the viscosity η of the medium. The proportionality coefficient depends on the shape of the particle. For example, for a sphere of radius r, it is $\gamma = 6\pi r \eta$. We will write in the following simply $\gamma = \eta$. We also call the diffusion coefficient ν (as in Eq. (7.37)) and any external force possibly acting on the particle $\vec{F}(\vec{x})$. We can, therefore, write the equation of dynamics in D dimensions as

$$m\ddot{\vec{x}} = -\eta\dot{\vec{x}} + \vec{F}(\vec{x}) + \vec{X}(t), \tag{7.48}$$

where $\vec{X}(t)$, according to Langevin, plays the role of a "fluctuating force", which encodes the randomness of diffusion due to collisions with the liquid molecules. We will discuss the term $\vec{X}(t)$, also known as "noise", in more detail in Section 7.6. For now, let us simply note that its mean will be zero and its covariance matrix is

$$\langle X_j(t) X_k(t') \rangle_{j,k} = \nu \delta_{j,k} \delta(t - t').$$

Since the Fourier antitransform of the Dirac delta $\delta(t - t')$ is a superposition of plane waves at all frequencies, this is called *white noise*, as with light.

If the liquid is particularly viscous, $\eta \gg 1$, the acceleration term will be irrelevant: $|m\ddot{\vec{x}}| \ll |\eta\dot{\vec{x}}|$. This is called the *overdamping limit*. Equation (7.48) thus reduces to a first-order differential equation with the addition of a stochastic term:

$$\frac{d\vec{x}}{dt} \simeq \frac{\vec{F}(\vec{x})}{\eta} + \frac{\sqrt{\nu}}{\eta}\vec{\xi}(t). \tag{7.49}$$

This is our first example of a *stochastic* differential equation, the *Langevin equation*, which we will further discuss in Section 7.6. The white noise $\vec{\xi}(t)$ that appears here, $\xi_k(t) \equiv X_k(t)/\sqrt{\nu}$, has covariance

$$\langle \xi_j(t) \xi_k(t') \rangle = \delta_{j,k} \delta(t - t').$$

It is the noise of a process of zero average velocity and unitary diffusion constant.

We are now interested in the representation of the stochastic process in terms of its probability density, i.e., the solution of a Fokker–Planck equation. In particular,

---

[3] The motion is named after the botanist Robert Brown (1773–1858), a pioneer in the use of the microscope, who observed that very small particles coming out of the microcavities/vacuoles of pollen grains suspended in water followed an agitated and irregular motion [55].

since the stochastic fluctuations are encoded in the diffusion term, we are interested in identifying the drift velocity to be included in the Fokker–Planck equation, which in the overdamping limit results in

$$\vec{v} = \frac{\vec{F}(\vec{x})}{\eta}\,.$$

The corresponding Fokker–Planck equation for the probability distribution of the position of the particle subject to Brownian motion in the strong viscosity limit eventually reads

$$\frac{\partial P(\vec{x},t)}{\partial t} = -\frac{1}{\eta}\,\vec{\nabla}\cdot(\vec{F}P(\vec{x},t)) + \frac{\nu}{2}\,\nabla^2 P(\vec{x},t)\,. \tag{7.50}$$

This is called the Smoluchowski equation[4] [56].

If the external force is conservative, we have $\vec{F}(\vec{x}) = -\vec{\nabla}U(\vec{x})$. In the absence of diffusion the motion of the particle would be directed towards a minimum of potential energy $U(\vec{x})$. In the presence of diffusion the particle can move away from the minimum and, in the case of the potential $U(\vec{x})$ possessing several minima, it can cross a potential barrier and explore positions $\vec{x}$ corresponding to another minimum. Equation (7.50) in the conservative case becomes

$$\frac{\partial P(\vec{x},t)}{\partial t} = \frac{\nabla^2 U(\vec{x})}{\eta}\,P(\vec{x},t) + \frac{\vec{\nabla}U(\vec{x})\cdot\vec{\nabla}P(\vec{x},t)}{\eta} + \frac{\nu}{2}\,\nabla^2 P(\vec{x},t)\,. \tag{7.51}$$

Solving the equation for $P(\vec{x},t)$ requires theoretical equipment that is beyond the contents of this book, if only slightly. We can focus, however, on the stationary solution $p_0(\vec{x})$, i.e., the limit function of $P(\vec{x},t)$ for $t \to \infty$. This coincides with the solution of Eq. (7.51), with no time dependence,

$$\frac{\partial p_0(\vec{x})}{\partial t} = 0\,,$$

and with appropriate boundary conditions. Among the boundary conditions that guarantee the existence and uniqueness of a stationary solution are the following:

- reflective conditions at the walls of the liquid container in which the particle is agitated so that the probability current – see (7.40) – is zero at the barriers; and
- so-called *natural* boundary conditions, when the probability density and current vanish at infinity,

$$\lim_{|\vec{x}|\to\infty} p_0(\vec{x}) = \lim_{|\vec{x}|\to\infty} J(\vec{x}) = 0 \iff \lim_{|\vec{x}|\to\infty} p_0(\vec{x}) = \lim_{|\vec{x}|\to\infty} \frac{\partial p_0(\vec{x})}{\partial x_\alpha} = 0\,. \tag{7.52}$$

With these boundary conditions, it can be shown that the stationary solution exists, is unique, and has the *potential* form

$$p_0(\vec{x}) \propto e^{-\Phi(\vec{x})}\,. \tag{7.53}$$

---

[4] The diffusive term in the original Smoluchowski formulation has coefficient $\nu = 2k_B T/\eta$, but we will see in a moment that this is exactly what happens at equilibrium in our case.

Before proceeding with the demonstration, let us dwell for a moment on the meaning of the "potential"[5] $\Phi(\vec{x})$. We expect that for a particle of potential energy $U(\vec{x})$, suspended in a viscous liquid, subject to the collisions of many molecules in equilibrium in a thermal bath of temperature T, the probability density of being in a given volume $d\vec{x}$ for long times tends to the Boltzmann–Gibbs distribution of the canonical ensemble [57]

$$p_0(\vec{x}) = \frac{e^{-\beta U(\vec{x})}}{Z},$$

$$Z \equiv \int d\vec{x}\, e^{-\beta U(\vec{x})}, \tag{7.54}$$

where $\beta = 1/(k_B T)$ is the inverse temperature. If this were the stationary solution, it would be $\Phi(\vec{x}) = \beta U(\vec{x})$, precisely proportional to the potential energy.

7.5.1 The Stokes–Einstein Relation

Let us assume that the solution is a well-normalized probability density and satisfies the necessary boundary conditions, such as, for example, Eq. (7.52). By inserting (7.54) into the Fokker–Planck equation (7.51), or into the probability current definition, it is easy to verify that it satisfies both the equation and the natural conditions:

$$-\frac{\vec{\nabla} U}{\eta}\, p_0(\vec{x}) - \frac{\nu}{2}\, \vec{\nabla} p_0(\vec{x}) = 0,$$

as long as one has

$$\nu = \frac{2 k_B T}{\eta}.$$

The latter formula is a particular realization of one of Einstein's most brilliant results. It is a general relation valid for systems at equilibrium at temperature T between the viscosity coefficient η and the diffusion coefficient ν:

$$\eta\, \nu \propto k_B T. \tag{7.55}$$

Equation (7.55) is called the *Stokes–Einstein relation* and, knowing the proportionality factor due to the shape of the particle, it allows the Boltzmann constant to be measured by observations on the diffusive motion of macroscopic bodies. In the particular case of spherical particles of radius r, we have, for example, that[6]

$$\nu\, \eta = \frac{k_B T}{3\pi r}. \tag{7.56}$$

---

[5] A stationary solution can also be obtained, e.g., by imposing periodic boundary conditions. In this case, one has a non-zero current flow. The solution is stationary but will not be of the potential type, and this is beyond the scope of our discussion. Interested readers may refer to the texts [30, 56].

[6] There is a factor of 2 difference in the definition of our diffusion coefficient ν from the covariance of the random walker displacement and the original Einstein definition of the diffusion constant D: $\nu = 2D$.

7.5.2 Uniqueness of Fokker–Planck Stationary Solution

Let us now turn to the proof of the existence and uniqueness of the stationary solution. Let us return to the notation of Eq. (7.37) where the external force field $\vec{F}$ is simply denoted by $\vec{v}$, more generically called the drift field. It is enough to rescale $\vec{F}/\eta = \vec{v}$ in Eq. (7.50). In the specific case of conservative forces, this means rescaling the potential energy $U(\vec{x})/\eta \to U(\vec{x})$.

We now want to show that if there is a choice of the partition function Z in (7.54) such that

$$\int d^D x \, p_0(\vec{x}) = \int d^D x \, \frac{e^{-\beta U(\vec{x})}}{Z} = \int d^D x \, \frac{e^{-(2/v)U(\vec{x})}}{Z} = 1 \,,$$

then the distribution $p_0(\vec{x})$ is the only time-independent solution and for any initial condition[7] we have that

$$\lim_{t \to \infty} P(\vec{x}, t) = p_0(\vec{x}) \,. \tag{7.57}$$

Solutions $P(\vec{x}, t)$ of this type, which are not stationary but tend towards a stationary solution for long times, are called *homogeneous solutions*. For this to be guaranteed, it is enough to assume that the potential $U(\vec{x})$ goes to infinity sufficiently quickly when $|\vec{x}|$ goes to infinity.

Proof The basic idea of the proof consists in transforming the Fokker–Planck equation,

$$\frac{\partial P(\vec{x}, t)}{\partial t} = -\vec{\nabla} \cdot [\vec{v} \, P(\vec{x}, t)] + \frac{v}{2} \nabla^2 P(\vec{x}, t) \equiv H_{\mathrm{FP}} P(\vec{x}, t) \,,$$

whose operator $H_{\mathrm{FP}} \equiv [- \vec{\nabla} \cdot \vec{v} + (v/2)\nabla^2]$ is not self-adjoint, into an equation with a self-adjoint operator.

This can be achieved by multiplicatively separating the time-dependent part of the probability density as

$$P(\vec{x}, t) = p_0(\vec{x})^{1/2} \rho(\vec{x}, t) \,. \tag{7.58}$$

Starting from the Fokker–Planck equation for systems with conservative forces, simple though somewhat tedious computations show that $\rho(\vec{x}, t)$ satisfies the equation

$$\frac{\partial \rho}{\partial t} = \frac{v}{2} \Delta \rho - \mathcal{V}(\vec{x})\rho \equiv -H\rho \,, \tag{7.59}$$

where

$$H \equiv \mathcal{V}(\vec{x}) - \frac{v}{2}\Delta \,, \tag{7.60}$$

with

$$\mathcal{V}(\vec{x}) \equiv \frac{1}{2v}|\vec{\nabla} U|^2 - \frac{1}{2}\Delta U(\vec{x}) \,, \tag{7.61}$$

---

[7] We do not consider here the case of singular potentials.

is a self-adjoint, otherwise called Hermitian, operator.[8] It is easy to verify that

$$H p_0(\vec{x})^{1/2} = 0, \tag{7.62}$$

i.e., the function $p_0(\vec{x})^{1/2}$ is an eigenfunction of the Hamiltonian H with null eigenvalue. The function $p_0(\vec{x})^{1/2}$ is real and positive, and the Hamiltonian H is self-adjoint; a known theorem (a consequence of the variational formulation) states that the wave function of the ground state must be positive (see e.g., ref. [58]). Since no two positive wave functions can exist orthogonal to each other, $p_0(\vec{x})^{1/2}$ must be the eigenfunction of the ground state. If we denote by λ_n and $\psi_n(\vec{x})$ the eigenvalues and eigenfunctions of the operator H, we obtain that the formal solution of (7.59) is

$$\rho(\vec{x}, t) = e^{-tH} \rho(\vec{x}, 0). \tag{7.63}$$

This can be computed using the expansion in the basis of the eigenfunctions $\{\psi_n\}$ as

$$\rho(\vec{x}, t) = \sum_n c_n \exp(-t\lambda_n) \psi_n(\vec{x})$$

with

$$c_n = \int d^D x \, \rho(\vec{x}, 0) \psi_n(\vec{x}).$$

As demonstrated in Appendix 7.B.2, the real eigenvalues of the operator H are all non-negative. From (7.62) we know that the smallest eigenvalue is $\lambda_0 = 0$ and the others are all positive: $\lambda_0 = 0 < \lambda_1 < \lambda_2 < \cdots$.

In the limit in which t tends to infinity, therefore, we obtain

$$\rho(\vec{x}, t) = \psi_0(\vec{x}) + O(e^{-\lambda_1 t}). \tag{7.64}$$

From (7.58) we eventually have that the probability density

$$P(x, t) = \sqrt{p_0(\vec{x})} \, \rho(\vec{x}, t)$$

for long times decays exponentially to

$$p_0(\vec{x}) = \psi_0(\vec{x})^2, \tag{7.65}$$

which is unique since the ground state ψ_0 is unique. The fact that there is a null eigenvalue of the Hermitian operator associated with the Fokker–Planck differential operator guarantees the existence and uniqueness of the stationary solution. ∎

7.5.3 Fokker–Planck and Schrödinger Equations

If we consider the transformation (7.58), the Fokker–Planck equation applied to $\rho(\vec{x}, t)$ transforms into (7.59). This is formally identical to a Schrödinger equation in pure imaginary time $\tau \equiv -\imath \hbar t$, for a particle of mass $m_S = \hbar^2/\nu$ in a potential $\mathcal{V}(\vec{x})$ which is exactly the same as the one defined in the formula (7.61):

---

[8]   For fans, it is a Sturm–Liouville operator – we will sometimes call it a Hamiltonian.

$$\imath\hbar\frac{\partial\phi(\vec{x},t)}{\partial\tau} = \left[-\frac{\hbar^2}{2m}\nabla^2 + \mathcal{V}(\vec{x})\right]\phi(\vec{x},\tau),$$

where we introduced the wave function $\phi(\vec{x},\tau) \equiv \rho(\vec{x},\imath\tau/\hbar)$.

If the potential energy of the original system $U(\vec{x})$ goes quickly to infinity, the effective potential $\mathcal{V}_S(\vec{x})$ also goes quickly to infinity, and the Schrödinger equation has a purely discrete spectrum. This analogy can be very useful in cases where the solutions of the Schrödinger equation are known. In terms of the eigenvalues and eigenfunctions of the operator H, indeed, it is possible to reconstruct $\rho(\vec{x},t)$ and the entire homogeneous solution of the Fokker–Planck equation in the case where the stationary limit exists.

Depending on the problem, in some cases it will be easier to consider the solution in the Schrödinger potential (7.61) (e.g., in the case of an infinite hole), instead of working with the Fokker–Planck equation with the original potential $U(\vec{x})$ of the external force. In one case the two potentials turn out to be equal. This is the case of motion under a harmonic potential $U(\vec{x}) = k|x|^2/2$, or a linear elastic force (the drift, linear in $\vec{x}$), which is also called the Ornstein–Uhlenbeck process. This can be verified by using the formula (7.61).

We note that, compared to the case of the random walker with traps, cf. Section 7.4, here it is the function $\rho(\vec{x},t)$, the time-dependent part of the homogeneous solution, that satisfies the Schrödinger equation. In the case of traps, instead, the wave function in the Schrödinger equation is for the entire probability density function $P(\vec{x},t)$. This is a very different phenomenon: in the presence of traps – which are also called *absorbing states* as we shall see in Chapter 10 (Sections 10.3.3 and 10.4.1) – the paths are sooner or later absorbed and a stationary probability distribution *does not exist*. As we showed in Section 7.4, indeed, $\mathcal{P}(t) = \int d^D x \, P(\vec{x},t) \to 0$, as $t \to \infty$.

7.6 Langevin Equation

The purpose of this section is to write an evolution law directly for the trajectory $x(t)$, where x is, for example, the position of a particle undergoing Brownian motion in a homogeneous and isotropic medium. Suppose that at time t_0 the particle is at point $\vec{x}_0$ and we compute the probability of finding the particle at point $\vec{x}_0 + \vec{\Delta x}$ at time $t_0 + \Delta t$. In the homogeneous case the drift velocity is constant and the Fokker–Planck equation is

$$\frac{\partial P}{\partial t} = \frac{\nu}{2}\Delta P - \vec{v}\cdot\vec{\nabla}P, \tag{7.66}$$

whose solution, conditional on the initial condition $\vec{x}(t_0) = \vec{x}_0$, is given by

$$P(\vec{x}_0 + \vec{\Delta x}, t_0 + \Delta t \mid \vec{x}_0, t_0) = \frac{1}{(2\pi\nu\Delta t)^{D/2}}\exp\left(-\frac{|\vec{\Delta x} - \vec{v}\Delta t|^2}{2\nu\Delta t}\right). \tag{7.67}$$

In other words, the displacement $\vec{\Delta x}$ has a Gaussian distribution around the mean value of the deterministic motion $\vec{v}\Delta t$, with standard deviation $(v\Delta t)^{1/2}$. We note that the distribution exclusively depends on the time interval Δt and the coordinate interval $\vec{\Delta x}$ and not on the start or the end values.

7.6.1 Wiener Processes

For a single displacement in D dimensions we can write the single coordinate *sample trajectory* extracted from the distribution (7.67) as

$$\Delta x_\alpha = v_\alpha \Delta t + v^{1/2}\Delta W_\alpha, \quad \alpha = 1,\ldots,D, \tag{7.68}$$

where we have introduced a vector of random variables $W_\alpha(t)$ whose increments ΔW_α are Gaussian with null average and variance Δt. They are called Wiener processes and their distribution is the solution of the Fokker–Planck equation of zero drift velocity and unitary diffusion constant,

$$P(\vec{\Delta W},t) = \frac{1}{(2\pi\Delta t)^{D/2}} \exp\left(-\frac{|\vec{\Delta W}|^2}{2\Delta t}\right). \tag{7.69}$$

Equation (7.68) is called the *Langevin equation* and describes the motion of a particle diffusing in a velocity field $\vec{v}$. Formally, in the limit $\Delta t \to 0$ we could write, as we have already seen in (7.49),

$$\frac{d\vec{x}}{dt} = \vec{v}(\vec{x}) + v^{1/2}\frac{d\vec{W}}{dt} = v(x) + v^{1/2}\vec{\xi}(t), \tag{7.70}$$

by defining a Gaussian-distributed "function" $\xi(t)$, with mean value 0 and covariance

$$\langle \xi_\alpha(t_1)\xi_\beta(t_2)\rangle = \delta_{\alpha\beta}\delta(t_1 - t_2). \tag{7.71}$$

It can, in fact, be easily verified that if (7.71) holds true, the increment of the variable W_α,

$$\Delta W_\alpha(t_0) = \int_{t_0}^{t_0+\Delta t} dW_\alpha(t) = \int_{t_0}^{t_0+\Delta t} \xi_\alpha(t)dt,$$

has variance $\langle \Delta W_\alpha(t_0)^2\rangle = \Delta t$ and that the increments at different times are uncorrelated:

$$\langle \Delta W_\alpha(t_0)\Delta W_\alpha(t_1)\rangle = 0$$

for $|t_0 - t_1| > \Delta t$. This is a result consistent with the fact that the joint distribution of a $W(t)$ process at different times factorizes into the distributions (7.69) of the single time intervals:

$$P(W_n,t_n;W_{n-1},t_{n-1};\ldots;W_2,t_2;W_1,t_1;W_0,t_0) = \prod_{k=1}^{n} \frac{1}{(2\pi\Delta t_k)^{1/2}} \exp\left(-\frac{(\Delta W_k)^2}{2\Delta t_k}\right), \tag{7.72}$$

where (for pure convenience) we have set $D = 1$ and defined the abbreviations $W_k \equiv W(t_k)$, $\Delta W_k \equiv W(t_k) - W(t_{k-1})$, and $\Delta t_k \equiv t_k - t_{k-1}$.

The Langevin equation correctly describes the motion of particles of small radius in a viscous liquid in the presence of external forces F. In such a case we can set $\vec{v}(x) = \mu\vec{F}(x)$, where μ, called *mobility*, is a constant that depends on the particle radius and is proportional to the inverse of the viscosity of the medium, $\mu \sim 1/\eta$, as we have seen[9] in Section 7.5.

As we have seen for Brownian motion, where the liquid is at equilibrium at temperature T, and the force is conservative, equal to $-\vec{\nabla}U(x)$, the stationary probability density distribution is $p_0(x) = \exp[-U(x)/k_BT]/Z$ and the diffusion coefficient ν of the particle is related to the temperature and mobility of the liquid by the Einstein relation

$$\mu = \frac{\nu}{2k_BT}.$$

On the left-hand side of this equation we have the response of the particle to some force in a viscous liquid (the mobility) and on the right-hand side we have the thermally induced fluctuations of the particle position, expressed in terms of diffusion, i.e., the covariance of the particle displacement.[10] We will then have that in a time t the particle will travel a distance of the order of

$$|x(t)| \simeq \sqrt{\nu t} = \sqrt{2\mu k_B T t}.$$

This observation tells us that Eq. (7.70) is purely formal, as there is no function, worthy of the name, that has the following required properties:

1. $x(t)$ is of order $t^{1/2}$ and $x(t)$ is not a differentiable function[11] at the origin; and
2. equation (7.71) implies a variance $\langle \xi(0)^2 \rangle = \infty$, and therefore $\xi(t)$ should always be infinite.

It is possible to regularize the function $\delta(t_1 - t_2)$ by writing

$$\langle \xi(t_1)\xi(t_2) \rangle \simeq \frac{1}{(2\pi\epsilon)^{1/2}} \exp\left(-\frac{(t_1 - t_2)^2}{2\epsilon}\right), \tag{7.73}$$

and subsequently sending ϵ to zero. A mathematically consistent treatment can be obtained by considering finite intervals of time Δt and only carrying out the limit $\Delta t \to 0$ at the end. Stochastic integration, however, as opposed to "deterministic" integration, *à la* Riemann, requires specifying the method followed in doing the discrete sums before taking the continuum limit. A detailed discussion of this problem requires the study of stochastic differential calculus [30, 60] and it is beyond the limits of our present analysis. We do, however, want to give an idea of the issue by giving a specific example in Section 7.6.3.

---

[9] If we consider spherical particles of radius r, the mobility is $\mu = 1/(6\pi r\eta)$ and the Stokes–Einstein relation (7.56) becomes, in the notation of this chapter, $\mu = D/(k_BT)$, originally known as the Einstein relation, in which the diffusion constant was $D = 2\nu$, to keep the notation of this chapter.

[10] This is an example of the fluctuation–dissipation theorem [56, 59].

[11] Technically $x(t)$ is a Hölder function of order 1/2. In general, the function $x(t)$ is Hölder of order α if $x(t + \Delta t) - x(t) = O(\Delta^\alpha)$. Differentiable functions are Hölder of order 1.

7.6.2 Autocorrelation of the Wiener Process

Before proceeding, we need to know the time autocorrelation function $\langle W(t_1)W(t_2)\rangle$ of a process $W(t)$, whose distribution is (7.72), in dimension $D = 1$. We choose for now $t_2 > t_1 > t_0$, where t_0 is the initial time and the initial condition is $W(t_0) = W_0$. Explicitly writing the two times correlation function we have

$$\langle W(t_1)W(t_2)\rangle = \int dW_1\, dW_2\, W_1 W_2 P(W_2, t_2; W_1, t_1; W_0, t_0)\,. \tag{7.74}$$

As already noted, the joint distribution of $W(t)$ factors into the distributions of the individual intervals ΔW. To exploit this property and greatly simplify the computation, we rewrite $W(t_1)$ and $W(t_2)$ as functions of the intervals:

$$W(t_1) = W(t_1) - W(t_0) + W(t_0) = \Delta W_1 + W_0\,, \tag{7.75}$$

$$W(t_2) = W(t_2) - W(t_1) + W(t_1) = \Delta W_2 + W(t_1) = \Delta W_2 + \Delta W_1 + W_0\,, \tag{7.76}$$

from which we obtain the expression

$$W(t_1)W(t_2) = \Delta W_2\, \Delta W_1 + \Delta W_2 W_0 + (\Delta W_1)^2 + \Delta W_1 W_0 + W_0^2\,, \quad t_2 > t_1\,.$$

Given that the correlations of the products of different intervals factorize and that $\langle \Delta W_k\rangle = 0$, the following holds for the autocorrelation function,

$$\langle W(t_1)W(t_2)\rangle = \langle(\Delta W_1)^2\rangle + W_0^2 = t_1 - t_0 + W_0^2\,, \quad t_2 > t_1\,, \tag{7.77}$$

where we used the fact that the variance of $\Delta W_1 = W(t_1) - W_0$ is $t_1 - t_0$. Similarly, we can proceed in the case $t_1 > t_2 > t_0$ by simply swapping $1 \leftrightarrow 2$. At the end we find

$$\langle W(t_1)W(t_2)\rangle = \min(t_1, t_2) - t_0 + W_0^2\,. \tag{7.78}$$

7.6.3 Stochastic Integration

Formally, one can construct a theory of stochastic integration from the sum over finite intervals, thus avoiding the problem of derivatives of non-derivable functions:

$$\int_{t_0}^{t} f(t')\, dW(t') \equiv \lim_{n\to\infty} \sum_{i=0}^{n-1} f(\tau_i)[W(t_{i+1}) - W(t_i)]\,, \tag{7.79}$$

where $t_n = t$, and

$$\tau_i \in [t_i, t_{i+1}]\,.$$

Apart from the non-trivial question of deciding which type of limit to adopt since, strictly speaking, the integral over a stochastic process is also a random variable (see Sections 3.1.1 and 3.1.2, and Appendix 3.I), we shall also see that the mean value of the integral changes with the choice of τ_i, on which we compute the function within the interval $\Delta t_i \equiv t_{i+1} - t_i$. To give a concrete example, let us consider the stochastic integral

$$\int_{t_0}^{t} W(t') \, dW(t') = \lim_{n\to\infty} \sum_{i=0}^{n-1} W(\tau_i)[W(t_{i+1}) - W(t_i)] \tag{7.80}$$

and compute the mean of the discrete sum using the result (7.78) of the time autocorrelation of W:

$$\sum_{i=0}^{n-1} \langle W(\tau_i)[W(t_{i+1}) - W(t_i)]\rangle = \sum_{i=0}^{n-1} [\langle W(\tau_i)W(t_{i+1})\rangle - \langle W(\tau_i)W(t_i)\rangle]$$

$$= \sum_{i=0}^{n-1} [\min(\tau_i, t_{i+1}) - \min(\tau_i, t_i)] = \sum_{i=0}^{n-1} (\tau_i - t_i). \tag{7.81}$$

Now let us set the intermediate time equal to $\tau_i = \alpha t_{i+1} + (1-\alpha)t_i$, so that we can vary the position of τ_i at will by varying $\alpha \in [0, 1]$. By substituting in (7.81) and (7.80) we will have

$$\left\langle \int_{t_0}^{t} W(t') \, dW(t') \right\rangle = \lim_{n\to\infty} \sum_{i=0}^{n-1} (\tau_i - t_i) = \alpha(t - t_0). \tag{7.82}$$

Changing α changes the mean value of the integral! We understand, therefore, that a stochastic integral is *defined* by the rule for choosing τ_i in the discrete interval. Different prescriptions on where to compute the function in the interval will give different results. Since the phenomena do not change (and neither does the Fokker–Planck equation), this means that the same stochastic process will be described by different stochastic differential equations if we use different integration prescriptions. The most famous prescriptions are those of Ito ($f(\tau_i) = f(t_i)$) and Stratonovich ($f(\tau_i) = [f(t_i) + f(t_{i+1})]/2$), but we refer the reader to texts dedicated to the theory of stochastic processes to discover their evolution, e.g., [30, 60].

Mathematical Appendices

7.A Fourier Transform on a Discrete Lattice

In crystallography [61] some periodic lattices are classified as *Bravais lattices*. In a Bravais lattice the arrangement and orientation of sites appear the same from whichever site you look at it. This feature *defines* the Bravais lattice and is equivalent to saying that in D dimensions there exists a set of non-coplanar D vectors, $\vec{v}^{\,(1)}, \vec{v}^{\,(2)}, \ldots, \vec{v}^{\,(D)}$ called *primitives*, in terms of which the position vector of each site can be written. The generic lattice point $\vec{m}$ will be a combination of them:

$$\vec{m} = \sum_{\alpha=1}^{D} c_\alpha \vec{v}^{(\alpha)}. \tag{7.83}$$

Now we can look at the behavior of a plane wave of wave vector $\vec{\kappa}$,

$$e^{\imath \vec{\kappa} \cdot \vec{m}},$$

in a system of points $\vec{m}$ composing a Bravais lattice. A plane wave oscillates in space with its own period, which, in general, will be different from that of the Bravais lattice, except for a particular set of values of $\vec{\kappa} = \vec{q}$ such that

$$e^{\imath \vec{q} \cdot (\vec{m}+\vec{d})} = e^{\imath \vec{q} \cdot \vec{d}} \quad \Longrightarrow \quad e^{\imath \vec{q} \cdot \vec{m}} = 1 \tag{7.84}$$

for each vector $\vec{d}$ on the periodic lattice. The set of vectors $\vec{q}$, in turn, defines a lattice, called the *reciprocal lattice*. But even in the reciprocal lattice $\vec{q}$, the arrangement and orientation of the lattice sites is independent of the site from which it is viewed, and therefore, also in the reciprocal lattice, having identified a set of D primitive vectors $u^{(\alpha)}$, one can always express each point as a combination

$$\vec{q} = \sum_{\alpha=1}^{D} b_\alpha \vec{u}^{(\alpha)}. \tag{7.85}$$

Concisely, the reciprocal lattice is also a Bravais lattice. To avoid confusion, we will call the starting lattice $\vec{m}$ the *direct lattice*. Obviously from formula (7.84) we have that the reciprocal lattice of the reciprocal lattice is the direct lattice.

Let us take a simple hypercubic lattice in D dimensions with the same step a in each direction $\hat{x}_\alpha$. In this case the primitive vectors are simply the versors of the Cartesian axes $\hat{x}_\alpha$ of the D-dimensional space multiplied by the step:

$$v^{(1)} = a\hat{x}_1, \quad v^{(2)} = a\hat{x}_2, \quad \ldots, \quad v^{(D)} = a\hat{x}_D,$$

where $\hat{x}^{(\alpha)} \cdot \hat{x}^{(\gamma)} = \delta_{\alpha\gamma}$. Crystals generally have much more complicated structures, but for our purposes studying this case is sufficient.

According to the formula (7.84) the set of D primitive vectors of the reciprocal lattice, of step a_R, is given by

$$u^{(\alpha)} = a_R \hat{x}_\alpha = \frac{2\pi}{a} \hat{x}_\alpha.$$

Indeed, since the coefficients of the projections in (7.83) and (7.85) are nothing but the Cartesian coordinates along the axes in lattice step units ($c_\alpha = n_\alpha$ and $b_\alpha = v_\alpha$, with $\vec{n}, \vec{v} \in \mathbb{Z}^D$), we have

$$1 = \exp\left(\imath \sum_{\alpha=1}^{D} b_\alpha \vec{u}^{(\alpha)} \cdot \sum_{\gamma=1}^{D} c_\gamma \vec{v}^{(\gamma)}\right) = \exp\left(\imath \sum_{\alpha,\gamma}^{1,D} a n_\alpha a_R v_\gamma \, \hat{x}_\alpha \cdot \hat{x}_\gamma\right)$$

$$= \exp\left(\imath a a_R \sum_{\alpha=1}^{D} n_\alpha v_\alpha\right), \tag{7.86}$$

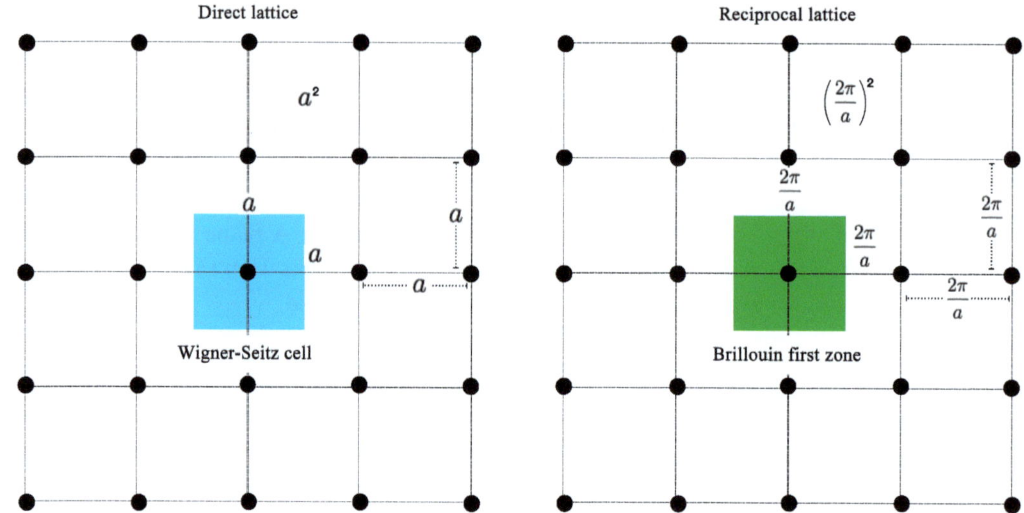

Figure 7.1 Direct (left) and reciprocal (right) Bravais lattices and related Wigner–Seitz cells. The Wigner–Seitz cell of the reciprocal lattice is also called the first Brillouin zone.

and since $\sum_{\alpha=1}^{D} n_\alpha \nu_\alpha$ is an integer, the lattice step of the reciprocal lattice must be

$$a_R = \frac{2\pi}{a}\,.$$

In the direct Bravais lattice, one can define a primitive cell with the same symmetry as the entire lattice: for example, the region of space around a point that is closer to that point than to any other point (Figure 7.1). This is called the *Wigner–Seitz cell* and is invariant for translations of length equal to the lattice step along the lattice primitive vectors. In our hypercubic lattice the Wigner–Seitz cell is a hypercubic cell of side a and volume a^D. In the reciprocal lattice, this primitive cell transforms into a hypercubic cell of side $2\pi/a$ and volume $(2\pi/a)^D$. This hypercube, which is the Wigner–Seitz primitive cell of the reciprocal lattice (space $\vec{q}$), is called the *first Brillouin zone.*

If the lattice step is 1, the first Brillouin zone in one direction will be 2π long. The direct lattice is periodic of period 1 and the reciprocal lattice of period 2π along each Cartesian axis. This is the case we discussed for the random walk on the lattice in Section 7.1.1 .

If the lattice step tends to 0, in the reciprocal space the first Brillouin zone will occupy an infinite volume. This is the case when we return to continuous Fourier transforms, and it is the limit in which we derived the Fokker–Planck equations in Section 7.3.

With these definitions in mind, we first consider the distribution of the position $\vec{x}$ of sites in a Bravais lattice $\vec{m} = a\vec{n}$, with $\vec{n} \in \mathbb{Z}^D$:

$$\rho(\vec{x}) = \sum_{\vec{n}} \delta^{(D)}(\vec{x} - a\vec{n}) = \prod_{\alpha=1}^{D} \sum_{n_\alpha} \delta(x_\alpha - an_\alpha)\,. \tag{7.87}$$

A function $F(\vec{x})$ on a lattice will, therefore, have values only for those values of $\vec{x}$ distributed with $\rho(\vec{x})$. Its D-dimensional Fourier transform (3.76) will, therefore, be

$$\tilde{F}_{\text{lattice}}(\vec{q}) = \int d^D x \, e^{\iota \vec{q}\cdot\vec{x}} \rho(\vec{x}) F(\vec{x}) = \sum_{\vec{n}} \prod_{\alpha=1}^{D} \int dx_\alpha \, e^{\iota q_\alpha x_\alpha} \delta(x_\alpha - a n_\alpha) F(\vec{x})$$

$$= \sum_{\vec{n}} \prod_{\alpha=1}^{D} e^{\iota a q_\alpha n_\alpha} F(a\vec{n}) = \sum_{\vec{n}} e^{\iota a \vec{q}\cdot\vec{n}} F(a\vec{n}) \,.$$

We will omit the subscript "lattice" from now on. By construction $\tilde{F}(\vec{q})$ is periodic in the reciprocal lattice and, as mentioned above, the primitive cell periodicity is the first Brillouin zone. If we want to invert the transform, we can, therefore, express the function at a point in the direct lattice as a combination of the contributions in the first Brillouin zone of the reciprocal lattice, i.e.,

$$F(a\vec{n}) = \left(\frac{a}{2\pi}\right)^D \int_B d^D q \, e^{-\iota a \vec{q}\cdot\vec{n}} \tilde{F}(\vec{q}) , \tag{7.88}$$

where in each direction α one has $q_\alpha \in [-\pi/a, \pi/a]$ (or any translation of the $2\pi/a$ long interval).

We take as a function on a lattice the Kronecker delta $\delta^D_{\vec{n},\vec{n}'} = \prod_{\alpha=1}^{D} \delta_{n_\alpha, n'_\alpha}$. Its discrete Fourier transform is

$$\sum_{\vec{n}} e^{\iota a \vec{q}\cdot\vec{n}} \delta^D_{\vec{n},\vec{n}'} = e^{\iota a \vec{q}\cdot\vec{n}'} , \tag{7.89}$$

and the antitransform is

$$\delta_{\vec{n},\vec{n}'} = \left(\frac{a}{2\pi}\right)^D \int_B d^D q \, e^{-\iota a \vec{q}\cdot(\vec{n}-\vec{n}')} . \tag{7.90}$$

With this property, we can verify that the transform of the antitransform is the original function:

$$F(a\vec{n}) = \left(\frac{a}{2\pi}\right)^D \int d^D q \, e^{-\iota a \vec{q}\cdot\vec{n}} \tilde{F}(\vec{q}) = \left(\frac{a}{2\pi}\right)^D \int d^D q \, e^{-\iota a \vec{q}\cdot\vec{n}} \sum_{\vec{m}} e^{\iota a \vec{q}\cdot\vec{m}} F(a\vec{m})$$

$$= \sum_{\vec{m}} \left(\frac{a}{2\pi}\right)^D \int d^D q \, e^{-\iota a \vec{q}\cdot(\vec{n}-\vec{m})} F(a\vec{m}) = \sum_{\vec{m}} \delta^{(D)}_{\vec{n},\vec{m}} F(a\vec{m}) = F(a\vec{n}) . \tag{7.91}$$

7.B Derivation and Properties of the Fokker–Planck Equation

7.B.1 Continuum Limit of the Random Walker

We start from the recursion equation (7.36) for the probability density on a hypercubic lattice of step a with time interval a^2:

$$P_{(a)}(\vec{x}, t + a^2) = \sum_{\vec{\ell}} J_{(a)}(\vec{x} - \vec{\ell}a, \vec{\ell}) P_{(a)}(\vec{x} - \vec{\ell}a, t) . \tag{7.92}$$

We expand in series for small a, up to second order. The left-hand side is simply

$$P_{(a)}(\vec{x},t) + a^2 \frac{\partial P_{(a)}(\vec{x},t)}{\partial t}\bigg|_{\vec{x},t} + O(a^4).$$

To perform the expansion on the right-hand side, we first define the auxiliary function

$$F(\vec{x} - a\vec{\ell}) \equiv J_{(a)}(\vec{x} - \vec{\ell}a,\, \vec{\ell})P_{(a)}(\vec{x} - \vec{\ell}a,\, t) \tag{7.93}$$

whose Taylor series expansion for small $\vec{\ell}$ will be

$$F(\vec{x} - a\vec{\ell}) = F(\vec{x}) - a \sum_{\alpha=1}^{D} \frac{\partial F(\vec{x} - a\vec{\ell})}{\partial x_\alpha}\bigg|_{\vec{x},t}$$

$$\ell_\alpha + \frac{a^2}{2} \sum_{\alpha,\beta}^{1,D} \frac{\partial^2 F(\vec{x} - a\vec{\ell})}{\partial x_\alpha \partial x_\beta}\bigg|_{\vec{x},t} \ell_\alpha \ell_\beta + O(a^3). \tag{7.94}$$

Summing the first term of the expansion over the values of $\vec{\ell}$, we simply have

$$\sum_{\vec{\ell}} J_{(a)}(\vec{x},\vec{\ell})P_{(a)}(\vec{x},t) = P_{(a)}(\vec{x},t), \tag{7.95}$$

due to the probability normalization property J. Summing the first derivative term over $\vec{\ell}$ gives the contribution

$$-a \sum_{\alpha=1}^{D} \frac{\partial}{\partial x_\alpha} \left(\sum_{\vec{\ell}} J_{(a)}(\vec{x} - \vec{\ell}a,\, \vec{\ell})\ell_\alpha P_{(a)}(\vec{x} - \vec{\ell}a,\, t) \right)\bigg|_{\vec{x},t}$$

$$= -a \sum_{\alpha=1}^{D} \frac{\partial}{\partial x_\alpha} [c_\alpha^{(a)}(\vec{x})P_{(a)}(\vec{x} - \vec{\ell}a,\, t)]\bigg|_{\vec{x},t}$$

$$= -a^2 \sum_{\alpha=1}^{D} \frac{\partial}{\partial x_\alpha} [v_\alpha^{(a)}(\vec{x})P_{(a)}(\vec{x},t)] + O(a^3),$$

where we have used the definition (7.33) of the drift velocity field. The second derivative term gives the contribution

$$\frac{a^2}{2} \sum_{\alpha,\beta}^{1,D} \frac{\partial^2}{\partial x_\alpha \partial x_\beta} \left(\sum_{\vec{\ell}} J_{(a)}(\vec{x} - \vec{\ell}a,\, \vec{\ell})\ell_\alpha \ell_\beta P_{(a)}(\vec{x} - \vec{\ell}a,\, t) \right)\bigg|_{\vec{x},t}$$

$$= \frac{a^2}{2} \sum_{\alpha,\beta}^{1,D} \frac{\partial^2}{\partial x_\alpha \partial x_\beta} [v_{\alpha\beta}^{(a)}(\vec{x})P_{(a)}(\vec{x},t)] + O(a^3),$$

where we have used the definition (7.35) of the diffusion tensor neglecting higher-order terms. All terms are at order a^2.

Putting the above contributions together, simplifying a^2, and sending $a \to 0$ yields the Fokker–Planck equation for the probability density of stochastic continuous-time processes:

$$\frac{\partial P(\vec{x},t)}{\partial t} = -\sum_{\alpha=1}^{D}\frac{\partial}{\partial x_\alpha}[v_\alpha(\vec{x})P(\vec{x},t)] + \frac{1}{2}\sum_{\alpha,\beta}^{1,D}\frac{\partial^2}{\partial x_\alpha \partial x_\beta}[\nu_{\alpha\beta}(\vec{x})P(\vec{x},t)]$$

$$= -\vec{\nabla}\cdot[\vec{v}(\vec{x})P(\vec{x},t)] + \frac{1}{2}\vec{\nabla}\vec{\nabla}[\hat{\nu}(\vec{x})P(\vec{x},t)].$$

Let us examine three special cases below:

- *Homogeneous stochastic process.* If processes are homogeneous in space, the probability of a move on a lattice step does not depend on the starting position: $J(\vec{m} - \vec{\ell}, \vec{\ell}) = J(\vec{\ell})$. By rescaling the lattice step, this also implies that $J_{(a)}(\vec{x} - \vec{\ell}a, \vec{\ell}) = J_{(a)}(\vec{\ell})$, so that the diffusion tensor (7.34) will not depend on position: $\nu_{\alpha\beta}(\vec{x}) = \nu_{\alpha\beta}$. We note that in this case the velocity will also be independent of position: $\vec{v}(\vec{x}) = \vec{v}$. The Fokker–Planck equation reduces to

$$\frac{\partial P(\vec{x},t)}{\partial t} = -\vec{v}\cdot\vec{\nabla}P(\vec{x},t) + \frac{1}{2}\hat{\nu}\vec{\nabla}\vec{\nabla}P(\vec{x},t)$$

$$= -\sum_{\alpha=1}^{D}v_\alpha\frac{\partial P}{\partial x_\alpha} + \frac{1}{2}\sum_{\alpha,\beta}^{1,D}\nu_{\alpha\beta}\frac{\partial^2 P(\vec{x},t)}{\partial x_\alpha \partial x_\beta}. \tag{7.96}$$

- *Isotropic stochastic process.* If processes are isotropic in space, the probability of displacement on a lattice in a step does not depend on the direction of displacement: $J(\vec{m}-\vec{\ell}, \vec{\ell}) = J(\vec{m}-\vec{\ell}, |\vec{\ell}|)$. This implies that $J_{(a)}(\vec{x}, \vec{\ell}) = J_{(a)}(\vec{x}, |\vec{\ell}|)$. Consequently the drift velocity and the diffusion tensor will not depend on the direction and the tensor will be diagonal: $\nu_{\alpha\beta}(\vec{x}) = \delta_{\alpha\beta}\nu(\vec{x})$. In this case the Fokker–Planck equation takes on the expression

$$\frac{\partial P(\vec{x},t)}{\partial t} = -\nabla\cdot[\vec{v}(\vec{x})P(\vec{x},t)] + \frac{1}{2}\nabla^2[\nu(\vec{x})P(\vec{x},t)]. \tag{7.97}$$

- *Homogeneous and isotropic process.* If both conditions hold, then we have a single diffusion constant valid everywhere and in any direction, $\nu_{\alpha\beta}(\vec{x}) = \delta_{\alpha\beta}\nu$, and a uniform velocity in space. The Fokker–Planck equation then becomes (7.37),

$$\frac{\partial P(\vec{x},t)}{\partial t} = -\vec{v}\cdot\nabla P(\vec{x},t) + \frac{\nu}{2}\nabla^2 P(\vec{x},t). \tag{7.98}$$

7.B.2 Fokker–Planck Operator Spectral Analysis

7.B.2.1 Stationary Potential Solution

In order for a stationary solution $p_0(\vec{x})$ to exist, the probability current must be constant, as we can observe from the formulation (7.39) of the Fokker–Planck equation as a continuity equation:

$$-\vec{\nabla}\cdot\vec{J} = \frac{\partial p_0}{\partial t} = 0. \tag{7.99}$$

In cases where the boundary conditions are reflective or natural barriers,[12] this constant is actually zero and the stationary solution corresponds to a current of zero probability everywhere in space. This is a special and very important case of the stationary solution of the Fokker–Planck equation which is found by imposing, see (7.40),

$$\vec{J} = \vec{v}p_0(\vec{x}) - \frac{v}{2}\vec{\nabla}p_0(\vec{x}) = 0 \quad \Longrightarrow \quad \vec{\nabla}p_0(\vec{x}) = \frac{2}{v}\vec{v}p_0(\vec{x}). \tag{7.100}$$

On solving this equation we obtain the distribution

$$p_0(\vec{x}) \propto \exp[-\Phi(\vec{x})], \tag{7.101}$$

$$\Phi(\vec{x}) \equiv \frac{2}{v}\int_{\vec{x}_0}^{\vec{x}} d\vec{x}' \cdot \vec{v}(\vec{x}'). \tag{7.102}$$

Taking the gradient of the "potential" function $\Phi(\vec{x})$, cf. (7.102), we find the relation

$$\vec{v} = -\frac{v}{2}\vec{\nabla}\Phi(\vec{x})$$

for the drift velocity of the Fokker–Planck equation. In the case of Brownian motion, see Section 7.5, the drift term $\vec{v}(\vec{x})$ is proportional to a conservative force field, $\vec{v}(\vec{x}) = -\vec{\nabla}U(\vec{x})$, and the function $\Phi(\vec{x})$ turns out to be precisely the potential energy of the Brownian particle. This is why the probability density (7.101) is called the *potential* stationary solution of the Fokker–Planck equation.

7.B.2.2 Fokker–Planck and Symmetric Operators

The Fokker–Planck equation can be concisely written by defining the non-symmetric operator

$$H_{\mathrm{FP}}[\vec{x}] \equiv -\vec{\nabla}\cdot\vec{v} + \frac{v}{2}\nabla^2 \tag{7.103}$$

as

$$\frac{\partial P(\vec{x},t)}{\partial t} = H_{\mathrm{FP}}[\vec{x}]P(\vec{x},t).$$

In general, in the non-stationary case, we can always express the probability current in terms of the function Φ, as

$$\vec{J}(\vec{x},t) \equiv \vec{v}P(\vec{x},t) - \frac{v}{2}\vec{\nabla}P(\vec{x},t) = -\frac{v}{2}\vec{\nabla}\Phi(\vec{x})P(\vec{x},t) - \frac{v}{2}\vec{\nabla}P(\vec{x},t)$$

$$= -\frac{v}{2}e^{-\Phi(\vec{x})}\vec{\nabla}(e^{\Phi(\vec{x})}P(\vec{x},t)). \tag{7.104}$$

We can then show that the differential operator H_{FP} is linked to the Hermitian operator H defined in Eqs. (7.60) and (7.61) by the relation

$$H = \frac{1}{\sqrt{p_0(\vec{x})}}H_{\mathrm{FP}}\sqrt{p_0(\vec{x})} = e^{\Phi(\vec{x})/2}H_{\mathrm{FP}}e^{-\Phi(\vec{x})/2}. \tag{7.105}$$

---

[12] *Natural* boundary conditions impose that current and probability density tend to zero at infinity, see (7.52).

To demonstrate this we can take the form of the operator H_{FP} as a function of the probability current,

$$H_{\mathrm{FP}}P = -\vec{\nabla}\cdot\vec{J} = \frac{v}{2}\vec{\nabla}\cdot[e^{-\Phi(\vec{x})}\vec{\nabla}(e^{\Phi(\vec{x})}P)]\,, \tag{7.106}$$

transform it according to (7.105), and apply it to a generic function ψ:

$$H\psi = e^{\Phi(\vec{x})/2}H_{\mathrm{FP}}e^{-\Phi(\vec{x})/2}\psi = e^{\Phi(\vec{x})/2}\frac{v}{2}\vec{\nabla}\cdot[e^{-\Phi(\vec{x})}\vec{\nabla}(e^{\Phi(\vec{x})/2}\psi)]$$

$$= \frac{v}{2}\left[-\frac{|\vec{\nabla}\Phi|^2}{2} + \frac{\nabla^2\Phi}{2} + \nabla^2\right]\psi\,.$$

In general, this H operator is self-adjoint. If we then take $\vec{v} = -\vec{\nabla}U(\vec{x})$ as in Section 7.5, $\vec{\nabla}\Phi(\vec{x}) = (2/v)\vec{\nabla}U(\vec{x})$, and obtain just the H operator of (7.60), then

$$H = e^{U(\vec{x})/v}H_{\mathrm{FP}}e^{-U(\vec{x})/v}\,.$$

7.B.2.3 Eigenvalues and Eigenfunctions of the Fokker–Planck Operators

Using the above formulations, we analyze the spectral properties of the Fokker–Planck operator. We call $\phi_n(\vec{x})$ and λ_n^{FP} the eigenfunctions and eigenvalues of the operator H_{FP}. We call $\psi_n(\vec{x})$ and λ_n the eigenfunctions and eigenvalues of the operator H. Thus

$$H_{\mathrm{FP}}\phi_n(\vec{x}) = -\lambda_n^{\mathrm{FP}}\phi_n(\vec{x})\,, \tag{7.107}$$

$$H\psi_n(\vec{x}) = -\lambda_n\psi_n(\vec{x})\,. \tag{7.108}$$

By multiplying (7.107) on the left by $e^{\Phi(\vec{x})/2}$ we obtain the chain of relations

$$-\lambda_n^{\mathrm{FP}}e^{\Phi(\vec{x})/2}\phi_n(\vec{x}) = e^{\Phi(\vec{x})/2}H_{\mathrm{FP}}e^{-\Phi(\vec{x})/2}e^{\Phi(\vec{x})/2}\phi_n(\vec{x})$$

$$= He^{\Phi(\vec{x})/2}\phi_n(\vec{x}) = -\lambda_n e^{\Phi(\vec{x})/2}\phi_n(\vec{x})\,,$$

from which we observe that the eigenfunctions of H and H_{FP} are linked by

$$\psi_n(\vec{x}) = e^{\Phi(\vec{x})/2}\phi_n(\vec{x}) \tag{7.109}$$

and, hence, the operators H and H_{FP} have the same eigenvalues. Since H is self-adjoint they will, therefore, all be real.

With the help of the expression (7.106), we will now show that they are also all non-negative. We start with the integral

$$\int d\vec{x}\,\phi_n(\vec{x})e^{\Phi(\vec{x})}H_{\mathrm{FP}}\phi_n(\vec{x}) = \frac{v}{2}\int d\vec{x}\,\phi_n(\vec{x})e^{\Phi(\vec{x})}\vec{\nabla}\cdot[e^{-\Phi(\vec{x})}\vec{\nabla}(e^{\Phi(\vec{x})}\phi_n(\vec{x}))]$$

$$= -\frac{v}{2}\int d\vec{x}\,[\vec{\nabla}(e^{\Phi(\vec{x})}\phi_n(\vec{x}))]^2 e^{-\Phi(\vec{x})} \leq 0\,. \tag{7.110}$$

In the last step we integrated by parts and the edge terms are null for all boundary conditions for which a potential stationary solution exists. Using the eigenfunction relation (7.109) and the definition (7.105) of the operator H, the integral (7.110) reads

$$\int d\vec{x}\, e^{-\Phi(\vec{x})/2}\psi_n(\vec{x})e^{\Phi(\vec{x})}H_{\mathrm{FP}}e^{-\Phi(\vec{x})/2}\psi_n(\vec{x})$$

$$= \int d\vec{x}\, \psi_n(\vec{x})H\psi_n(\vec{x}) = -\lambda_n \int d\vec{x}\, |\psi_n(\vec{x})|^2 = -\lambda_n \,,$$

from which it follows that $\lambda_n \geq 0$, $\forall n$. The equality only holds for the stationary solution. The eigenvalue $\lambda_0 = 0$ corresponds to the eigenfunction of the ground state of H_{FP}, which is nothing other than the stationary Fokker–Planck solution: $\phi_0(\vec{x}) = p_0(\vec{x}) = e^{-\Phi(\vec{x})}$. Furthermore, $\lambda_0 = 0$ also corresponds to the eigenfunction of the ground state of H, $\psi_0(\vec{x}) = e^{-\Phi(\vec{x})/2} = \sqrt{p_0(\vec{x})}$. Indeed, we recall that in the special "potential–potential" case of Section 7.5.2, where $\vec{v} = -\nabla U(\vec{x})$, we showed that $\psi_0(\vec{x}) = \sqrt{p_0(\vec{x})}$ satisfies Eq. (7.62): $H\psi_0(\vec{x}) = 0$.

Generating Functions and Chain Reactions

8.1 Generating Functions

Before tackling the study of other random phenomena, let us look in detail at a tool that allows us to deal with complicated problems in a simple manner. We have already seen something like this in the constructive proof of the central limit theorem carried out in Section 3.2.1. Here we will give a more general and in-depth presentation of the so-called generating functions.

The generating function $F(s)$ associated with a series $f(n)$, $n \in \mathbb{N}$, is an analytic function of the variable s with domain in a region of the complex plane containing the origin,

$$F(s) = \sum_{n=0}^{\infty} f(n)s^n . \tag{8.1}$$

The series is convergent for sufficiently small s, $|s| < 1$, if (as we always assume) $f(n)$ does not grow faster than an exponential in n. In the case where $\sum_n f(n)$ is bounded, the uniform convergence domain includes at least the edge of the convergence disk: $|s| \leq 1$. This is the case for probability distributions, where n labels the event and $f(n)$ its probability. If $f(n)$ decays fast enough, then $F(s)$ can also converge outside the $|s| = 1$ disk. Often the function can be analytically continued beyond the convergence domain of the series.

In the case of a probability distribution, it is possible to derive all its moments from the generating functions using differentiation instead of integrating (summing). First of all, the generating function evaluated at $s = 1$ is the normalization condition (1.1), whereby

$$F(1) = \sum_{n=0}^{\infty} f(n) = 1 . \tag{8.2}$$

If we take the first derivative and evaluate it at $s = 1$ we obtain the mean value

$$F'(1) = \sum_{n=1}^{\infty} nf(n) = \langle n \rangle = \mu . \tag{8.3}$$

The second derivative gives

$$F''(1) = \sum_{n=2}^{\infty} n(n-1)f(n) = \langle n^2 \rangle - \langle n \rangle ,$$

so the variance is obtained from the combination

$$\sigma^2 = \langle (n - \langle n \rangle)^2 \rangle = F''(1) + F'(1) - [F'(1)]^2 , \tag{8.4}$$

and so on for higher-order moments. For example, we can compute the generating function of the Poisson distribution, cf. (2.31), $p(n|\lambda) = e^{-\lambda}\lambda^n/n!$, as

$$P(s) = \sum_{n=0}^{\infty} p(n|\lambda)s^n = e^{-\lambda}\sum_{n=0}^{\infty} \frac{(s\lambda)^n}{n!} = e^{\lambda(s-1)} .$$

We note that, in this case, where $p(n) \sim 1/n!$, the series also converges beyond the disk $|s| = 1$. We can check the normalization, mean value, and variance of the distribution, see Section 2.3, with the relations of the generating functions of probability distributions.

$$P(1) = 1 ,$$
$$\mu = P'(1) = \lambda e^{\lambda(s-1)}\big|_{s=1} = \lambda ,$$
$$\sigma^2 = P''(1) + P'(1) - P'(1)^2 = [\lambda^2 + \lambda]e^{\lambda(s-1)}\big|_{s=1} - \lambda^2 e^{2\lambda(s-1)}\big|_{s=1} = \lambda .$$

Another example is the generating function of the probability of counting, according to the binomial distribution,

$$b(N, K, p) = \left(\begin{array}{c} N \\ K \end{array} \right) p^K q^{N-K} ,$$

a number K of Bernoulli events of probability p in N attempts (with $q = 1 - p$), which equals

$$B_N(s) = \sum_{K=0}^{\infty} \left(\begin{array}{c} N \\ K \end{array} \right) p^K q^{N-K} s^K = (ps + q)^N , \quad |s| \leq 1 ,$$

and is precisely the expansion of a binomial. We can easily check the normalization, mean value, and variance by applying the formulas (8.2), (8.3), and (8.4).

8.1.1 Generating Functions, and Fourier and Laplace Transforms

On the edge of the disk in the complex plane $|s| = 1$, we can set $s = \exp(\imath q)$, with $q \in [-\pi, \pi]$ (first Brillouin zone), and the generating function along the unitary circumference turns out to be the Fourier transform of a discrete distribution function $p(n)$, on a one-dimensional discrete lattice, see (7.5),

$$\tilde{P}(q) \equiv P(e^{\imath q}) = \sum_{n=0}^{\infty} p(n)e^{\imath q n} .$$

We note that the generating function of a generic sequence is still well defined for $|s| < 1$ even in those cases where its Fourier transform may have convergence problems on the disk. If we turn to the continuous case and express the probability

$p(n) \to dp(x) = p(x)dx$ in terms of the probability density $p(x)$, the generating function coincides with the Fourier transform:

$$\tilde{P}(q) = \int_{-\infty}^{\infty} dx\, p(x)e^{\iota q x}\,.$$

As we have already seen in the Taylor expansion for the proof of the central limit theorem, cf. Section 3.2.1 and Appendix 3.C.1, the moments of $p(x)$ are given by derivatives

$$\langle x^k \rangle = (-\iota)^k \frac{d}{dq^k} \tilde{P}(q) \Big|_{q=0}\,.$$

As an alternative, instead of generating the moments of a distribution by taking the limit for $s \to 1$ of a function defined along the convergence disk $|s| = 1$ ($s \to 1$ corresponds to $q \to 0$), we can start from a generating function defined on a domain s along the real axis between 0 and 1: i.e., $s \equiv e^{-h}$, with $h \in [0, \infty)$ (so the limit $s \to 1$ corresponds to the limit $h \to 0$). From (8.1) we thus achieve the (bilateral) Laplace transform, see Appendix 3.D, which we write directly in the continuous notation as

$$\hat{P}(h) \equiv P(e^{-h}) = \int_{-\infty}^{\infty} dx\, e^{-hx} p(x)\,, \quad h \in [0, \infty)\,.$$

Again, all moments are obtained by deriving the Laplace transform and evaluating it at $s = 1$, which corresponds to $h = 0$:

$$\langle x^k \rangle = (-1)^k \frac{d}{dh^k} \hat{P}(h) \Big|_{h=0}\,. \tag{8.5}$$

To avoid the annoying power of the minus sign, the generating function of the moments of the random variable as Laplace transform can equivalently be defined as

$$\hat{P}(z) = \langle e^{zX} \rangle = \int_{-\infty}^{\infty} dx\, e^{zx}(x)\,, \quad \text{with} \quad z \in (-\infty, 0]\,. \tag{8.6}$$

The latter formulation will be used in Chapter 12 to prove the central limit theorem for correlated random variables.

8.1.2 Inverse Generating Functions

As for the special cases of transforms, there is an inversion formula for generating functions:

$$f(k) = \frac{1}{2\pi\iota} \oint \frac{ds}{s}\, s^{-k} F(s)\,. \tag{8.7}$$

The integral is carried out on a path around the origin in a counterclockwise direction and does not enclose any singularities of the function $F(s)$. The proof is an immediate consequence of the residue theorem: the integral on a closed circuit around a singularity of order $k + 1$ at $s = 0$ of the function $F(s)/s^{k+1}$ is

$$\oint ds\, \frac{F(s)}{s^{k+1}} = 2\pi\iota \operatorname{Res}(F(s)s^{-k-1}, 0) = 2\pi\iota\, \frac{1}{k!} \frac{d^k F(s)}{ds^k} \Big|_{s=0} = 2\pi\iota f(k)\,. \tag{8.8}$$

In the case of probability distributions, the integration path can be brought along the circle $|s| = 1$, and by defining $s = e^{iq}$ the formula (8.7) becomes the Fourier antitransform. If, on the other hand, we set $s = e^{-h}$, Eq. (8.7) coincides with the inverse Laplace transform, defined in Appendix 3.D.

8.1.3 Tail and Cumulative Probability Generating Functions

If $p(n)$ is the probability that the random variable X defined on the integers takes value $X = n$, we call the probability $c(n)$ that $X > n$ the *tail probability*:

$$c(n) \equiv \sum_{i=n+1}^{\infty} p(i)\,.$$

This is well defined as a probability for each n, being a number between 0 and 1, but it is not a probability distribution over the values n. There is a relationship between the generating functions of the probability distribution and of the tail probabilities sequence that will become important in Chapter 9, when dealing with recurrent events. As we know, $P(s) = \sum_{n=0}^{\infty} p(n)s^n$ converges in $|s| \leq 1$. On the other hand, $C(s) = \sum_{n=0}^{\infty} c(n)s^n$ converges for $|s| < 1$ because each term is finite ($c(n) \leq 1$) but their sum is not necessarily so.

Theorem. (Tail probability generating function) *The tail probability theorem says that, if $|s| < 1$, then*

$$C(s) = \frac{1 - P(s)}{1 - s}\,. \tag{8.9}$$

Proof   To prove this we take

$$(1 - s)C(s) = \sum_{n=0}^{\infty} c(n)s^n - \sum_{m=0}^{\infty} c(m)s^{m+1} = c(0) + \sum_{k=1}^{\infty} (c(k) - c(k-1))s^k$$

$$= 1 - p(0) - \sum_{k=1}^{\infty} p(k)s^k = 1 - P(s)\,, \tag{8.10}$$

where in the last line we have used the definition of the queue probability to write

$$c(k) - c(k-1) = -p(k) \quad \text{if} \quad k \geq 1$$

and $c(0) = 1 - p(0)$. $\blacksquare$

It is a simple exercise to realize that the generating function $\Sigma(s)$ of the cumulative probability distribution,

$$\sigma(n) \equiv \sum_{i=0}^{n} p(i)\,, \tag{8.11}$$

is related to the generatrix of $\{p\}$ by the complementary relation to that (8.9) of the tails:

$$\Sigma(s) = \frac{P(s)}{1 - s}.\qquad(8.12)$$

8.1.4 Generating Function of the Convolution

As we have seen for Fourier transforms in Appendix 3.C, we consider the sum of random variables

$$Z = X + Y,$$

where X and Y are independent random variables that take non-negative integer values. If the values x, y, and z of the variables X, Y, and Z are realized with probabilities $a(x)$, $b(y)$, and $c(z)$, respectively, and if we call $A(s)$, $B(s)$, and $C(s)$ the respective generating functions, after a simple computation we find that

$$C(s) = A(s)B(s).\qquad(8.13)$$

In fact, the value z of Z can be obtained by the occurrence of several joint events of x and y:

$$(x = 0, y = z), \quad (x = 1, y = z - 1), \quad \ldots, \quad (x = z - 1, y = 1), \quad (x = z, y = 0).$$

All these events are mutually exclusive and, therefore, the probability $c(z)$ that the sum variable Z takes the value z is written as

$$c(z) = a(0)b(z) + a(1)b(z - 1) + \cdots + a(z - 1)b(1) + a(z)b(0) = \sum_{n=0}^{z} a(n)b(z - n),$$

$$(8.14)$$

which is a convolution product between the distributions $\{a\}$ of X and $\{b\}$ of Y:

$$\{c\} = \{a\} * \{b\}.$$

The generating function of the distribution of the sum variable Z will then be

$$C(s) = \sum_{k=0}^{\infty} c(k)s^k = \sum_{k=0}^{\infty}\sum_{n=0}^{k} a(n)b(k - n)s^k = \sum_{n=0}^{\infty}\sum_{k=n}^{\infty} a(n)b(k - n)s^k$$

$$= \sum_{n=0}^{\infty}\sum_{m=0}^{\infty} a(n)b(m)s^{n+m} = A(s)B(s),\qquad(8.15)$$

where we set $k = n + m$.

As an exercise, we can use the convolution property to compute the distribution of the sum variable of two Poisson-distributed variables, $\{p(\lambda_1)\}$ and $\{p(\lambda_2)\}$, respectively. Given the variable $k = k_1 + k_2$, its distribution is

$$p^{(2)}(k|\lambda_1, \lambda_2) = p(k|\lambda_1) * p(k|\lambda_2).$$

If we denote by

$$P_{1,2}(s) = \sum_k p(k|\lambda_{1,2})s^k$$

the generating functions of the two Poisson variables, the generating function of the sum variable turns out to be

$$P^{(2)}(s) = \sum_{k=0}^{\infty} p^{(2)}(k)s^k = P_1(s)P_2(s) = e^{\lambda_1(s-1)}e^{\lambda_2(s-1)} = e^{(\lambda_1+\lambda_2)(s-1)}.$$

In summary, defining $\lambda = \lambda_1 + \lambda_2$, the distribution of the sum variable $k = k_1 + k_2$ is still Poisson with expectation value the sum of the expectation values:

$$p^{(2)}(k \,|\, \lambda = \lambda_1 + \lambda_2) = e^{-\lambda}\frac{\lambda^k}{k!}.$$

If, instead of the sum of two variables, we consider the sum of several variables $X_1, X_2, \ldots, X_N$ with distributions $\{a^{(1)}\}, \{a^{(2)}\}, \ldots, \{a^{(N)}\}$, the distribution of their sum $Z = \sum_{j=1}^{N} X_j$ will be the convolution

$$\{c\} = \{a^{(1)}\} * \{a^{(2)}\} * \cdots * \{a^{(N)}\}, \tag{8.16}$$

whence $C(s) = \prod_{j=1}^{N} A^{(j)}(s)$. If the summed variables are all identically distributed we have

$$\{c\} = \{a\} * \{a\} * \cdots * \{a\} \equiv \{a\}^{*N},$$
$$C(s) = [A(s)]^N.$$

We can get there faster. Let us take a random variable X with integer values n, distributed with $\{p\}$. The function s^X will surely be well defined for $|s| \leq 1$ and take values s^n. The generating function of $\{p\}$ can therefore be thought of as

$$P(s) = \sum_{n=0}^{\infty} p(n)s^n = \langle s^X \rangle, \tag{8.17}$$

where the expected value of a generic function of the variable X is

$$\langle f(X) \rangle = \sum_{n=0}^{\infty} p(n)f(n). \tag{8.18}$$

With this notation it is immediate to see that, if $Z = \sum_{j=1}^{N} X_j$ is the sum of independent, identically distributed variables, then

$$C(s) = \langle s^Z \rangle = \prod_{j=1}^{N} \langle s^{X_j} \rangle = [P(s)]^N. \tag{8.19}$$

8.1.5 Composite Probability Theorem

The composite probability theorem applies to the sum of a random number of random variables.

Theorem. (Composite probability generating function) *Let $P(s)$ and $Q(s)$ be the generating functions of two probability distributions $\{p\}$ and $\{q\}$. If a number k is the sum of m independent numbers $k_i \in \mathbb{N}$, i.e.,*

$$k = \sum_{i=1}^{m} k_i \,, \tag{8.20}$$

where the values of k_i are i.i.d. random variables distributed with $\{p\}$, and also the number m of variables in the sum is a random variable, whose distribution is $\{q\}$, then the generating function $R(s)$ of the probability distribution $r(k)$ of k is given by

$$R(s) = Q(P(s))\,. \tag{8.21}$$

Proof To prove it, we first note that the simultaneous occurrence of two different values of m is impossible. The variable m, therefore, labels a complete system of self-excluding events and, therefore, we can write the probability that the value of the sum is k as

$$r(k) = \sum_{m=0}^{\infty} p(k|m)q(m)\,. \tag{8.22}$$

Here $p(k|m)$ is the probability that the sum of a fixed number m of variables is equal to k. Since all k_i in (8.20) have the same distribution $\{p\}$, $p(k|m)$ is equal to the mth convolution

$$p(k|m) = \{p\}^{*m}\,.$$

We can then move on to the generating functions using the convolution property, whereby the generating function of the sum is given by the product of the generating functions. Let us call

$$P_m(s) = \sum_{k=0}^{\infty} p(k|m)s^k = [P(s)]^m$$

the generating function of the distribution of the variable k sum of m (fixed) random variables. In the last step, we apply Eq. (8.19).

The generating function of the probability distribution of the values of the sum of a random number m of identically distributed random variables k_i eventually turns out to be

$$R(s) \equiv \sum_{k=0}^{\infty} r(k)s^k = \sum_{k=0}^{\infty}\sum_{m=0}^{\infty} p(k|m)q(m)s^k = \sum_{m=0}^{\infty} q(m)[P(s)]^m = Q(P(s))\,. \tag{8.23}$$

In the last step we used the property that, by definition, the generating function of a well-normalized distribution has $|P(s)| \leq 1$ and, therefore, the series $Q(P(s))$ exists. $\blacksquare$

The use of generating functions makes it possible to write the result for an apparently very complicated problem in a very simple and compact manner. The theorem of the generating function of composite probabilities is very useful for studying the problem of *branching* processes. These are processes in which a given parent element at each step of the dynamics can branch into several other children elements (or disappear). Another name we will use for branching processes is *chain reactions*.

8.2 Chain Reactions

Let us apply the formalism of generating functions to the chain reactions we are going to define here. We consider a system defined for discrete times, denoted by the index k, e.g., generations of a population. The internal state of the system can be characterized by a number $n(k)$, which often refers to the number of elements of the system at time k, i.e., a population. Suppose that in a time unit each one of the elements of the system can transmute into m elements, with a probability $p(m)$. The total number of elements at time $k + 1$ will, therefore, be given by

$$n(k + 1) = \sum_{i=1}^{n(k)} m(i),\tag{8.24}$$

where each $m(i)$ is independently extracted with probability $p(m)$.

For simplicity, we are considering discrete time. Thus all births of new elements, for example, are synchronized. The formalism can be extended to continuous time, non-synchronized events in various possible ways. This is a complication that we will deal with in Chapter 13 for simple cases. In any case, we lose very little in qualitative terms.

This type of phenomenon is called a *branching process* or *chain reaction*. To better clarify what we are dealing with, here are three rather different areas of applicability of the theory we are going to develop.

- The quantity $n(k)$ represents the number of neutrons in a fissile material. After a unit of time, each neutron may have produced, due to collisions, m neutrons (including itself) with probability $p(m)$. The neutron may also be absorbed without producing any more neutrons, or it may be lost into the surrounding environment: both these events contribute to the event 0, with probability $p(0)$.
- The quantity $n(k)$ represents the number of pairs of rabbits (or the number of amoebae) and $p(m)$ is the probability that a pair of rabbits gives birth to other m pairs of rabbits (or an amoeba divides itself into m amoebae) in one generation. Compared to the real world, we are making various approximations: for example, that rabbits are born in pairs of different sexes (which is not always true), there are no effects due to the age of the rabbits, and there are no limiting factors related to the total number of rabbits. Despite all these approximations, this type of model can be considered a reasonable description to capture the qualitative behavior of the system, at least in certain situations.

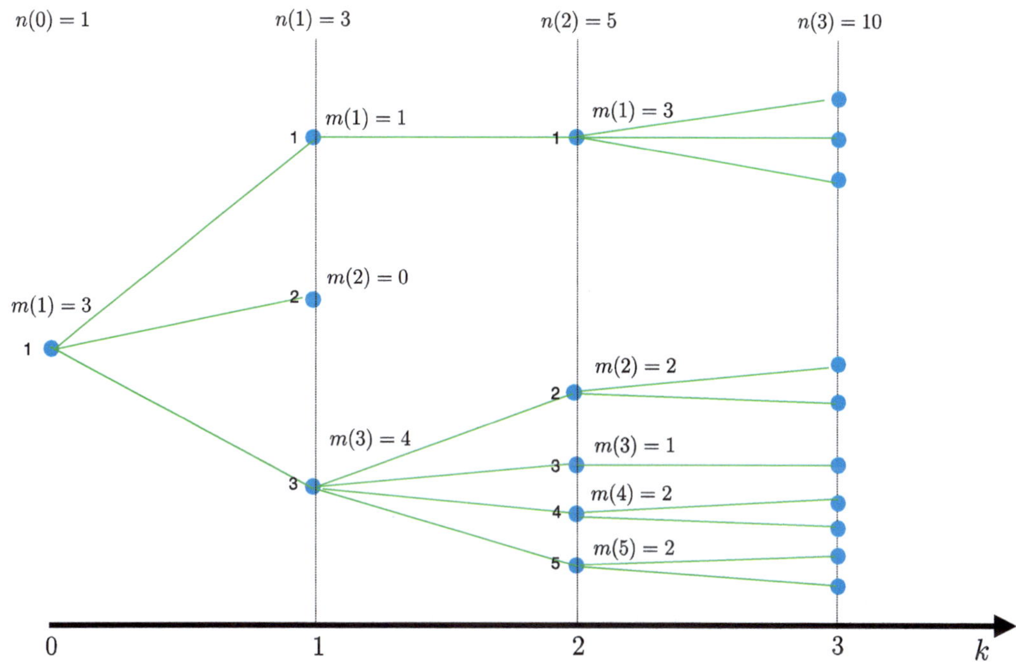

Figure 8.1   Illustration of the initial steps of a branching process.

- The quantity $n(k)$ represents the number of patients infected with a given virus and $p(m)$ the probability that an infected individual infects another m individuals in a given probing time interval. The possible epidemic spread depends on how many subjects susceptible to the virus can be infected on average by those already infected. Here, too, we make several approximations: namely we are neglecting that in reality the probability will also depend on self-limiting factors that are triggered by the spread of the disease, because, for instance, over time, someone may recover and develop antibodies, on the mitigation actions taken (protective devices, tracing, isolation), and on the possible expansion of vaccines and medicines. The theory we shall study concerns the case in which the probability of the spread of the virus is stationary over time.

Let us start at the beginning, when we have the progenitor of the population: $n(0) = 1$ is the initial population, see Figure 8.1. We call $m(1)$ the offspring of this one element, which will be a random integer number with probability distribution $p(m)$; $p(0)$ is the probability of becoming extinct, $p(1)$ that of remaining alive without reproducing, and $p(m > 1)$ that of reproducing and/or multiplying. At generation $k = 1$, the population will then be $n(1) = m(1)$. At the following generation, $k = 2$, the population will be the descendants of all the elements $j = 1, \ldots, n(1)$ of the previous generation:

$$n(2) = \sum_{j=1}^{n(1)} m(j).$$

Iterating for a general number of steps k, we obtain Eq. (8.24) for the population size dynamics. We want to compute the probability distribution $p^{(k)}(n)$ of having a population of n elements at the kth generation. In particular, we are interested in the behavior of $E^{(k)} = p^{(k)}(0)$, representing the extinction probability, or the behavior of the complementary survival probability $S^{(k)} = 1 - E^{(k)}$. At this point, it is useful to state a fundamental theorem and discuss its proof.

8.2.1 Chain Reactions Fundamental Theorem

Theorem. (Chain reactions) *Let the average value of the population after one generation be*

$$\mu = \sum_{m=0}^{\infty} mp(m), \tag{8.25}$$

and let k be the generation step. Then the following hold:

- *if $\mu \leq 1$, the reaction dies out with probability 1 as $k \to \infty$; and*
- *if $\mu > 1$, the reaction has a non-zero probability of self-sustaining for infinite time.*

There are three possible cases, as we will derive in a short while, by proving the theorem.

1. The survival probability $S^{(k)}$ tends to zero very quickly as k increases, more exactly exponentially: $S^{(k)} \sim e^{-k}$. In this case, the reaction is called sub-critical.
2. The quantity $S^{(k)}$ tends to a positive non-zero value and the average number of elements in the population grows exponentially with generations k. In this case, the reaction is called super-critical. The quantity $S^{(k)}$ is not identically equal to 1, as there is always the possibility that the reaction dies out during the first steps, but the extinction rate $T_k = E^{(k)} - E^{(k-1)}$ tends to zero exponentially. Only in this situation can the chain reaction be self-sustaining, once triggered. It is evident that, due to exponential growth, n will soon become extremely large and the assumptions about the absence of self-limiting factors will no longer be true.
3. The quantity $S^{(k)}$ tends to zero with k, but more slowly than an exponential, more precisely $S^{(k)} \sim 1/k$. This situation is called critical and in parameter space separates the sub-critical case from the super-critical case.

8.2.2 Proof of the Chain Reactions Fundamental Theorem

To prove the theorem and outline the three scenarios, we need to obtain an exact expression of the probability $p^{(k)}(n)$ of having n elements at time k and its generating function $P^{(k)}(s)$.

Proof   If we assume that we have only one element at time 0, we obtain that the elements of the generation at time 1 are distributed with $p(m)$ and the generating function at time 1 is given by

$$P(s) \equiv \sum_{m} p(m)s^{m} ,$$

with mean value

$$\left.\frac{dP(s)}{ds}\right|_{s=1} = \mu. \tag{8.26}$$

At time 2, $n(2) = \sum_{j=1}^{n(1)} m(j)$ is given by the sum of a random number (the number $n(1)$ of the ancestor population) of random variables (the descendants $m(j)$). The composite probability theorem, presented in Section 8.1.5, applies to distributions of such doubly random sums. In this case, the number of addends and the value of each addend are distributed in the same way, with $\{p\}$. Equation (8.21) in that theorem tells us that, at time $k = 2$,

$$P^{(2)}(s) = P(P(s)). \tag{8.27}$$

Continuing to iterate we obtain the recursion formula

$$P^{(k)}(s) = P(P^{(k-1)}(s)) = P^{(k-1)}(P(s)). \tag{8.28}$$

Formula (8.28), together with the property (8.26), guarantees that

$$P^{(k)}(1) = 1, \quad \left.\frac{dP^{(k)}(s)}{ds}\right|_{s=1} = \mu^k. \tag{8.29}$$

The first of equations (8.29) assures that the probability is well normalized, while the second equation tells us that the mean value of n at time k is simply given by

$$\langle n \rangle_k = \mu^k. \tag{8.30}$$

At this point, it is easy to see, as we shall shortly demonstrate, that the sub-critical, critical, and super-critical cases correspond to the cases where μ is, respectively, less than, equal to, or greater than 1. Indeed, since we can bound from above the survival probability at step k as

$$\mu^k = \langle n \rangle_k = \left.\frac{dP^{(k)}}{ds}\right|_{s=1} = \sum_{n=1}^{\infty} n p^{(k)}(n) \geq \sum_{n=1}^{\infty} p^{(k)}(n) = S^{(k)}, \tag{8.31}$$

it follows that, for $\mu < 1$, $S^{(k)} \lesssim e^{-k|\ln \mu|}$ goes exponentially to zero.

In general, we can proceed in a systematic way. We first observe that:

- the value $P^{(k)}(0)$ of the generating function is the extinction probability $E^{(k)} = 1 - S^{(k)}$;
- the recurrence relation (8.28) implies

$$E^{(k)} = P(E^{(k-1)}); \tag{8.32}$$

and

- since P is an increasing function, the quantities $E^{(k)}$ as k grows form an increasing monotonic sequence (the extinction is irreversible) bounded above by 1 and, therefore, there is necessarily a limit $E^\star \leq 1$ for this sequence when $k \to \infty$.

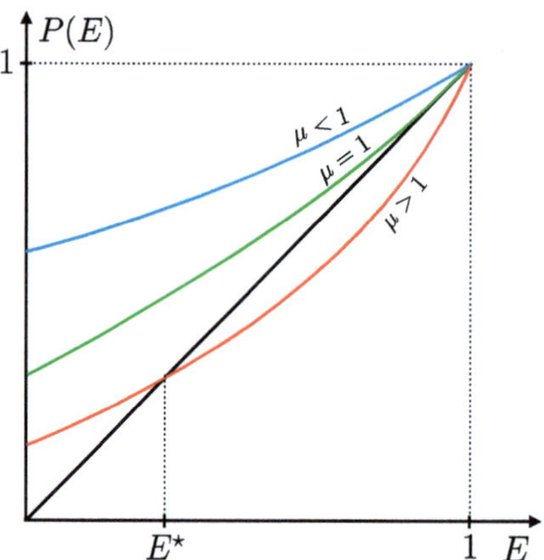

Figure 8.2 Graphical solution of the extinction probability equation for average branching value: $\mu < 1$ (top blue curve), $\mu = 1$ (middle green curve), and $\mu > 1$ (bottom red curve).

Putting everything together, we find that the extinction probability must satisfy the following asymptotic equation as $k \to \infty$:

$$E = P(E). \tag{8.33}$$

If the equation has a unique solution in the range $E \in [0, 1]$, we have found the value of E; otherwise, we need to think again. Let us study the function $P(s)$. We first note that it is a convex function for non-negative real s. Indeed

$$\frac{d^2 P}{ds^2} = \sum_{m=2}^{\infty} m(m-1)s^{m-2} p_m \geq 0. \tag{8.34}$$

A convex function is always greater than its tangent at any point, and can be intersected by a straight line at most in two points.

Since $P(1) = 1$, it is evident that $E = 1$ is a solution of Eq. (8.33); see Figure 8.2. We only need to see if there can also be another one. If $P'(1) = \mu < 1$ the curve $P(E)$ is above its tangent at $E = 1$ (the line $y = \mu(E - 1) + 1$) and, therefore, cannot have intersections with the line $y = E$ in $0 \leq E < 1$. The same is true for $\mu = 1$ (excluding the trivial case where $P(s) = s$).

Conversely, for $\mu > 1$ we obtain that $P(E) < E$ for $E \lesssim 1$, while for $E \gtrsim 0$ we have that $P(E) > E$, if we exclude the trivial case $P(0) = 0$. For continuity, the equation $P(E) = E$ must, therefore, have a solution $E^{\star}$ in the open interval $(0, 1)$. Since the function $P(E)$ is increasing, $P(0) < P(E^{\star})$ and, therefore, we will have that

$$E^{(1)} = P(E^{(0)}) = P(0) < P(E^{\star}) = E^{\star}.$$

The same reasoning also implies that $E^{(1)} < E^{(2)} < E^{\star}$. In short, all $E^{(k)}$ form an increasing sequence with k, all remaining strictly smaller than $E^{\star} < 1$:

$$E^{(1)} < E^{(2)} < \cdots < E^{(k)} < \cdots < E^{\star} < 1. \tag{8.35}$$

Their limit for $k \to \infty$ must therefore be $E^{\star}$, and cannot be 1.

It remains to solve the issue of how quickly the limit is reached. Let us call E_∞ the solution of the equation $E = P(E)$, i.e., the limit of the sequence $E^{(k)}$. For large k the difference $E^{(k)} - E_\infty = \Delta_k$ will be small. We can then expand the extinction probability in powers of Δ_k in Eq. (8.32). If we stop at first order we obtain

$$E^{(k+1)} = P(E^{(k)}) = E_\infty + P'(E_\infty)(E^{(k)} - E_\infty) + O((E^{(k)} - E_\infty)^2),$$

from which

$$\Delta_{k+1} = A\Delta_k + O(\Delta^2), \quad \text{with} \quad A \equiv \left.\frac{dP(E)}{dE}\right|_{E=E_\infty}.$$

In both non-critical cases, one has $A < 1$. In the sub-critical case $A = P'(1) = \mu < 1$ by construction. In the super-critical case, the above equations imply that $A = P'(E^\star) < 1$. Thus the approach to the asymptotic limit occurs exponentially:

$$\Delta_k = \Delta_0 A^k = -E_\infty e^{-k/\tau}, \tag{8.36}$$

with decay time

$$\tau = -\frac{1}{\ln A}.$$

The only delicate case is the critical case where $A = P'(E_\infty = 1) = \mu = 1$, τ diverges, and the behavior is no longer exponential. In this, case we have to expand to second order,

$$\Delta_{k+1} = \Delta_k + B\Delta_k^2 + O(\Delta_k^3), \tag{8.37}$$

where

$$B = \left.\frac{1}{2}\frac{d^2 P(E)}{dE^2}\right|_{E=1} = \frac{1}{2}(\langle n^2\rangle - \langle n\rangle) = \frac{1}{2}(\langle n^2\rangle - \langle n\rangle^2) = \frac{\sigma^2}{2}, \tag{8.38}$$

with the expectation value computed from the probability distribution $\{p\}$. We have made it explicit that in the critical regime $\mu = \langle n\rangle = \langle n\rangle^2 = 1$. The solution of Eq. (8.37) for large k is given by

$$\Delta_k = -\frac{1}{Bk} = -\frac{2}{\sigma^2 k} \quad \Longrightarrow \quad E_k = 1 - \frac{2}{\sigma^2 k}. \tag{8.39}$$

The interpretation of the above formula is clear. The mean value of n is constant and is equal to 1. However, since the variance σ^2 is not zero, there are fluctuations. The fluctuations produce rare events (survival), with probability certainly much smaller than 1 as $k \gg 1$, and typical events (extinction), with probability $1 - 2/(\sigma^2 k)$. The larger the variance of $\{p\}$, the faster the probability of extinction tends to 1. Conversely, in the limit $\sigma^2 \to 0$, the population will always be equal to its average, $n = \mu = 1$, and there will be no extinction. Even a small fluctuation induces certain extinction of the reaction, but in an infinitely long time. $\blacksquare$

A detailed study of the behavior of the asymptotic distribution is possible, but is beyond the scope of this book.

Recurrent Events

9.1 Definitions and Examples

Let us consider a system evolving in discrete time labeled by an integer step index n. To begin with, let us consider the case in which the event can occur at any time n. The event is said to be a *recurrent event* if the probability of occurring at time n depends only on the time interval since a previous occurrence. If we call this time k, the recurrence probability p depends only on $n - k$, and the times of all the events that occurred before k play no role in predicting the future after the time k. In other words, an event is recurrent if its probability is invariant under time translation on discrete steps.

Like the random walker, whose probability of moving to a new position depends only on the position the traveller occupies at the previous time step, and like chain reactions, in which the probability of a population depends only on the population of the previous generation, recurrent events are an example of correlated processes without memory. You only need to know the last recurrence of the event to (stochastically) predict the future. All history prior to the last occurrence of the event is irrelevant.

9.1.1 Examples of Recurrent Events

9.1.1.1 Head or Tail, True or False, Even or Odd … Balance

Let us take a sequence of Bernoulli events (true or false, heads or tails, …) described by a sequence of variables $x_i = \pm 1$ (true $= 1$, false $= -1$). We call "recurrent event $\mathcal{E}$" the point in the sequence where the number of true Bernoulli events equals the number of false Bernoulli events. In order to characterize $\mathcal{E}$, we define a sum variable at step n:

$$S_n = \sum_{k=1}^{n} x_k \, . \tag{9.1}$$

At time $n = 0$ (before any statement is made) one will have $S_0 = 0$. The recurrent event $\mathcal{E}$ is, in our case, signaled by $S_n = N_{\text{true}} - N_{\text{false}} = 0$. For example, in the unfolding of a talk show, which we can idealize as a sequence of true or false statements, there will be times when half of the statements uttered so far will be false and half true, hence $S_n = 0$. The interval between two such events is the recurrence time. The recurrence time is a random variable itself but it can only be even: this event has a periodicity equal to 2, meaning that it cannot occur if the time is not a multiple of 2. The same happens when rolling dice and betting on even or odd: the event defined by the fact that after n throws the number of even values obtained is equal to the number of odd ones can only occur for n even, and hence has periodicity 2.

Always assuming the formalization of the Bernoulli sequence, we now define a recurrent event $\mathcal{E}$ as the moment when the number of times a true statement is asserted is a multiple M of the number of false statements, $N_{\text{true}} = MN_{\text{false}}$. This event occurs at each step n for which the sum variable S_n takes on the value $(M-1)N_{\text{false}}$. If $M = 1$ we return to the previous case. We note that for the event to occur, at the very least, a sequence of M true and 1 false must occur. The event can, therefore, only recur every multiple of $M + 1$ steps: it is periodic of periodicity $M + 1$.

9.1.1.2 The Simple Queue

Let us take a post office counter, whose status is represented by a Boolean variable S: free $S = 0$, and busy $S = 1$. Let us take the number of waiting users, represented by an integer-valued variable $X = 0, 1, 2, \ldots$. We now define the recurrent event as the (heavenly) situation in which the counter is free and there is no queue: $\mathcal{E} \equiv \{S = 0, X = 0\}$. If at the first time step a first customer arrives, they will directly go to the counter, and we have $\{S = 1, X = 0\}$. If at the next step a second customer arrives while the first is still busy, you will have $\{S - 1, X = 1\}$, and so on. When the queue empties and the counter becomes free, we have a recurrence of $\mathcal{E}$. This event has no time constraint for its recurrence and is called *aperiodic*.

9.1.1.3 The Return of the Random Walker

The return of a random walker on a lattice to its starting point after a certain number of steps N is another example of a recurrent event. Interestingly, the probability $P(\vec{0}, N)$ of returning to the start, see (7.14), depends on the size D of the lattice. The greater the number of directions D along which the walker can move, the greater the number of paths to walk, and, moving randomly, the more difficult it becomes to return to the starting point.

9.2 Classification of Recurrent Events

9.2.1 Periodic and Aperiodic Events

The first distinction we must make, therefore, is between periodic and aperiodic events. Here the name is deceptive. An event is periodic only if it can occur at fixed time intervals, not if it always happens at fixed time intervals. A sunny Sunday can occur only at multiple seven-day intervals, but the sun need not necessarily shine every Sunday: this event "Sunday of sunshine" is said to be periodic (probably a more suitable name would be *periodicized* instead of periodic). An event that has no precise recurrence time constraints is called aperiodic.

The study of periodic events can easily be reduced to the study of non-periodic events. If M is the periodicity, the probability that the event recurs at time n is

$$p(n) \begin{cases} = 0 & \text{if } n \bmod(M) \neq 0 , \\ \geq 0 & \text{if } n \bmod(M) = 0 . \end{cases} \tag{9.2}$$

To eliminate all times at which p is identically zero, it is sufficient to consider events occurring at times Mk and take k as the new time step index: $p(n = kM) \to p(k)$.

9.2.2 Persistent and Transient Events

We denote by $p(k)$ the probability of the event occurring at time k. Conventionally we assume that the event occurs at time zero, i.e., that time starts from a certain occurrence of the event. We will, therefore, have that $p(0) = 1$.

We observe that the sequence of probabilities $\{p(k)\}$ that the event occurs at all time steps k does not represent a probability distribution. Indeed, the event may recur at more than one time step in the sequence and, therefore, the recurrence of the event at a given step is not mutually exclusive with the recurrence at another step.

There are two possibilities for the event.

- The probability $p(k)$ goes to zero so fast with k that the sum of probabilities over time remains finite when the number of steps N goes to infinity:

$$\lim_{N \to \infty} \sum_{k=0}^{N} p(k) < \infty . \tag{9.3}$$

In this case, we speak of a *transient* event, as it happens a finite number of times.
- The probability $p(k)$ is such that its sum diverges:

$$\sum_{k=0}^{\infty} p(k) = \infty . \tag{9.4}$$

In this case, we speak of a *returning, persistent,* or *certain* event, as it certainly happens an infinite number of times.

In this latter persistent case, we can make a further distinction.

- The case where the mean probability over time is zero:

$$\lim_{N \to \infty} \frac{1}{N} \sum_{k=0}^{N} p(k) = 0 , \tag{9.5}$$

and the event is called *null persistent*. An example is given by $p(k) \propto k^{-1/2}$ for $k \gg 1$.

- The case where the same quantity is non-zero:

$$\lim_{N \to \infty} \frac{1}{N} \sum_{k=0}^{N} p(k) > 0, \tag{9.6}$$

and the event is called *non-null returning*, or *non-null persistent*. In the case where it is also aperiodic, it is called an *ergodic recurrent event*. This occurs, for example, if $p(k) \to$ constant > 0 when $k \to \infty$.
- The last theoretical possibility is that the limit in (9.5) above does not exist, but we will show in Section 9.4 that this possibility cannot occur.

To fix ideas a little, let us think, for example, of the random walker in Chapter 7. Let us put the origin of the hypercubic lattice at the starting point and look at the probability of the recurrent event to return to the origin in k steps, $P(\vec{0}, k)$, given by the formula (7.14): the dependence on the number of steps goes as $1/k^{D/2}$. This implies that for $D \geq 3$ the condition (9.3) for transient events is fulfilled, while for $D < 3$ the divergence (9.4) occurs. The return of the random walker is a transient event in three or more dimensions, while it is persistent in one or two dimensions. Furthermore, in $D = 2$ dimensions, the condition (9.5) is verified, while for $D = 1$ the condition (9.6) is true: the return is a null persistent event in two dimensions and a non-null persistent event in one dimension.

9.3 Fundamental Relations of Recurrent Events

9.3.1 First Return Probability

A fundamental role in recurrent events is played by the *first return probability*, i.e., the probability $f(n)$ that the event occurs for the first time after n time steps since the last recurrence. For example, $f(n)$ is the probability that an event occurs for the first time at time $k + n$ if it occurred at time k. Expressing this more pedantically, $f(n)$ is the probability that the event occurs at time k, that it occurs at time $k + n$, and that it does not occur at any time between k and $k + n$. Conventionally, we can, therefore, set $f(0) = 0$. It may also happen that an event no longer occurs after a certain time N. In this case we will have that $f(n > N) = 0$.

Recurrent events with different times of first return are self-exclusive and, therefore, the sequence $\{f(n)\}$ is a probability distribution of the event's recurrence time intervals. The first return time n is the time between the last recurrence and the next, thus it is an interval between two consecutive recurrences. This is why the normalization condition applies:

$$f \equiv \sum_{n=1}^{\infty} f(n) \leq 1, \tag{9.7}$$

where equality holds when it is always possible (sooner or later) for the event to recur. Normalization is incomplete, on the other hand, if there is a finite probability $f_\infty \equiv 1 - f$ that the event will never occur again. If

$$f = 1 \tag{9.8}$$

the event will therefore be persistent and we will see that the condition is equivalent to (9.4), whereas if

$$f < 1 \tag{9.9}$$

the event can only occur a finite number of times and is therefore transient, as expressed in the formula (9.3) in terms of the return probability $p(n)$.

If the event is persistent, the series $\{f(n)\}$ represents the distribution of the first return time, which takes integer values, and its expected value

$$\mu \equiv \langle n \rangle = \sum_{n=0}^{\infty} f(n)n \tag{9.10}$$

is the mean (first) recurrence time.

As we can easily guess, the two probabilities p and f are not independent of each other. Let us try to formalize our intuition by considering a specific sequence. Given discrete times $n > 0$ and $k \in (0, n]$, we see the case in which the first recurrence of the event occurs at step k, with probability $f(k)$, and subsequently the event also (not *only*) recurs after $n - k$ steps, with probability $p(n - k)$. This sequence is a composite event that occurs with probability $f(k)p(n - k)$. As k changes, composite events of this kind are all mutually exclusive and by Kolmogorov's fourth axiom (see Section 1.2.2) the probability of the event recurring at step $n > 0$ is, therefore, the sum of their probabilities:

$$p(n) = \sum_{k=1}^{n} f(k)p(n - k) = \sum_{k=0}^{n} f(k)p(n - k), \quad n \geq 1, \tag{9.11}$$

where the second equality follows from the fact that $f(0) = 0$. We recognize at first glance that the formula (9.11) is a convolution operation: $\{p\} = \{f\} * \{p\}$.

We can make Eq. (9.11) also valid for $n = 0$ by adding an additional term for $p(0) = 1$:

$$p(n) = \sum_{k=0}^{n} f(k)p(n - k) + \delta_{n,0}. \tag{9.12}$$

The above equations allow us to compute p, given f, recursively from $p(1)$. In the same way, we can compute f from p. In fact, the formula above can be written as

$$f(n) = p(n) - \sum_{k=0}^{n-1} f(k)p(n - k) - \delta_{n,0}. \tag{9.13}$$

Much of the theory of recurrent events is based on the basic equation (9.12) of the convolution of $\{p\}$ and $\{f\}$ and on the fact that both p and f are non-negative quantities. An arbitrary choice of p would naturally lead to negative f as well. On the other hand, we notice that any choice of f, whose sum is less than or equal to 1, produces non-negative p. So if we want to design a recurrent event, we have no constraints on the first return probabilities, but we have very strong ones on p. For example, from Eq. (9.13) we have

$$f(0) = p(0) - 1 = 0,$$

$$f(1) = \frac{p(1)}{2},$$

$$f(2) = \frac{p(2)}{2} - \frac{p(1)^2}{2},$$

which implies that, if $p(1) = 1/2$, one would need to have $p(2) \geq 1/4$ in order to have $f(2) > 0$.

9.3.2 Generating Functions of Recurrent Events Probability

The relationship between $\{p\}$ and $\{f\}$ can be written in a simple and compact form if we introduce their generating functions $P(s)$ and $F(s)$:

$$P(s) = \sum_{k=0}^{\infty} p(k)s^k, \tag{9.14}$$

$$F(s) = \sum_{k=0}^{\infty} f(k)s^k. \tag{9.15}$$

Here $\{p\}$ and $\{f\}$ are always sequences of finite numbers (less than or equal to 1). Consequently, $P(s)$ and $F(s)$ are analytic functions in the open disk $|s| < 1$. Furthermore, as we have seen, $f(k)$ represents mutually exclusive events and constitutes a probability distribution, so

$$F(1) = \sum_{k=0}^{\infty} f(k) = f \leq 1, \tag{9.16}$$

where the "less than" part is realized if there is a probability that the event will never occur. This implies that $|F(s)| \leq 1$ also on the convergence circumference $|s| = 1$ and the convergence region is larger than that of $P(s)$. The convolution relation (9.12) translates for the generating functions in a simple way as

$$P(s) = P(s)F(s) + 1 \quad \Longrightarrow \quad P(s) = \frac{1}{1 - F(s)}. \tag{9.17}$$

From this relation we immediately derive the connection between the conditions (9.3) and (9.9) for transient events and the conditions (9.4) and (9.8) for persistent ones. It is evident, in fact, that

$$P(1) = \sum_{k=0}^{\infty} p(k) < \infty \quad \Longleftrightarrow \quad F(1) < 1.$$

The event, then, is transient if and only if $F(1) < 1$: the probability to have a first return is less than 1. Conversely, the event is persistent if $F(1) = 1$: the probability of having a recurrence is 1, and the event will recur in the future with certainty.

From the above definitions it also follows that an event is periodic with periodicity M if $f(n)$ and $p(n)$ are equal to zero when n is not a multiple of M. If f and p have different periodicities, their least common multiple is indicated as the periodicity of the event.

9.3.3 Lack of Memory and First Return Properties

To emphasize the lack of memory in an otherwise correlated process such as a recurrent event, it can be illustrative to consider a description solely in terms of first return probability. Let us begin with the probability of an event occurring for the second time at time n, the *second return* probability,

$$f^{(2)}(n) = \sum_{k=1}^{n} f(k)f(n-k). \tag{9.18}$$

This is the convolution of two first return probabilities: $\{f^{(2)}\} = \{f\} * \{f\} = \{f\}^{*2}$. Iterating the relation, one will have $\{f^{(3)}\} = \{f^{(2)}\} * \{f\} = \{f\}^{*3}$ and so on until the generic probability of the rth return, for which the convolution $\{f^{(r)}\} = \{f^{(r-1)}\}*\{f\} = \{f\}^{*r}$ applies:

$$f^{(r)}(n) = \sum_{k=1}^{n} f^{(r-1)}(k)f(n-k). \tag{9.19}$$

By definition, $f^{(r)}(n = T^{(r)})$ is the probability of the time $T^{(r)}$ of r subsequent recurrences. The fact that it is the convolution of r first return probabilities corresponds to saying that the random variable $T^{(r)}$ is the sum

$$T^{(r)} = \sum_{a=1}^{r} T_a^{(1)}, \tag{9.20}$$

where the first recurrence times $T_a^{(1)}$ are all drawn independently of each other, each one with probability $f(T_a^{(1)})$. We denote by $T_1^{(1)}$ the time of the first return, by $T_2^{(1)}$ the time elapsed between the first and second returns, and so on. The strength of the theory of recurrent events lies precisely in this reduction of a generic situation into the sum of independent random variables.

In terms of generating functions, it follows from (9.19) that the generating function of the rth return probability reads

$$F^{(r)}(s) = \sum_{k=0}^{\infty} f^{(r)}(k)s^k = [F(s)]^r, \tag{9.21}$$

from which $F^{(r)}(1) = [F(1)]^r = f^r$. For a transient recurrent event ($f < 1$) the probability of the rth return becomes smaller and smaller as the number of recurrences increases, and the event can, hence, only occur a finite number of times because at infinity the probability is zero. In contrast, for a persistent event $f^r = f = 1, \forall r$, and the rth is certain at any r, also infinite.

Keeping in mind formula (9.21) and considering the fact that for $|s| < 1$ we can bound from above the absolute value of the first return probability generator as

$$|F(s)| < \sum_{k=1}^{\infty} f(k)|s|^k < \sum_{k=1}^{\infty} f(k) = f \leq 1, \tag{9.22}$$

we can rewrite the formula (9.17) as a series expansion of $F(s)$ and, using (9.21), obtain

$$P(s) = \frac{1}{1 - F(s)} = \sum_{r=0}^{\infty} [F(s)]^r = \sum_{r=0}^{\infty} F^{(r)}(s).$$

Moving from generators to distributions, this relation implies that

$$\sum_{n=0}^{\infty} p(n)s^n = \sum_{r=0}^{\infty} \sum_{n=0}^{\infty} f^{(r)}(n)s^n \implies \sum_{n=0}^{\infty} \left[p(n) - \sum_{r=1}^{\infty} f^{(r)}(n) \right] s^n = 0,$$

and, hence, for $s > 0$ one has

$$p(n) = \sum_{r=1}^{n} f^{(r)}(n) \quad \forall n,$$

where in the upper limit of the summation we have replaced ∞ with n because the probability of r returns in $n < r$ steps is zero (we also have omitted the term of $r = 0$ returns). The probability that the event recurs in n steps is equal to the sum of the disjoint instances of recurrence in n steps for the first time, or for the second, or for the third, or, ..., for the nth.

9.4 Limit Probability Theorem for Recurrent Events

In this section, we want to show the following theorem.

Theorem. (Limit probability theorem for recurrent events) *For non-null persistent and aperiodic recurrent events, the probability $p(n)$ for large n satisfies*

$$\lim_{n \to \infty} p(n) = \frac{1}{\mu}, \tag{9.23}$$

where μ is the mean recurrence time, cf. Eq. (9.10).

We note that the theorem actually holds also in the case of null persistent events, for which the normalization is $f = 1$ but the mean recurrence time is $\mu \to \infty$. Consistently, indeed, we find that $\lim_{n \to \infty} p(n) = 0$.

We now propose a proof of the theorem of the limit probability of recurrence for aperiodic events, i.e., of period 1, for which one can have $f(n) > 0$ at every step, but the result is easily generalizable to the periodic case, as we shall see in Section 9.4.1. An alternative proof, under the same assumptions, is that given by Feller [28].

Proof   For the proof of the theorem we consider a series $p(n)$ of all non-negative and finite elements and its generating function $P(s) = \sum_n p(n)s^n$. Using two theorems on the convergence of series, one of *Abelian* type and one of *Tauberian* type, due to Godfrey Harold Hardy and John Edensor Littlewood [62, 63], we can write the relation

$$\lim_{n \to \infty} p(n) = \lim_{s \to 1^-} (1 - s)P(s). \tag{9.24}$$

In order not to burden the discussion, the proofs of the theorems underlying Eq. (9.24) are given in Appendix 9.A.

For s close to 1, we can expand the generating function of the probability distribution of first return times into a series:

$$F(s) = 1 - \mu(1 - s) + \tfrac{1}{2}F''(1)(1 - s)^2 + O((1 - s)^3), \tag{9.25}$$

where we have assumed, for simplicity of proof, that the second derivative of the generating function of the first return probabilities is finite,

$$F''(1) = \sum_{k=2}^{\infty} f(k)k(k - 1) < \infty. \tag{9.26}$$

We recall that the variance of the first return times is $F''(1) - \mu^2 + \mu$ and, thus, this will also be finite if (9.26) holds. In this case, expanding (9.17) in a Taylor series near $s = 1$, we obtain that

$$P(s) = \frac{1}{1 - F(s)} = \frac{1}{\mu(1 - s)} + \frac{F''(1)}{2\mu^2} + O(1 - s), \tag{9.27}$$

where $O(1-s)$ denotes terms going to zero for s tending[1] to 1. Plugging this expression into the relation (9.24) then gives the expected result (9.23).   ∎

9.4.1 Limit Theorem Generalization to Periodic Case

In the case of periodic events of periodicity $M > 1$, M is the greatest common divisor of the steps n for which we can have $f(n = kM) > 0$. In all other cases, $n \pmod M \neq 0$, one will have $f(n) = 0$. The theorem can also be generalized to periodic events: simply rescaling the times. We start by writing the identities

$$F(s) = \sum_{n=1}^{\infty} f(n)s^n = \sum_{k=1}^{\infty} f(kM)s^{kM}, \tag{9.28}$$

$$\mu = F'(1) = \sum_{k=1}^{\infty} f(kM)kM. \tag{9.29}$$

---

[1] If the series (9.26) had been divergent, there would have been additional singularities at $s = 1$, but these singularities would have been less strong than a pole of order 1.

We can eliminate the null terms in the power series of the generating function by rescaling the variable $\zeta \equiv s^M$, so as to obtain the generating functions of the series at times when the probabilities are not compulsorily null:

$$F_M(\zeta) = F(\zeta^{1/M}) = \sum_{k=1}^{\infty} f(kM)\zeta^k, \tag{9.30}$$

$$P_M(\zeta) = P(\zeta^{1/M}) = \sum_{k=0}^{\infty} p(kM)\zeta^k. \tag{9.31}$$

The mean first return time in this rescaled counting of time is also rescaled by M:

$$F'_M(1) = \sum_{k=1}^{\infty} kf(kM) = \frac{F'(1)}{M} = \frac{\mu}{M}, \tag{9.32}$$

and we can then generalize Eq. (9.23) in the limit probability theorem of recurrent events to periodic events by considering only sequences of probabilities at multiple times of periodicity. This gives

$$\lim_{n \to \infty} p(nM) = \frac{1}{F'_M(1)} = \frac{M}{\mu}. \tag{9.33}$$

9.4.2 Periodicity and Singularities of the Generating Function $P(s)$

We know that in the unit disk $|s| < 1$ the generating functions $F(s)$ and $P(s)$ are analytic and that $F(s)$ also converges on the circumference $|s| = 1$. It also follows that for $|s| < 1$ the bound (9.22) from above applies and in the case of persistent events we have, therefore, $|F(s)| \leq 1$.

We want to show below that if the events are periodic, the singularities of $P(s)$ in (9.27) are not only in $s = 1$, real, but at all points along the unit circumference in the complex plane where $F(s) = 1$. To this end, it is useful to consider the generating function exclusively on the edge of the convergence disk, of coordinate $s = e^{iq}$, $q \in [-\pi, \pi]$, i.e., the Fourier transform on the discrete lattice (see Appendix 7.A),

$$\tilde{F}(q) = F(e^{iq}), \tag{9.34}$$

and use the inverse formula,

$$f(n) = \frac{1}{2\pi} \int_{-\pi}^{\pi} dq\, e^{-inq}\, \tilde{F}(q). \tag{9.35}$$

It is easy to realize that, since all $f(k)$ are positive or null, the relation $F(\exp(iq)) = 1$ can hold only if all non-zero terms of the series

$$\sum_{k=1}^{\infty} f(k)e^{ikq}$$

have the same phase q. Indeed, we have that the real part is

$$0 = \text{Re}[F(\exp{(\iota q)}) - 1] = \text{Re}\left[\sum_{k=1}^{\infty} f(k)(\exp{(\iota k q)} - 1)\right] = \sum_{k=1}^{\infty} f(k)[\cos{(kq)} - 1].$$

The last expression is the sum of negative or null terms and can only be zero if all its terms are null:

$$f(k)[\cos{(kq)} - 1] = 0 \quad \forall k \in \mathbb{N}. \tag{9.36}$$

Either the probability of first return is zero or $\cos{(kq)} = 1$ (or both).

We then understand that a singularity of $\tilde{P}(q) = [1 - \tilde{F}(q)]^{-1}$ for $q \neq 0$ is only possible in the periodic case. Indeed, we first note that q must be of the form $2\pi/M$, with M integer, otherwise $\cos{(kq)} - 1$ can never be zero and, consequently, all the first return probabilities $f(k)$ should be zero, which is not possible by hypothesis (their sum is equal to 1).

If q is of the form $2\pi/M$, the probability of first return $f(k)$ can be non-zero only for k being a multiple of M. We are, therefore, in the periodic case, in which, on the convergence disk, the transform $\tilde{F}(q)$ is equal to 1 for

$$q = 0, \ \frac{2\pi}{M}, \ \frac{4\pi}{M}, \ \ldots, \ \frac{2(M-1)\pi}{M}.$$

These are the values of the singularities of $P(e^{\iota q})$ on the convergence disk $s = e^{\iota q}$. In the aperiodic case ($M = 1$) the only possible singularity of $\tilde{P}(q)$ is at $q = 0$, i.e., a null phase ($s = 1$).

9.4.3 Long-Time Approach to Limit Probability

Since the periodic case is easily taken into account by restricting the study to times that are multiples of M, we now consider the aperiodic case only. Let us define the series $g(n)$ as the deviation of $p(n)$ from its asymptotic value:

$$g(n) \equiv p(n) - \frac{1}{\mu}. \tag{9.37}$$

The function $G(s)$ generating the moments of $g(n)$ is then given by

$$G(s) = P(s) - \frac{1}{\mu(1 - s)}. \tag{9.38}$$

From (9.27) we see that the function $G(s)$ is finite at $s = 1$, where one has

$$G(1) = \sum_{n=1}^{\infty} g(n) = \frac{F''(1)}{2\mu^2}, \tag{9.39}$$

and it does not diverge on the remainder of the unit circle. It is a finite, norm-constrained L^2 function along the unit circle where $\tilde{G}(q) \equiv G(\exp{(\iota q)})$:

$$\int_{-\pi}^{\pi} dq \, |\tilde{G}(q)|^2 < \infty. \tag{9.40}$$

Because of one of the fundamental theorems of the Fourier transform we hence have

$$\sum_{n=1}^{\infty} g(n)^2 = \frac{1}{2\pi} \int dq \, |\tilde{G}(q)|^2 < \infty , \tag{9.41}$$

demonstrating that $g(n)$ tends to zero for large n. The non-negativity of $f(n)$ played a crucial role in the proof.

Mathematical Appendices

9.A Two Theorems for Divergent Series Summation

In this appendix we aim to prove the following two theorems connecting the limit of a series for times going to infinity, and the order of divergence of its generating function on the edge of the disk of convergence. We shall restrict ourselves to considering aperiodic distributions and, therefore, look only at the divergence for real $s \to 1$.

We shall refer to the first theorem as the *Abelian* theorem (see, e.g., ref. [63]) or the *direct* theorem. The other theorem, which we shall call the *Tauberian* theorem (see, e.g., ref. [62]), is almost the inverse of the first one.

Theorem. (Abelian theorem) *If $p(n) \to C > 0$ for $n \to \infty$ and if $P(s) \equiv \sum_k p(k)s^k$, then, for $s \to 1$, the following limit exists:*

$$\lim_{s \to 1} P(s)(1 - s) \to C . \tag{9.42}$$

Theorem. (Tauberian theorem) *If for $s \to 1$ the generating function diverges as $P(s) \sim C/(1 - s)$ and if $p(k) \geq 0$ for each k, then, for $n \to \infty$, the following holds:*

$$\lim_{n \to \infty} \frac{1}{n} \sum_{k=0}^{n} p(k) = C . \tag{9.43}$$

In (9.42) and (9.43) C is the same constant, i.e., for positive and limited series the two limits lead to the following relation valid in both directions:

$$\lim_{n \to \infty} \frac{1}{n} \sum_{k=0}^{n} p(k) = \lim_{s \to 1} (1 - s) \sum_{k=0}^{\infty} p(k)s^k . \tag{9.44}$$

This last equation is the relation on which our proof of the asymptotic probability theorem of recurrent events, set out in Section 9.4, is based.

Before moving on to the proofs of the theorems, let us carry out a small preliminary exercise. Let us define the sum of probabilities

$$\sigma(n) \equiv \sum_{k=0}^{n} p(k) , \tag{9.45}$$

whose generating function satisfies the relation

$$\Sigma(s) = \sum_{n=0}^{\infty} \sigma(n)s^n = \sum_{n=0}^{\infty}\sum_{k=0}^{n} p(k)s^n = \sum_{k=0}^{\infty}\sum_{n=k}^{\infty} p(k)s^n$$

$$= \sum_{k=0}^{\infty}\sum_{j=0}^{\infty} p(k)s^{j+k} = \sum_{k=0}^{\infty} p(k)s^k \sum_{j=0}^{\infty} s^j = \frac{P(s)}{1-s}. \tag{9.46}$$

In the case in which $p(k)$ is a probability distribution, this coincides with the result (8.12) derived from the proof of the tail probability theorem. Indeed, in that case, $\sigma(n)$ is the cumulative probability, the complement of the tail probability defined in (8.9), $c(n) = 1 - \sigma(n)$, and their generating functions thus satisfy the relation $\Sigma(s) + C(s) = 1/(1-s)$ consistent with Eq. (8.10) in the theorem.

Let us now demonstrate the Abelian or direct theorem.

9.A.1 Proof of the Generating Function Divergence Theorem

Let us first see a general theorem on the asymptotic behavior of series and their generating functions near the circle of convergence $|s| = 1$.

Theorem. (Theorem for the asymptotic behaviour of series and their generating functions) *Given the non-negative and limited series $\{p(n)\}$ and $\{q(n)\}$, with $p(n) < \infty$ and $q(n) < \infty$, $\forall n$, and their generating functions*

$$P(s) \equiv \sum_{k=1}^{\infty} p(k)s^k \quad and \quad Q(s) \equiv \sum_{k=1}^{\infty} q(k)s^k \,,$$

convergent for $s \in [0, 1)$ and divergent in $s = 1$, then

$$p(n) \underset{n\to\infty}{\sim} Cq(n) \quad \Longrightarrow \quad P(s) \underset{s\to 1}{\sim} CQ(s). \tag{9.47}$$

Here the symbol $\sim$ indicates that the quantity on the left "scales like" the one on the right in the limit marked below the symbol. We now give the proof of (9.47).

Proof The limitation hypothesis results in the property that, given an $\epsilon > 0$, there exists an N large enough such that for every $n > N$ there occurs $|p(n) - Cq(n)| < \epsilon q(n)$ (remember that $q(n)$ is limited by hypothesis). We therefore have

$$|P(s) - CQ(s)| = \left| \sum_{k=0}^{\infty} [p(n) - Cq(n)]s^n \right|$$

$$\leq \left| \sum_{n=0}^{N} [p(n) - Cq(n)]s^n \right| + \left| \sum_{n=N+1}^{\infty} [p(n) - Cq(n)]s^n \right|$$

$$\leq \sum_{n=0}^{N} |p(n) - Cq(n)|s^n + \epsilon \sum_{n=N+1}^{\infty} q(n) \leq \sum_{n=0}^{N} |p(n) - Cq(n)| + \epsilon Q(s).$$

$$\tag{9.48}$$

At this point we choose a sufficiently large s, i.e., we take an arbitrarily small $\delta > 0$ such that, for $s > 1 - \delta$, $Q(s)$ satisfies, for each ϵ, the inequality

$$\epsilon Q(s) \geq \sum_{n=0}^{N} |p(n) - Cq(n)|, \quad \text{if } s > 1 - \delta. \tag{9.49}$$

This is always possible because by hypothesis $Q(s)$ diverges for $s \to 1$ while the sum on the right-hand side is taken over a finite number N of terms. For these values of s we can then perform the additional estimation from above in formula (9.48):

$$|P(s) - CQ(s)| \leq 2\epsilon Q(s), \quad \text{for } s > 1 - \delta. \tag{9.50}$$

So, provided we are arbitrarily close to $s = 1$ ($s \to 1$), for arbitrarily small ϵ, it holds that

$$P(s) \sim CQ(s), \quad \text{for } s \to 1, \tag{9.51}$$

completing the proof of the theorem. ∎

We now apply Eq. (9.47) to the cumulative series

$$\sigma(n) = \sum_{k=0}^{n} p(k) \quad \text{and} \quad \tau(n) = \sum_{k=0}^{n} q(k). \tag{9.52}$$

Proof of Abelian theorem　We observe that if $p(k) \sim Cq(k)$ when $k \to \infty$, then $\sigma(n) \sim C\tau(n)$ as $n \to \infty$. Using the result (9.46) on the generating function of the cumulative series we can rewrite their generating functions as

$$\Sigma(s) = \frac{P(s)}{1 - s} \quad \text{and} \quad T(s) = \frac{Q(s)}{1 - s},$$

which are convergent for $s < 1$. Because of Eq. (9.47) we hence have that

$$\frac{P(s)}{1 - s} \sim C \frac{Q(s)}{1 - s} \quad \text{for } s \to 1. \tag{9.53}$$

As a final step, we eventually focus on the specific series $q(n) = 1$ for each $n > 0$, whence $\tau(n) = n$ and $T(s) = s/(1 - s)^2$. Thus the above formula becomes

$$\frac{P(s)}{1 - s} \underset{s \to 1}{\sim} \frac{C}{(1 - s)^2}. \tag{9.54}$$

Equation (9.47) implies, therefore, that for $s \to 1$

$$P(s) \underset{s \to 1}{\sim} \frac{C}{1 - s}, \tag{9.55}$$

and we obtain the proof of the Abelian theorem (direction $\Rightarrow$ of the equality (9.44)),

$$\lim_{n \to \infty} p(n) = C \quad \Longrightarrow \quad \lim_{s \to 1} P(s)(1 - s) = C, \tag{9.56}$$

as a special case of Eq. (9.47) of the general theorem. ∎

9.A.2 Proof of the Hardy–Littlewood Theorem on the Limit of a Series

The inverse theorem is not trivial, in general. Indeed, to assume that a function $P(s)$ diverges in $s = 1$ as $1/(1 - s)$ (in order to have a finite limit $\lim_{s \to 1} P(s)(1 - s) = C < \infty$) is not enough to have a finite limit of $p(n)$ when $n \to \infty$. Let us take as a counter-example the function

$$P(s) = \frac{1}{(1 + s)^2(1 - s)}.$$

(9.57)

The limit $\lim_{s \to 1} P(s)(1-s) = 1/4$, but if we write it as a generating function we obtain

$$P(s) = \frac{1}{(1 + s)^2(1 - s)} = \sum_{n=0}^{\infty} (n + 1)(s^{2n} - s^{2n+1}),$$

(9.58)

and the terms of the series oscillate

$$p(k) = \begin{cases} k/2 + 1 & \text{if } k \,(\text{mod } 2) = 0, \\ -k/2 - 1 & \text{if } k \,(\text{mod } 2) \neq 0, \end{cases}$$

(9.59)

without having a definite limit for $k \to \infty$. In order to state the inverse theorem we must, therefore, assume that the terms of the series are non-negative, as in the assumptions of the direct theorem.

9.A.2.1 Hardy and Littlewood's Theorem

In this case we can state the inverse theorem of Hardy and Littlewood (1914) [62].

Theorem. (Inverse Hardy–Littlewood theorem) *(i) If the function*

$$P(s) = \sum_{n=0}^{\infty} p(n)s^n < \infty, \quad s < 1,$$

(9.60)

satisfies the limit

$$\lim_{s \to 1} P(s)(1 - s) = C < \infty,$$

(9.61)

and (ii) if $p(n) \geq 0$, $\forall\, n$, then

$$\lim_{n \to \infty} \frac{1}{n} \sum_{k=0}^{n} p(k) = C.$$

9.A.2.2 About the Weierstrass Theorem

Before proving the Hardy–Littlewood theorem, let us recall the Weierstrass theorem, which states that any continuous function with no singularity can be approximated by a sequence of polynomials. In particular, in a given finite domain, a function $\phi(s)$ that has no singularities can always be bound from above by a polynomial $M(s)$ and from

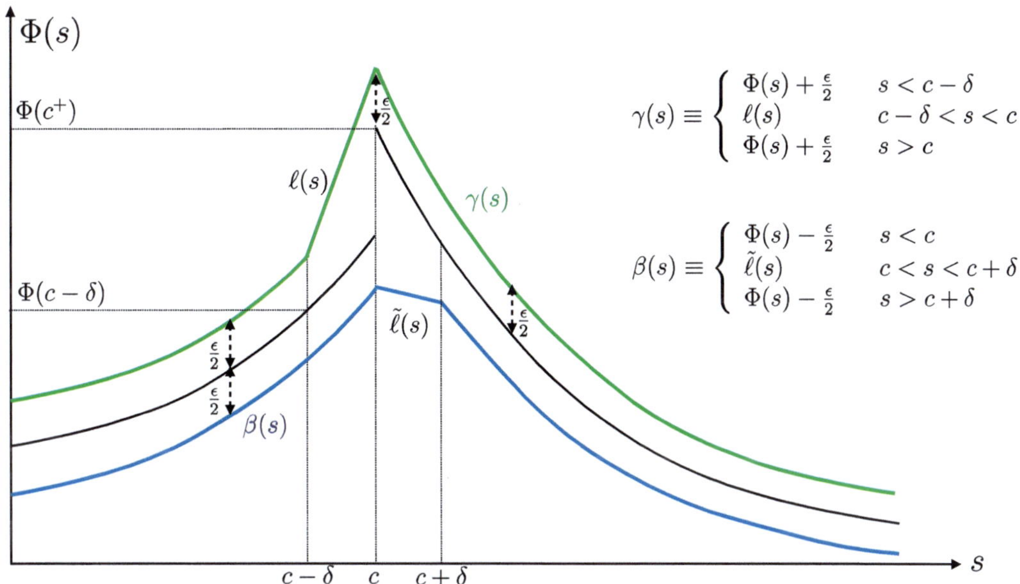

Figure 9.1 Encasing a discontinuous function $\phi(s)$ with continuous functions $\beta(s)$ and $\gamma(s)$ such that $\beta(s) < \phi(s) < \gamma(s)$, by application of Weierstrass' sandwich theorem of polynomials (9.62).

below by a polynomial $m(s)$. More formally, given $\phi \in C^0$ in $s \in [0,1)$ and given $\epsilon > 0$, it is always possible to find two polynomials $M(s)$ and $m(s)$ such that

$$m(s) \leq \phi(s) \leq M(s),$$

$$\int_0^1 ds\,[\phi(s) - m(s)] \leq \epsilon \quad \text{and} \quad \int_0^1 ds\,[M(s) - \phi(s)] \leq \epsilon. \tag{9.62}$$

However, it is not strictly necessary for the function to be continuous, as long as it is limited and has isolated singularities, in order to be able to implement these bounds with polynomials. Let us take a function with a discontinuity at a point $c \in (0,1)$ and, for illustration, take $\phi(c^+) > \phi(c^-)$ (see Figure 9.1). Let us take $\delta > 0$ and define the function

$$\gamma(s) \equiv \begin{cases} \phi(s)+\epsilon/2, & s < c-\delta, \\ \ell(s), & c-\delta < s < c, \\ \phi(s)+\epsilon/2, & s > c, \end{cases} \tag{9.63}$$

where $\ell(s)$ is a linear interpolation between $\phi(c-\delta)+\epsilon/2$ and $\phi(c^+)+\epsilon/2$:

$$\ell(s) = \phi(c-\delta) + \epsilon/2 + (s-c+\delta)[\phi(c^+) - \phi(c-\delta)]/\delta.$$

For a sufficiently small δ it will always be possible to find a polynomial $M(s)$ such that

$$M(s) \geq \gamma(s) \geq \phi(s), \tag{9.64}$$

to which the Weierstrass theorem for the continuous function $\gamma(s)$ can be applied for a given $\epsilon/2$,

$$\int_0^1 ds\,[M(s) - \gamma(s)] \le \epsilon/2, \tag{9.65}$$

and from which we obtain the bound from above of the distance from the discontinuous function $\phi(s)$,

$$\int_0^1 ds\,[M(s) - \phi(s)] \le \epsilon. \tag{9.66}$$

With identical philosophy we can define a continuous function less than $\phi(s)$ as

$$\beta(s) \equiv \begin{cases} \phi(s) - \epsilon/2, & s < c, \\ \tilde{\ell}(s), & c < s < c + \delta, \\ \phi(s) - \epsilon/2, & s > c + \delta, \end{cases} \tag{9.67}$$

where $\tilde{\ell}(s)$ is the linear interpolation between $\phi(c^-) - \epsilon/2$ and $\phi(c + \delta) - \epsilon/2$. When δ is small enough, one can always find a polynomial $m(s)$ such that

$$m(s) \le \beta(s) \le \phi(s), \qquad \int_0^1 ds\,[\beta(s) - m(s)] \le \epsilon/2, \tag{9.68}$$

and hence such that

$$\int_0^1 ds\,|\phi(s) - m(s)| \le \epsilon. \tag{9.69}$$

In summary, we can also bound from above and below, with suitable polynomial limited functions, a function $\phi(s)$ with isolated discontinuities.

9.A.2.3 Proof of Hardy and Littlewood's Tauberian Theorem

With this result, we can move on to the proof of Hardy and Littlewood's theorem attributed to Karamata [63]. Let us first prove an intermediate step:

For each polynomial $Z(x) = \sum_k \mu_k x^k$ we have

$$\lim_{s \to 1} (1 - s) \sum_{n=0}^{\infty} p(n)s^n Z(s^n) = C \int_0^1 Z(x)\,dx. \tag{9.70}$$

Proof of intermediate step. We start with the single monomial $Z(x) = x^k$ for which we have

$$\lim_{s \to 1} (1 - s) \sum_{n=0}^{\infty} p(n)s^n s^{nk} = \lim_{s \to 1} \frac{1 - s}{1 - s^{k+1}} (1 - s^{k+1}) \sum_{n=0}^{\infty} p(n)(s^{k+1})^n.$$

Using the first assumption (9.61) of the inverse theorem, with s^{k+1} as the variable, we obtain that

$$\lim_{s \to 1} (1 - s) \sum_{n=0}^{\infty} p(n)s^n s^{nk} = \lim_{s \to 1} \frac{1 - s}{1 - s^{k+1}} C = \frac{C}{k + 1} = C \int_0^1 dx\,x^k.$$

Adding up several monomials, we have that for any polynomial $Z(x)$ Eq. (9.70) holds.

We now take the arbitrary function $\phi(s)$, continuous or discontinuous, and the relations (9.62). We first take the bound from above with polynomial $M(x)$,

$$
\lim_{s\to 1}(1-s)\sum_{n=0}^{\infty} p(n)s^n\phi(s^n) \leq \lim_{s\to 1}(1-s)\sum_{N=0}^{\infty} p(n)s^n M(s^n)
$$

$$
= C\int_0^1 dx\, M(x) < C\int_0^1 dx\,[\phi(x)+\epsilon], \qquad (9.71)
$$

where we have used the condition that $M(s)-\phi(s)$ can be at most ϵ for the second inequality of (9.62) to be satisfied. Similarly, we bound from below as

$$
\lim_{s\to 1}(1-s)\sum_{n=0}^{\infty} p(n)s^n\phi(s^n) \geq \lim_{s\to 1}(1-s)\sum_{N=0}^{\infty} p(n)s^n m(s^n)
$$

$$
= C\int_0^1 dx\, m(x) > C\int_0^1 dx\,\phi(x) - C\epsilon, \qquad (9.72)
$$

where we implemented that, in order to satisfy the third inequality of (9.62), $\phi(s)-m(s)$ cannot exceed ϵ. By sending $\epsilon\to 0$ in (9.71) and in (9.72) we obtain the relation (9.70) also for ϕ:

$$
\lim_{s\to 1}(1-s)\sum_{n=0}^{\infty} p(n)s^n\phi(s^n) = C\int_0^1 dx\,\phi(x). \qquad (9.73)
$$

To finish the proof, let us consider a specific function ϕ that suits our case:

$$
\phi(x) = \begin{cases} 0 & \text{if } 0 \leq x < 1/e, \\ 1/x & \text{if } 1/e \leq x < 1. \end{cases} \qquad (9.74)
$$

On the left-hand side of (9.73) the summation becomes

$$
\sum_{n=0}^{\infty} p(n)s^n\phi(s^n) = \sum_{n=0}^{N(s)} p(n) = \sigma(N(s)),
$$

where $N(s)$ is given by the condition $s^n \geq 1/e$, and thus

$$
n \leq N(s) \equiv \frac{1}{\ln s^{-1}}.
$$

Its inverse is

$$
s(N) = e^{-1/N}.
$$

The right-hand side of (9.73) is

$$
C\int_0^1 dx\,\phi(x) = C\int_{e^{-1}}^1 dx\,\frac{1}{x} = C,
$$

and finally (9.73) implies that

$$
\lim_{s\to 1}(1-s)\sigma(N(s)) = C
$$

or, equivalently, it implies that for $s \rightarrow 1$ we have $\sigma(N(s)) \sim C/(1-s)$. Expressing $s = s(N) = e^{-1/N}$, we see that the limit for $s \rightarrow 1$ corresponds to the limit for $N \rightarrow \infty$ and we obtain

$$\lim_{s \rightarrow 1} (1-s)\sigma(N(s)) = \lim_{s(N) \rightarrow 1} (1-s(N))\sigma(N) = \lim_{N \rightarrow \infty} \frac{\sigma(N)}{N} = C. \tag{9.75}$$

Therefore, provided that the terms of the series $p(n)$ are non-negative (assumption (ii) of the theorem) we obtain

$$\lim_{s \rightarrow 1} (1-s)P(s) = C \implies \lim_{n \rightarrow \infty} p(n) = C. \tag{9.76}$$

This completes the proof of the intermediate step needed for the proof of Hardy and Littlewood's Tauberian theorem. ∎

Using both equations (9.56) and (9.76) we have proved the relation (9.44).

Markov Chains

10.1 Examples of Markov Chains

By Markov *chains* we mean random dynamical processes of discrete variables at discrete time steps that have the peculiarity of being described by knowing the probability of transition from one value of the variable to another in a time step and simply concatenating these probabilities over time. We will often call the values of the variables "states". Knowing the distribution of states at a given time step, we only need to know the probability of transition in a time step to another state to determine the current state of the system. It does not matter how one arrived at the starting state: the previous history does not condition the future step. To learn (stochastically) the evolution of the system it is completely irrelevant to remember what happened before. The distribution of the values of the variables at the previous step will, in turn, also be given by the distribution of a yet previous step and the related transition probability. And so on, backwards forming a chain. Markov chains are a very general mathematical structure that contain as special cases many of the systems we have studied so far. Before going into the formal details of the theory of Markov chains, indeed, let us first characterize this type of phenomenon that is already well known to us.

10.1.1 Random Walker

In Chapter 7 we studied the random walker on a lattice. We saw that the probability $P(\vec{m}, N)$ that the walker is at the site $\vec{m}$ at the time step N satisfies Eq. (7.25), which we report here for ease of reference:

$$P(\vec{m}, N) = \sum_{\vec{\ell}} J(\vec{m} - \vec{\ell}, \vec{\ell}) P(\vec{m} - \vec{\ell}, N - 1), \tag{10.1}$$

where $P(\vec{m} - \vec{\ell}, N - 1)$ is the probability of being at site $\vec{m} - \vec{l}$ at the previous step and $J(\vec{m} - \vec{\ell}, \vec{\ell})$ is the probability of making a transition from $\vec{m} - \vec{\ell}$ to $\vec{m}$ in one time step. We note that we only summed over the probabilities at the previous step $N - 1$. The probability of being at $\vec{m}$ depends only on the probabilities of being somewhere else at the previous time step. It does not depend on what happened before $N - 1$.

10.1.2 Chain Reactions

In Chapter 8 we studied chain reactions, also called branching processes. A population $n(N)$ at the time step (generation) N depends on the population $n(N-1)$ at the time step $N-1$ and on how many descendants m_α each member $\alpha = 1, \ldots, n(N-1)$ of that population yields in one generation:

$$n(N) = \sum_{\alpha=1}^{n(N-1)} m_\alpha(N-1).$$

Both m_α and n are random variables (and have the same distribution). We have seen that the probability distribution of the population evolves according to the equation

$$P(n(N)) = \sum_{m=1}^{\infty} p(n(N)|m)P(m), \tag{10.2}$$

where the index m represents the number of elements of the predecessor generation, $m = n(N-1)$. Here $p(n|m)$ denotes the probability of having a population of n elements conditioned to sum over the descendants (of random number) of a population of m elements, or, in other words, the probability that the sum of m random variables is equal to the value n, or, again, in "Markovian terms" it is the probability of transition from a population of m elements to a population of n elements in one generation. Nothing depends on the properties of the generations preceding the $N-1$ steps or generations.

10.1.3 Recurrent Events

In Chapter 9, we finally analyzed the probability of a recurrent event and showed that the probability of an event recurring at time n depends on the probability of the time $k < n$ of the last previous recurrence. In other words, $p(n)$ is the convolution of p to all previous times with the probability of first return:

$$p(n) = \sum_{k=0}^{n-1} f(n-k)p(k). \tag{10.3}$$

Here the "state" is the time of recurrence, the dynamic step is played by subsequent recurrences, and the first return probability plays the role of the transition probability from a recurrence time to another recurrence time in a subsequent recurrence. The only thing that matters is the last time k at which the event recurred. All returns prior to k do not help determining $p(n)$.

These phenomena recalled above are three examples of chains of events (position in space, size of a population, and return at a given time) that are correlated (to the previous position, the previous population, or the previous recurrence) but are *memory-less*. They are called Markov chains.

As we will formalize shortly, each Markov chain is characterized by an absolute probability of the states, such as, for example,

Table 10.1 Summary of the Markov chains studied so far (incognito) in Chapters 7, 8, and 9, each with its own notation.

Basic symbols of a Markov chain	Random walker	Chain reaction	Recurrent event
state α	lattice position	population size	recurrence time
dynamic step t	step on the lattice	generation	subsequent recurrence
state absolute probability α: $p(\alpha(t))$	$P(\vec{\alpha}, t)$	$p(\alpha(t))$	$p(\alpha)$
transition probability $\alpha \to \beta$: $P_{\alpha,\beta}$	$J(\vec{\alpha}, \vec{\beta} - \vec{\alpha})$	$p(\beta(t) \mid \alpha(t-1))$	$f(\beta - \alpha)$

- $P(\vec{n}, N)$ to be at $\vec{n}$ at time N, for the random walker,
- $p(n(N))$ to have a population $n(N)$ at time N, for a chain reaction, and
- $p(n)$ to return at time n, for a recurrent event,

and by a transition probability between states, such as, for example,

- $J(\vec{k}, \vec{n} - \vec{k})$ to go from $\vec{k}$ to $\vec{n}$ in one step, for the random walker,
- $p(n|k)$ to generate a population of n elements from a population of k elements in one generation, in a chain reaction, and
- $f(n - k)$ to have a single recurrence at time n starting from time k, for a recurrent event.

All these processes are summarized in Table 10.1.

10.2 General Properties of Markov Chains

Let us take a system in which time is described by a discrete variable, n. Let us further assume that the state of the system is also described by an integer variable k. The dynamics of the system will then be described by a function $k(n)$ indicating the state k of the system at time n.

10.2.1 Transition Probability and Conditional Probability

Let us consider the evolution of the system, which we will assume to be probabilistic. We assume that the probability that the system is in the state $k(n + 1)$ at time $n + 1$

depends only on the state of the system at time n, $k(n)$, and not on the preceding history at steps $n-1$, $n-2$, ... In this case, the system dynamics is said to form a *Markov chain*. In other words, a Markov chain is a system with a number of numerable internal states, stochastically evolving over time in discrete steps. The transition probability does not depend on the previous history, except the most recent one, implying that the system is *memory-less*. We have already noted, recalling the branching processes, for example, that in place of transitions between states we can, equivalently, speak of the realization of events conditional on the occurrence of other events.

10.2.2 Stochastic Matrix

We have seen that a fundamental role in the description of Markov chains is played by the probability of transition from the state $k(n)$ to the state $i(n+1)$ in a discrete dynamic step: $P_{k(n),i(n+1)}$. This is equivalent to the probability of the occurrence of an event i at step $n+1$ conditional on the occurrence of the event k at the previous step:

$$P_{k(n),i(n+1)} \equiv p(i(n+1)\,|\,k(n))\,.$$

As a consequence of the fact that the probability must be normalized to 1, the matrix P must satisfy the following relation for any starting k:

$$\sum_i P_{k,i} = 1, \quad \forall k, \tag{10.4}$$

where the sum over i extends to all states of the chain, whether they are finite or infinite. In the future, in all cases where we do not specify limits, the sums over the states will extend to all possible states. In general, a matrix (finite or infinite) whose elements are all non-negative and satisfy the relation (10.4) is called a (row) *stochastic matrix* as it is the transition probability matrix for a Markov chain. Conversely, any non-negative matrix that satisfies Eq. (10.4) can be associated with a Markov chain.

10.2.3 Joint Probability of a Chain

More formally, a sequence of states is called a Markov chain if the probability of a specific sequence of N states is exclusively described in terms of:

- the absolute probability distribution of being in a given initial state k at step 0, i.e., $p_k(0)$, with $\sum_k p_k = 1$; and
- the transition probability between two states $k(n)$ and $k(n+1)$ in one step, which can be seen as the probability of the system to evolve into the state $k(n+1)$ conditional on being in the state $k(n)$, i.e., $P_{k(n),k(n+1)} = p(k(n+1)\,|\,k(n))$, for which the normalization (10.4) applies.

The reverse is also true: a probability distribution of initial conditions and a stochastic matrix *completely define* a Markov chain.

The previous relations imply that the joint probability of following a given history for n steps, starting from the state $k(0)$ at time 0 to the state $k(n)$, is simply given by the concatenation:

$$p(k(0), k(1), k(2), \ldots, k(n)) = p(k(0)) P_{k(0),k(1)} P_{k(1),k(2)} \cdots P_{k(n-1),k(n)} . \tag{10.5}$$

Let us take $n = 2$. The joint probability of being in the states $k(0)$ at time 0, $k(1)$ at time 1, and $k(2)$ at time 2 in a Markov chain is given by

$$p(k(0), k(1), k(2)) = p(k(0)) P_{k(0),k(1)} P_{k(1),k(2)} . \tag{10.6}$$

If we marginalize over all intermediate states $k(1)$, we obtain the joint probability of being in $k(0)$ at time 0 and in $k(2)$ after 2 steps, which we can write as

$$\sum_j p(k(0), j, k(2)) = p(k(0)) \sum_j P_{k(0),j} P_{j,k(2)} = p(k(0)) P^{(2)}_{k(0),k(2)} ,$$

where we have defined the probability $P^{(2)}$ of transition in two steps. If we start from n, rather than from 0, it reads

$$P^{(2)}_{k(n),k(n+2)} \equiv \sum_j P_{k(n),j} P_{j,k(n+2)} = p(k(n+2) \,|\, k(n)). \tag{10.7}$$

10.2.4 Stationary Markov Chain

If the elementary stochastic matrix P does not depend on absolute time n, its elements are time-translation-invariant,

$$P_{k(n),k(n+1)} = P_{k(n+m),k(n+1+m)} ,$$

for each m (possibly also negative). In this case the Markov chain is called stationary. The random walkers, branching processes, and recurrent events that we have studied in previous chapters are all stationary chains of events. In this book we will always consider stationary chains.

10.2.4.1 Dynamic Iteration of the Transition Probability

In the presence of stationarity, the indices in the formula (10.7) lose their dependence on absolute time n and this reduces to

$$P^{(2)}_{i,k} = \sum_j P_{i,j} P_{j,k} = (P^2)_{i,k},$$

which is nothing more than the product of the stochastic matrix with itself. In general, by repeating the procedure, we arrive at the identity[1]

---

[1] Beware of the notation: by $k(n-1)$ we mean the state of the chain at time $n-1$, while k_{n-1} is just the $(n-1)$th summation index in a sum over several indices of the states of the chain.

$$P_{i,j}^{(n)} = \sum_{k_1,k_2,\ldots,k_{n-1}} P_{i,k_1} P_{k_1,k_2} \cdots P_{k_{n-1},j} = (P^n)_{i,j}\,. \tag{10.8}$$

The multiplication of n stochastic matrices is still a stochastic matrix because the normalization still applies:

$$
\begin{aligned}
\sum_j P_{i,j}^{(n)} &= \sum_{k_1,k_2,\ldots,k_{n-1}} P_{i,k_1} P_{k_1,k_2} \cdots P_{k_{n-2},k_{n-1}} \sum_j P_{k_{n-1},j} \\
&= \sum_{k_1,k_2,\ldots,k_{n-2}} P_{i,k_1} P_{k_1,k_2} \cdots P_{k_{n-3},k_{n-2}} \sum_{k_{n-1}} P_{k_{n-2},k_{n-1}} = \cdots = 1\,.
\end{aligned} \tag{10.9}
$$

10.2.4.2 Dynamic Iteration of Absolute Probability

Given the simple evolution rule (10.8) of the transition matrix, the absolute probability of being in a given state k at step n iterates as

$$p_k(n) = \sum_j p_j(n-1)P_{j,k} = \cdots = \sum_j p_j(0)P_{j,k}^{(n)}\,. \tag{10.10}$$

10.2.5 The Chapman–Kolmogorov Equation

From the iteration properties in the steps of the stochastic matrix of the Markov chain, we can write the equation for the probability of transition from a state j to a state k in $n + m$ steps:

$$P_{j,k}^{(n+m)} = (P^{n+m})_{j,k} = \sum_l (P^n)_{j,l}(P^m)_{l,k} = \sum_l P_{j,l}^{(n)} P_{l,k}^{(m)}\,, \tag{10.11}$$

which is the discrete state and discrete time step form of the so-called Chapman–Kolmogorov equation for stochastic processes.

10.3 Further Examples of Markov Chains

10.3.1 Recurrent Events, Again

True, we have already discussed this, but after introducing the stochastic matrix we want to try to give a formally more explicit representation of the sequence of recurrent events as a Markov chain. Let us therefore consider a Markov chain with states S_k, $k = 0, 1, \ldots$, and take, for example, as a recurrent event $\mathcal{E} = \{$be in state S_0 of the chain$\}$. We can describe the dynamics of the recurrent event $\mathcal{E}$ using a Markov chain with the following stochastic matrix:

$$P = \begin{pmatrix} & S_0 & S_1 & S_2 & S_3 & & \\ f_1 & f_2 & f_3 & f_4 & \cdots & \cdots \\ 1 & 0 & 0 & 0 & \cdots & \cdots \\ 0 & 1 & 0 & 0 & \cdots & \cdots \\ 0 & 0 & 1 & 0 & \cdots & \cdots \\ 0 & 0 & 0 & 1 & \cdots & \cdots \\ \cdots & \cdots & \cdots & \cdots & \cdots & \cdots \end{pmatrix}. \qquad (10.12)$$

The elements of row 0 are the first return probabilities, $P_{0,k} = f_{k+1}$, of a persistent (otherwise known as certain or returning) event. The array elements in the lower rows are all null except for

$$P_{k-1,k-2} = P_{k-2,k-3} = \cdots = P_{2,1} = P_{1,0} = 1.$$

The normalization property of the stochastic matrix $P_{j,k}$ is trivially satisfied for $j > 0$ (a single non-zero element equal to 1). For $j = 0$ we have

$$\sum_{k=0}^{\infty} P_{0,k} = \sum_{k=1}^{\infty} f_k = 1$$

because we are considering a persistent recurrent event. If the first step leads from S_0 to S_{k-1}, in $k-1$ subsequent steps the chain is bound to make the transitions $k-1 \to k-2$, $k-2 \to k-3, \ldots, 1 \to 0$ with probability 1. The event S_0 is therefore a (persistent) recurring event and the return times k have distribution $\{f_k\}$. It goes without saying that the same return dynamics can be represented for any state S_k.

10.3.2 Two-State Markov Chain

Let us take a Markov chain with only two states, 0 and 1, such as that reproduced in Figure 10.1. We denote $P_{0,1} = \gamma$ and $P_{1,0} = \alpha$. For the normalization condition, we will have $P_{0,0} = 1 - \gamma$ and $P_{1,1} = 1 - \alpha$. We denote by $p_k(n)$, with $k = 0, 1$, the probability of being in the state k at step n.

At each step n the normalization is $p_0(n) + p_1(n) = 1$. We can write the equations of the two-state chain dynamics from formula (10.10) as

$$p_k(n) = \sum_{j=0,1} p_j(n-1)P_{j,k} = p_0(n-1)P_{0,k} + p_1(n-1)P_{1,k}, \quad k = 0,1. \qquad (10.13)$$

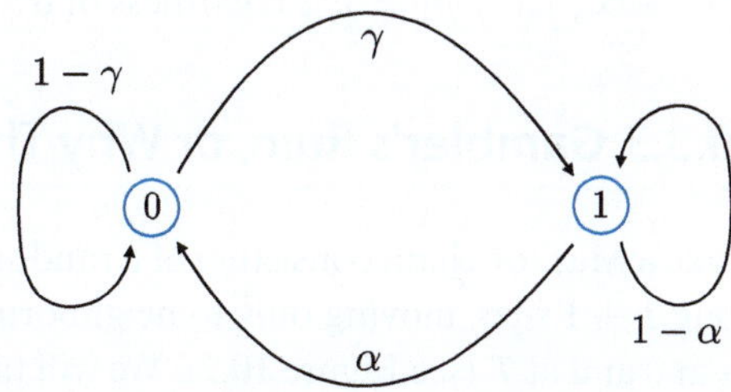

Figure 10.1   Two-state Markov chain.

In the non-trivial case, $\alpha, \gamma \in (0, 1)$, we iterate $p_0(n)$ starting from n steps:

$$
\begin{aligned}
p_0(n) &= p_0(n-1)P_{0,0} + p_1(n-1)P_{1,0} \\
&= p_0(n-1)(1-\gamma) + [1 - p_0(n-1)]\alpha \\
&= \alpha + (1 - \gamma - \alpha)p_0(n-1) \\
&= \alpha + (1 - \gamma - \alpha)[\alpha + (1 - \gamma - \alpha)p_0(n-2)] \\
&= \alpha + (1 - \gamma - \alpha)\{\alpha + (1 - \gamma - \alpha)[\alpha + (1 - \gamma - \alpha)p_0(n-3)]\} \\
&= \alpha + \alpha(1 - \gamma - \alpha) + \alpha(1 - \gamma - \alpha)^2 + (1 - \gamma - \alpha)^3 p_0(n-3) \\
&= \alpha + \alpha(1 - \gamma - \alpha) + \alpha(1 - \gamma - \alpha)^2 + \cdots \\
&\quad + \alpha(1 - \gamma - \alpha)^{n-1} + (1 - \gamma - \alpha)^n p_0(0) \\
&= \alpha \sum_{k=0}^{n-1} (1 - \gamma - \alpha)^k + (1 - \gamma - \alpha)^n p_0(0).
\end{aligned}
$$

(10.14)

(10.15)

There are two borderline cases. For $\alpha = \gamma = 1$ there is a continuous cycle between 0 and 1. If we start from 0, at odd steps the system will certainly be in 1, and at even steps it will certainly be in 0 (and vice versa if we start from 1). If $\alpha = \gamma = 0$, one has $P_{0,0} = P_{1,1} = 1$ and it is never possible to exit from the two states. These states are called *absorbing*.

For all other cases, $1 - \alpha - \gamma \in (-1, 1)$, we obtain the solution

$$
p_0(n) = \frac{\alpha}{\gamma + \alpha} + (1 - \alpha - \gamma)^n \left[p_0(0) - \frac{\alpha}{\gamma + \alpha} \right],
\tag{10.16}
$$

starting from the initial condition $p_0(0)$. The limiting distribution $(n \to \infty)$ is, therefore,

$$
p_0 = \alpha/(\alpha + \gamma), \quad p_1 = \gamma/(\alpha + \gamma),
$$

and loses its dependence on the initial condition. Furthermore, if we choose the stochastic transition matrix with $\gamma = \alpha$, the Eq. (10.16) reduces to

$$
p_0(n) = \frac{1}{2} + (1 - 2\alpha)^n \left[p_0(0) - \frac{1}{2} \right],
\tag{10.17}
$$

and for $n \to \infty$, $p_0 = p_1 = 1/2$ regardless of α.

10.3.3 Gambler's Ruin, or Why There Is a Zero in Roulette

Let us take a Markov chain consisting of a random walker that can move in one dimension along $T + 1$ sites, moving only to neighboring states in a single step, between two barriers at 0 and at T (see Figure 10.2). We will take absorbing barriers: the state 0 and the state T have the peculiarity that, once they are reached, they cannot be left. This absorption is represented by the elements of the transition matrix $P_{0,0} = P_{T,T} = 1$.

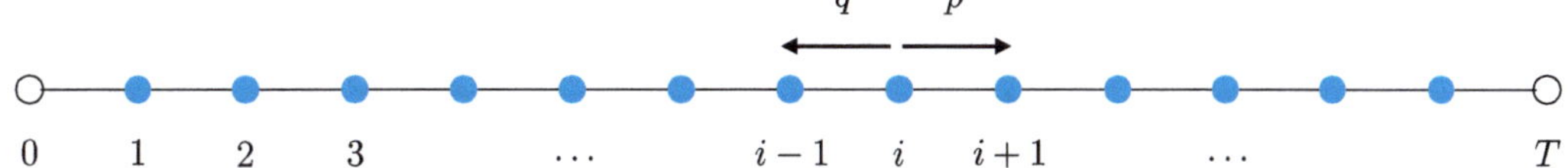

Figure 10.2 Markov chain with transitions to *nearest neighbors* and absorbing barriers at extremes 0 and T.

We denote the probability of moving forward by one step with $P_{i,i+1} = p$ and the probability of moving backward with $P_{i,i-1} = q$, with $p + q = 1$. Eventually, the stochastic matrix of this Markov chain will have the form

$$P \equiv \begin{pmatrix} 1 & 0 & 0 & 0 & 0 & \cdots & 0 & 0 & 0 \\ q & 0 & p & 0 & 0 & \cdots & 0 & 0 & 0 \\ 0 & q & 0 & p & 0 & \cdots & 0 & 0 & 0 \\ 0 & 0 & q & 0 & p & \cdots & 0 & 0 & 0 \\ \vdots & \vdots & \vdots & \vdots & \vdots & \ddots & \vdots & \vdots & \vdots \\ 0 & 0 & 0 & 0 & 0 & \cdots & q & 0 & p \\ 0 & 0 & 0 & 0 & 0 & \cdots & 0 & 0 & 1 \end{pmatrix}. \tag{10.18}$$

Using this Markov chain we can compute the probability of ending up in the *absorbing state* 0 starting from a state $i > 0$. *Mutatis mutandis* we can rephrase this in terms of gambling and ask what is the probability of a gambler going broke if they start with i euros and with each bet they can win or lose 1 euro. In addition, the gambling house will exclude the player from making new bets once a certain T euros ceiling has been reached.

Let us call $\mathcal{E}$ the event of running out of money and let us denote by $\mathcal{R}_i$ the probability of $\mathcal{E}$ occurring by starting the game with i euros. We will say that the player's pool at step 0 (the initial position of the random walker) is $x(0) = i$, thus

$$\mathcal{R}_i \equiv \mathrm{Prob}(\mathcal{E}|x(0) = i).$$

The origin of the times does not really matter since $\mathcal{E}$ is defined for any time interval. It is not the event of failure in n steps, only of failure, sooner or later. Consequently, the probability $\mathcal{R}_i$ is the probability of ruin starting from i euros at any time, or equivalently of failing having i euros at some time.

Two cases are already given. The first is $\mathcal{R}_0 = 1$: if you begin without money, with certainty you will remain without it. The other case is $\mathcal{R}_T = 0$: one does not start the game with a stake equal to the ceiling, so one cannot go broke.

For the probabilities of the intermediate cases, we can write an iterative formula. We denote by $p(x(n + 1) = k \,|\, x(n) = j)$ the conditional probability of having k euros at step $n + 1$ if we have j euros at step n, which is equal to the probability of transition $P_{j,k}$ from the state j to the state k in the matrix (10.18). We then write the following identity:

$$\mathcal{R}_i = \mathcal{R}_{i+1}p(x(1) = i + 1 \,|\, x(0) = i) + \mathcal{R}_{i-1}p(x(1) = i - 1 \,|\, x(0) = i)$$
$$= \mathcal{R}_{i+1}p + \mathcal{R}_{i-1}q\,, \tag{10.19}$$

which expresses the fact that the probability of going bankrupt having i euros is equal to the probability of winning 1 euro (i.e., $p(x(1) = i + 1 \,|\, x(0) = i) = P_{i,i+1} = p$) and going bankrupt starting with $i + 1$ added to the probability of the disjoint event of losing 1 euro (i.e., $p(x(1) = i - 1 \,|\, x(0) = i) = P_{i,i-1} = q$) and going broke starting with $i - 1$ euros. Let us repeat again that $\mathcal{R}_i$ is not the probability of having i euros (i.e., of being in the state i), but it is the probability of going bankrupt starting from i euros.

An elementary solution of Eq. (10.19) will have the form $\mathcal{R}_k = A^k$. The value of the base A can be found by checking (10.19). We find two possible values $A = 1$ and q/p. In the case of $q \neq p$ the general solution is found as a linear combination of the two elementary solutions:

$$\mathcal{R}_k = c_1 + c_2 \left(\frac{q}{p} \right)^k. \tag{10.20}$$

The values of the constants c_1 and c_2 are found by imposing conditions on the boundaries, $\mathcal{R}_0 = 1$ and $\mathcal{R}_T = 0$:

$$c_1 = \frac{1}{1 - (q/p)^T}, \qquad c_2 = -\frac{(q/p)^T}{1 - (q/p)^T}, \tag{10.21}$$

whence

$$\mathcal{R}_k = \frac{(p/q)^{T-k} - 1}{(p/q)^T - 1}. \tag{10.22}$$

If, on the other hand, $q = p = 1/2$, the elementary solution of the type A^k gives a single solution ($A = 1$). To formulate the general solution, we then use another elementary solution of the form $\mathcal{R}_k = k$, whence $\mathcal{R}_k = c_1 k + c_2$. By imposing conditions on the edges, we finally obtain

$$\mathcal{R}_k = 1 - \frac{k}{T}. \tag{10.23}$$

We can see that if the game is fair ($p = q = 1/2$), such as heads or tails, or even and odd, the probability of going broke increases linearly as the player's stack gets smaller, which is very reasonable. If a player starts with 500 euros the probability of going broke if the cap is $T = 1000$ euros is $\mathcal{R}_{500} = 1/2$.

Let us see what happens if the game is not fair. For example, suppose the player bets on red in roulette. There are 18 red boxes, 18 black boxes, and one green (the zero). So $p = 18/37$, $q = 19/37$, and $p/q = 18/19$. If we take the same case as before, a sum of $k = 500$ to bet and a ceiling of $T = 1000$ imposed by the croupier, the formula (10.22) gives us the probability $\mathcal{R}_{500} = 1 - 2 \times 10^{-12}$. A difference in probability between winning and losing of $1/37$ changes the probability of ruin from $1/2$ to 1! In order to have a probability equal to the probability of heads or tails when betting on red or black in roulette, you must have at least 987 euros to bet (out of a 1000 euros ceiling).

What happens if we remove the cap? If the game is fair, we see from Eq. (10.23), with $T \to \infty$, that $\mathcal{R}_k = 1$, $\forall k$. *A fortiori*, ruin is certain if $p < q$, as confirmed by

Eq. (10.22) when sending $T \to \infty$. If, on the other hand, $p > q$, again from (10.22) we have $\mathcal{R}_k = (q/p)^k$. If there is no above absorbing state, there can be, in this case, a finite probability equal to $1 - (q/p)^k$ of being able to play forever.

In general, in all of the above examples, the property of depending only on the previous step is evident, which *identifies* Markov chains: they are *correlated processes* in which the probability of ending up in a state depends on the probability of the previous state, but they are *without memory*, as everything concerning the history prior to the occurrence of the previous event does not count, it is forgotten.

10.4 Classification of Markov Chains

In order to simplify the study of Markov chains, it is convenient to introduce a first classification. To a Markov chain we can associate a graph, see Figures 10.3 and 10.4, in such a way that the states (or events) of the chain are the nodes of the graph and the allowed transitions are indicated with lines oriented from one node to another.

10.4.1 Types of Markov Chains

10.4.1.1 Decomposable and Indecomposable Chains

A Markov chain is said to be decomposable if its states can be divided into two subsets $\mathcal{A}$ and $\mathcal{B}$ such that it is not possible to make transitions from $\mathcal{A}$ to $\mathcal{B}$ and vice versa. In this case, the matrix P can be thought of as the tensor product of two matrices acting on two independent spaces. A Markov chain is said to be indecomposable if it is not decomposable. The study of a decomposable chain can be separated into the separate study of the two Markov chains acting separately in $\mathcal{A}$ and in $\mathcal{B}$. Very often indecomposable chains will, then, be considered. Each Markov chain can be decomposed into a numerable sum of indecomposable chains (see Section 10.5.5).

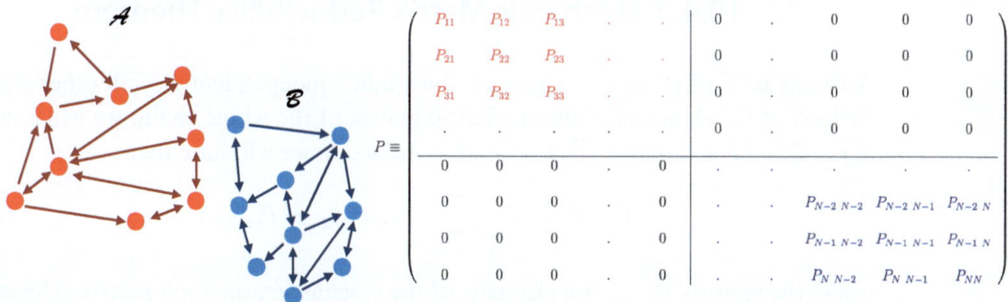

Figure 10.3   Markov chain of N states, decomposable into two subchains $\mathcal{A}$ and $\mathcal{B}$, and its stochastic matrix.

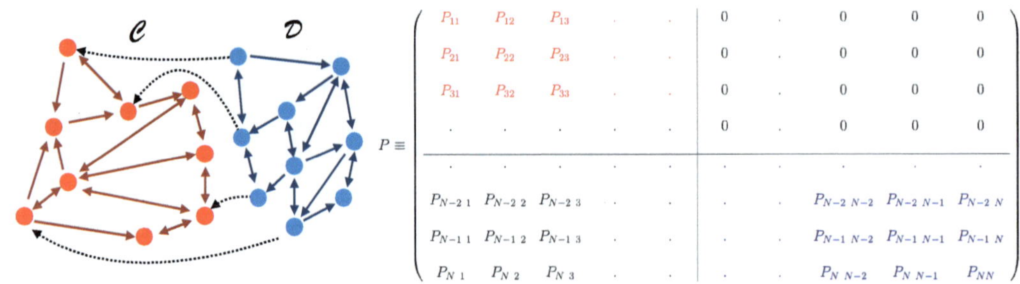

Figure 10.4 Markov chain composed of two subsets of states $\mathcal{C}$ and $\mathcal{D}$ reducible to the subchain $\mathcal{C}$ and its stochastic matrix.

10.4.1.2 Reducible and Irreducible Chains

A chain is said to be reducible if the states can be divided into two subsets $\mathcal{C}$ and $\mathcal{D}$ such that it is not possible to make transitions from $\mathcal{C}$ to $\mathcal{D}$ while transitions from $\mathcal{D}$ to $\mathcal{C}$ can take place. In this case, we can define a Markov subchain restricted to the set $\mathcal{C}$. Of course, a chain is irreducible if it is not reducible. In general, a Markov chain may contain one or more irreducible subchains.

10.4.1.3 Accessible State, Absorbing State, and Closed Set of States

A state k is said to be *accessible* from the state j if there exists a number of steps n such that the transition probability $P_{j,k}^{(n)} > 0$: sooner or later $j \to k$ is a possible transition.

A subset of states $\mathcal{C}$ to which it is possible to access from states outside $\mathcal{C}$ but from which it is not possible to access any state outside $\mathcal{C}$ is called a *closed set*. If no closed set exists in a chain, it means that the entire chain is the closed set and it is an irreducible and indecomposable chain. *In an irreducible Markov chain each state is accessible to each other.*

There can be a closed set consisting of a single state. This is called the absorbing state (see Section 7.4, the random walk with traps, or Section 10.3.3, the gambler's ruin), for which $P_{k,k} = 1$: once the state k is reached, one remains in k with certainty.

10.4.2 Stochastic Matrix Reducibility Theorem

A closed set $\mathcal{C}$ of states in a chain can be studied independently of all other states. If we look at the elements of the stochastic matrix of the whole chain, we will have that if $j \in \mathcal{C}$ and $k \notin \mathcal{C}$ then $P_{j,k}^{(n)} = 0$ for all n. In detail, we will have that

$$0 = P_{j,k}^{(n)} = \sum_{i_1,\dots,i_{n-1}} P_{j,i_1} P_{i_1,i_2} \cdots P_{i_{n-1},k}, \tag{10.24}$$

where the various P_{i_{r-1},i_r} are elements of the stochastic transition matrix. The expression above will be null because in the n steps sooner or later the transition between a state $i_{r-1} \in \mathcal{C}$ and a state i_r outside the closed set will be proposed, and thus at least the

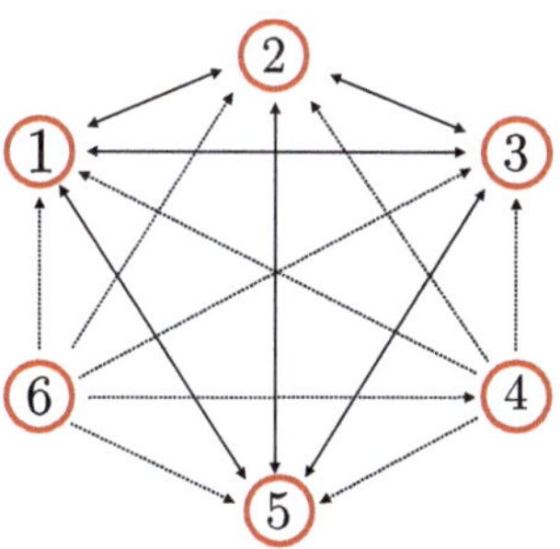

Figure 10.5 Example of a reducible Markov chain. The states $1, 2, 3, 5 \equiv C_4$ form an irreducible closed subchain: each of them is accessible to each of the others, and vice versa. The states 4 and 6, on the other hand, are not part of a closed set: from these two states, one can reach the states in the set C_4 but they are not reachable from any of the states in C_4.

element P_{i_{r-1},i_r} will be null. The opposite is not true in general, see Figure 10.4: from a state outside the closed set, one can make a transition to an internal state, $P_{k,j}^{(n)} \geq 0$.

At this point we can state the reducibility theorem for the stochastic matrix of a reducible Markov chain consisting of $M > N$ states (possibly also with $M \to \infty$).

Theorem. (Stochastic matrix reducibility theorem) *A reducible Markov chain of stochastic matrix P contains at least one closed set C_N of N states. The other states belong to the complement $\bar{C}_N$. Removing from P all rows containing the probabilities of transition to states belonging to C_N from states in $\bar{C}_N$ and all columns containing the probabilities of transition from states in the closed set to states in the complement (the latter are all null) yields a reduced $N \times N$ matrix, $P(C_N)$, which is still stochastic and which characterizes the irreducible Markov chain exclusively consisting of the closed set C_N.*

Take for example the reducible Markov chain reproduced in Figure 10.5. The stochastic matrix P of transition probabilities between states is

$$P \equiv \begin{pmatrix} 0 & P_{12} & P_{13} & 0 & P_{15} & 0 \\ P_{21} & 0 & P_{23} & 0 & P_{25} & 0 \\ P_{31} & P_{32} & 0 & 0 & P_{35} & 0 \\ P_{41} & P_{42} & P_{43} & 0 & P_{45} & 0 \\ P_{51} & P_{52} & P_{53} & 0 & 0 & 0 \\ P_{61} & P_{62} & P_{63} & P_{64} & P_{65} & 0 \end{pmatrix}, \tag{10.25}$$

where we explicitly set to zero the probabilities of transitions absent in Figure 10.5. The reduced 4×4 matrix of transitions between states of the irreducible subchain C_4 then holds:

$$P(C_4) = \begin{pmatrix} 0 & P_{12} & P_{13} & P_{15} \\ P_{21} & 0 & P_{23} & P_{25} \\ P_{31} & P_{32} & 0 & P_{35} \\ P_{51} & P_{52} & P_{53} & 0 \end{pmatrix}. \tag{10.26}$$

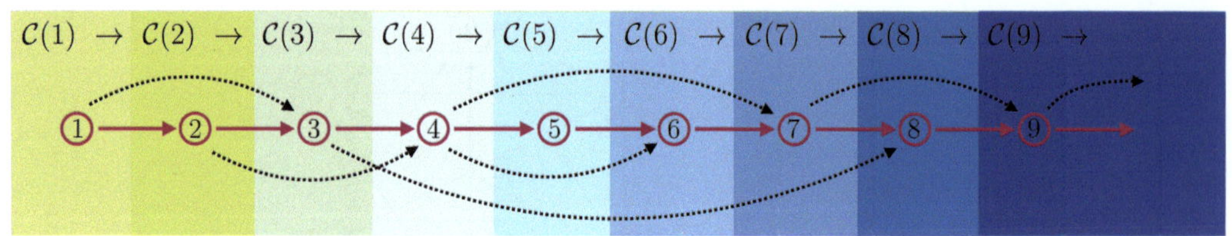

Figure 10.6   Example of state closures of the Markov chain defined in (10.27). With the symbol $\mathcal{C}(k) \rightarrow$ we denote the closure of state k (open to the right).

If we remove the rows and columns of transitions to and from states outside of the closed set $\mathcal{C}_4 \equiv \{1, 2, 3, 5\}$, the normalization conditions of each row do not change, because from the rows of transitions from states into $\mathcal{C}_4$ we only remove null elements. Consequently $P(\mathcal{C}_4)$ is still stochastic and the Chapman–Kolmogorov equation (10.11) applies. We will review and prove this property in Section 10.5.5, and in Appendix 10.C as part of more general theorems.

10.4.3 State Closure

The set "closure of a state" is defined to be the smallest closed set of states that contains the state in question. In other words, the closure of a state consists of all the states that are accessible from that state (and from which it is accessible). For some types of states, which we will define shortly, their closure is an irreducible set. Hence, we will have the same closure for all the states in the irreducible set. Other types of states, on the other hand, have more complicated closures, which may contain the closures of other states (even an infinite number of closures) that, in turn, *do not* contain them. Take, for instance, the chain defined by the stochastic matrix (see Figure 10.6)

$$
\begin{aligned}
P_{j,k} &= 0, && \text{if } k \le j, \\
P_{j,k} &> 0, && \text{if } k = j+1, \\
P_{j,k} &\ge 0, && \text{if } k > j+1.
\end{aligned}
\tag{10.27}
$$

There can only be forward transitions, which is why the closure of j will consist of j and all the $k > j$ states of the Markov chain. This set also contains the closure of $j+1$, which does not contain j but does contain the normalization of $j+2$, and so on.

10.4.4 States Classification

We have already seen that, given an arbitrary state k and the defined event "the Markov chain is in state k at time n", this satisfies all the conditions for being a recurrent event at time n. In particular, the no-memory condition for the Markov chain implies the memory-less condition typical of recurrent events. The quantity $P_a^{(n)} \equiv P_{a,a}^{(n)}$ satisfies all the conditions in the definition of recurrent events. Each state a can be classified according to the nature of the associated recurrent event and thus according to the properties of the first return. We will have periodic and aperiodic events, transient events, and persistent events, the latter divided into null events and non-null events (called ergodic if they are also aperiodic).

The use of the results on recurrent events obtained in the previous chapter allows a very simple approach (due to Feller [28]) to Markov chains, both for chains with a finite number of states as well as for infinite chains. Some results specifically valid for finite chains can be found in Sections 10.5.2 and 10.7.2.

For the sake of completeness, we report here the generalized classification of states of Markov chains in terms of the transition probability $P_{j,k}$ and the first transition probability, or *first passage*, $f_{j,k}$ between states j and k.

10.4.4.1 Periodic and Aperiodic States

The state j is periodic of periodicity M if

$$P_{j,j}^{(n)} = 0, \quad \forall n \mid n \, (\mathrm{mod}\, M) \neq 0. \tag{10.28}$$

An aperiodic state is a periodic state of periodicity 1. We stress that it may happen that j is aperiodic and that one has $P_{j,j}^{(n)} = 0$ for some n.

10.4.4.2 Persistent and Transient States

We define the *first passage* or *first transition* probability from j to k in n steps, i.e., the probability of going for the first time from j to k in n steps, $f_{j,k}^{(n)}$, which generalizes the first return probability of a recurrent event, studied in Chapter 9. At step zero, $f_{j,k}^{(0)} = 0$ for all k because in zero steps one cannot go (or return) anywhere for the first time.

We then define the probability of the first transition $f_{j,k}$ from j to k in any number of steps. Since the first transitions at different times are all disjoint events, this is given by the sum

$$f_{j,k} \equiv \sum_{n=1}^{\infty} f_{j,k}^{(n)}. \tag{10.29}$$

In this notation, where $f_{j,j}$ is the probability of the first return of the event j, a state j is classified to be one of the following.

- Transient, if $f_{j,j} < 1$.
- Persistent, if $f_{j,j} = 1$.

  In the latter case, the return of the event j is certain but its mean recurrence time

$$\mu_j = \sum_{\ell=1}^{\infty} \ell f_{j,j}^{(\ell)} \tag{10.30}$$

  could be infinite, with qualitative implications on the behavior of the represented system. This is why persistent states are, in turn, sub-classified into

  – null when $\mu_j = \infty$;
  – non-null when $\mu_j < \infty$.

If a non-null persistent state is also aperiodic, this is called an *ergodic state*, as its occurrence is a certain event in any finite time (with no limit cycles, no multiplicity constraints of a given period).

10.4.4.3 Transition Probability and First Passage Probability

Let us consider, for simplicity, aperiodic states. One can reformulate the classification in persistent and transient states in terms of the behavior of the transition probability $P_{j,k}^{(n)}$ through the convolution relation

$$P_{j,k}^{(n)} = \sum_{\ell=1}^{n} f_{j,k}^{(\ell)} P_{k,k}^{(n-\ell)} = f_{j,k} * P_{j,k}, \quad \forall n \geq 1, \tag{10.31}$$

with $P_{k,k}^{(0)} = 1$. Given this convolution product, in order to proceed we define the generating functions of the sequences $\{P_{i,j}^{(n)}\}$ and $\{f_{i,j}^{(n)}\}$:

$$\mathcal{F}_{i,j}(s) \equiv \sum_{n=1}^{\infty} f_{i,j}^{(n)} s^n, \tag{10.32}$$

$$\mathcal{P}_{i,j}(s) \equiv \sum_{n=0}^{\infty} P_{i,j}^{(n)} s^n, \tag{10.33}$$

and look at their relationship. Due to the convolution property, the generating function of the right-hand side of (10.31) is the product of the just defined generating functions. Since $f_{j,k}^{(0)} = 0$, however, the generator of the convolution is defined from $n = 1$, so on the left-hand side we will have to take away the $n = 0$ contribution:

$$\sum_{n=1}^{\infty} P_{i,j}^{(n)} s^n = \mathcal{P}_{i,j}(s) - P_{i,j}^{(0)} = \mathcal{F}_{i,j}(s)\mathcal{P}_{i,j}(s). \tag{10.34}$$

In particular for $i = j$ (for which $P_{j,j}^{(0)} = 1$) we have

$$\mathcal{P}_{j,j}(s) = \frac{1}{1 - \mathcal{F}_{j,j}(s)}, \tag{10.35}$$

which is nothing other than (9.17) for the event j. At this point, using the previous relation computed in $s = 1$, we can reclassify the states as follows:

- Transient

$$\mathcal{P}_{j,j}(1) = \sum_{n=0}^{\infty} P_{j,j}^{(n)} < \infty,$$

    whereby we have

$$\lim_{n \to \infty} P_{j,j}^{(n)} = 0. \tag{10.36}$$

- Persistent

$$\mathcal{P}_{j,j}(1) = \sum_{n=0}^{\infty} P_{j,j}^{(n)} = \infty.$$

In the latter case, applying the limit probability theorem that holds for non-null persistent recurrent events, demonstrated in Section 9.4, and the definition of mean recurrence time of state j, i.e., $\mu_j = F'_{j,j}(1)$, we have

$$\lim_{n\to\infty} P_{j,j}^{(n)} = \frac{1}{\mu_j}, \quad \forall j. \tag{10.37}$$

The return to a j state of a Markov chain, indeed, is nothing more than a recurrent event, with its own mean time of recurrence μ_j. We can, therefore, complete the classification of persistent states as

– null persistent

$$\mathcal{P}_{j,j}(1) = \infty, \quad \lim_{n\to\infty} P_{j,j}^{(n)} = 0; \tag{10.38}$$

– non-null persistent

$$\mathcal{P}_{j,j}(1) = \infty, \quad \lim_{n\to\infty} P_{j,j}^{(n)} = \frac{1}{\mu_j}. \tag{10.39}$$

Let us now see how the transition probability $P_{j,k}^{(n)}$ between different states behaves in the discrete time limit $n \to \infty$.

10.4.5 Limit Transition Probability to a Persistent State

Theorem. *The asymptotic time limit of the transition probability between two ergodic states a and b is*

$$\lim_{n\to\infty} P_{ab}^{(n)} = \frac{1}{\mu_b}. \tag{10.40}$$

Proof   We start with the convolution formula (10.31), which we reproduce here for convenience:

$$P_{a,b}^{(n)} = \sum_{m=1}^{n} f_{a,b}^{(m)} P_{b,b}^{(n-m)}. \tag{10.41}$$

Let us assume that the states a and b are aperiodic and that all elements of the series can be non-zero. In the periodic case, we will have null elements whenever the number of steps is not a multiple of the periodicity, but we have seen in Section 9.4.1 that it is sufficient to rescale the times over the periodicity in order to consider a periodic state as aperiodic. What we will say is also true, therefore, for periodic start and arrival states with the same period.

The first passage probabilities are probabilities of exclusive events and we consider persistent states for which

$$f_{a,b} \equiv \sum_{m=1}^{\infty} f_{a,b}^{(m)} = 1 \,.$$

We notice that the series $\{f\}$ is absolutely convergent ($\{f\} \in \ell^1$). We have just reviewed Eq. (9.23) in the theorem for the limit probability of recurrent events rewritten in (10.37) for states of a Markov Chain, whereby, for example,

$$\lim_{n \to \infty} P_{b,b}^{(n-m)} = \frac{1}{\mu_b} \,.$$

When we take the limit $n \to \infty$ in Eq. (10.41) we formally obtain that

$$\lim_{n \to \infty} P_{a,b}^{(n)} = \sum_{m=1}^{\infty} f_{a,b}^{(m)} \lim_{n \to \infty} P_{b,b}^{(n-m)} = \frac{1}{\mu_b} \,,$$

where the limit inside the sum of a number of terms going to infinity is mathematically allowed as $\{f\} \in \ell^1$ (and therefore $\{f * P\} \in \ell^1$), as shown in Appendix 10.A.   ■

10.5 Irreducible Markov Chains Fundamental Theorems

We saw in Section 10.4.2 that all reducible chains contain at least one closed set that is itself a Markov chain and is irreducible. Its stochastic matrix is a reduction of the stochastic matrix of the original chain. In general, as we will formalize better in a moment, all reducible chains can be reduced into a number of irreducible subchains plus a set of states (transients) that do not belong to any closed set.

We can therefore concentrate on studying the properties of irreducible Markov chains (IMCs). To this end, we state and prove a series of interconnected and sometimes equivalent theorems valid under the irreducibility hypothesis. For ease of reference, we list them below.

1. The states of an IMC are all of the same type (Section 10.5.1).
2. In a finite IMC, the states are all non-null persistent (Section 10.5.2).
3. A persistent state belongs to an IMC (Section 10.5.3).
4. In an ergodic IMC, the probability of a state does not depend on the initial conditions (Section 10.5.4).
5. There is a unique division of the states of a Markov chain into one single set of transient states and one or more IMCs of persistent states (Section 10.5.5).

We begin to analyze the properties of irreducible Markov chains with the first theorem.

10.5.1 All States of an Irreducible Markov Chain Are of the Same Type

By type of state, we mean periodic or aperiodic, transient or persistent, and non-null persistent or null persistent.

Theorem. *The states of an IMC are all of the same type.*

Proof   For the proof we consider two states a and b of the chain. The irreducibility condition implies that b is accessible from a, i.e., that there exists an integer step m such that $P_{a,b}^{(m)}$ is non-zero.

To see this better we consider the set $\mathcal{A}(m)$ consisting of all the states of the chain accessible from a in m steps and we define the set $\mathcal{A} = \bigcup_m \mathcal{A}(m)$ of all states accessible from a (sooner or later). It is immediate to see that there can be no transitions between an element of $\mathcal{A}$ and an element that does not belong to $\mathcal{A}$. Take, for example, a generic state d that is accessible in one step from an element $c \in \mathcal{A}$. By definition of $\mathcal{A}$, the state c is accessible from a after some n: $c \in \mathcal{A}(n)$. Thus d will be accessible in $n + 1$ steps from a: $d \in \mathcal{A}(n + 1) \in \mathcal{A}$.

The condition of irreducibility implies that the set $\mathcal{A}$ coincides with the chain itself, thus any state b belongs to $\mathcal{A}$ and, consequently, there exists a number m such that $P_{a,b}^{(m)}$ is strictly non-zero. In the same way, there exists a number n such that $P_{b,a}^{(n)}$ is strictly non-zero. For brevity, we will say $P_{a,b}^{(m)} = \alpha > 0$ and $P_{b,a}^{(n)} = \beta > 0$.

In general, the probability of starting from a and returning to a in $l + m + n$ steps will be greater than or equal to the probability of the system going from a to b in m steps, returning to b in l steps, and then going from b to a in n steps; see Figure 10.7. In fact, the possible existence of other alternative paths from a to a can only increase the probability of returning to a. We therefore have the inequality

$$P_{a,a}^{(\ell+m+n)} \geq P_{a,b}^{(m)} P_{b,b}^{(\ell)} P_{b,a}^{(n)} = \alpha\beta\, P_{b,b}^{(\ell)} . \tag{10.42}$$

In the same way, swapping $a \leftrightarrow b$, we get

$$P_{b,b}^{(\ell+m+n)} \geq \alpha\beta\, P_{a,a}^{(\ell)} , \tag{10.43}$$

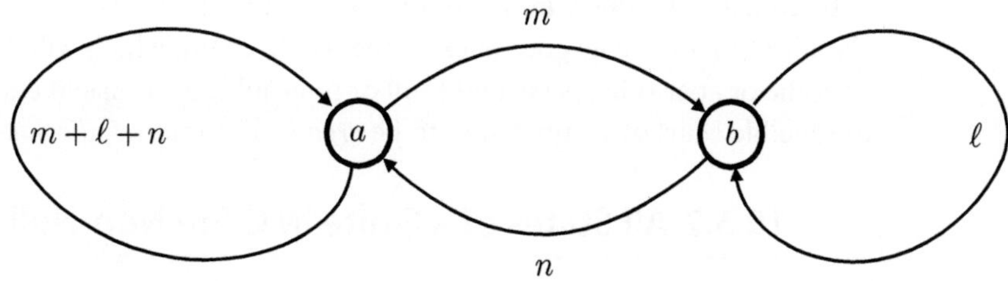

Figure 10.7   Scheme for the proof of the theorem in Section 10.5.1. Transition $a \rightarrow b \rightarrow a$ consisting of transition from a to b in m steps, followed by return to b in l steps, and $b \rightarrow a$ in n steps. In the text, the inequality (10.42) is compared with a more generic return to a ($a \rightarrow a$) in $m + l + n$ steps.

where we recall that both α and β are strictly non-zero probabilities and, therefore, $0 < \alpha\beta \leq 1$. The two previous relations imply that the asymptotic behaviors for large l of $P_{a,a}^{(\ell)}$ and $P_{b,b}^{(\ell)}$ are the same. Consequently, we will have

$$\sum_{\ell=0}^{\infty} P_{a,a}^{(\ell)} \sim \sum_{\ell=0}^{\infty} P_{b,b}^{(\ell)}, \tag{10.44}$$

where $\sim$ denotes the behavior neglecting a finite number of terms. Then the series will either be both convergent series, and a and b will be both transient, or both divergent, and then the states will be both persistent. In the latter case, since also

$$\lim_{\ell\to\infty} P_{a,a}^{(\ell)} = \lim_{\ell\to\infty} P_{b,b}^{(\ell)}, \tag{10.45}$$

a and b will be both null or non-null persistent.

Furthermore, it is also true that two states in an irreducible chain must both be periodic, with the same periodicity, or also both aperiodic (of periodicity 1). Let us demonstrate this.

If a is periodic of periodicity M, by definition, there exists a multiple ℓ of M such that $P_{a,a}^{(\ell)} > 0$. From (10.42) with $\ell = 0$ we also have

$$P_{a,a}^{(m+n)} \geq \alpha\beta P_{b,b}^{(0)} = \alpha\beta > 0. \tag{10.46}$$

Since a is periodic of periodicity M, one must have $(m + n) \bmod M = 0$: $m + n$ is a multiple of the periodicity. Again, for the periodicity of a using (10.43), we have that

$$P_{b,b}^{(\ell+m+n)} \geq \alpha\beta P_{a,a}^{(\ell)} > 0, \quad \text{for } \ell\,(\bmod M) = 0. \tag{10.47}$$

It follows that if a is periodic then $(\ell+m+n)\,(\bmod M) = 0$, because $m+n$ is a multiple of M, and b will also be periodic of periodicity M.

Finally, we can also easily prove the further property that

$$P_{a,b}^{(\ell+m)} \geq P_{a,b}^{(m)} P_{b,b}^{(\ell)} = \alpha P_{b,b}^{(\ell)} \tag{10.48}$$

and the transition probability between different states has the same behavior as the probability of returning to the same state. ∎

In conclusion, since, in an irreducible chain, all events have the same nature, the classification of events goes back to the whole chain, which will, therefore, be called a periodic or aperiodic, persistent (null or non-null), or transient chain. In particular, an irreducible Markov chain is said to be *ergodic* if it is *aperiodic non-null persistent*.

10.5.2 All States of a Finite IMC Are Non-Null Persistent

Theorem. *In a finite IMC, the states are all non-null persistent.*

Proof A corollary of the previous theorem implies that in an *irreducible finite chain* it is not possible for all states to be transient or null persistent. Indeed, the condition of probability normalization implies that

$$\sum_{b=1}^{N} P_{a,b}^{(n)} = 1, \tag{10.49}$$

where the sum over b extends over a finite number N of states. We can then exchange the limit $n \to \infty$ with the summation over the states and obtain:

$$1 = \lim_{n \to \infty} \sum_{b=1}^{N} P_{a,b}^{(n)} = \sum_{b=1}^{N} \lim_{n \to \infty} P_{a,b}^{(n)}. \tag{10.50}$$

As we have seen, by the result (10.40) of the limit probability of Markov chains for non-null persistent states, we have

$$\lim_{n \to \infty} P_{a,b}^{(n)} = \lim_{n \to \infty} P_{b,b}^{(n)} = \frac{1}{\mu_b} > 0,$$

while for null persistent states, see (10.38), or for transient states, see (10.36), every term in the sum on the right-hand side of Eq. (10.50) is zero. Since they must sum to 1, all states will be non-null persistent. ∎

As a sub-corollary, we note that, in this case, the average recurrence frequencies $\{\mu^{-1}\}$ compose a well-normalized probability distribution of states. We will return to this property in Section 10.6.

10.5.3 Persistent State Closure Theorem

We see and prove another theorem.

Theorem. *A persistent state belongs to an IMC.*
  Also, for a given persistent state p:

- *there exists a unique irreducible closed set C_p containing it; and*
- *for each pair of states belonging to C_p, the first passage in both directions is a certain event, $f_{a,b} = f_{b,a} = 1$.*

Proof   If an unknown state i belongs to the closure of p (see (10.4.1)) it means that i is accessible from p. By definition there is at least one time step n for which the probability of going from p to i for the first time is strictly positive: $f_{p,i}^{(n)} = \alpha > 0$.

The probability of going from p to i for the first time in any number of steps is $f_{p,i} = \sum_{n=1}^{\infty} f_{p,i}^{(n)}$. Similarly, we shall denote by $f_{i,p}$ the probability of going from i to p for the first time in any number of steps. The probability of the complementary event, i.e., never going from i to p, will therefore be $1 - f_{i,p}$.

With this terminology, we can write the probability $R_{\searrow p}$ of leaving p never to return to it as:

$$R_{\searrow p} \equiv \sum_{i} f_{p,i}(1 - f_{i,p}) = \sum_{n=1}^{\infty} \sum_{i} f_{p,i}^{(n)}(1 - f_{i,p}) > \alpha(1 - f_{i,p}). \tag{10.51}$$

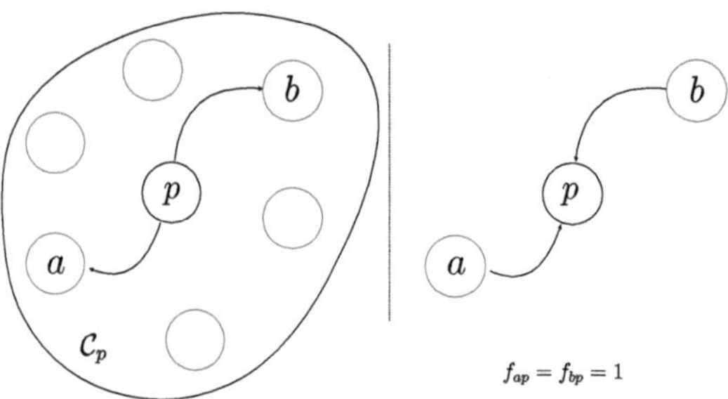

Figure 10.8 Left: The set C of states accessible from the persistent state p. Right: From any state accessible by p, it is certain to return to p, because this is persistent.

However, by hypothesis, p is persistent, i.e., the chain certainly returns to p, and so, by definition, one must have $R_{\searrow p} = 0$, which implies

$$\alpha(1 - f_{i,p}) = 0 \quad \Longrightarrow \quad f_{i,p} = 1.$$

In summary, if p is persistent, then p is accessible from any i state (whose type we ignore) of its closure. Let us recall that in the classification of Section 10.4.4 we saw that a persistent state p is defined by the property $f_{p,p} = 1$. We now see that the persistent nature of p also implies that $f_{i,p} = 1, \forall i$.

In Figure 10.8, in the left-hand panel, we show an example of the set C_p of states accessed by p. Let us take two states $a, b \in C_p$. We have just seen, and we draw it in the right-hand panel of Figure 10.8, that the mere fact that p is persistent implies $f_{a,p} = f_{b,p} = 1$. Whatever the nature of a and b, the state p is certainly reached by both a and b a first time (sooner or later). This leads to the conclusion that the state a and the state b are also accessible to each other. All the states in C_p are accessible to each other, just because of p being persistent, and we thus have $f_{a,b} = f_{b,a} = 1$. We saw in Section 10.4.1 that a Markov chain in which every state is accessible from every other state is an irreducible chain and we saw in Section 10.5.1 that in an IMC all states are of the same type, so all states in the closed set C_p are persistent as p. ∎

Let us make a few considerations related to this property of persistent states. First, the closure of a persistent state can only be an IMC. This means that *no transient state can ever be reached by a persistent state*. It also means that, starting from an arbitrary persistent state, every other state of the closure will be visited with certainty, i.e., in the IMC any event is certain regardless of the starting point.

To further confirm this we propose, for pedagogical purposes, a direct proof of this property with the following theorem.

10.5.4 Theorem of Initial Conditions Independence in an IMC

Theorem. *In an ergodic IMC, the probability of a state does not depend on the initial conditions.*

Proof   To prove that in an ergodic irreducible Markov chain an arbitrary event b is certain regardless of the starting point a, we prove the equivalent statement that the probability of *never crossing b* starting from any a state in n steps tends to zero when n tends to infinity.

We have already seen in Section 10.5.1 that under the assumptions of an aperiodic IMC there exists an m such that the probability $P_{a,b}^{(m)}$ of going from a to b in m steps is non-zero. *A fortiori* there will exist an $\ell \leq m$ such that the first passage probability $f_{a,b}^{(\ell)} \geq \gamma > 0$ of going from a to b in ℓ steps without passing through a in the intermediate step is non-zero.

Suppose by contradiction that the probability $R_{a\smallsetminus b}^{(m)}$ of returning to a in m steps without passing through b does not go to zero when $m \to \infty$. We define the event $\mathcal{E}_{m,\ell}$ to go from a to a in m steps without ever going through b, and then to go from a to b in ℓ steps without ever passing through a:

$$\mathcal{E}_{m,\ell} \equiv \left\{ a \underset{\substack{\searrow\\ \smallsetminus b}}{\xoverset{m}{\longrightarrow}} a \underset{\substack{\searrow\\ \smallsetminus a}}{\xoverset{\ell}{\longrightarrow}} b \right\}.$$
$$\qquad\qquad R_{a\smallsetminus b}^{(m)} \quad f_{a,b}^{(\ell)}$$

In other words, m is the last time the system passed through a before transiting into the b state in ℓ steps. It is evident that events characterized by different m times are self-excluding and therefore the probability

$$Q(n) = \sum_{m=0}^{n} R_{a\smallsetminus b}^{(m)} f_{a,b}^{(\ell)} \tag{10.52}$$

that at least one of these events will be realized in n steps is given by the sum of the probabilities to realize each trajectory. Using the hypotheses, we have the inequality

$$\sum_{m=0}^{n} R_{a\smallsetminus b}^{(m)} f_{a,b}^{(\ell)} \geq \gamma \sum_{m=0}^{n} R_{a\smallsetminus b}^{(m)}. \tag{10.53}$$

Since we assumed (in contradiction) that $R_{a\smallsetminus b}^{(m)}$ does not tend to zero when $m \to \infty$, the result of the summation diverges when m goes to infinity, and thus the right-hand side of the previous equation becomes larger than 1 for m sufficiently large. This is not possible since $Q(n)$ is a probability. The assumption we have made is contradictory. It therefore follows that the probability $R_{a\smallsetminus b}^{(m)}$ of not passing at least once through b starting from a tends to zero when the time m tends to infinity, and therefore the passage for b is a certain event regardless of the starting a state.   ∎

10.5.5 Unique Partition Theorem for Markov Chain States

We saw in Section 10.5.3 that persistent states form irreducible closed sets and that each persistent state belongs to a unique irreducible closed set. Consequently, we have noted that no transient state can ever be reached by a persistent state. What can we say about reducible chains? What role do transient states play?

Theorem. *There is a unique division of the states of a Markov chain into one single set of transient states and one or more IMCs of persistent states.*

Proof Let us take a Markov chain with N states of which $M < N$ constitute a closed set C. If we order them by first labeling the M states, with $i = 1, \ldots, M, M + 1, \ldots, N$ the $\mathbb{P}(C)$ matrix of transition probabilities between the states of the closed set is still a stochastic matrix, as shown in Section 10.4.2, because if $i \leq M$ and $j > M$, then $P_{i,j} = 0$ and $\sum_{j=1}^{N} P_{i,j} = \sum_{j=1}^{M} P_{i,j} = 1$. We will then have the following block decomposition for the stochastic matrix

$$
P = \begin{pmatrix} \mathbb{P}(C) & 0 \\[2mm] \underbrace{\mathbb{F}}_{M} & \underbrace{\mathbb{P}(\bar{C})}_{N-M} \end{pmatrix} \begin{array}{l} \rbrace M \\[2mm] \rbrace N - M \end{array},
\tag{10.54}
$$

where $\mathbb{F}$ is the matrix of transitions from states outside the closed set to states inside C, and $\mathbb{P}(\bar{C})$ is the matrix of transitions between states not belonging to C and, therefore, belonging to its complement $\bar{C}$. Multiplying (10.54) by itself after n steps we arrive at the expression

$$
P^n = \begin{pmatrix} \mathbb{P}^n(C) & 0 \\[2mm] \mathbb{F}^{(n)} & \mathbb{P}^n(\bar{C}) \end{pmatrix},
\tag{10.55}
$$

from which we observe that $\mathbb{P}^{(n)}(C)$ remains stochastic across iterations and the closed component of the Markov chain evolves independently of the other states $k > M$. Furthermore, the transition matrix between the complement states also follows the same recursion law for the stochastic matrix as C: $\mathbb{P}^{(n)}(\bar{C}) = \mathbb{P}^n(\bar{C})$. However, $\mathbb{P}^n(\bar{C})$ is not a stochastic matrix, since the elements of $\mathbb{F}^{(n)}$ are not null (nor do they have a simple recursion law).

The closed set on which $\mathbb{P}^{(n)}(C)$ acts may itself be reducible into several irreducible closed subsets: $C = \bigcup_\alpha C_\alpha$. In this case the matrix $\mathbb{P}(C)$ simply decomposes into diagonal blocks $\mathbb{P}(C_\alpha)$ independent of each other, with all null elements outside the blocks. In any case, see Section 10.5.3, a persistent state will belong to only one of the irreducible subsets C_α.

It also follows from this decomposition that if we consider the set $\bar{C}$ of events that do not belong to any of the irreducible subchains, this must necessarily be composed of transient events.

We have thus proved the theorem of the unique partition of the states of a Markov chain. ∎

The states of any Markov Chain can be uniquely partitioned into a set of disjoint $\{T, C_1, \ldots, C_K\}$ sets such that:

- the set T contains all and only the transient states; and
- the sets $C_\alpha, \alpha = 1, \ldots, K$, are irreducible and contain only persistent states.

10.6 Balance Equation for Ergodic Irreducible Markov Chains

Let us consider an irreducible chain of aperiodic persistent states that we will, hence, call an ergodic chain.[2] We saw in Section 10.4.5 that the probability of going from the state a to the state b in a number of steps tending to infinity does not depend on the starting point a, and we denote this stationary probability by the symbol

$$u_b \equiv \lim_{n \to \infty} P_{a,b}^{(n)} = \lim_{n \to \infty} P_{b,b}^{(n)} = \frac{1}{\mu_b}. \tag{10.56}$$

We now want to show that the following results hold:

1. the quantities u_b satisfy the following linear equation, called the *balance* relation,

$$\sum_a u_a P_{a,b} = u_b\,; \tag{10.57}$$

2. the solution of Eq. (10.57) is unique, if we have $\{u\} \in \ell^1$, i.e., $\sum_b |u_b| < \infty$; and
3. the solution $\{u\}$ is a well-normalized probability distribution, i.e., $\sum_b u_b = 1$.

The proofs we will give of the previous theorems are obviously false for an (infinite) chain in which all states are transient, see formula (10.36), or persistent null, see formula (10.39). As in the case of Section 10.4.5, formally the proof of the previous theorems can be carried out by trivially exchanging limits for infinite summations. This operation is not always lawful and becomes legal only in the case of ergodic chains. In any case, however, the result (10.56) is still valid even for transient and persistent null states, with $u_b = 0$ for any b.

10.6.1 Balance Equation Theorem

Proof of result 1 To prove (10.57) we consider the following Chapman–Kolmogorov equation, see the formula (10.11):

$$P_{c,b}^{(n+1)} = \sum_{a=1}^{\infty} P_{c,a}^{(n)} P_{a,b}\,. \tag{10.58}$$

Because of row normalization for a stochastic matrix, we have $\sum_{a=1}^{\infty} P_{c,a}^{(n)} = 1$. Since the transition probabilities are $P_{a,b} \leq 1$, this implies

$$\sum_{a=1}^{\infty} P_{c,a}^{(n)} P_{a,b} \leq \sum_{a=1}^{\infty} P_{c,a}^{(n)} = 1$$

---

[2] The periodic case can be easily tackled with the same technique by observing the system only at times that are a multiple of the period, as done in Section 9.4.1.

and the series in a, $P^{(n)}_{c,a} P_{a,b}$, is in ℓ^1. Taking the asymptotic limit of (10.58) we can thus bring the limit inside the sum on the right-hand side and obtain

$$u_b = \lim_{n \to \infty} P^{(n+1)}_{c,b} = \sum_{a=1}^{\infty} \lim_{n \to \infty} P^{(n)}_{c,a} P_{a,b} = \sum_{a=1}^{\infty} u_a P_{a,b} , \qquad (10.59)$$

which is the balance equation (10.57) and result 1 is proved. $\blacksquare$

We can also provide a slightly alternative proof, not using the ℓ^1 convergence. We first introduce a small lemma.

Lemma. *If the generic quantities $G_a^{(n)}$ are not negative, then the following inequality holds:*

$$\lim_{n \to \infty} \sum_{a=1}^{\infty} G_a^{(n)} \geq \sum_{a=1}^{\infty} \lim_{n \to \infty} G_a^{(n)} . \qquad (10.60)$$

Proof   Indeed, we can write

$$\lim_{n \to \infty} \sum_{a=1}^{\infty} G_a^{(n)} \geq \lim_{n \to \infty} \sum_{a=1}^{N} G_a^{(n)} = \sum_{a=1}^{N} \lim_{n \to \infty} G_a^{(n)} ,$$

and by sending $N \to \infty$ in the above formula we obtain (10.60). $\blacksquare$

Alternative proof of result 1.   Applying (10.60) in the lemma on the limit of Eq. (10.58) we have

$$u_b = \lim_{n \to \infty} P^{(n+1)}_{c,b} \geq \sum_{a=1}^{\infty} \lim_{n \to \infty} P^{(n)}_{c,a} P_{a,b} = \sum_{a=1}^{\infty} u_a P_{a,b} ,$$

i.e.,

$$u_b - \sum_{a=1}^{\infty} u_a P_{a,b} \geq 0.$$

However, since $\sum_b P_{a,b} = 1$, the sum over b of the left-hand side of this inequality is zero. Since the sum of non-negative terms can only be zero if all terms are zero, we obtain Eq. (10.57). $\blacksquare$

10.6.2 Uniqueness Theorem for the Balance Equation Solution

Proof of result 2.   Suppose that there exists a $\{v\} \in \ell^1$ series such that

$$\sum_a v_a P_{a,b} = v_b .$$

Applying the previous equation n times we have that

$$\sum_{a=1}^{\infty} v_a P^{(n)}_{a,b} = v_b .$$

Let us take the limit $n \to \infty$. Since the sum is absolutely convergent, we can take the limit inside the sum. We then have that

$$v_b = \lim_{n \to \infty} \sum_{a=1}^{\infty} v_a P_{a,b}^{(n)} = \sum_a v_a u_b. \tag{10.61}$$

Hence, $v_b \propto u_b$ and the solution of (10.57) is unique (proving result 2). ∎

Proof of result 3. Eventually, summing (10.61) over b and dividing by $\sum_{a=1}^{\infty} v_a$, we obtain that $\sum_{a=1}^{\infty} u_a = 1$. We have therefore also proved the result 3, that $\{u\}$ is a well-normalized probability distribution. ∎

10.7 Finite Chains and Spectral Decomposition

We have just seen that, in an irreducible Markov chain, the limit probability distribution of states $\{u\} \equiv \vec{u}^{\,T} = (u_1, u_2, u_3, \dots)$ is solution of the system of balance equations. Written in vector terms, this can be read as an equation for the eigenvalues of the non-Hermitian matrix P,

$$\vec{u}^{\,T} P = \lambda \vec{u}^{\,T},$$

with $\vec{u}$ the *left* eigenvector of the matrix P associated with the eigenvalue $\lambda = 1$.

10.7.1 Ranking Pages

Before going into the formal details of this representation (see Section 10.7.2), let us look at an application of the balance equation in the case of a Markov chain with a finite number N of states. Let us take a set of sites on the World Wide Web, all containing a given group of words, and ask ourselves, of all the sites containing that given group of words, which ones are the most representative. We take N sites ($N = 5$ in the example in Figure 10.9) connected to each other by directional links, from the site A that has a link to the site B, weighted with a probability $P_{A,B}$. The sum of the probabilities of all the possible links of A to other sites containing the same group of words will be equal to $\sum_k P_{A,k} = 1$: the P matrix of links is a stochastic matrix and the sites are representable as states of a Markov chain. Matrix P is non-symmetric, hence non-Hermitian.

Let us take as an initial working hypothesis that all links on a page have the same probability, ignoring eye-catching graphics, strategic positioning, pop-ups, etc. With this assumption, the stochastic matrix of the mini Web, represented in Figure 10.9 (left), is

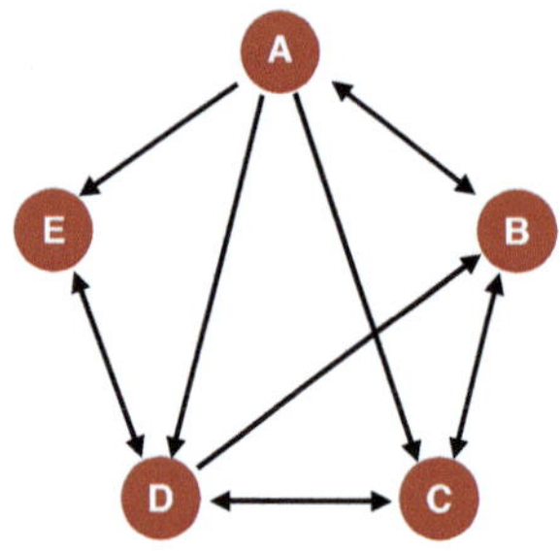
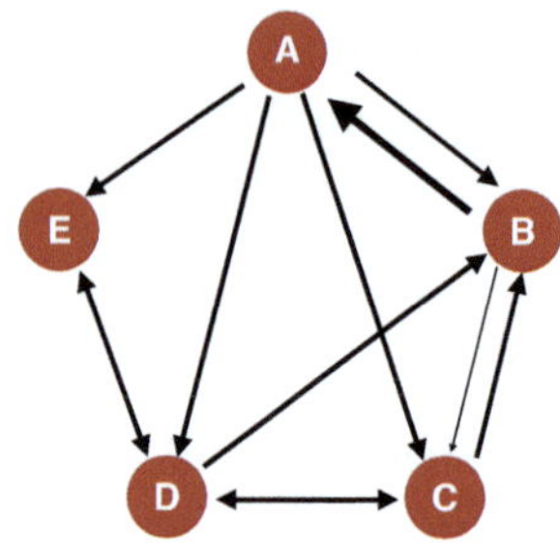

Figure 10.9 Example of a set of five sites containing a given group of words with their links. The arrows indicate that the site has a link to another site. On the left, all links from one site are uniform; on the right, links from site B have different weights.

$$
P = \begin{array}{c} \\ \\ \\ \\ \\ \\ \end{array}
\begin{array}{ccccc}
A & B & C & D & E \\
\end{array}
\left(\begin{array}{ccccc}
0 & 1/4 & 1/4 & 1/4 & 1/4 \\
1/2 & 0 & 1/2 & 0 & 0 \\
0 & 1/2 & 0 & 1/2 & 0 \\
0 & 1/3 & 1/3 & 0 & 1/3 \\
0 & 0 & 0 & 1 & 0
\end{array}\right)
\begin{array}{c}
A \\ B \\ C \\ D \\ E
\end{array} .
\tag{10.62}
$$

In descending order, the eigenvalues of P are[3]

$$
\{\lambda_1, \lambda_2, \lambda_3, \lambda_4, \lambda_5\} = \left\{1, \frac{\sqrt{3}-1}{4}, 0, -\frac{1}{2}, -\frac{\sqrt{3}+1}{4}\right\},
\tag{10.63}
$$

and the left eigenvector corresponding to the eigenvalue $\lambda_1 = 1$ is $\vec{u}^T = (4/33, 8/33, 8/33, 9/33, 4/33)$. Ranking the sites by importance we, therefore, obtain that

Ranking	Page	Weight
1	D	$0.\overline{27}$
2	B, C	$0.\overline{24}$
3	A, E	$0.\overline{12}$

showing that D is the most relevant site among those containing the group of searched words – it is the one to which the most links are directed.

If we release the assumption of uniformity of the link weights and, for example, take $P_{B,A} = 3/4 > P_{B,C} = 1/4$ (see Figure 10.9, right), page A will clearly become more relevant than C, but the effect has consequences on the whole ordering. In this case, the limit probability distribution of the sites becomes $\vec{u}^T = (0.172, 0.230, 0.190, 0.273, 0.134)$ and the ranking results in

---

[3] We notice that the largest eigenvalue is 1 and that all eigenvalue have $|\lambda| \leq 1$.

Ranking	Page	Weight
1	D	0.273
2	B	0.23
3	C	0.19
4	A	0.172
5	E	0.134

10.7.2 Spectral Decomposition of Non-Hermitian Matrices

In the case of a Markov chain with a finite number of states, one can obtain very accurate results using the spectral theorem for non-Hermitian matrices. Sometimes the techniques we use here can also be extended to the infinite case (and this will be done in the next chapter in a special case), but this extension is not always possible.

Let us consider the case of a chain with a finite number N of states and suppose that the matrix P has N distinct eigenvalues λ, which are the zeros of the characteristic polynomial $\det(P - \lambda) = \det(P^T - \lambda)$, where P^T is the transposed matrix of P. The case of degenerate eigenvalues is slightly more complicated and will not be dealt with here, but we mention that the Perron–Frobenius theorem (see Appendix 10.C) guarantees that, for a certain set of matrices (*irreducible* matrices with all non-negative real elements), the largest eigenvalue is real, positive, and unique [64, 65, 66, 67, 68]. Moreover, the eigenvectors associated with the largest eigenvalue have all positive elements. The theorem, which we report in Appendix 10.C, and which was proved by Oskar Perron in 1907 [69] and by Georg Frobenius in 1912 [70], is a simple but extremely useful linear algebra theorem for stochastic processes. In fact, the stochastic non-negative matrices of irreducible Markov chains belong to the set of matrices for which the Perron–Frobenius theorem holds. In this case, the maximal eigenvalue turns out to be equal to 1, as we have unconsciously already verified in the example in Section 10.7.1 of the page-ranking chain, and as we can immediately verify in the example in Section 10.3.2 of the two-state chain and, perhaps less immediately, in all the examples of Sections 10.1 and 10.3. We anticipate that if the chain is reducible, the uniqueness thesis falls. In particular, there will be an eigenvalue equal to 1 for each irreducible component of the reducible chain.

We introduce the N right ($\vec{d}$) and left ($\vec{s}$) eigenvectors that satisfy the following equations:

$$\sum_k P_{i,k} d_k(m) = \lambda(m) d_i(m), \quad \forall i, \tag{10.64}$$

$$\sum_i s_i(m) P_{i,k} = \lambda(m) s_k(m), \quad \forall i, \tag{10.65}$$

where we recall that the eigenvalues $\lambda(m)$ will, in general, be complex numbers since P is non-Hermitian.

As can be easily demonstrated (see Appendix 10.B), the right and left eigenvectors corresponding to different eigenvalues are orthogonal to each other, and we can impose the condition of orthonormalization,

$$\sum_k s_k(\ell)d_k(m) = \vec{s}(\ell) \cdot \vec{d}(m) = \delta_{\ell m}, \tag{10.66}$$

where $\vec{a} \cdot \vec{b}$ denotes the scalar product between the vectors $\vec{a}$ and $\vec{b}$. The vectors $\vec{d}$ and $\vec{s}$ are sometimes called a skew basis. Indeed, each vector $\vec{v}$ can be written as

$$\vec{v} = \sum_m (\vec{v} \cdot \vec{s}(m))\vec{d}(m) = \sum_m (\vec{v} \cdot \vec{d}(m))\vec{s}(m). \tag{10.67}$$

In the skew basis the matrix P can be decomposed (Appendix 10.B) as

$$P_{i,k} = \sum_m d_i(m)\lambda(m)s_k(m). \tag{10.68}$$

In the same way we obtain that the nth power of the matrix is given by

$$(P^n)_{i,k} = \sum_m d_i(m)\lambda(m)^n s_k(m). \tag{10.69}$$

If we denote by $\lambda(1)$ the *spectral radius* of the matrix P, i.e., the eigenvalue largest in modulus (assuming it is the only one for the moment), we obtain that, for large n,

$$(P^n)_{i,k} \simeq d_i(1)\lambda(1)^n s_k(1). \tag{10.70}$$

From here, it is clear that eigenvalues in modulus larger than 1 cannot exist, otherwise the probability would diverge for large n, rather than being limited by 1. From (10.64) it is straightforward to verify that the matrix normalization condition, $\sum_k P_{i,k} = 1, \forall i$, implies that the vector with components $d_k = 1$ is the right eigenvector with eigenvalue 1. We thus obtain that in the case when there is only one eigenvalue of spectral radius 1, i.e., $\lambda(1) = 1$, we have

$$\lim_{n \to \infty} (P^n)_{i,k} = s_k(1). \tag{10.71}$$

Therefore, from Eq. (10.56), we have that the components of the left eigenvector of the eigenvalue 1 coincide with the limit distribution of the Markov chain: $\vec{s}(1) = \vec{u}$. More precisely, considering also the sub-dominant terms, we have

$$(P^n)_{i,k} = u_k + d_i(2)s_k(2)\lambda(2)^n + O(\lambda(3)^n), \tag{10.72}$$

where we have ordered the eigenvalues in order of decreasing modulus. Given that

$$\lambda(2)^n = e^{n \log(\lambda(2))} = e^{n \log(|\lambda(2)|)+\imath n \arg(\lambda(2))} = e^{-n/\tau}e^{\imath n \arg(\lambda(2))}, \tag{10.73}$$

the quantity

$$\tau \equiv -\frac{1}{\log(|\lambda(2)|)} \tag{10.74}$$

is also called the chain correlation time, since only for $n \gg \tau$ are the initial conditions forgotten.

In the general, periodic case, cf. Eq. (10.93) and Sec. 10.C.1, where there are M eigenvalues of spectral radius 1, of the form $\lambda(j) = e^{\imath \arg(\lambda(j))}$, we have that, for large n,

$$(P^n)_{i,k} \simeq \sum_j d_i(j)\exp\{\imath n \arg(\lambda(j))\}s_k(j). \tag{10.75}$$

That is, $M - 1$ terms oscillate and only one has a non-zero limit in distribution, the term corresponding to the *spectral limit* of P, i.e., the eigenvalue with the largest real part: $\lambda(1) = 1$.

10.7.2.1 Spectral Properties of a Stochastic Matrix

The structure of chains, decomposable or indecomposable, reducible or irreducible, with periodic or aperiodic, transient or persistent states, is faithfully reflected in the structure of the eigenvalues and eigenvectors. In particular, the degeneracy of eigenvalues equal to the spectral limit $\lambda = 1$ tells us how many irreducible subchains there are, and, in an irreducible matrix block, eigenvalues of spectral radius $|\lambda| = 1$ give us information about the periodicity. We will later rigorously deepen this topic, which is called Perron–Frobenius theory, see Appendix 10.C, but here we will mention the correspondence.

Reducible or decomposable chains correspond to stochastic matrices with more than one real eigenvalue equal to the spectral limit 1, one for each irreducible or indecomposable component according to the theorem of the unique partition of states reported in Section 10.5.5. The case in which the eigenvalue 1 is non-degenerate corresponds to an irreducible chain.

In the case of aperiodic chains, there is only one eigenvalue whose radius is equal to the spectral limit, $|\lambda| = \lambda = 1$, one for each irreducible component. In contrast, for periodic chains of periodicity M, there are M eigenvalues of radius 1 per irreducible component, given by the formula

$$\lambda(j) = \exp\left(\frac{2\pi i (j - 1)}{M}\right), \quad j = 1, \ldots, M, \tag{10.76}$$

where the phase $\arg(\lambda(j))$ takes values that are multiples of $2\pi/M$, as discussed in Section 9.4.1 in the case of recursive events and as a consequence of the second part of the Perron–Frobenius theorem, cf. Eq. (10.93).

It is not easy to extend this discussion to the case of infinite chains in all generality because there is no extension of the spectral theorem for non-symmetric limited operators in the infinite-dimensional case. Note that in infinite irreducible chains, it is possible that the correlation time τ is infinite, i.e., that the asymptotic limit $\{u\}$ of the probabilities is reached more slowly than an exponential.

10.7.2.2 Two-State Chain Revisited with Spectral Decomposition

To fix the ideas in a simple case, let us take the exercise carried out in Section 10.3.2 of the two-state chain, 0 and 1, defined by the stochastic matrix

$$P \equiv \begin{pmatrix} 1 - \gamma & \gamma \\ \alpha & 1 - \alpha \end{pmatrix} \tag{10.77}$$

of determinant $\det P = 1 - \alpha - \gamma$. For $\alpha, \gamma \in (0, 1)$ the matrix is irreducible, so we already know that the largest eigenvalue, which we will call λ_0 here, is equal to 1. From the determinant, we know, therefore, that the other eigenvalue is $\lambda_1 = 1 - \alpha - \gamma$. The reader can verify that the two eigenvalues are those found by solving $\det(P - \lambda I) = 0$. Equation (10.72) is in this case simply (note that the numbering now starts from 0)

$$P_{i,k}^{(n)} = u_k + d_i(1)s_k(1)\lambda(1)^n, \quad \text{with } i, k = 0, 1.$$

The asymptotic probability is found by solving the balance equation (10.57) $\vec{u}^T P = \vec{u}^T$:

$$(u_0 \quad u_1) \begin{pmatrix} 1 - \gamma & \gamma \\ \alpha & 1 - \alpha \end{pmatrix} = (u_0 \quad u_1)$$

$$\implies \quad (u_0 \quad u_1) \begin{pmatrix} -\gamma & \gamma \\ \alpha & -\alpha \end{pmatrix} = (0, \quad 0), \tag{10.78}$$

which gives $u_1 = (\gamma/\alpha)u_0$. Together with the normalization condition $u_0 + u_1 = 1$, we obtain the stationary probabilities

$$u_0 = \frac{\alpha}{\alpha + \gamma} \quad \text{and} \quad u_1 = \frac{\gamma}{\alpha + \gamma}. \tag{10.79}$$

The eigenvectors of the skew basis can be found from the left and right equations of the eigenvalues, $P\vec{d}(1) = \lambda(1)\vec{d}(1)$ and $\vec{s}(1)P = \lambda(1)\vec{s}(1)$, respectively. Alternatively, knowing $\vec{d}(0)^T = (1, 1)$ and $\vec{s}(0) = \vec{u}$, we can use the orthonormality conditions of the skew basis to determine $\vec{s}(1)$ and $\vec{d}(1)$:

$$\vec{s}(0) \cdot \vec{d}(1) = \vec{s}(1) \cdot \vec{d}(0) = 0,$$

$$\vec{s}(1) \cdot \vec{d}(1) = 1.$$

In any case, after a quick algebraic count, we obtain the relations

$$s_0(1)d_0(1) = \frac{\gamma}{\alpha + \gamma} \quad \text{and} \quad s_1(1)d_1(1) = \frac{\alpha}{\alpha + \gamma}. \tag{10.80}$$

With formula (10.80) we obtain the elements of the stochastic matrix at step n:

$$P_{00}^{(n)} = u_0 + d_0(1)s_0(1)\lambda(1)^n = \frac{\alpha}{\alpha + \gamma} + \frac{\gamma}{\alpha + \gamma}(1 - \alpha - \gamma)^n,$$

$$P_{11}^{(n)} = u_1 + d_1(1)s_1(1)\lambda(1)^n = \frac{\gamma}{\alpha + \gamma} + \frac{\alpha}{\alpha + \gamma}(1 - \alpha - \gamma)^n,$$

$$P_{01}^{(n)} = 1 - P_{00}^{(n)} = \frac{\gamma}{\alpha + \gamma} - \frac{\gamma}{\alpha + \gamma}(1 - \alpha - \gamma)^n,$$

$$P_{10}^{(n)} = 1 - P_{11}^{(n)} = \frac{\alpha}{\alpha + \gamma} - \frac{\alpha}{\alpha + \gamma}(1 - \alpha - \gamma)^n.$$

The probability of being in the state 0 at step n is thus written as

$$\begin{aligned} p_0(n) &= p_0(0)P_{00}^{(n)} + (1 - p_0(0))P_{10}^{(n)} \\ &= \frac{\alpha}{\alpha + \gamma} - \frac{\alpha}{\alpha + \gamma}(1 - \alpha - \gamma)^n + p_0(0)(1 - \alpha - \gamma)^n \\ &= \frac{\alpha}{\alpha + \gamma} + \left[p_0(0) - \frac{\alpha}{\alpha + \gamma} \right](1 - \alpha - \gamma)^n, \end{aligned}$$

which coincides with the solution (10.16) of the stochastic dynamics of Section 10.3.2, without performing any recursion.

10.8 Non-Markov Chains

In general, most processes cannot be described just in terms of transition probability and the absolute probability of the initial condition. The joint probability $p(k(0), k(1), \ldots, k(n))$ of a given chain of events k in discrete time n is not decomposable as in (10.5) into the product $P_{k(i),k(i+1)}$ of transition probabilities from one state to another in one step. Indeed, the rule (1.40) for multiplication of conditional probability applies, of which the Markov rule is a drastic simplification. All processes for which this special simplification is invalid are generically called non-Markovian.

In several cases it is possible to "Markovianize" a chain by expanding the space of variables. A non-Markov chain in one variable can become a Markov chain in two variables under certain conditions. Reversing the argument, a non-Markov chain can be seen as the projection onto a subspace of Markovian dynamics in a larger space.

Let us take the example of a symmetric random path in one dimension. As we know, it is a random Markov process X that takes values $k = -L, \ldots, -1, 0, 1, \ldots, L$ on the integers (L can also be infinite), whose stochastic matrix is

$$P_{i,k} = \frac{1}{2}\delta_{k,i+1} + \frac{1}{2}\delta_{k,i-1}. \tag{10.81}$$

As proposed by Nico van Kampen [71], we modify the dynamics by favoring the tendency to continue along the same direction. If, on the previous move, the walker went in a given direction, then they will have a probability $p > 1/2$ of continuing in that direction and a probability $q < 1/2$ of going back. So the probability of a transition at step $n + 1$ no longer depends only on which position X the random walker occupies at step n but also depends on *remembering* which position X the walker came from at step $n - 1$. The joint probability of the whole dynamics can now decompose at most into products of terms of the type $p(k(n + 1) \,|\, k(n), k(n - 1))$, and $X(n)$ is no longer a Markov chain.

We can, however, represent the same dynamics by means of a two-component vector variable $\vec{X}$:

$$\vec{X}(n) \equiv \begin{pmatrix} X_1(n) \\ X_2(n) \end{pmatrix} = \begin{pmatrix} X(n - 1) \\ X(n) \end{pmatrix}.$$

If we look at the transition of this two-dimensional variable, there are two events that can occur in one step:

- the system moves (when the original walker makes two moves in the same direction), with probability p; or
- the system does not move (in the original system the walker moves and comes back), with probability q.

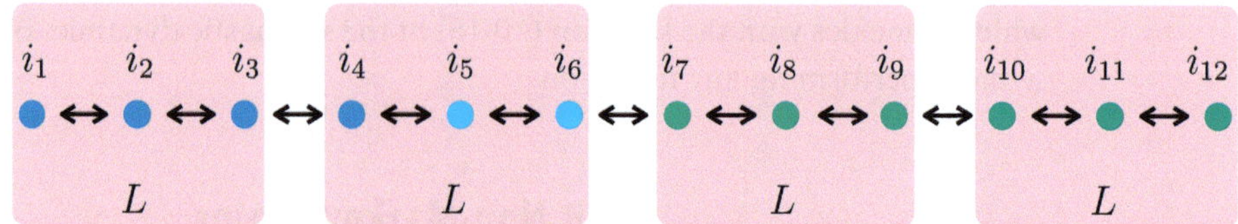

Figure 10.10 Example of a non-Markov chain. States i_k are grouped into boxes of L states and the variables depend on the box and not on the state. Variable changes do not follow Markovian dynamics.

The step of the dynamics in the variable $\vec{X}$ taking values $\vec{k} = \{k_1, k_2\}$ can be illustrated as

$$
\begin{array}{ccc}
i_1 & & i_2 \\
\bullet \longrightarrow & \bullet \longrightarrow & \bullet \\
& k_1 & k_2
\end{array}
$$

and can be written through the simple transition probability in one step from the (two-dimensional) state $\vec{i}$ to the state $\vec{k}$,

$$P_{\vec{i},\vec{k}} = \delta_{k_1,i_2}(p\delta_{k_2-k_1,i_2-i_1} + q\delta_{k_2,i_1}), \tag{10.82}$$

which is the stochastic matrix of a Markov chain for $\vec{X}$.

We can take another example starting once again with the one-dimensional random walker X taking values on the sites i. We group the sites into boxes of length L and consider the dynamics between events that consist of being inside a box. In practice, we roll up the process of random walking between sites into a process between boxes (but still moving one site at a time), as in Figure 10.10.

We define a "box" random variable Y_L, taking the value $k(n) = [i(n)/L]$ at step n, where $[\ldots]$ denotes the integer part. If $L > 1$ we do not know exactly where the walker is inside the box, so we cannot evaluate its probability of staying in the box or going to another box just as a simple transition probability from one box to another. We need to know when and how the walker entered the starting box and all the possible dynamics within the box (invisible to us). The dynamics between boxes, therefore, requires knowledge of Y_L at the previous steps, and the chain of events is no longer Markov. In this example, it is explicitly shown how the mechanism of *coarse graining*, i.e., putting together different positions in the box, leads from a Markov to a non-Markov chain.

The same can be said of the extraction from an urn of balls of different types, characterized by a given weight. If the random variable to observe is the total weight of all extracted balls, the process is not Markovian. It becomes so if we consider as a random variable the vector whose elements represent the weight of the different types of balls.

In general, if we start from a Markov chain and reduce the event space, losing some information, the resulting chain may not be a Markov chain, and, on the contrary, starting from a non-Markov chain, this may become equivalent to a Markov chain if we expand the event space.

10.8.1 Continuous Non-Markov Processes

Turning from discrete to continuous, in both variables and time, as we did for the case of the random walker in Sections 7.3–7.6, deriving the Fokker–Planck and the Langevin equations, a Markovian process satisfies a first-order stochastic differential equation, called the Langevin equation, such as Eq. (7.48). In Section 7.5 we saw that this is an approximation by large viscosity, the so-called overdamped approximation, of Eq. (7.49). The latter is a second-order stochastic differential equation, which we report here for convenience in one dimension,

$$\ddot{x} + \eta\dot{x} = F(x) + \sqrt{\nu}\xi(t),$$

with

$$\langle \xi(t) \rangle = 0, \quad \langle \xi(t)\xi(t') \rangle = \delta(t - t'),$$

whose solution is a non-Markovian process. However, it is possible also in this case to extend the variable space by defining a vector variable (x, v) for which the equation becomes a system

$$\begin{cases} v = \dot{x}, \\ \dot{v} = -\eta v + F(x) + \sqrt{\nu}\xi(t), \end{cases}$$

of first-order stochastic differential equations (only the second one is actually stochastic), whose solution $(x(t), v(t))$ is, instead, a Markov process, which can be expressed in terms of the probability density of transition $p(x, v, t \mid x_0, v_0, t_0)$ from state (x_0, v_0) to state (x, v). This, in turn, is the solution of the Fokker–Planck equation for Brownian motion (see Sections 7.3 and 7.5), otherwise called the Smoluchowski equation.

Mathematical Appendices

10.A Limit-Sum Exchange under Series Absolute Convergence

We here want to elaborate a bit on the last step in the proof of the limit probability theorem in Section 10.4.5 and explicitly show why this is allowed and how the hypothesis of absolute convergence comes into play. The first thing is to justify the limit

$$\lim_{n\to\infty} \sum_{m=1}^{n} f_{a,b}^{(m)} P_{b,b}^{(n-m)} = \lim_{n\to\infty} \sum_{m=1}^{\infty} f_{a,b}^{(m)} P_{b,b}^{(n-m)}. \tag{10.83}$$

Trivially, one could put every $P^{(n<0)}$ to zero before taking the limit, but it can be seen that it is sufficient for $P^{(n<0)}$ to be probabilities (i.e., ≤ 1) to guarantee this step. Indeed, by adding and removing the elements of the series up to ∞, using (10.31) we can write

$$\left| \lim_{n \to \infty} P_{a,b}^{(n)} - \lim_{n \to \infty} \sum_{m=1}^{\infty} f_{a,b}^{(m)} P_{b,b}^{(n-m)} \right| = \lim_{n \to \infty} \sum_{m=n+1}^{\infty} f_{a,b}^{(m)} P_{b,b}^{(n-m)} \leq \lim_{n \to \infty} \Delta(n) ,$$

where we have defined

$$\Delta(n) \equiv \sum_{m=n+1}^{\infty} f_{a,b}^{(m)} . \tag{10.84}$$

Since $\sum_{\ell=1}^{\infty} f_{a,b}^{(\ell)} = 1$, we have $\lim_{n \to \infty} \Delta(n) = 0$ and (10.83) applies.

The second thing is to rewrite the sum of infinite terms as a finite sum of the first M terms plus the infinite sum of the remaining ones:

$$P_{a,b}^{(n)} = \sum_{m=1}^{M} f_{a,b}^{(m)} P_{b,b}^{(n-m)} + \sum_{m=M+1}^{\infty} f_{a,b}^{(m)} P_{b,b}^{(n-m)} \leq \sum_{m=1}^{M} f_{a,b}^{(m)} P_{b,b}^{(n-m)} + \Delta(M) .$$

Thus, by sending $n \to \infty$ and passing the limit into the finite sum, we obtain

$$\lim_{n \to \infty} P_{a,b}^{(n)} - \sum_{m=1}^{M} f_{a,b}^{(m)} u_b \leq \Delta(M)$$

and sending $M \to \infty$ results in Eq. (10.40).

10.B Stochastic Matrix Spectral Decomposition

10.B.1 Spectral Decomposition of Hermitian Matrices

Before dealing with the spectral decomposition of non-Hermitian matrices such as the stochastic matrices of Markov chains, we would like to recall the properties of Hermitian matrices. A Hermitian matrix $\mathbb{A}$ can always be written as a function of its eigenvalues and eigenvectors as

$$\mathbb{A} = \mathbb{V} \Lambda \mathbb{V}^T ,$$

where Λ is the diagonal matrix of eigenvalues, $\Lambda_{j,k} = \delta_{j,k} \lambda(k)$, and $\mathbb{V}$ is the matrix of column eigenvectors

$$\mathbb{V} = \left(\vec{v}(1) \; \middle| \; \vec{v}(2) \; \middle| \; \ldots \; \middle| \; \ldots \; \middle| \; \ldots \; \middle| \; \vec{v}(N) \right) , \tag{10.85}$$

with $V_{j,k} = v_j(k)$ the jth component of the kth eigenvector. The eigenvectors form an orthonormal basis:

$$\vec{v}(j) \cdot \vec{v}(k) = \delta_{j,k} \, .$$

The spectral decomposition of the matrix $\mathbb{A}$ can thus be written as

$$A_{j,k} = \sum_{\ell,m} V_{j,\ell}\lambda(\ell)\delta_{\ell,m}(V^T)_{m,k} = \sum_{m} V_{j,m}\lambda(m)V_{k,m} = \sum_{m} v_j(m)\lambda(m)v_k(m). \quad (10.86)$$

10.B.2 Spectral Decomposition of Non-Hermitian Matrices

Stochastic matrices, however, are non-Hermitian and there is no orthonormal basis for the decomposition.

A non-Hermitian $N \times N$ matrix P has, in general, N eigenvalues. The eigenvalues are the zeros of $\det(P - \lambda I)$. We initially assume that they are all simple zeros, otherwise it is a bit messy. The zeros can be complex, but to avoid complications we also assume they are real (an approximation that is correct in the case of reversible Markov chains, as we will see in the next chapter, in Section 11.2).

The right eigenvector $\vec{d}$ of a non-Hermitian matrix $\mathbb{P}$, satisfying

$$\mathbb{P}\vec{d}(j) = \lambda(j)\vec{d}(j), \quad (10.87)$$

is not, in general, orthonormal to the other right eigenvectors $\vec{d}(j) \cdot \vec{d}(k) \neq \delta_{j,k}$. Similarly for the left eigenvector $\vec{s}$, i.e., the solution of

$$\vec{s}(j)^T \mathbb{P} = \lambda(j)\vec{s}(j)^T, \quad (10.88)$$

we have $\vec{s}(j) \cdot \vec{s}(k) \neq \delta_{j,k}$.

Let us define the following matrices:

- $\Lambda_{j,k} \equiv \delta_{j,k}\lambda(j)$, the diagonal matrix of the eigenvalues;
- $D_{j,k} \equiv d_j(k)$, the column-wise right eigenvalue matrix, where $d_j(k)$ denotes the jth element of the kth eigenvector, associated with $\lambda(k)$; and
- $S_{j,k} \equiv s_k(j)$, the row-wise eigenvector matrix, where $s_k(j)$ denotes the kth element of the jth eigenvector, associated with $\lambda(j)$.

In terms of these matrices, we can rewrite the formula (10.87) for all eigenvectors

$$\sum_{b} P_{i,b}d_b(k) = \lambda(k)d_i(k) = \sum_{b} d_i(b)\delta_{b,k}\lambda(b) \quad \Longrightarrow \quad \sum_{b} P_{i,b}D_{b,k} = \sum_{b} D_{i,b}\Lambda_{b,k}$$

in the matrix form $\mathbb{P}\mathbb{D} = \mathbb{D}\Lambda$. Multiplying this expression on the left by $\mathbb{S}$ gives $\mathbb{S}\mathbb{P}\mathbb{D} = \mathbb{S}\mathbb{D}\Lambda$. Similarly, we can rewrite (10.88) as

$$\sum_{b} s_b(i)P_{b,k} = \lambda(i)s_k(i) = \sum_{b} \delta_{i,b}\lambda(b)s_k(b) \quad \Longrightarrow \quad \sum_{b} S_{i,b}P_{b,k} = \sum_{b} \Lambda_{i,b}S_{b,k},$$

i.e., $\mathbb{S}\mathbb{P} = \Lambda\mathbb{S}$. If we multiply on the right by $\mathbb{D}$ we have $\mathbb{S}\mathbb{P}\mathbb{D} = \Lambda\mathbb{S}\mathbb{D}$, from which it follows that the matrix $\mathbb{S}\mathbb{D}$ commutes with the diagonal matrix Λ, which implies that

$\mathbb{S}\mathbb{D}$ is diagonal and that, under appropriate normalization, $\mathbb{S}\mathbb{D} = \mathbb{D}\mathbb{S} = \mathbb{I}$. We have, therefore, the orthonormalization of the skew basis

$$\sum_{b=1}^{N} s_b(j)d_b(k) = \vec{s}(j) \cdot \vec{d}(k) = \delta_{j,k}\,. \tag{10.89}$$

With this result, multiplying $\mathbb{S}\mathbb{P} = \Lambda\,\mathbb{S}$ on the left by $\mathbb{D}$, we obtain $\mathbb{P} = \mathbb{D}\,\Lambda\,\mathbb{S}$, i.e., the spectral decomposition

$$P_{i,k} = \sum_{j,l}^{1,N} D_{i,j}\Lambda_{j,l}S_{l,k} = \sum_{l=1}^{N} d_i(l)\lambda(l)s_k(l)\,. \tag{10.90}$$

Eventually, multiplying $\mathbb{P} = \mathbb{D}\,\Lambda\,\mathbb{S}$ iteratively n times by $\mathbb{P}$ (left or right is the same) gives the nth iteration decomposition of the stochastic matrix:

$$P_{i,k}^{(n)} = \sum_{l=1}^{N} d_i(l)\lambda(l)^n s_k(l)\,. \tag{10.91}$$

10.C Perron–Frobenius Theorem

Before stating this general theorem for non-negative matrices, let us give a few definitions that may be useful. We will define and manipulate mathematical objects, which, for the most part, have already been seen in the course of this chapter on Markov chains. The purpose of this appendix is to formalize some concepts and to rigorously investigate some mechanisms in the case of finite chains.

If we have a spectrum of radius-ordered eigenvalues $|\lambda_1| \geq \cdots \geq |\lambda_N|$ of a generic non-Hermitian matrix $\mathbb{P}$, the eigenvalue λ_1 with the largest real part is called the *spectral limit*,

$$\sup_{\lambda} \mathrm{Re}\,[\lambda] \equiv \lambda_1\,,$$

while the largest modulus available among all the eigenvalues,

$$\sup_{\lambda} |\lambda| \equiv \rho(\mathbb{P})\,,$$

is called the *spectral radius*.

A (not necessarily square, even rectangular) matrix is called positive if all its elements are positive. It is written as $\mathbb{P} > 0$. A matrix is called non-negative if all its elements are non-negative. It is written as $\mathbb{P} \geq 0$. In particular, these definitions apply to vectors, which are $1 \times N$ matrices.

A non-negative matrix is said to be larger than another non-negative matrix, $\mathbb{P} > \mathbb{Q}$, if the matrix $\mathbb{P} - \mathbb{Q} > 0$ is positive.

A matrix obtained by permuting rows (or columns) of the identity matrix is a *projection matrix*. Since the symbols p, P, $\mathcal{P}$, and $\mathbb{P}$ are abundantly used herein, we will use $\mathbb{G}$ for the projection matrix. With the projection matrix we can project the matrix $\mathbb{P}$ into another matrix,

$$\mathbb{Q} = \mathbb{G}\,\mathbb{P}\,\mathbb{G}^{-1}.$$

The matrix $\mathbb{Q}$ constructed in this way is called the *cogredient matrix* of $\mathbb{P}$.

A matrix is *reducible* if there exists a cogredient of it of the following form:

$$\begin{pmatrix} \mathbb{A} & \mathbf{0} \\ \mathbb{B} & \mathbb{C} \end{pmatrix} \tag{10.92}$$

(with $\mathbb{A}$, $\mathbb{B}$, $\mathbb{C}$, and $\mathbf{0}$ block matrices, of which $\mathbf{0}$ is a matrix of all null elements), i.e., if there exists a projection matrix $\mathbb{G}$ that transforms it into this form; see Section 10.4.2 for the reducibility of Markov chains. We note that a matrix of all positive elements can never be projected into this form, so all positive matrices are irreducible. Non-negative matrices are irreducible when they are not reducible.

We now enunciate the Perron–Frobenius theorem on non-negative matrices.

Theorem. (Perron–Frobenius theorem on non-negative matrices) *This is in two parts.*

1. *If a matrix $\mathbb{P}$ is non-negative and irreducible then:*

    (a) *there exists a single eigenvalue λ_1 realizing the spectral limit that coincides with the spectral radius $\rho(\mathbb{P})$ (in simple terms, we are saying that the eigenvalue with the largest real part is unique, real, and positive);*

    (b) *there is a positive right eigenvector $\vec{d}_1$ (and a corresponding positive left eigenvector $\vec{s}_1$) associated with λ_1, i.e., $\mathbb{P}\,\vec{d}_1 = \lambda_1 \vec{d}_1$ (and $\vec{s}_1^{\,T}\mathbb{P} = \lambda_1 \vec{s}^{\,T}$); and*

    (c) *there is no other positive right (or left) eigenvector of $\mathbb{P}$, associated with any other eigenvalue, that is not parallel to $\vec{d}_1$ (or $\vec{s}_1$).*

2. *If $\mathbb{P}$ is non-negative, irreducible, and has M complex eigenvalues of modulus equal to the spectral limit λ_1, which we denote by*

$$\lambda_\alpha = \lambda_1 e^{\iota q_\alpha}, \quad \text{with} \quad 0 = q_1 < q_2 < \cdots < q_M < 2\pi\,,$$

   then these eigenvalues are all simple roots of the equation

$$\lambda^M = \lambda_1^M. \tag{10.93}$$

There are numerous proofs of this theorem in its one hundred years of existence, and we refer to some of the texts where they are given, in different areas and with different techniques [64, 65, 66, 67, 68, 72]. Here we limit ourselves to making some remarks on the consequences of the theorem in the case of a particular sub-case of non-negative matrices: stochastic matrices.

10.C.1 Application to Stochastic Matrices of Markov Chains

First of all, the normalization condition

$$\sum_k P_{i,k} = 1\,, \quad \forall\, i = 1,\ldots,N\,, \tag{10.94}$$

can be read as a right eigenvector equation $\mathbb{P}\vec{1} = \vec{1}$ associated to the eigenvalue $\lambda = 1$, where $\vec{d} = \vec{1}$ has elements all equal to 1. If $\mathbb{P}$ is irreducible, the Perron–Frobenius theorem guarantees that there is only one strictly positive right eigenvector and it is the one associated with the eigenvalue equal to the spectral limit. Hence, for stochastic matrices of irreducible Markov chains, we derive that the maximum eigenvalue and its right eigenvector are

$$\lambda(1) = 1, \quad \vec{d}(1) = \vec{1}.$$

What is the left eigenvector $\vec{s}(1)^T$? If we take the balance equation (10.57) $\vec{u}^T = \vec{u}^T\mathbb{P}$, and recall that its solution is unique, this can be read as the left eigenvector equation for the eigenvalue $\lambda = 1$. Thus

$$\vec{s}(1) = \vec{u},$$

the vector whose elements constitute the stationary probability distribution of the irreducible Markov chain of which $\mathbb{P}$ is the stochastic matrix.

The second part of the theorem is very relevant in the case of periodic Markov chains. Equation (10.93) is in fact equivalent to

$$e^{i q_\alpha M} = 1,$$

whose solutions are (10.76):

$$q_\alpha = \frac{2\pi(\alpha - 1)}{M}, \quad \alpha = 1, \ldots, M.$$

Numerical Simulations

11.1 Inverse and Reversible Markov Chains

We have variously seen that the asymptotic probability distribution for irreducible and aperiodic Markov chains is the solution of the balance equation, (10.57), $u_a = \sum_b u_b P_{b,a}$. From the point of view of an eigenvalue equation, the limit distribution $\{u\}$ is the left eigenvector of the stochastic non-Hermitian matrix P, relative to $\lambda = 1$.

As already noted in Section 10.2 we can consider the transition probability between two states in n steps as a conditional probability:

$$P_{a,b}^{(n)} = p(b(n + n_0)\,|\,a(n_0)) = \frac{p(a(n_0)\,|\,b(n + n_0))p(b(n + n_0))}{p(a(n_0))}, \qquad (11.1)$$

for which Bayes' formula applies. Since we consider stationary Markov chains, the stochastic matrices of n-step transitions depend only on the difference in steps n separating a from b, and not on the absolute times n_0. Assuming we are already in an ergodic regime, for example, taking $n_0 \to \infty$, the absolute probability of being in one state, a or b, does not depend on the time step, but is equal to the stationary probability. Consequently, in (11.1) the *a priori* probability is $p(b(n + n_0)) = u_b$ and the marginal probability is $p(a(n_0)) = u_a$.

Let us now call $Q_{b,a}^{(n)} \equiv p(a(n_0)\,|\,b(n + n_0))$ the probability of the inverse transition in time from the state b at step $n + n_0$ to the state a at step n_0. Rewriting Eq. (11.1) in the stationary limit ($n_0 \gg 1$), we have, therefore,

$$u_a P_{a,b}^{(n)} = u_b Q_{b,a}^{(n)}, \quad \forall n. \qquad (11.2)$$

For $n = 1$, we have

$$u_a P_{a,b} = u_b Q_{b,a}. \qquad (11.3)$$

From (11.3), using the balance equation (10.57), it easily follows that the matrix Q of inverse transitions is also a stochastic matrix:

$$\sum_a u_a P_{a,b} = u_b \sum_a Q_{b,a} \quad \Longrightarrow \quad \sum_a Q_{b,a} = 1.$$

Furthermore, with (11.2) it follows immediately that the limit probability theorem applies for the matrix Q of the inverse chain, as well:

$$\lim_{n \to \infty} Q_{b,a}^{(n)} = \lim_{n \to \infty} P_{a,b}^{(n)} \frac{u_a}{u_b} = u_a.$$

In the particular case of time-reversal-invariant Markov chains, called *reversible* chains, we have $Q_{a,b} = P_{a,b}$ and thus the relation (11.3) becomes

$$u_a P_{a,b} = u_b P_{b,a}\,, \tag{11.4}$$

called the *detailed balance* relation.

We notice that if we sum this relation over all states a, we obtain the balance relation (10.57), valid for all irreducible ergodic Markov chains. However, the opposite is not true: the *detailed* balance is only satisfied in a reversible irreducible chain.

11.2 Detailed Balance for Reversible Markov Chains

As we shall see in the next section, the detailed balance relation is of particular importance when Markov chains are to be applied in the numerical evaluation of complicated integrals (typically, as we shall see later, of high dimensionality) and for the numerical simulation of the dynamics of statistical mechanics systems.

As we know from Chapter 10, in particular Section 10.7.2, P is a non-Hermitian matrix and its eigenvalues are, in general, complex. In the case of reversible chains, it is possible to construct a symmetric matrix associated with the matrix P, and to obtain further information on the spectrum of the eigenvalues of the stochastic matrix. To this end, it is convenient to introduce the quantities

$$\zeta_i = \sqrt{u_i}\,, \tag{11.5}$$

always positive, which satisfy the detailed balance relations

$$\zeta_i^2 P_{i,k} = \zeta_k^2 P_{k,i}\,. \tag{11.6}$$

It is straightforward to verify that the condition (11.6) is equivalent to imposing that the matrix $P_{i,k}$ can be written as

$$P_{i,k} = \zeta_i^{-1} M_{i,k} \zeta_k\,, \tag{11.7}$$

where $M_{i,k}$ is a symmetric matrix. Summing the relation (11.6) over k we obtain that

$$\sum_k \zeta_k^2 P_{k,i} = \zeta_i^2 \sum_k P_{i,k} = \zeta_i^2\,, \tag{11.8}$$

which is nothing other than the balance equation (10.57). The detailed balance relation implies that, unless $\sum_k \zeta_k^2$ is divergent, ζ_k^2 is the asymptotic probability distribution of the state k of the chain.

11.2.1 Analogy with Continuous Reversible Dynamics

In a short diversion, we notice that the procedure is exactly the same as the one we used in Section 7.5.2 to pass from the Fokker–Planck equation

$$\frac{\partial P(\vec{x},t)}{\partial t} = H_{\mathrm{FP}} P(\vec{x},t)\,,$$

whose differential operator

$$H_{FP} = -\vec{\nabla} \cdot \vec{v} + \frac{v}{2}\nabla^2$$

is not symmetric, to an evolution equation of the type (7.59):

$$\dot{\rho} = -H\rho,$$

where $H \equiv \mathcal{V}(\vec{x}) - (v/2)\nabla^2$ is a symmetric operator (we have taken license to call it self-adjoint). To go from one to the other we set (7.58), which we repeat for convenience:

$$P(\vec{x}, t) = \sqrt{p_0(\vec{x})}\, \rho(\vec{x}, t).$$

In that case the relationship between the stationary probability density $p_0(\vec{x})$, the solution of the stationary Fokker–Planck equation $H_{FP}\, p_0(\vec{x}) = 0$, and the ground-state eigenfunction $\psi_0(\vec{x})$ of the operator H, the solution of $H\psi_0(\vec{x}) = 0$, was given by the formula (7.65):

$$p_0(\vec{x}) = \psi_0^2(\vec{x}).$$

In the reversible dynamics of a process with discrete steps and discrete states, i.e., a chain, where the state $\vec{x}$ is represented by the index k, we have now set the analogous relation (11.5)

$$u_k = \zeta_k^2.$$

We will see that, indeed, the sequence $\vec{\zeta}$ is nothing other than the eigenvector $\vec{\psi}(1)$ of the matrix M corresponding to the eigenvalue 1.

11.2.2 A Symmetric Matrix for a Reversible Markov Chain

As in the case of a stochastic continuous process, where we introduced a symmetric operator H associated to the non-self-adjoint operator H_{FP}, in the case of a chain we can define a symmetric matrix M associated with the non-Hermitian stochastic matrix P.

The transition probabilities after n steps can be easily obtained. A simple computation shows that

$$P_{i,k}^{(n)} = \zeta_i^{-1}(M^n)_{i,k}\zeta_k. \tag{11.9}$$

The nth power of the matrix M can be obtained using the spectral theorem for symmetric matrices:

$$(M^n)_{i,k} = \sum_m \psi_i(m)\lambda^n(m)\psi_k(m), \tag{11.10}$$

where

$$\sum_k M_{i,k}\psi_k(m) = \lambda(m)\psi_i(m), \tag{11.11}$$

the eigenvalues $\lambda(m)$ are all real, and the eigenvectors $\vec{\psi}$ form an orthonormal basis:

$$\vec{\psi}(\ell) \cdot \vec{\psi}(m) = \delta_{\ell,m}.$$

In the case of a Markov chain with a finite number of states, the matrix M is a finite-dimensional matrix. If we recall the spectral decomposition of the stochastic matrix for finite chains in the skew basis of right eigenvectors $\vec{d}(m)$ and left eigenvectors $\vec{s}(m)$ presented in Section 10.7.2,

$$P_{i,k} = \sum_m d_i(m)\lambda(m)s_k(m),$$

by using the relation (11.7) to M we obtain

$$P_{i,k} = \sum_m \zeta_i^{-1}\psi_i(m)\lambda(m)\psi_k(m)\zeta_k,$$

so that in the case of reversible chains we can identify the components of the right and left eigenvectors as

$$d_i(m) = \frac{\psi_i(m)}{\zeta_i}, \quad s_k(m) = \psi_k(m)\zeta_k,$$

and observe that P and M have the same eigenvalues. Thus, the eigenvalues of the stochastic non-Hermitian matrix of a reversible chain are all real.

By the theorem in Section 10.5.2, for finite chains, the states are all non-null persistent and, if the chain is also aperiodic, by the result in Section 10.4.5 of the stationary probability, we have that, for each k,

$$\lim_{n\to\infty} P_{i,k}^{(n)} = \zeta_k^2. \tag{11.12}$$

However, from the formula (11.9) we also know that

$$\lim_{n\to\infty} P_{i,k}^{(n)} = \lim_{n\to\infty} \zeta_i^{-1} M_{i,k}^n \zeta_k = \zeta_i^{-1}\psi_i(1)\psi_k(1)\zeta_k, \tag{11.13}$$

where we have used the spectral decomposition (11.10) of the matrix M.

It follows that

$$\zeta_i \zeta_k = \psi_i(1)\psi_k(1)$$

and the eigenvalue equal to 1, which, according to the Perron–Frobenius theorem (Appendix 10.C), exists and is unique because the stochastic matrix is irreducible, corresponds to the eigenvector $\vec{\psi}(1) = \vec{\zeta}$. Since the chain is ergodic, all other eigenvalues must be smaller than 1 and their contribution to the transition probability vanishes in the limit $n \to \infty$. If we sort the eigenvalues in decreasing order, $1 = \lambda(1) > \lambda(2) > \cdots > \lambda(N-1) > \lambda(N)$, we obtain, for large n,

$$P_{i,k}^{(n)} = [\psi_k(1)]^2 + [\lambda(2)]^n \frac{\psi_i(2)}{\psi_i(1)}\psi_k(1)\psi_k(2) + O([\lambda(3)]^n). \tag{11.14}$$

Notice that the sum over k of the term proportional to $\lambda(2)$ is automatically null due to the orthonormality property of the eigenvectors $\vec{\psi}$, and this applies to all subsequent orders in the expansion.

The type of algebraic manipulation that we have used is very similar to that introduced for diffusion processes for conservative forces, which are gradients of potential, in Section 7.5. This fact should not come as a surprise, since a diffusion process on a

Table 11.1 Comparison of the properties of the probability and its stationary limit (FP sol. = Fokker–Planck solution; DB sol. = detailed balance solution) in the continuous diffusion process and in the discrete Markov chain.

	Continuous process	Discrete chain
Stationary solution and eigenstate	$p_0(\vec{x}) = \psi_0^2(\vec{x})$	$u_k = \zeta_k^2$
Homogeneous probability expansion	$P(\vec{x}, t \mid \vec{x}_i, 0)$ $\simeq p_0(\vec{x})e^{-\lambda_0 t} + c(\vec{x}, \vec{x}_i)e^{-\lambda_1 t}$	$P_{i,k}^{(n)}$ $\simeq u_k e^{-n\|\ln \lambda(1)\|} + c_{i,k} e^{-n\|\ln \lambda(2)\|}$
Stationarity condition and ground-state eigenvalue	If $\lambda_0 = 0$, $\exists$ FP sol. s.t. $\lim_{t \to \infty} P(\vec{x}, t \mid \vec{x}_i, 0) = p_0(\vec{x})$	If $\lambda(1) = 1$, $\exists$ DB sol. s.t. $\lim_{n \to \infty} P_{i,k}^{(n)} = u_k$

lattice is a reversible Markov chain: in the limit where the lattice step and the time step tend to zero, the detailed balance condition for the Markov chain corresponding to the diffusion process implies that a stationary solution of the type $p_0(\vec{x}) = e^{-\Phi(\vec{x})}$ exists, where $\Phi(\vec{x})$ is called the potential function. We saw in Section 7.5 a particular example of this type of stationary solution, where the drift term of the Fokker–Planck equation (proportional to the force) was the gradient of a potential energy. In this case, the potential function of the stationary solution is the true potential energy, hence the name. In Table 11.1 we summarize the analogous properties of homogeneous and stationary probability distributions in the continuous case and in the discrete case.

In the infinite-dimensional case, the formula (11.14) for the behavior at a large number of time steps might still be extended, but one must consider the possible existence of a continuous spectrum since the number of eigenvalues will be infinite and the domain of their values is limited between 0 and 1. In the case where the continuous spectrum includes the value 1, or 1 is an accumulation point of the discrete spectrum, the probabilities $P_{i,k}^{(n)}$ continue to tend to the stationary value u_k when $n \to \infty$ (the result in Section 10.4.5 of the stationary probability was proved for infinite chains), but slower than an exponential, e.g., as an inverse power.

The results presented in this section are useful for constructing Markov chains leading to a chosen form of the asymptotic probability: they will be used, particularly in the next section, to construct processes that will allow us to efficiently compute high-dimensional integrals.

It is interesting to mention Einstein's relations [73, 74] for the transition probabilities between two states i and j with energy $E(i)$ and $E(j)$, respectively, in a system in equilibrium at temperature T:

$$P_{i,j} = \exp\{-\beta[E(j) - E(i)]\}P_{j,i}, \tag{11.15}$$

where $\beta = (k_B T)^{-1}$, and k_B is the Boltzmann constant. These relations coincide with those of the detailed balance, if the limit probability u_i is proportional to $\exp(-\beta E(i))$.

In other words, Einstein's relations on the stochastic matrix of a reversible Markov chain guarantee that the asymptotic probability distribution is given precisely by the Boltzmann–Gibbs distribution

$$u_i = \frac{e^{-\beta E(i)}}{Z}, \quad Z = \sum_i e^{-\beta E(i)}.$$

11.3 Monte Carlo Method

11.3.1 Integrals in Few and in Many Dimensions

The Monte Carlo method that we will discuss here proves to be of great utility in the computation of high-dimensional integrals. This is a common situation in the study of physical systems, where the reference scale for a macroscopic system is Avogadro's number, of order 10^{23}.

Let us begin, however, by considering the simple one-dimensional integral

$$G \equiv \int_{x_A}^{x_B} g(x)\,dx\,, \tag{11.16}$$

where $g(x)$ is a regular function, i.e., twice differentiable. Typically, a numerical computation is performed by means of a quadrature method, more or less refined. One starts by choosing equispaced values of x_i on the support (advanced methods may use values of x_i chosen from more complex considerations). We then decompose the integral into N infinitesimal contributions, of width 2ϵ,

$$G \simeq \sum_i^N \int_{x_i-\epsilon}^{x_i+\epsilon} g(x)\,dx\,, \tag{11.17}$$

where $N \sim 1/\epsilon$, and approximate the single contribution between $-\epsilon$ and $+\epsilon$ according to some quadrature formula [42]. For example, the trapezium rule gives

$$G_\epsilon \equiv \int_{-\epsilon}^{+\epsilon} g(x)\,dx = \epsilon[g(\epsilon) + g(-\epsilon)] + O(\epsilon^3)\,, \tag{11.18}$$

while the Cavalieri–Simpson rule gives

$$G_\epsilon \equiv \int_{-\epsilon}^{+\epsilon} g(x)\,dx = \frac{\epsilon}{3}[g(\epsilon) + 4g(0) + g(-\epsilon)] + O(\epsilon^5)\,. \tag{11.19}$$

We immediately note that the error on the integral is

$$\mathrm{Err}(G) = \begin{cases} O(N\epsilon^3) = O(N^{-2}), & \text{trapezium}\,, \\ O(N\epsilon^5) = O(N^{-4}), & \text{Cavalieri–Simpson}\,, \end{cases}$$

which we can summarize as $N^{-\eta}$, with $\eta = 2$ for the trapezium rule or $\eta = 4$ for the Cavalieri–Simpson rule. $\mathrm{Err}(G)$ therefore decreases quite rapidly with the number of points approximating the continuous interval in one dimension (we are, of course, assuming that the integrand function g has a reasonably regular behavior).

An idea that at first sight may seem peculiar is to choose values x_i at random, uniformly distributed in the function support $x \in [x_A, x_B]$. This is the basic idea of the method called *Monte Carlo* (for reasons that can be deduced by our savvy readers), which leads to results of much greater interest than might appear from a first assessment. Having, therefore, chosen the values $\{x_i\}$ at random, we approximate the integral as

$$G \simeq \frac{1}{N} \sum_{i=1}^{N} g(x_i). \tag{11.20}$$

What will be the error in the estimate of the integral approximated in this way? Since we have randomly chosen the values of x_i, it is reasonable to consider $g(x_i)$ as a random variable and use the central limit theorem for our estimate in the limit of large N. If we define the empirical averages

$$\bar{g} \equiv \frac{1}{N} \sum_{i=1}^{N} g(x_i) \quad \text{and} \quad \overline{g^2} \equiv \frac{1}{N} \sum_{i=1}^{N} g(x_i)^2, \tag{11.21}$$

the mean square deviation of $g(x_i)$ will be

$$\sigma_g \equiv \sqrt{\overline{g^2} - \bar{g}^2},$$

and the error on our estimate (11.20) of G will be

$$\text{Err}(G) = \sigma_G \simeq \frac{\sigma_g}{N^{1/2}}. \tag{11.22}$$

The error is of order $N^{-1/2}$ (to be compared with N^{-2} for the trapezium quadrature and N^{-4} for the Cavalieri–Simpson quadrature). This is clearly an unexciting result, since for our one-dimensional integral the Cavalieri–Simpson rule is much more effective. It is, however, a result destined to become remarkable in the case of multi-dimensional integrals..

Before evaluating and comparing the different approaches to computing multi-dimensional integrals, we note that the error is proportional to σ_g, the mean square deviation of the function g: the smaller σ_g, the smaller the error on the integral of g. This consideration will be the origin of the importance sampling approach of the Monte Carlo method, which we will discuss in Section 11.3.2.

Let us now analyze the behavior of an integral in D dimensions:

$$\int \prod_{\alpha=1}^{D} dx_\alpha \, g(\{x\}) = \int d^D x \, g(\vec{x}). \tag{11.23}$$

Here D will be large (on our computers we often have to deal with $D \simeq 10^8$): this is a common situation in physics, where we are interested in describing a system with many degrees of freedom.

Suppose we use some deterministic integration rule, approximating the integration domain with N total points. Each one of the D dimensions will be divided into $N^{1/D}$ intervals of spacing $\epsilon \sim N^{-1/D}$. For example, 1024 points in $D = 10$ dimensions correspond to a hypercubic lattice of side 2.

An elementary cell will have volume $\sim \epsilon^D$. The error on the integral on an elementary cell will be $O(\epsilon^{D+\eta})$, where η is a small number, of order 1, which, as we have seen, depends on the detail of the quadrature rule used. The total error committed on the integral will therefore be $O(N\epsilon^{D+\eta})$, which, using the scaling law for ϵ, is of order $N^{-\eta/D}$. Since we are interested in integrals of high dimensionality, this scaling law is dramatically slow and the computation of the integral cannot be performed using this approach: the scaling of the error with N is too slow.

In the Monte Carlo method, on the other hand, the error still decreases as $N^{-1/2}$, independently of D. Obviously, one must take into account prefactors, which may be large, but which we will assume, anyway, constant in N. For the classical trapezium rule, for example, the fact that $\eta = 2$ implies that the naive Monte Carlo method we have described here is asymptotically better than deterministic integration starting from $D = 4$. The Cavalieri–Simpson method ($\eta = 4$) is outperformed by the Monte Carlo method starting from D somewhat larger. It is clear, however, that, since we are interested in values of D that allow us to approximate situations potentially dominated by an Avogadro's number of degrees of freedom, *the game is worth the candle* (it's worth the risk).

However, all that glitters is not gold. Suppose we consider the integral on the D-dimensional vector variable $\vec{x}$ of the function

$$g(\vec{x}) = \exp\left\{-\frac{D}{2}\sum_{\alpha=1}^{D} x_\alpha^2\right\},$$

that is,

$$\int_{r<R} d^D x \, \exp\left\{-\frac{D}{2}\sum_{\alpha=1}^{D} x_\alpha^2\right\} = S(D) \int_0^R dr \, r^{D-1} \exp\left\{-\frac{D}{2}r^2\right\}$$

$$= S(D) \int_0^R \frac{dr}{r} \, \exp\left\{-D\left[\frac{1}{2}r^2 - \log(r)\right]\right\}, \quad (11.24)$$

where $r^2 = |\vec{x}|^2$,

$$S(D) = \frac{\pi^{D/2}}{\Gamma(D/2+1)}$$

is the surface area of the unit sphere in D dimensions, and we consider the case $R > 1$. The maximum of the integral is at $r = 1$ and, for large D, the leading contribution comes from this region, as known from Laplace's maximum method discussed in Appendix 2.C. But if we sample the points $\vec{x}_i$ on which to compute the integral with the Monte Carlo method uniformly, for large D most of the points are concentrated in the region $r \approx R$, close to the surface of the hypersphere. For this reason the integration of (11.24) with the Monte Carlo method only starts to work when the number of sampled points is at least of order R^D such that we have a non-negligible probability of placing at least some of the points in the relevant region $r \simeq 1$. Obviously, for large D we face disaster.

As we shall see, this is a well-known phenomenon in statistical mechanics: when the space of configurations has a very high number of dimensions, the probability distribution is often concentrated in a small region of phase space, which may not be the most probable by uniformly sampling random values in the domain of the definition of the integral. To bypass this difficulty, it is necessary to use the method of *importance sampling*, described in the next section.

11.3.2 Importance Sampling

In the previous section, we noted the fact that the error made when computing an integral using the Monte Carlo method decreases when σ_g decreases, i.e., when the function to be integrated has a smooth limited behavior (in the case of a constant, obviously, a single representative point chosen at random allows us to reduce the error to zero). The problem seems difficult to solve, since σ_g is an intrinsic feature of the function $g(x)$ and cannot be modified. The appropriate answer is, in this case, to perform a transformation that leads us to integrate a different function, characterized by smaller fluctuations. To do this, we must modify the probability measure on which we perform the integral, which, presumably, will no longer be uniform. Let us consider a probability density $\pi(x)$, such that

$$\int_{x_A}^{x_B} \pi(x)\,dx = 1 \quad \text{and} \quad \pi(x) > 0 \quad \forall x \in [x_A, x_B], \tag{11.25}$$

and rewrite (11.16) as

$$G = \int_A^B \tilde{g}(x)\pi(x)\,dx, \tag{11.26}$$

where we have defined the function

$$\tilde{g}(x) \equiv \frac{g(x)}{\pi(x)}. \tag{11.27}$$

Computing (11.26) means choosing the values $\{x_i\}$ no longer with a uniform probability but according to $\pi(x)\,dx$. This is the idea behind the so-called *importance sampling*. In realistic cases, we cannot exactly compute this measure, but the use of the Markov chain techniques that we have just learned allows us to implement algorithms that sample the values of x generated by the probability density $\pi(x)$.

Now the error on the integral will be dominated by the variance of $\tilde{g}(x)$ instead of by σ_g. If we can choose a weight function $\pi(x)$ that is large where $g(x)$ is large and small where $g(x)$ is small, the new function $\tilde{g}(x)$ will be a smooth function, its variance will be small, and the error in the estimate of the integral will also be small.

In simple one-dimensional examples, it is possible to exactly compute the probability $\pi(x)\,dx$ of the values $x_i \in [x, x + dx)$. In general, and in almost all interesting cases, this is not possible, though. Instead, we will use the techniques discussed above that enable the construction of a Markov chain that converges for a large number of steps to the stationary measure $\pi(x)\,dx$. We will introduce the classical Metropolis algorithm that allows this operation to be carried out in a simple way.

In statistical mechanics, this formulation comes about very naturally. Recall that in this case the integrals to be computed are typically the expectation values of an observable function O of the configurations C of the system (list of positions, momenta, angular momenta, spins, etc., of many particles) of the form

$$\langle O \rangle \equiv \frac{\int \mathcal{D}C\, e^{-\beta H[C]} O[C]}{Z}, \quad Z \equiv \int \mathcal{D}C\, e^{-\beta H[C]}, \tag{11.28}$$

where with C we symbolically denote the relevant degrees of freedom of the physical system under consideration, with $\mathcal{D}C$ the multivariate differential, and with H the Hamiltonian functional determining the interaction properties between the elementary components of the theory. For example, for a system composed of N particles in three dimensions, each with coordinates x_i, y_i, and z_i, the multi-dimensional differential has the form

$$\mathcal{D}C = \prod_{i=1}^{N} dx_i\, dy_i\, dz_i\,,$$

and $H[C] = H[\vec{x}, \vec{y}, \vec{z}\,]$ depends on the $3N$ variables characterizing the system's degrees of freedom. It is clear that a natural choice for the weight factor is, in this case, precisely the so-called Boltzmann–Gibbs distribution

$$\pi[C] = \frac{e^{-\beta H[C]}}{\int \mathcal{D}C\, e^{-\beta H[C]}}\,, \tag{11.29}$$

while the rest of the integrand is $\tilde{g}[C] = O[C]$.

The fact that in statistical mechanics, for large N, the expectation value normally depends on an extremely small area of the phase space makes the Monte Carlo method with importance sampling very efficient: to have a good estimate of the integral (11.28), it is sufficient to sample a small part of the values that the variables C can take, corresponding to the minima of the Hamiltonian.

11.3.3 Markov Chain Monte Carlo: Metropolis Algorithm

We must, therefore, analyze the problem of how we choose to sample the medium in such a way that allows us to reproduce the chosen measurement. We specialize our discussion to a form of the type (11.28), which is, as we have discussed, typical of problems in statistical mechanics. In this case, we are not able, of course, to exhibit the explicit form of the measure, and we are willing to use our knowledge of Markov chains, learned in Chapter 10 and Section 11.1.

To do this we write the detailed balance relation (11.4) in the case of the integral (11.28), whereby the stationary probability for the state k, i.e., the configuration $C^{(k)}$

of the values of the degrees of freedom, is (11.29):

$$u_k = \pi[C^{(k)}].$$

We obtain

$$e^{-\beta H_i} P_{i,k} = e^{-\beta H_k} P_{k,i}, \tag{11.30}$$

where we have abbreviated $H_i \equiv H[C^{(i)}]$. An irreducible Markov chain in the configuration space of the system that satisfies (11.30) – and is, therefore, ergodic – reconstructs the equilibrium Boltzmann–Gibbs probability distribution (11.29).

This point of view defines the integration algorithm, which we called Monte Carlo, as a dynamic procedure of transitions between states of an ergodic and reversible Markov chain: it is the method also called *Markov Chain Monte Carlo*. The dynamics takes place in discrete time steps defined by the Markovian numerical chain simulation algorithm, in which time may have more or less to do with real physical time.

We, therefore, study a particular Monte Carlo algorithm developed by Nicholas Metropolis, Arianna and Marshall Rosenbluth, and Augusta and Edward Teller in 1953 [75], which is nowadays called the *Metropolis* algorithm. We proceed by decomposing the transition probability between two configurations into (i) the *proposal* of a state update and (ii) the possible *acceptance* of such an update.

(i) Starting from a system configuration, which we denote by $C = C_i$, we propose to change it to the trial configuration, denoted by C_k, requiring the probability $S_{i,k}$ of proposing the exchange $C_i \to C_k$ to be equal to the probability $S_{k,i}$, i.e., the probability of proposing the reverse exchange $C_k \to C_i$. Depending on the number of states k accessible from i, we can take, for example, a uniform distribution. This may also include the possibility of proposing the starting state i itself, depending on the algorithm.

   Typically, in a model of variables defined on a lattice, the change of a single site variable is proposed. The way in which C_k is proposed influences the efficiency of the algorithm, and its speed of convergence at equilibrium. In an essentially heuristic way, a good choice is considered to be a C_k such that the acceptance probability of the change (see item (ii)) is not too far from $1/2$. The reason is quite intuitive: unlikely changes, although potentially very large, slow down convergence (as very frequent, though presumably too small, changes).

(ii) We accept such an update proposal with a probability $A_{i,k}$ satisfying the detailed balance equation (11.30) and guaranteeing that the stationary probability distribution $\pi(C)$ is the Boltzmann–Gibbs distribution, Eq. (11.29). Such an acceptance probability is proportional to the transition probability, $A_{i,k} \propto P_{i,k}$, so, from (11.30) we have

$$\frac{A_{i,k}}{A_{k,i}} = e^{-\beta(H_k - H_i)}.$$

Given C_i and C_k we then compute the variation of the Hamiltonian function H that appears in the exponential, and the values of $A_{i,k}$ and $A_{k,i}$ will depend on the relative values of the Hamiltonian in the C_i and C_k configurations. The Metropolis algorithm is, eventually, defined by the choice

$$
A_{i,k} = \begin{cases} e^{-\beta(H_k - H_i)}, & \text{if } H_k > H_i\,, \\ 1, & \text{if } H_k < H_i\,. \end{cases}
\tag{11.31}
$$

If the change from C_i to C_k decreases H, we accept the change with certainty, and move to the new configuration $C = C_k$. If, instead, the function H increases as a result of the proposed change ($H_k > H_i$), we do not reject the change *a priori*, as one would do in a convergence procedure to a local minimum, but

(a) we compute the value of $A_{i,k} = e^{-\beta(H_k - H_i)}$,
(b) we extract a random number ρ with uniform probability in $[0, 1)$, and
(c) if $\rho < A_{i,k}$ we accept the change and set $C = C_k$, whereas if $\rho > A_{i,k}$ we reject the change and set $C = C_i$.

It is interesting to note the following. Once the integration procedure is translated into a dynamics that suitably samples the phase space, we note that it might be possible to introduce a continuous dynamics, e.g., via the Langevin equation discussed in Section 7.6. In this sense, the Metropolis method can be seen as an extremely astute discretization of the Langevin equation, which, while using a finite integration step, converges to the exact result for the asymptotic distribution without introducing any systematic error. This is in contrast to a direct discretization of a differential equation (and even more so of a stochastic differential equation), where the correct result is obtained only in the limit of infinitesimal increments [76, 77]. This statement can be made more quantitative by showing that, when infinitesimal increments are chosen for the elementary variables, the Metropolis algorithm leads to the same Fokker–Planck equation as obtained from the Langevin equation.

In the Monte Carlo method we only have to deal with statistical errors, which afflict our results only insofar as we have finite statistics, and disappear in the limit where the number of simulation steps diverges. This difference between statistical error and systematic error, and the peculiarity, in this respect, of the Metropolis method, are facts of great importance, and the reader is invited to carefully think about them.

11.3.4 Ising Model

The Ising model is one of the highlights of statistical mechanics. It is a model that is simple to define but has a behavior of high complexity, which allows one to introduce and discuss in detail ideas related to the criticality of the system and to the so-called *phase transitions*, where, starting from local interaction laws between microscopic components, collective behaviors emerge. The Ising model schematizes the behavior of a magnetic material and succeeds in describing the mechanism by which a material spontaneously magnetizes, at low temperatures, even in the absence of a magnetic field.

We shall define here, as a typical example for a Monte Carlo study, the Ising model in two spatial dimensions (i.e., where our elementary magnets live on a two-dimensional square lattice): this model is exactly solvable, as first demonstrated by Onsager in 1944 (see also [57]), as opposed to the model, physically more relevant, defined in three spatial dimensions. The latter has no exact solution so far and can only be studied using approximate methods or, indeed, through Monte Carlo simulation methods. The results obtained through Monte Carlo numerical simulations in the 2D case can then be compared to the exact results, making it possible to analyze the method in detail through, for example, the study of the behavior of statistical errors, of the effects of finite volume, and of the convergence of the Markov chain of system configurations to equilibrium (i.e., the stationary limit under time reversal invariance).

Consider a square lattice of linear size L and volume $N = L^2$. Each site has $2 \times D = 4$ nearest neighbors, which we label with the integer pair (x, y), where $x = 0, 1, \ldots, L-1$, and $y = 0, 1, \ldots, L-1$. The dynamic variables σ representing the elementary magnets live on the sites of the lattice, and can take on only the two values ± 1 (local magnetization of type plus or type minus):

$$\sigma_{x,y} = \pm 1 . \tag{11.32}$$

They are called Ising spins. Typically, the system is defined with periodic boundary conditions (minimizing systematic errors due to the finiteness of the lattice), although boundary conditions of a different type are admissible. The function defining the interaction between two elementary magnets is the Hamiltonian H,

$$H \equiv -J \sum_{\langle i\, j \rangle}^{1,N} \sigma_{x_i, y_i} \sigma_{x_j, y_j} , \tag{11.33}$$

where $J > 0$ (we will set it equal to 1 without loss of generality) and the sum runs over the pairs $\langle i\, j \rangle$ of nearest neighbors of the lattice:

$$\sqrt{(x_i - y_i)^2 + (x_j - y_j)^2} = 1 .$$

At zero temperature, the system is not subject to thermal agitation, and stabilizes in its minimum energy state. The minimum energy is obtained when the spins are all parallel, all with the value $+1$ (positive magnetization state) or -1 (negative magnetization state). In contrast, at non-zero temperatures, the system is described by a probability distribution of configurations, which becomes wider at higher temperatures. If we define an inverse temperature $\beta \equiv T^{-1}$ (we are ignoring multiplicative constants J/k_B, which we set to one), the fundamental laws of statistical mechanics tell us that the probability of a given configuration of the elementary variables will be

$$P(\{\sigma\}) = \frac{e^{-\beta H[\sigma]}}{Z} . \tag{11.34}$$

Macroscopic quantities (such as the magnetization or susceptibility of the system) will be expressed by the values of expectation $\langle \cdot \rangle$ on this probability measure, of the form

$$\langle O \rangle \equiv \sum_{\{\sigma\}} \frac{e^{-\beta H[\sigma]}}{Z} O[\sigma] , \tag{11.35}$$

where the sum runs over all possible 2^N configurations $\{\sigma\}$ of Ising spins and the *partition function Z* is such that $\langle 1 \rangle = 1$. The task of a Monte Carlo simulation is therefore to compute expectation values, i.e., integrals of the type (11.35).

In a realistic simulation, on computers available today, the typical lattices have linear sizes of the order of 10^3 (or even much more if one looks for the record[1]), but for the first tests, a lattice of linear size $L = 10$ is more than sufficient.

The Metropolis algorithm is very simple to apply to this model. The spins are changed one at a time. The lattice can be swept deterministically, following the sites in lexicographic order or by choosing the site to be changed randomly, uniformly with probability $1/N$. We emphasize that, in the first case, detailed balance is not satisfied at the level of the change of the single site variable, because if we call a a given configuration and b the same configuration a with the spin j reversed, if one has $S_{a,b} = 1$ then $S_{b,a} = 0$ (the spin subsequently proposed for the flip in lexicographic order is the spin $j + 1$) and $S_{a,b} \neq S_{b,a}$. However, the simpler balance equation (10.57) is satisfied. Though less stringent, this condition can be proven to be sufficient for convergence to equilibrium.[2] In the second case, the probability of proposing the configuration b moving from a corresponds to extracting a given spin j to be inverted, which has probability $1/N$. The inverse consists of starting from the configuration b (which is the configuration a with the spin j upside down) and extracting the spin j, an event that has, again, probability $1/N$ to occur. Hence, $S_{a,b} = S_{b,a}$ and in this case detailed balance is satisfied.

We propose for the change the site variable $\sigma_{x,y}$, which we call σ_1: the new value of the variable $\sigma_{x,y}$, which will be obtained as a result of the update procedure, will be called σ_2. A good choice for the trial value is $\sigma_t = -\sigma_1$, symmetric as it should be. Also choosing $\sigma_t = \pm 1$ with probability $1/2$ is a legitimate choice. The first choice does not include the starting configuration; the second one does, and is less efficient. We will adopt the first. We then compute

$$\Delta H = -(\sigma_t - \sigma_1) \sum_{\substack{x'=x\pm 1 \\ y'=y\pm 1}} \sigma_{x',y'} = 2\sigma_1 \sum_{\substack{x'=x\pm 1 \\ y'=y\pm 1}} \sigma_{x',y'}, \tag{11.36}$$

where the sum goes over the four nearest neighbors of (x, y). Now we follow the procedure mentioned before: If ΔH is negative we accept the change, setting $\sigma_2 = \sigma_t$. If, instead, the energy increases, we compute the factor $f = e^{-\beta \Delta H}$, and compare it with a random number ρ drawn with uniform probability in the interval $[0, 1)$. If $f > \rho$, then we accept the change and set $\sigma_2 = \sigma_t$; otherwise we reject the proposed change and leave $\sigma_2 = \sigma_1$.

Interestingly, the binary character of the Ising spin variables allows particularly efficient codes to be written. For example, it is not necessary to compute at each step the exponential in f. Its exponent can, indeed, take only a few different integer values,

---

[1] In recent work, the following have been achieved: $2d$ grids of linear size $L = 83\,968$ on an artificial intelligence integrated circuit built for machine learning in neural networks, called a *Tensor Processing Unit* (TPU) [78].

[2] It can be shown that the balance equation is sufficient to guarantee that the dynamics of the Metropolis method tends to the Boltzmann–Gibbs distribution for not excessively small systems [79].

because one can only have $\Delta H = -4, -2, 0, 2, 4$, and in the case of zero or negative values we do not need to compute f, because acceptance is certain. The remaining values

$$f = \begin{cases} e^{-2\beta}, & \Delta H = 2, \\ e^{-4\beta}, & \Delta H = 4, \end{cases}$$

can be computed once and stored in a vector.

Furthermore, since the elementary variables can be stored in a single bit, it is possible to store in a word of 64 bits, for example, 64 copies of the system. In this way it is possible to contemporaneously compute, basically at the cost normally required to follow the dynamics of a single system, 64 copies of the system (this is the so-called *multi-spin coding* method).

It is not appropriate here to enter into a discussion of the very interesting physical phenomena that occur in a system that is so simple to define. It is enough to mention that, for a precise value of T, namely, the *critical temperature T_c*, the system generates spatial correlations of infinite range: this has repercussions on the dynamics that experiences a *critical* slowing down.

Finally, we note that the Metropolis algorithm is only one of several algorithms that can be used. In the case of the Ising model, for example, the probability of a spin can be computed exactly when the other spins are fixed. We can, for example, compute what is the choice of σ_2 that satisfies the Boltzmann–Gibbs distribution if the other spins are already distributed according to Boltzmann–Gibbs. This algorithm is called the *heat bath* algorithm (because it is equivalent to considering the chosen spin as embedded in a thermal bath of neighboring spins), and leads to choosing the new spin with probability

$$P(\sigma_2 = +1) = \frac{e^{-\beta \sum' \sigma}}{e^{-\beta \sum' \sigma} + e^{+\beta \sum' \sigma}}, \qquad P(\sigma_2 = -1) = 1 - P(\sigma_2 = +1), \qquad (11.37)$$

where the primed sum indicates that the sum goes over the nearest neighbors of the considered spin:

$$\sum{}' \sigma \equiv \sum_{\substack{x'=x\pm1 \\ y'=y\pm1}} \sigma_{x',y'}.$$

The heat bath algorithm has the same dynamic properties of the Metropolis one.

Correlated Events

In this chapter we aim to study the extension of the central limit and large deviations theorems to the case of correlated random variables.

At first we will consider two concrete cases, one more general and one more specific, which we have explored in previous chapters: finite Markov chains and recurrent events. In each of the two cases, we will be able to obtain explicit results for the central limit behavior after a long and exact computation.

Next we will characterize and study correlation functions in detail and show how the same qualitative behavior for a large number of variables can be obtained from general considerations on correlation functions, and we will prove the central limit theorem and the large deviations theorem for correlated events.

In the last part of the chapter we shall emphasize the contact points between the probability theory of correlated events with the statistical mechanics theory of a system in the thermodynamic limit of infinite volume.

12.1 Central Limit Distribution for Finite Markov Chains

We consider a Markov chain of states denoted by k and a random variable $a(k)$ defined on the states of the chain. We designate with k_t the state of the chain at the time step t. We propose to compute the probability distribution of the sum of random variables defined on the states of the chain:

$$A(n) = \sum_{t=1}^{n} a(k_t).$$

$$(12.1)$$

We call k_0 the initial condition, the state of the chain at time zero. To compute the probability distribution of $A(n)$, we should consider the generating function $F_n(z)$ of its moments, see formula (8.6), which we will write in the related case in terms of the joint probability density, conditional on the initial condition, of all the states visited in the n steps over which we sum $a(k_t)$:

$$F_n(z) \equiv \langle \exp\{z A(n)\} \rangle = \sum_{k_1, k_2, \ldots, k_n} p(k_1, k_2, \ldots, k_n \mid k_0) \exp\left\{ z \sum_{t=1}^{n} a(k_t) \right\}. \quad (12.2)$$

In other words, the average is over all possible paths $\{k_1, k_2, \ldots, k_n\}$ along the states of the chain in n steps. If we use the probability factorization property in a Markov chain,

$$p(k_1, k_2, \ldots, k_n \mid k_0) = p(k_n|k_{n-1}) \cdots p(k_2|k_1)p(k_1|k_0) = \prod_{t=1}^{n} P_{k_{t-1},k_t} \,,$$

and introduce the matrix element

$$Q_{i,k}(z) = p(k|i)\exp\{za(k)\} = P_{i,k}\exp\{za(k)\}, \tag{12.3}$$

we can rewrite

$$F_n(z) = \sum_{k_1,k_2,\ldots,k_n} \prod_{t=1}^{n} [P_{k_{t-1},k_t}\exp\{za(k_t)\}] = \sum_{k} (Q(z)^n)_{k_0,k} \,, \tag{12.4}$$

where we denote by $(Q(z)^n)_{i,k}$ the matrix elements of the nth power of the matrix Q.

The argument proceeds in much the same way as in Section (10.7.2). Let us consider the case of an irreducible ergodic reversible chain with a finite number N of states and suppose that the matrix Q has N non-degenerate real eigenvalues, which are the zeros of the characteristic polynomial $\det(Q(z) - \lambda)$. We introduce the N right (d) and left (s) eigenvectors, satisfying the following equations:

$$\begin{aligned}
\sum_{k} Q_{i,k}(z)d_k(i) &= \lambda(i,z)d_i(i)\,, \\
\sum_{i} s_i(j)Q_{i,k}(z) &= \lambda(j,z)s_k(j)\,.
\end{aligned} \tag{12.5}$$

The right eigenvectors are orthogonal to the left eigenvectors corresponding to different eigenvalues, and we can impose the orthonormalization condition of the skew basis:

$$\sum_{k} d_k(j)s_k(m) = \delta_{j,m}\,. \tag{12.6}$$

With the spectral decomposition of non-Hermitian matrices, see Section 10.7.2, the matrix Q can be written as

$$Q_{i,k}(z) = \sum_{j} d_i(j)\lambda(j,z)s_k(j)\,. \tag{12.7}$$

In the same way, we obtain that the nth power of the matrix is given by

$$(Q(z)^n)_{i,k} = \sum_{j} d_i(j)\lambda(j,z)^n s_k(j)\,. \tag{12.8}$$

If we denote by $\lambda(1, z)$ the largest eigenvalue (spectral limit) of the non-negative matrix $Q(z)$, we obtain that, for large n, the generating function reads

$$F_n(z) = \sum_{k=1}^{N} (Q(z)^n)_{k_0,k} = d_{k_0}(1)\sum_{k=1}^{N} s_k(1)\lambda(1,z)^n + O(\lambda(2,z)^n)$$

$$\simeq \lambda(1,z)^n = \exp\left[n\log\lambda(1,z)\right]. \tag{12.9}$$

The contributions of the other eigenvalues are exponentially small with n, and they go to zero as

$$\left|\frac{\lambda(k > 1, z)}{\lambda(1, z)}\right|^n \simeq \exp\left(-n \log \frac{\lambda(1, z)}{|\lambda(k, z)|}\right) \xrightarrow[n \to \infty]{} 0.$$

Asymptotically, therefore, the generator (12.9) is factorized and the distribution of $A(n)$ is apparently the same as the sum of n independent and identically distributed variables with probability density $p(x)$ and generating function

$$\int dx \, p(x) \exp(xz) = \exp\{\log[\lambda(1, z)]\} = \lambda(1, z). \tag{12.10}$$

Where did the correlation go? To see this, we must explicitly compute $\lambda(1, z)$ and the moments generated by it.

If the chain, as we initially assumed, is irreducible, the stochastic matrix P, which coincides with $Q(0)$, has only one eigenvalue $\lambda(1)$ equal to 1. In the ergodic case (hence aperiodic), this is also the only eigenvalue of spectral radius 1.

A matrix is said to be analytic in a certain variable if all its elements are analytic functions of that variable. A well-known theorem [80] states that, if a matrix is analytic in the variable z, its eigenvalues are also analytic functions of z, unless two or more eigenvalues coincide. Even in the case where there are some degenerate eigenvalues, though, the distinct eigenvalues remain analytic functions of z. For an irreducible chain this theorem implies that $\lambda(1, z)$ is an analytic function of z around $z = 0$ ($\lambda(1, 0) = \lambda(1) = 1$) and consequently the distribution of $A(n)$ turns out to have, for large n, mean and variance equal to

$$\langle A \rangle = n \frac{\partial}{\partial z} \log \lambda(1, z)\Big|_{z=0},$$

$$\langle A^2 \rangle - \langle A \rangle^2 = n \frac{\partial^2}{\partial z^2} \log \lambda(1, z)\Big|_{z=0}. \tag{12.11}$$

It is immediate to see that this behavior, proportional to n, implies that the variance of $A(n)/\sqrt{n}$, see, e.g., the variable in Eq. (3.18) defined in the central limit case for independent events, is finite:

$$\frac{\langle (A(n) - \langle A(n) \rangle)^2 \rangle}{n} = O(1), \quad n \gg 1,$$

and that the higher-order cumulants of the variable $A(n)/\sqrt{n}$ tend to zero for $n \to \infty$. As we have seen in Appendix 3.C.3, dedicated to the generating function of the cumulants of a Gaussian variable, this is sufficient to guarantee that the distribution of $A(n)$ tends to a Gaussian distribution, as the central limit theorem states in general.

To compute the variance (12.11), we can use perturbation theory [58]. In the case of a non-Hermitian matrix H with real elements,[1] with eigenvalues $E(j)$, $j = 1, \ldots, N$, we can use the Dirac notation for the skew basis. The left eigenvectors will be the bra $\langle j| \equiv \vec{s}(j)$, and the right eigenvectors will be the ket $|k\rangle \equiv \vec{d}(k)$, so

---

[1] The formulas for a Hermitian matrix are familiar to those who have studied quantum mechanics. Nevertheless, we recall them in Appendix 10.B.

$$\langle j|k\rangle = \sum_{a=1}^{N} s_a(j)d_a(k) = \delta_{j,k}\,,$$

$$H|j\rangle = \left\{\sum_{b=1}^{N} H_{a,b}d_b(j)\right\} = E(j)\vec{d}(j) = E(j)|j\rangle\,,$$

$$\langle j|H = \left\{\sum_{a=1}^{N} s_a(j)H_{a,b}\right\} = E(j)\vec{s}(j) = \langle j|E(j)\,,$$

$$\langle j|H|k\rangle = \sum_{a,b}^{1,N} s_a(j)H_{a,b}d_b(k)\,. \tag{12.12}$$

In the case where the eigenvalues of H are not degenerate, the eigenvalues $E(j,\epsilon)$ of the matrix $H + \epsilon\Delta$ are given, for small ϵ, by the formula [58]

$$E(j,\epsilon) \doteq E(j) + \epsilon\langle j|\Delta|j\rangle + \epsilon^2 \sum_{k\neq j} \frac{\langle j|\Delta|k\rangle\langle k|\Delta|j\rangle}{E(j) - E(k)} + O(\epsilon^3)\,. \tag{12.13}$$

We now return to our Markov chain. We abbreviate $\lambda(j) = \lambda(j,0)$, define the matrices $\hat{a}$ and $\widehat{a^2}$ of elements, respectively, as

$$(\hat{a})_{i,k} = \delta_{i,k}a(i) \quad \text{and} \quad (\widehat{a^2})_{i,k} = \delta_{i,k}a(i)^2,$$

and recall that in the braket notation the eigenvalue equations read as

$$\langle m|P = \lambda(m)\langle m| \quad \text{and} \quad P|m\rangle = \lambda(m)|m\rangle.$$

Expanding in z the analytic matrix $Q(z)$ in Eq. (12.3) around $z = 0$, we obtain

$$Q(z) = P + zP\hat{a} + \frac{z^2}{2}P\widehat{a^2} + O(z^3)\,,$$

from which, applying the perturbation theory formula (12.13), we derive the expansion in z for the maximum eigenvalue of the perturbed system:

$$\lambda(1,z) = 1 + z\langle 1|\hat{a}|1\rangle + \frac{1}{2}z^2\langle 1|\widehat{a^2}|1\rangle + z^2 \sum_{j=2}^{N} \frac{\lambda(j)}{1 - \lambda(j)}\langle 1|\hat{a}|j\rangle\langle j|\hat{a}|1\rangle\,. \tag{12.14}$$

We then obtain, by means of (12.11), the mean and the variance of the sum variables

$$\langle A\rangle = n\langle 1|\hat{a}|1\rangle = n\sum_{k=1}^{N} a(k)u(k)\,,$$

$$\langle A^2\rangle - \langle A\rangle^2 = n\left\{\langle 1|\widehat{a^2}|1\rangle - \langle 1|\hat{a}|1\rangle^2 + 2\sum_{j=2}^{N} \frac{\lambda(j)}{1 - \lambda(j)}\langle 1|\hat{a}|j\rangle\langle j|\hat{a}|1\rangle\right\}$$

$$= n\left\{\sum_{k=1}^{N} a(k)^2 u(k) - \left[\sum_{k=1}^{N} a(k)u(k)\right]^2 + 2\sum_{j=2}^{N} \frac{\lambda(j)}{1 - \lambda(j)}\langle 1|\hat{a}|j\rangle\langle j|\hat{a}|1\rangle\right\}, \tag{12.15}$$

where we have denoted the kth component of $\langle 1|$ as $u(k) = s_k(1) = \lim_{t\to\infty} P_{i,k}^{(t)}$, as per the formulas (10.40), (10.56), and (10.71). We also recall that $|1\rangle = \vec{d}(1)$ is the positive eigenvector of components all equal to 1. We will see in Section 12.4.2 that the same result can be obtained without bothering with perturbation theory.

Let us now make a few observations. The correlation of the variables $a(k)$ comes into play in the variance, which we can write as

$$\sigma_{A(n)}^2 \equiv \langle A^2(n)\rangle - \langle A(n)\rangle^2 = n\left[\sigma_a^2 + \sum_{k,l}^{1,N} a(k)a(l)p(k,l)\right], \tag{12.16}$$

where we have defined the joint probability of the states k and l as

$$p(k,l) \equiv u(k)p(l|k) = u(k)\sum_{m=2}^{N} \frac{2\lambda(m)}{1-\lambda(m)} d_k(m)s_l(m) \simeq 2u(k)\frac{\lambda(2)}{1-\lambda(2)} d_k(2)s_l(2).$$

Notice the term $1 - \lambda(2)$ in the denominator. When $\lambda(2)$ is very close to 1, this term is essentially the inverse of the correlation time $\tau \equiv -1/\ln\lambda(2)$, cf. (10.74). Furthermore, in the case where the events are uncorrelated, the Markov chain decomposes into N single state closed sets, all with a single eigenvalue $\lambda(1) = 1$, so that the last term in Eqs. (12.15) and (12.16) is null and we obtain the known result for independent events (see Section 3.2).

If the Markov chain contains an infinite number of states, we can distinguish two cases. If the eigenvalue at $\lambda(1) = 1$ is isolated and the function $a(k)$ satisfies the appropriate conditions, there are no essential differences with the finite-dimensional case: $\lambda(1, z)$ is an analytic function around $z = 0$, and the distribution of $A(n)$ is Gaussian. If, instead, the eigenvalue $\lambda(1) = 1$ is not isolated, we can have very different behavior, and the distribution of $A(n)$ may be very different from a Gaussian.

12.2 Central Limit for Recurrent Events

In this section, we aim to compute the probability distribution of the variable $K(n)$, which is equal to the number of events that occurred in the first n steps of the discrete dynamics of a recurrent event (origin excluded).

We first compute the mean value and variance of recurrences $K(n)$. To this purpose we introduce a "counter" function $\epsilon(t)$ which is 1 if the event recurs at time t and is 0 otherwise. We have

$$K(n) = \sum_{t=1}^{n} \epsilon(t). \tag{12.17}$$

The expected value is then

$$\langle K(n)\rangle = \sum_{t=1}^{n} \langle \epsilon(t)\rangle = \sum_{t=1}^{n} p(t), \tag{12.18}$$

since $\langle \epsilon(t) \rangle$ is precisely the probability of the event occurring at time t. In the same way, we obtain for the second moment

$$\langle K^2(n) \rangle = \sum_{t_1,t_2}^{1,n} \langle \epsilon(t_1)\epsilon(t_2) \rangle = \sum_{t_1,t_2}^{1,n} p(t_1,t_2), \tag{12.19}$$

where $p(t_1,t_2)$ is the joint probability that the event occurs at both times t_1 and t_2. Because of the property $\epsilon(t)^2 = \epsilon(t)$ when the times are equal, $p(t,t)$ coincides with $p(t)$, while for different times we can use the fundamental expression for stationary Markovian events:

$$p(t_1,t_2) = \begin{cases} p(t_1), & \text{for } t_1 = t_2, \\ p(t_2 - t_1)p(t_1), & \text{for } t_2 > t_1, \\ p(t_1 - t_2)p(t_2), & \text{for } t_1 > t_2. \end{cases}$$

Eventually Eq. (12.19) can be rewritten as

$$\langle K^2(n) \rangle = \sum_{t=1}^{n} p(t) + 2 \sum_{t_1 < t_2}^{1,n} p(t_1)p(t_2 - t_1), \tag{12.20}$$

where the factor 2 comes from the symmetry by exchange of the indices $t_1 \neq t_2$.

If, for convenience, we conventionally assume that at time zero no event occurs, i.e., $p(0) = 0$ (and no longer is $p(0) = 1$, as in Chapter 9, see Section 9.2.2), we can rewrite the second term of Eq. (12.20) in the convolution form for the probability p,

$$2 \sum_{t_1 \leq t_2}^{1,n} p(t_1)p(t_2 - t_1) = 2 \sum_{t_2=1}^{n} \sum_{t_1=1}^{t_2} p(t_1)p(t_2 - t_1) = 2 \sum_{t_2=1}^{n} \left[p * p \right](t_2), \tag{12.21}$$

from which

$$\langle K^2(n) \rangle = \sum_{t=0}^{n} p(t) + 2 \sum_{t_2=1}^{n} p^{*2}(t_2). \tag{12.22}$$

We then introduce the generating function of the $p(n)$, assuming $p(0) = 0$:

$$\underline{P}(s) \equiv \sum_{n=0}^{\infty} p(n)s^n = \sum_{n=1}^{\infty} p(n)s^n. \tag{12.23}$$

Please pay attention to the fact that this generating function $\underline{P}(s)$ differs by 1 (the term in $n = 0$) from the definition of the generating function $P(s)$ given in formula (9.14). Its relation to the function $F(s)$ in (9.15), which generates the moments of the first return probability, is given by

$$\underline{P}(s) = \frac{F(s)}{1 - F(s)}, \tag{12.24}$$

rather than (9.17). We assume that the recurrent event is persistent and therefore that the limit of $p(t)$ exists when $t \to \infty$. We also know that the function $F(s)$ is sufficiently regular around $s = 1$, since the first return probabilities are well normalized and their expected value, the mean recurrence time μ, is well defined. In this case we can write

$$p(t) = p_\infty + g(t), \tag{12.25}$$

where $g(t)$ goes to zero quickly when $t \to \infty$. We introduce the generating function $G(s)$,

$$G(s) \equiv \sum_{n=1}^{\infty} g(n)s^n, \tag{12.26}$$

in terms of which the probability generator of the event is written[2] as

$$\underline{P}(s) = p_\infty \frac{s}{1-s} + G(s) = \frac{p_\infty}{1-s} + G(s) - p_\infty. \tag{12.27}$$

The function $G(s)$ is regular around $s = 1$ so we get the behavior

$$\underline{P}(s) \underset{s\to 1}{\sim} \frac{p_\infty}{1-s} + G(1) - p_\infty. \tag{12.28}$$

A simple computation shows that, for large n, (12.18) becomes

$$\langle K(n) \rangle = p_\infty n + \sum_{t=1}^{n} g(t) \underset{n\gg 1}{\sim} p_\infty n + \sum_{t=1}^{\infty} g(t) = p_\infty n + G(1), \tag{12.29}$$

where the last step is justified as $G(1)$ converges while $p_\infty n$ continues to grow for large n.

Alternatively, we can obtain the same result by calling upon the property of the generating function (8.11) of a sequence $\{b\}$. If its generating function is

$$B(s) = \sum_{t} b(t)s^t,$$

indeed, the generating function of the cumulative sequence of elements

$$\sigma(n) = \sum_{t=0}^{n} b(t) \tag{12.30}$$

is given by the formula (9.46), which we quote here for convenience:

$$\Sigma(s) = \frac{B(s)}{1-s}. \tag{12.31}$$

We then define the generator of the sum of $p(t)$, formula (12.18),

$$K_1(s) = \sum_{n=0}^{\infty} \langle K(n) \rangle s^n, \tag{12.32}$$

---

[2] As shown in (12.23) we can sum $p(n)s^n$ from $n = 1$ (first expression), or from $n = 0$ where $g(0) = p(0) - p_\infty = -p_\infty$ (directly second expression). The two results are obviously identical.

where the coefficient of the term in s^n is the mean value of the number of returns of the event at time n, and $\langle K(0)\rangle = p(0) = 0$. Using the relation (12.31) we write the generator of the sum, which for $s \to 1$ behaves as

$$K_1(s) = \frac{\underline{P}(s)}{1-s} \underset{s\to 1^-}{\sim} \frac{p_\infty}{(1-s)^2} + \frac{G(1)-p_\infty}{1-s} + O(1). \tag{12.33}$$

Since the behavior for $s \to 1^-$ is dominated by contributions at large n and

$$\frac{1}{(1-s)^2} = \sum_{n=0}^{\infty} (n+1)s^n, \tag{12.34}$$

we can rewrite (12.33) as

$$K_1(s) \underset{s\to 1^-}{\sim} \sum_{n=0}^{\infty} [p_\infty(n+1) + G(1) - p_\infty]s^n, \tag{12.35}$$

and we obtain the formula (12.29) for $n \gg 1$, once again.

We proceed in this way to determine the trend of $\langle K^2(n)\rangle$ for large n. We define

$$K_2(s) \equiv \sum_{n=1}^{\infty} \langle K^2(n)\rangle s^n, \tag{12.36}$$

which for (12.22) is nothing other than the generator of a cumulative combination of $\{p\}$ and $\{p*p\}$. Using the convolutional structure of the formula (12.22) we obtain that

$$K_2(s) = (1-s)^{-1}[\underline{P}(s) + 2\underline{P}(s)^2]. \tag{12.37}$$

The first term has already been estimated in (12.33). For the second term, the behavior for $s \to 1^-$ has the form

$$\frac{\underline{P}(s)^2}{1-s} \underset{s\to 1^-}{\sim} \frac{p_\infty^2}{(1-s)^3} + \frac{2[G(1)-p_\infty]p_\infty}{(1-s)^2} + O\left(\frac{1}{1-s}\right). \tag{12.38}$$

If we recall that

$$\frac{1}{(1-s)^3} = \sum_{n=0}^{\infty} \frac{(n+1)(n+2)}{2} s^n \tag{12.39}$$

and we keep only terms of order n and n^2 in the coefficients of s^n (we are interested in large n), putting together (12.33) and (12.38) we obtain that

$$K_2(s) \underset{s\to 1^-}{\sim} \sum_n [p_\infty n + p_\infty(n^2 + 3n) + 4p_\infty n(G(1) - p_\infty)]s^n. \tag{12.40}$$

Thus, for $n \gg 1$,

$$\langle K(n)^2\rangle \sim n^2 p_\infty^2 + np_\infty(1 - p_\infty) + 4nG(1)p_\infty. \tag{12.41}$$

By grouping (12.29) and (12.40), in the limit of large n, we obtain that the expected value of the frequency is given by

$$\lim_{n\to\infty} \frac{K(n)}{n} = p_\infty \tag{12.42}$$

and that the variance of the frequency is proportional to $1/n$. More precisely, we obtain that for n tending to infinity:

$$\lim_{n\to\infty} \frac{\langle K^2(n)\rangle - \langle K(n)\rangle^2}{n} = p_\infty(1 - p_\infty) + 2p_\infty G(1). \tag{12.43}$$

The first term of (12.43) is also present in the case of uncorrelated events: it is the variance of the mean of n Bernoulli events. The effect of time correlations in recurrent events, on the other hand, occurs in the second term.

We now propose to prove that for large n the distribution of $K(n)$ is Gaussian and, when possible, obtain formulas for large deviations. The complete proof is complicated and there are general methods to obtain it more simply, as we shall see in Section 12.4 when we will state and demonstrate the central limit theorem for correlated variables. Here we will only mention the salient points. The simplest method is to verify that the cumulants are those of a Gaussian distribution, as we did, for example, with the reweighting method in the analysis of experimental data in Section 5.4.3.

For example, the starting formula for the fourth moment is

$$\langle K^4(n)\rangle = \sum_{t_1,t_2,t_3,t_4}^{1,n} p(t_1, t_2, t_3, t_4). \tag{12.44}$$

This formula can be simplified by observing that

$$\langle K(n)(K(n) - 1)(K(n) - 2)(K(n) - 3)\rangle = \sum_{\substack{t_1,t_2,t_3,t_4 \\ t_i \neq t_j \\ \forall i,j=1,2,3,4}}^{1,n} p(t_1, t_2, t_3, t_4), \tag{12.45}$$

where the sum is restricted to the case where all times are different. The proof of formula (12.45) can be sketched in three steps.

1. Writing the formulas for the moments of $K(n)$ on the left-hand side up to the fourth ($n = 1, 2, 3, 4$), we obtain that the right-hand side of Eq. (12.45) must be equal to the expectation value of a fourth-order polynomial (with no zero-degree term) in the variable $K(n)$.
2. The equation is clearly true for $n = 1, 2, 3$ since both sides are zero and a simple computation shows that it is also true for $n = 4$.
3. The preceding conditions unambiguously fix the polynomial.

The argument can be easily generalized to the case of general n.

By exploiting the formula (12.45) and the Markovian property of recurrent events, whereby the joint probability of recurrences at times t_1, t_2, t_3, t_4 depends only on the differences between consecutive recurrences

$$p(t_1, t_2, t_3, t_4) = p(t_4 - t_3)p(t_3 - t_2)p(t_2 - t_1)p(t_1),$$

we can write

$$\langle K(n)(K(n)-1)(K(n)-2)(K(n)-3)\rangle = 4! \sum_{t_1<t_2<t_3<t_4}^{1,n} p(t_4-t_3)p(t_3-t_2)p(t_2-t_1)p(t_1).$$

We have, therefore, a four-fold convolution:

$$
\sum_{t_1 < t_2 < t_3 < t_4}^{1,n} p(t_4 - t_3)p(t_3 - t_2)p(t_2 - t_1)p(t_1)
$$

$$
= \sum_{t_4=1}^{n} \sum_{t_3=1}^{t_4} p(t_4 - t_3) \sum_{t_2=1}^{t_3} p(t_3 - t_2) \sum_{t_1=1}^{t_2} p(t_2 - t_1)p(t_1)
$$

$$
= \sum_{t_4=1}^{n} \sum_{t_3=1}^{t_4} p(t_4 - t_3) \sum_{t_2=1}^{t_3} p(t_3 - t_2)p^{*2}(t_2)
$$

$$
= \sum_{t_4=1}^{n} \sum_{t_3=1}^{t_4} p(t_4 - t_3)p^{*3}(t_3) = \sum_{t_4=1}^{n} p^{*4}(t_4) .
$$

The generator of the convolution $\{p^{*4}\}$ is $[\underline{P}(s)]^4$ and the generator of its sum up to the time n, see formula (12.31), is $(1-s)^{-1}[\underline{P}(s)]^4$. We then obtain the following expression for the generating function:

$$
K_4(s) \equiv \sum_{n=1}^{\infty} \langle K(n)(K(n) - 1)(K(n) - 2)(K(n) - 3)\rangle s^n = \frac{4!\underline{P}(s)^4}{1 - s} . \tag{12.46}
$$

We can introduce the following abbreviation of the mean of the quartic polynomial (12.45),

$$
\mathcal{K}_4(n) \equiv \langle K(n)(K(n) - 1)(K(n) - 2)(K(n) - 3)\rangle = \left\langle \prod_{a=0}^{3} (K(n) - a) \right\rangle, \tag{12.47}
$$

and its generalization for any similar polynomial of degree m,

$$
\mathcal{K}_m(n) \equiv \left\langle \prod_{a=0}^{m-1} [K(n) - a] \right\rangle. \tag{12.48}
$$

Analogous to the $m = 4$ case, the generating functions of the sequences $\{\mathcal{K}_m(n)\}$ in n are

$$
K_m(s) \equiv \sum_{n=1}^{\infty} \mathcal{K}_m(n)s^n = \frac{m![P(s)]^m}{1 - s} . \tag{12.49}
$$

We now introduce a real variable ρ and define the mean value of $\rho^{K(n)}$ on the recurrences (12.18),

$$
r(\rho, n) \equiv \langle \rho^{K(n)} \rangle , \tag{12.50}
$$

and its generating function,

$$
R(\rho, s) = \sum_{n=1}^{\infty} r(\rho, n)s^n .
$$

Since, from (12.48), we have

$$\left.\frac{\partial^m r(\rho,n)}{\partial\rho^m}\right|_{\rho=1} = \left\langle\left.\frac{\partial^m \rho^{K(n)}}{\partial\rho^m}\right|_{\rho=1}\right\rangle$$

$$= \left\langle K(n)(K(n)-1)\cdots(K(n)-m+1)\rho^{K(n)-m}\Big|_{\rho=1}\right\rangle = \mathcal{K}_m(n),$$

it follows that

$$\left.\frac{\partial^m R(\rho,s)}{\partial\rho^m}\right|_{\rho=1} = \sum_{n=1}^{\infty} \left.\frac{\partial^m r(\rho,n)}{\partial\rho^m}\right|_{\rho=1} s^n = \sum_{n=1}^{\infty} \mathcal{K}_m(n)s^n = K_m(s). \tag{12.51}$$

We note that the above formula can alternatively also be proved by means of Newton's formula:

$$\langle \rho^{K(n)}\rangle = \sum_m \frac{\mathcal{K}_m(n)}{m!}(\rho-1)^m. \tag{12.52}$$

By expanding the generator $R(\rho,s)$ in a Taylor series around the point $\rho=1$ and using (12.51) and (12.49), we obtain that

$$R(\rho,s) = \sum_{m=0}^{\infty} \frac{(\rho-1)^m}{m!} \left.\frac{\partial^m R(\rho,s)}{\partial\rho^m}\right|_{\rho=1} = \sum_{m=0}^{\infty} \frac{(\rho-1)^m}{m!} K_m(s)$$

$$= \sum_{m=0}^{\infty} \frac{(\rho-1)^m}{m!} \cdot \frac{m![\underline{P}(s)]^m}{1-s}$$

$$= \frac{1}{1-s}\frac{1}{1-(\rho-1)\underline{P}(s)} = \frac{1-F(s)}{1-s}\frac{1}{1-\rho F(s)}, \tag{12.53}$$

where in the last line we have used the relation (12.24). We note that the fixed pole at $s=1$ is canceled by the zero of the numerator ($F(1)=1$).

As we have already seen in the study of the Abelian and Tauberian theorems for the limits of discrete sequences, see Appendix 9.A, a detailed analysis shows that the behavior for large n of the mean $r(\rho,n)$ is connected to the poles in s of its generating function $R(\rho,s)$. In particular, the large n leading term determines the kind of singularity in the complex plane closest to the origin, as shown in Appendix 12.A.

Let us consider for simplicity the position of the pole on the positive real axis closest to the origin. It is easy to see that, if the function $R(\rho,s)$ has a simple pole at $s=s_0(\rho)$ with residue $a(\rho)$ (see, for example, Eq. (8.8)), the behavior for large n is of the form provided by (12.130), in Appendix 12.A:

$$r(\rho,n) \propto a(\rho)s_0(\rho)^{-n}.$$

The details of the computation, generalized to several, even complex, poles, can be found in Appendix 12.A, in particular in the formula (12.133). If $\rho \geq 1$ the value of $s_0(\rho)$ can be found by solving the equation of the zeros of the denominator (12.53):

$$\rho F(s) = 1. \tag{12.54}$$

This equation for ergodic recurrent events ($F(1) = 1$) has one and only one real solution strictly less than 1 ($s \in [0, 1)$), where the series defining the generating function $F(s)$ of the first return probabilities is convergent throughout the $|s| \leq 1$ disk. The previous equation could also have other solutions for complex s, but it is easy to see that, in the case of aperiodic events, the solution of the equation on the positive real axis is closer to the origin than any other possible complex or negative real-axis solutions (the reasoning is the same as that used in Section 9.4.1).

We hence obtain that, for large n,

$$\langle \rho^{K(n)} \rangle \propto s_0(\rho)^{-n} . \tag{12.55}$$

We have thus proved the fundamental estimate that allows us to compute the probability distribution of large deviations and to prove the central limit theorem.

We note that if the function $F(s)$ is not analytic around $s = 1$, with this method we obtain reasonable results only in the region $K(n) \geq p_\infty$, which corresponds to $\rho > 1$. For example, it is found that the large deviations theorem expressed in (4.3) is satisfied with a probability of the event counts that scales as

$$P(K(n)) \propto \exp\{nS(K(n))\} . \tag{12.56}$$

The general study of large deviations in the whole range $0 < K(n) < p_\infty$ is quite complicated and will not be addressed now for the specific case of recurrent events. In Section 12.4.1 we will see how to state and prove the central limit theorem and the large deviations theorem for correlated variables (such as recurrent events) in general.

We note *en passant* that if we set

$$z \equiv \log \rho \quad \text{and} \quad Q(z) \equiv \log \left[\frac{1}{s_0(e^z)} \right] , \tag{12.57}$$

the formula (12.55) can be rewritten as

$$\langle e^{zK(n)} \rangle \propto e^{nQ(z)} = [e^{Q(z)}]^n . \tag{12.58}$$

From the first relation we have that the cumulants of $K(n)$ are generated by the generating function $nQ(z) = \ln\langle e^{zK(n)} \rangle$ as

$$n(dQ/dz)\big|_{z=0} = \langle K(n) \rangle ,$$
$$n(d^2Q/dz^2)\big|_{z=0} = \langle K^2(n) \rangle - \langle K(n) \rangle^2 ,$$
$$\vdots$$

The fact that the mean $r(\rho, n)$, cf. (12.50), can be factorized into the nth power also means that, if there exists a probability density $p(x)$ (which is not evident in general) such that

$$\int dx \, p(x)e^{zx} = e^{Q(z)} , \tag{12.59}$$

then the probability distribution of the variable $K(n)$ of the number of events is the same as that of the sum of n independent (Boolean) variables with probability density $p(x)$.

If the singularity $s_0(\rho)$, and, consequently, the function $Q(z)$ in (12.57), is twice differentiable around $\rho = 1$ (or $z = 0$), and if both p_∞ and $G(1)$ are finite, the distribution of $K(n)$ is, then, a Gaussian distribution of variance

$$\langle K^2(n) \rangle - \langle K(n) \rangle^2 = n \frac{d^2 Q}{dz^2}\Bigg|_{z=0}, \tag{12.60}$$

and in this case the previous formulas allow us to study large deviations as well. On the contrary, if the function $s(\rho)$ is not analytic around $\rho = 1$, and, in particular, if it is not twice differentiable, the probability distribution will not be a Gaussian and it will be similar to, for example, the distributions of the sum of variables with a fat tail probability distribution studied in Section 3.3.

The central limit distribution for recurrent events, as well as for all Markov chains, see Section 12.1, can be constructed in a general way from the central limit theorem for correlated events, which we will state and prove, under certain assumptions, in Section 12.4. Before addressing the theorem, we will devote the next section to the definition and manipulation of correlation functions that generalize the moments (unconnected correlation functions) and cumulants (connected correlation functions) of non-factorizable distributions of many variables.

12.3 Connected Correlation Functions

12.3.1 Definitions

Let us consider a set of variables A_i, with $i = 1, \ldots, N$. We define the generalized moments of these variables with respect to a joint probability distribution $P(A_1, \ldots, A_N)$ as

$$M(n) = \left\langle \prod_{i=1}^{N} A_i^{n_i} \right\rangle, \tag{12.61}$$

where n_i are positive or null integers.

In the literature, these moments are called correlation functions. The quantity

$$R = \sum_{i=1}^{N} n_i$$

is the order of the correlation function. The name "correlation function" is partially abusive, as their value is not an immediate indicator of the correlation existing between these variables. To illustrate this, let us consider the case in which only two variables appear (order $R = 2$). We can define a matrix of correlation functions

$$M_{i,k} = \langle A_i A_k \rangle, \tag{12.62}$$

but the amount of correlation between the variables is shown by the elements of the covariance matrix,

$$C_{i,k} = \langle A_i A_k \rangle - \langle A_i \rangle \langle A_k \rangle . \tag{12.63}$$

The element i, k of the covariance matrix is, indeed, zero if the variables A_i and A_k are not correlated and their joint probability distribution factorizes. The combination of correlation functions of order 2 and order 1 (the expected values) making up the elements of the covariance matrix is a *connected* correlation function.

The connected correlation functions generalize the cumulants. They are constructed in such a way that, as soon as the probability distribution is factorized in any way, i.e., there are variables that are independent from the others, they are equal to zero.

Let us look at the relationships between connected and disconnected correlation functions for lower orders:

$$\begin{aligned}
\langle A_i \rangle_c &= \langle A_i \rangle , \\
\langle A_i A_k \rangle_c &= \langle A_i A_k \rangle - \langle A_i \rangle \langle A_k \rangle .
\end{aligned} \tag{12.64}$$

The connected correlation function of two variables coincides with the covariance matrix and can be rewritten as

$$\langle A_i A_k \rangle_c = \langle (A_i - \langle A_i \rangle)(A_k - \langle A_k \rangle) \rangle . \tag{12.65}$$

Similarly, the connected correlation function of three variables is given by

$$\langle A_i A_j A_k \rangle_c = \langle (A_i - \langle A_i \rangle)(A_j - \langle A_j \rangle)(A_k - \langle A_k \rangle) \rangle . \tag{12.66}$$

We emphasize, however, that beyond third order it is not true in general that the connected correlation function coincides with the disconnected correlation function of the fluctuations of the variable with respect to the mean.

For example, in the case of four variables with zero mean value, the expression is more complicated than for the previous orders (12.65) and (12.66). We have in fact

$$\langle ABCD \rangle = \langle ABCD \rangle_c + \langle AB \rangle_c \langle CD \rangle_c + \langle AC \rangle_c \langle BD \rangle_c + \langle AD \rangle_c \langle BC \rangle_c .$$

In this case, it is easy to see that if the probability distribution is factorized into a product, for example,

$$P(A, B, C, D) = P(A, B)P(C, D), \tag{12.67}$$

the related connected correlation function is equal to zero. Indeed, in this case, we have that $\langle ABCD \rangle = \langle AB \rangle \langle CD \rangle$, and $\langle AC \rangle = \langle A \rangle \langle C \rangle = 0$ and $\langle AD \rangle = \langle A \rangle \langle D \rangle = 0$.

We will see in the following section two methods for the construction of connected correlation functions of arbitrary order: van Kampen's construction method and the method of generating functions.

Before that, we observe here that at second and third order we have

$$\begin{aligned}
\langle A_i A_k \rangle &= \langle A_i A_k \rangle_c + \langle A_i \rangle_c \langle A_k \rangle_c, \\
\langle A_i A_j A_k \rangle &= \langle A_i A_j A_k \rangle_c + \langle A_i A_j \rangle_c \langle A_k \rangle_c + \langle A_i A_k \rangle_c \langle A_j \rangle_c + \langle A_k A_j \rangle_c \langle A_i \rangle_c \\
&\quad + \langle A_i \rangle_c \langle A_j \rangle_c \langle A_k \rangle_c .
\end{aligned} \tag{12.68}$$

Disconnected correlation functions can, therefore, be expressed in terms of a sum of products of connected correlation functions of lower order. This is a general property, with the following structure:

$$\left\langle \prod_{i=1}^{N} A_i^{n_i} \right\rangle = \left\langle \prod_{i=1}^{N} A_i^{n_i} \right\rangle_c + \cdots , \tag{12.69}$$

where the dots are sums of products of connected correlation functions of lower order.

12.3.1.1 Cumulants of a Single Random Variable

From the above definitions applied to the connected moments of the same variable, called *cumulants*, the expressions

$$\langle A^2 \rangle_c = \sigma^2 , \quad \langle A^3 \rangle_c = \sigma^3 S , \quad \langle A^4 \rangle_c = \sigma^4 (K - 3) , \tag{12.70}$$

immediately follow, where S and K are, respectively, the skewness (2.6) and the kurtosis (2.7). We can observe that for a Gaussian variable we find the relations of Section 2.4: the third and fourth cumulants are null ($S = 0$, $K = 3$) and, as we have seen in Appendix 3.C.3, so are, in general, all cumulants of order higher than the second.

12.3.2 Van Kampen's Construction

The van Kampen construction [60] can be used to determine the dependence of connected correlation functions on disconnected ones. To be able to represent a connected correlation function of N variables, which, for the time being, we consider all distinct, we take a sequence of N points (here, for illustrative purposes, we take $N = 6$):

$$\underbrace{\langle \cdot \quad \cdot \quad \cdot \quad \cdot \quad \cdot \rangle}_{N} , \quad N = 6 .$$

We then separate the points into $p + 1$ sub-sequences by inserting p barriers " $\rangle\langle$ ", such as $p = 2$ barriers as below:

$$\underbrace{\langle \cdot \quad \cdot \quad \cdot \rangle\langle \cdot \quad \cdot \rangle\langle \cdot \rangle}_{N} , \quad p = 2 .$$

We now distribute N variables instead of points so that all *distinct* expressions of the products of $p + 1$ correlations are represented:

$$\langle a_1 \, a_2 \, a_3 \rangle \langle a_4 \, a_5 \rangle \langle a_6 \rangle$$
$$\langle a_1 \, a_2 \, a_3 \rangle \langle a_4 \, a_6 \rangle \langle a_5 \rangle$$
$$\langle a_1 \, a_2 \, a_3 \rangle \langle a_5 \, a_6 \rangle \langle a_4 \rangle$$
$$\langle a_1 \, a_2 \, a_4 \rangle \langle a_3 \, a_5 \rangle \langle a_6 \rangle$$
$$\langle a_1 \, a_2 \, a_4 \rangle \langle a_3 \, a_6 \rangle \langle a_5 \rangle$$
$$\langle a_1 \, a_2 \, a_4 \rangle \langle a_5 \, a_6 \rangle \langle a_3 \rangle$$

$$\langle a_1\, a_2\, a_5\rangle\langle a_3\, a_4\rangle\langle a_6\rangle$$
$$\langle a_1\, a_2\, a_5\rangle\langle a_3\, a_6\rangle\langle a_4\rangle$$
$$\langle a_1\, a_2\, a_5\rangle\langle a_4\, a_6\rangle\langle a_3\rangle$$
$$\langle a_1\, a_2\, a_6\rangle\langle a_3\, a_4\rangle\langle a_5\rangle$$
$$\langle a_1\, a_2\, a_6\rangle\langle a_3\, a_5\rangle\langle a_4\rangle$$
$$\langle a_1\, a_2\, a_6\rangle\langle a_4\, a_5\rangle\langle a_3\rangle$$
$$\langle a_1\, a_3\, a_4\rangle\langle a_2\, a_5\rangle\langle a_6\rangle$$
$$\langle a_1\, a_3\, a_4\rangle\langle a_2\, a_6\rangle\langle a_5\rangle$$
$$\langle a_1\, a_3\, a_4\rangle\langle a_5\, a_6\rangle\langle a_2\rangle$$

$$\vdots$$

We then define the sum of all products of $p + 1$ distinct disconnected correlation functions: $C_p(a_1,\ldots,a_N)$. For example,

$$C_1(a_1, a_2, a_3) = \langle a_1 a_2\rangle\langle a_3\rangle + \langle a_1 a_3\rangle\langle a_2\rangle + \langle a_2 a_3\rangle\langle a_1\rangle\,.$$

Once all C_p have been computed, the connected correlation function of N variables as a function of the disconnected correlation functions is expressed by the combination

$$\langle a_1 a_2 \ldots a_N\rangle_c = \sum_{p=0}^{N-1} (-1)^p p!\, C_p(a_1,\ldots,a_N)\,. \tag{12.71}$$

We note that, in the case of correlation functions of variables with multiplicity other than 1, it is sufficient to consider all variables as separate. If the order is $R = \sum_{i=1}^{N} n_i$ we can always write

$$\left\langle \prod_{k=1}^{N} a_k^{n_k} \right\rangle = \left\langle \prod_{\ell=1}^{R} a_{k_\ell} \right\rangle,$$

where the indices $k_\ell = 1,\ldots,N$. For example $\langle a_1 a_2^3 a_3^2\rangle = \langle a_1 a_2 a_2 a_2 a_3 a_3\rangle$.

To better exemplify the procedure, let us take the example of computing $\langle a_1 a_2 a_3 a_4\rangle_c$.

- $p = 0$. The first term is simply the disconnected correlation function,

$$C_0(a_1, a_2, a_3, a_4) = \langle a_1 a_2 a_3 a_4\rangle\,.$$

- $p = 1$. The term with one division has two types of contributions, $\langle\cdot\rangle\langle\cdot\;\cdot\;\cdot\rangle$ and $\langle\cdot\cdot\rangle\langle\cdot\cdot\rangle$, so

$$C_1(a_1, a_2, a_3, a_4) = \langle a_1\rangle\langle a_2 a_3 a_4\rangle + \langle a_2\rangle\langle a_1 a_3 a_4\rangle + \langle a_3\rangle\langle a_1 a_2 a_4\rangle + \langle a_4\rangle\langle a_1 a_2 a_3\rangle$$
$$+ \langle a_1 a_2\rangle\langle a_3 a_4\rangle + \langle a_1 a_3\rangle\langle a_2 a_4\rangle + \langle a_1 a_4\rangle\langle a_2 a_3\rangle\,.$$

- $p = 2$. With two divisions, the four variables only give one possible contribution, of the type

$$C_2(a_1, a_2, a_3, a_4) = \langle a_1\rangle\langle a_2\rangle\langle a_3 a_4\rangle + \langle a_1\rangle\langle a_3\rangle\langle a_2 a_4\rangle + \langle a_1\rangle\langle a_4\rangle\langle a_2 a_3\rangle$$
$$+ \langle a_2\rangle\langle a_3\rangle\langle a_1 a_4\rangle + \langle a_2\rangle\langle a_4\rangle\langle a_1 a_3\rangle + \langle a_3\rangle\langle a_4\rangle\langle a_1 a_2\rangle\,.$$

- $p = 3$. This term $p = N - 1$ is the product of $N = 4$ expected values,

$$C_3(a_1, a_2, a_3, a_4) = \langle a_1 \rangle \langle a_2 \rangle \langle a_3 \rangle \langle a_4 \rangle \,.$$

Using the formula (12.71) we eventually have

$$\begin{aligned}
\langle a_1 a_2 a_3 a_4 \rangle_c = {} & C_0(a_1, a_2, a_3, a_4) - C_1(a_1, a_2, a_3, a_4) \\
& + 2C_2(a_1, a_2, a_3, a_4) - 6C_3(a_1, a_2, a_3, a_4) \,.
\end{aligned} \tag{12.72}$$

We will see in the next section another method to build connected correlation functions of arbitrary order in such a way that they are zero as soon as there are variables that are independent.

12.3.3 Generating Functions of Correlation Functions

A very important relationship links the generating function of the disconnected correlation functions to the generating function of the connected correlation functions. This relation will allow us to easily build correlation functions of any order. First, we must introduce the generating function in N variables, which is a natural extension of the generating function in a single variable that we have previously seen in Eq. (12.2).

We introduce N auxiliary variables z_i, each one conjugated to an A_i variable, and define the generating function of the correlation functions as

$$G(\mathbf{z}) \equiv \left\langle \exp\left(\sum_{i=1}^{N} z_i A_i \right) \right\rangle = \left\langle \prod_{i=1}^{N} \left(\sum_{n_i=0}^{\infty} \frac{z_i^{n_i}}{n_i!} A_i^{n_i} \right) \right\rangle = \prod_{i=1}^{N} \left(\sum_{n_i=0}^{\infty} \frac{z_i^{n_i}}{n_i!} \right) \left\langle \prod_{i=1}^{N} A_i^{n_i} \right\rangle . \tag{12.73}$$

We immediately see from the last relation that we can generate the correlation functions from the derivatives of the function $G(\mathbf{z})$ computed at the origin of the vector $\mathbf{z}$:

$$\left\langle \prod_{i=1}^{N} A_i^{n_i} \right\rangle = \prod_{i=1}^{N} \frac{\partial^{n_i}}{\partial z_i^{n_i}} G(\mathbf{z}) \Bigg|_{\mathbf{z}=0} . \tag{12.74}$$

There is, hence, a bi-univocal correspondence between the sets of all the correlation functions of N random variables A_i and the analytic functions of N variables z_i around the origin.

In the same way we define the generating function of the connected correlation functions as

$$C(\mathbf{z}) = \prod_{i=1}^{N} \left(\sum_{n_i=0}^{\infty} \frac{z_i^{n_i}}{n_i!} \right) \left\langle \prod_{i=1}^{N} A_i^{n_i} \right\rangle_c , \tag{12.75}$$

from which

$$\left\langle \prod_{i=1}^{N} A_i^{n_i} \right\rangle_c = \prod_{i=1}^{N} \frac{\partial^{n_i}}{\partial z_i^{n_i}} C(\mathbf{z}) \Bigg|_{\mathbf{z}=0} , \tag{12.76}$$

where conventionally we place $\langle 1 \rangle_c = 0$ or, equivalently, we exclude from the sum the term with all n_i equal to zero. Alternatively, one can rewrite the expression of $C(\mathbf{z})$ as a resummed expansion in the order R of the connected correlations. By inserting a Kronecker delta $\delta\big(R, \sum_{i=1}^{N} n_i\big)$ on the order and making some cosmetic manipulations, one can rewrite the generator of the connected correlations as a series ordered by the order of the correlation functions:

$$
\begin{aligned}
C(\mathbf{z}) &= \sum_{R=1}^{\infty} \sum_{\{n_1, n_2, \ldots, n_N\}}^{0, \infty} \delta\left(R, \sum_{i=1}^{N} n_i\right) \left[\prod_{i=1}^{N} z_i^{n_i}\right] \left[\prod_{i=1}^{N} \frac{1}{n_i!}\right] \left\langle \prod_{i=1}^{N} A_i^{n_i} \right\rangle_c \\
&= \sum_{R=1}^{\infty} \frac{1}{R!} \left[\prod_{\ell=1}^{R} \sum_{i_\ell=1}^{N}\right] \left[\prod_{\ell=1}^{R} z_{i_\ell}\right] \left\langle \prod_{\ell=1}^{R} A_{i_\ell} \right\rangle_c .
\end{aligned}
\tag{12.77}
$$

Here the square brackets around the products are solely intended as a guide for the eyes but are not necessary. In the case of $N = 1$, the two equivalent expressions (12.75) and (12.77) are also identical at first sight, with $R = n_1 = n$:

$$
C(\mathbf{z}) = \sum_{n=1}^{\infty} \frac{z^n}{n!} \langle A^n \rangle_c .
$$

A theorem that we will see shortly states that the two generating functions are related by the important functional relationship

$$
C(\mathbf{z}) = \ln\left[G(\mathbf{z})\right] .
\tag{12.78}
$$

In other words, this theorem states that, if we take the formula (12.78) as the definition of the generating functional of the related correlation functions, these functions satisfy the property of being null in the case of independent variables. Indeed, it is immediate to verify that if the variables can be divided into two groups of uncorrelated variables, the connected correlation functions of these variables are null. More precisely, let us consider the case in which the variables can be divided into two groups for $i \leq K$ (which we will denote with $A^{(1)}$) and $i > K$ (which we will denote with $A^{(2)}$) such that the probability $P(\{A\})$ factorizes into the product $P_1(\{A^{(1)}\}) \, P_2(\{A^{(2)}\})$. The factorization of probabilities implies that

$$
\begin{aligned}
G(\mathbf{z}) &= G_1(\mathbf{z}^{(1)}) G_2(\mathbf{z}^{(2)}) , \\
C(\mathbf{z}) &= C_1(\mathbf{z}^{(1)}) + C_2(\mathbf{z}^{(2)}) ,
\end{aligned}
\tag{12.79}
$$

where the functions G_1 and G_2 (like the functions C_1 and C_2) depend only on, respectively, the first ($\mathbf{z}^{(1)} \equiv \{z_1, z_2, \ldots, z_K\}$) and the second ($\mathbf{z}^{(2)} = \{z_{K+1}, z_{K+2}, \ldots, z_N\}$) groups of auxiliary variables.

The generating function of the connected correlation functions does not contain products of the $\mathbf{z}$ variables belonging to the two different groups and, therefore, the correlation functions of variables of the two groups are equal to zero. It remains to prove that the relation (12.78) between generating functions of connected and disconnected correlation functions that we have chosen is unique, but it is better not to deal with this too technical argument. It is conceptually easy, but a little laborious to verify that the formulas on correlation functions of order less than or equal to 4, which we saw in the previous section, are a consequence of the formula (12.78).

12.3.4 Generating Function of Multivariate Gaussian Variables Correlations

Another important property is that connected correlation functions of order greater than the second are zero for a multivariate Gaussian probability distribution (for simplicity with mean zero):

$$P(\{A\}) = \frac{1}{\sqrt{(2\pi)^N \det C}} \exp\left(-\frac{1}{2}\sum_{i,k}^{1,N} A_i (C^{-1})_{i,k} A_k\right). \tag{12.80}$$

Indeed, using the formula (2.59), we obtain that the generating function of N Gaussian-distributed quantities is

$$G(\mathbf{z}) = \left\langle \exp\left(\sum_{k=1}^{N} z_k A_k\right)\right\rangle = \exp\left(\frac{1}{2}\sum_{i,k}^{1,N} z_i C_{i,k} z_i\right), \tag{12.81}$$

where $C_{i,k}$ is, as usual, the covariance matrix. It follows from (12.78) that

$$C(\mathbf{z}) = \frac{1}{2}\sum_{i,k}^{1,N} z_i C_{i,k} z_k, \tag{12.82}$$

from which $\langle A_i A_k \rangle_c = C_{i,k}$ and the connected correlation functions of order higher than second are all zero.

The opposite is also true: a quadratic multivariate generating function is the generating function of a multivariate Gaussian distribution. Let us take the biquadratic form

$$C(\mathbf{z}) = \sum_{j=1}^{N} \mu_j z_j + \frac{1}{2}\sum_{i,k}^{1,N} z_i C_{i,k} z_k, \tag{12.83}$$

from which

$$G(\mathbf{z}) = \exp\left(\sum_{j=1}^{N} \mu_j z_j + \frac{1}{2}\sum_{i,k}^{1,N} z_i C_{i,k} z_k\right). \tag{12.84}$$

To compute $P(\{A\})$ we only need to carry out the anti-generator, which is none other than the inverse Laplace transform, or the Fourier transform if we perform the rotation in complex space $z_i = \iota q_i$, $\forall i$, and set $\tilde{G}(\mathbf{q}) = G(-\iota \mathbf{q})$:

$$P(\{A\}) = \frac{1}{(2\pi)^N} \int \prod_{i=1}^{N} dq_i \exp\left(-\iota \sum_{k=1}^{N} q_k A_k\right) \tilde{G}(\mathbf{q})$$

$$= \frac{1}{(2\pi)^N} \int \prod_{i=1}^{N} dq_i \exp\left(\iota \sum_{k=1}^{N} (\mu_k - A_k) q_k - \frac{1}{2}\sum_{i,k}^{1,N} q_i C_{i,k} q_k\right)$$

$$= \frac{1}{(2\pi)^{N/2}} \frac{1}{\det \mathbb{C}} \exp\left(-\sum_{i,k}^{1,N} (\mu_i - A_i)(C^{-1})_{i,k}(\mu_k - A_k)\right), \tag{12.85}$$

which is a multivariate Gaussian of N variables of mean $\vec{\mu}$ and covariance $\mathbb{C}$.

12.3.5 Cumulants Generating Function

In the case of a single variable, the generating function of the connected correlation functions is given by

$$C(z) = \ln\left(\int da\, P(a)\exp(za)\right). \tag{12.86}$$

This is a generator that we have already encountered in Chapter 4, Section 4.2, and before that in Chapter 3, in Appendix 3.C.3, in the latter case in the guise of a Fourier transform instead of Laplace. We have seen that it plays an extremely important role in proving the central limit theorem and in the study of large deviations for independent variables. For small z, expanding to the fourth order we have

$$C(z) = z\langle x\rangle + z^2 \frac{1}{2}\sigma^2 + z^3 \frac{1}{6}S\sigma^3 + z^4 \frac{K-3}{24}\sigma^4 + O(z^5). \tag{12.87}$$

At this point, we have in our hands all the tools to undertake, once again, the proofs of the central limit theorem and of the large deviations theorem, generalized to the case of correlated variables.

12.4 Central Limit and Large Deviations for Correlated Events

We can prove a theorem that generalizes the central limit theorem and allows us to obtain estimates of large deviations for sets of correlated variables.

Suppose we consider an infinite sequence of variables a_i with $i \in (-\infty, \infty)$ such that the correlation functions depend only on the difference of the indices.[3] For example, $\langle a_i\rangle$ does not depend on i and

$$\langle a_i a_k\rangle = f(i-k), \quad \langle a_i a_k\rangle_c = c(i-k). \tag{12.88}$$

In other words, we assume that the probability distribution is translation-invariant, or stationary. This means that, if we define the sequence $b_i = a_{i+m}$, the probability of the translated sequence b is the same as that of the sequence a for any value of m.

We want to estimate the probability distribution of

$$A(N) \equiv \sum_{i=1}^{N} a_i. \tag{12.89}$$

---

[3] This condition is not necessary and can be replaced by a much weaker one. For the sake of brevity, we will not discuss this issue here.

We assume that the system satisfies the so-called *clustering decomposition* property, which states that the connected correlation functions go to zero when the points on which they depend can be divided into two blocks whose distance tends to infinity. Some examples are

$$\lim_{k\to\infty}\langle a_i a_k\rangle_c = \lim_{k\to\infty}\langle a_i a_j a_k\rangle_c = \lim_{k\to\infty}\langle a_i a_j a_l a_k\rangle_c = \lim_{k\to\infty}\langle a_i a_j a_k a_{k+l}\rangle_c = 0.$$

The clustering decomposition can also be stated by saying that the disconnected correlation functions tend to factorize into the product of the correlation functions in each separated cluster. In other words, for large k we require that

$$\langle a_i a_k\rangle \simeq \langle a_i\rangle\langle a_k\rangle,$$
$$\langle a_i a_j a_k a_{k+l}\rangle \simeq \langle a_i a_j\rangle\langle a_k a_{k+l}\rangle. \tag{12.90}$$

The clustering decomposition property formalizes the statement that variables depending on distant points are uncorrelated.

To obtain the desired result for large N, i.e., a Gaussian distribution of $A(N)$, it is necessary to assume that the clustering occurs fast enough, e.g., exponentially, such that the connected correlation functions are soon very small. We shall see below, in the course of the proof of the central limit theorem, how quickly they must decay.

Let us consider the generating function of the moments of the quantity $A(N)$, defined as

$$\langle\exp(zA(N))\rangle = \left\langle\exp\left(z\sum_{i=1}^N a_i\right)\right\rangle \equiv G_N(z), \tag{12.91}$$

where by $G_N(z)$ we denote a function like $G(\mathbf{z})$, (12.73), but computed at $z_i = z$ for $1 \le i \le N$.

The theorem of Section 12.3.3 and the formula (12.77) allow us to write

$$G_N(z) = \exp\{C_N(z)\} = \exp\left\{\sum_{R=1}^\infty \frac{z^R}{R!}\left[\prod_{\ell=1}^R \sum_{i_\ell=1}^N\right]\left\langle\prod_{\ell=1}^R a_{i_\ell}\right\rangle_c\right\}. \tag{12.92}$$

The coefficient of order z^R is then given by the sum of connected correlation functions of order R. The previous formula was derived directly using the definition of a connected correlation function. If we stop at order z^2 we obtain

$$C_N(z) = zN\langle a\rangle + \frac{z^2}{2}\sum_{i,j}^{1,N}\langle a_i a_j\rangle_c + O(z^3) = zN\langle a\rangle + \frac{z^2}{2}\sum_{i=1}^N\sum_{j=1}^N c(i-j) + O(z^3). \tag{12.93}$$

When the connected correlation function $c(d)$ goes to zero fast enough for $d \to \infty$, for large N, neglecting terms of $O(1)$ with respect to N, we get

$$C_N(z) = zN\langle a\rangle + \frac{z^2}{2}\left[N\sum_{j=-\infty}^\infty c(j) + O(1)\right] = N\left[z\langle a\rangle + \frac{z^2}{2}\sum_{j=-\infty}^\infty\langle a_0 a_j\rangle_c\right] + O(1).$$
$$\tag{12.94}$$

To prove this, we rewrite (12.93) as

$$\sum_{i,j}^{1,N} c(i-j) = \sum_{i=1}^{N} \left[\sum_{j=-\infty}^{\infty} c(i-j) - \sum_{j=-\infty}^{0} c(i-j) - \sum_{j=N+1}^{\infty} c(i-j) \right]. \qquad (12.95)$$

The first sum brings us the leading extensive contribution (i.e., proportional to N) to (12.94):

$$\sum_{i=1}^{N} \sum_{j=-\infty}^{\infty} c(i-j) = N \sum_{j=-\infty}^{\infty} c(j) = N \sum_{j=-\infty}^{\infty} \langle a_0 a_j \rangle_c. \qquad (12.96)$$

Alternatively, by defining $d = |i - j|$, we can also write

$$\sum_{j=-\infty}^{\infty} c(i-j) = c(0) + 2 \sum_{d=1}^{\infty} c(d) = \langle a_0^2 \rangle + 2 \sum_{d=1}^{\infty} \langle a_0 a_d \rangle_c, \qquad (12.97)$$

which is converging as long as $c(d)$ decays faster than an inverse power $d^{-\alpha}$ with $\alpha > 1$.

The last two summations in formula (12.95) are, however, quantities of order 1 when $N \to \infty$, if the correlation function $c(d)$ decays fast enough when $d \to \infty$. Let us see how fast, by examining, for example, the second term. Defining $d = i - j$, it can be rewritten as

$$\sum_{i=1}^{N} \sum_{j=-\infty}^{0} c(i-j) = \sum_{i=1}^{N} \sum_{d=i}^{\infty} c(d) = \sum_{i=1}^{\infty} \sum_{d=i}^{\infty} c(d) - \sum_{i=N+1}^{\infty} \sum_{d=i}^{\infty} c(d)$$

$$= \sum_{d=1}^{\infty} \sum_{i=1}^{d} c(d) - \sum_{d=N+1}^{\infty} \sum_{i=N+1}^{d} c(d)$$

$$= \sum_{d=1}^{\infty} d\, c(d) - \sum_{d=N+1}^{\infty} (d - N) c(d). \qquad (12.98)$$

If $c(d) \sim d^{-\alpha}$, with $\alpha > 2$, the first term converges for every N while the second term tends to zero for large values of N. The same argument can be made for the last term. Finally, we note that, even if $\alpha = 2$ in (12.98), the first sum on the right-hand side becomes

$$\sum_{d=1}^{N} d\, c(d) \underset{N \gg 1}{\sim} \ln N,$$

being the maximum distance between N variables of order N. So this term will be divergent for large N, but much smaller than the order N of (12.96), and its contribution to the cumulant computation of $A(N)$ is negligible in the central limit.

We can follow the same reasoning for the higher orders in (12.93) and eventually obtain:

$$\log \langle \exp\{z A(N)\} \rangle = C_N(z) = N h(z) + O(1), \qquad (12.99)$$

where the function $h(z)$ can be formally written as

$$h(z) = \sum_{R=1}^{\infty} \frac{z^R}{R!} h_R \,,$$

with

$$h_1 = \langle a \rangle \,, \tag{12.100}$$

$$h_2 = \sum_{j=-\infty}^{\infty} \langle a_0 a_j \rangle_c \,, \tag{12.101}$$

$$h_3 = \sum_{j_1=-\infty}^{\infty} \sum_{j_2=-\infty}^{\infty} \langle a_0 a_{j_1} a_{j_2} \rangle_c \,, \tag{12.102}$$

$$h_R = \left[\prod_{\ell=1}^{R-1} \sum_{j_\ell=-\infty}^{\infty} \right] \left\langle a_0 \prod_{l=1}^{R-1} a_{j_\ell} \right\rangle_c \,. \tag{12.103}$$

The sum over the points j_ℓ is convergent due to the sufficiently fast decrease of the connected correlation functions. In the definition of the function $h(z)$, the radius of convergence of the series is not evident. In most cases, the series has a finite radius of convergence, but it is possible to construct instances where the radius of convergence is zero.

We can use the standard arguments of Section 3.2 to show that the probability distribution of the quantity

$$w_N = \frac{A(N) - \langle A(N) \rangle}{\sqrt{N}}$$

becomes a Gaussian for large N with variance

$$\langle w_N^2 \rangle_c = h_2 = \sum_{j=-\infty}^{\infty} \langle a_0 a_j \rangle_c \,,$$

while cumulants of order R greater than 2,

$$\langle w_N^R \rangle_c \simeq \frac{h_R}{N^{R/2-1}} \,, \quad R > 2, \tag{12.104}$$

tend to zero for N large.

For the limit distribution of the empirical average in an experiment,

$$\bar{a} = \frac{A(N)}{N} \,,$$

we will have a Gaussian of average $\langle A \rangle$ and variance

$$\sigma_{\bar{a}}^2 = \langle (\bar{a})^2 \rangle_c = \left\langle \left(\frac{A(N)}{N} \right)^2 \right\rangle_c = \frac{1}{N} \sum_{j=-\infty}^{\infty} \langle a_0 a_j \rangle_c = \frac{\sigma_a^2}{N} + \frac{2}{N} \sum_{d=1}^{\infty} \langle a_0 a_d \rangle_c \,. \tag{12.105}$$

The last identity clearly expresses what was studied in detail in Chapter 6, namely, that neglecting the correlation of the data leads to an underestimation of the experimental errors.

12.4.1 Large Deviations for Correlated Variables

Finally, thanks to the result (12.99), we can study large deviations as well, generalizing the results of Chapter 4 obtained in the case of independent variables.

Indeed, the cumulants generator $C_N(z)$ is nothing other than the generalized form for correlated variables of the cumulant generating function $R(h)$ we defined in (4.14). The expression

$$C_N(z) = \ln \int \prod_{i=i}^{N} da_i \, p(a_1, \dots, a_N) \exp\left(z \sum_{i=1}^{N} a_i \right) = Nh(z) \qquad (12.106)$$

is the logarithm of the N-dimensional bilateral Laplace transform of the joint probability distribution of the N correlated variables a_i. For (12.99) it is equal[4] to $Nh(z)$. The probability distribution of the extremely large values, $A(N) = \lambda N$, is obtained by considering the detailed properties of the whole function $h(z)$. The large deviations distribution

$$d_N(\lambda) \sim e^{NS(\lambda)}$$

is hence obtained from the Legendre transform of $h(z)$,

$$S(\lambda) = h_N(\bar{z}(\lambda)) - \bar{z}(\lambda)\lambda \,, \qquad (12.107)$$

where the function $\bar{z}(\lambda)$ is the solution to the saddle point equation

$$\frac{dh(z)}{dz} = \lambda \,. \qquad (12.108)$$

We will return to the theory of large deviations for correlated events when we study its application to statistical mechanics, in Section 14.5.

12.4.2 Central Limit for Markov Chains, Revisited

We can now apply the theorem to the case of an irreducible aperiodic Markov chain with a finite number of states, which we explicitly analyzed in Section 12.1. In this case, we will find that the quantities

$$\left\langle \prod_{j=1}^{N} a_{i_j} \prod_{k=1}^{M} a_{l_k+t} \right\rangle - \left\langle \prod_{j=1}^{N} a_{i_j} \right\rangle \left\langle \prod_{k=1}^{M} a_{l_k+t} \right\rangle$$

tend to zero as $t \to \infty$ exponentially fast, i.e., proportionally to $\lambda(2)^t = e^{-t|\ln \lambda(2)|}$, where $\lambda(2)$ is the second eigenvalue (with modulus $|\lambda(2)| < 1$) of the stochastic matrix.

---

[4] It should be noted that here h is the function, z is the conjugate variable, and A is the sum of the random variables, whereas in Chapter 4 R was the function, h was the conjugate Laplace variable, and z was the sum of the random variables.

The correlation functions therefore go to zero exponentially in time as $e^{-t/\tau}$, with characteristic correlation time $\tau = -1/\ln|\lambda(2)|$.

In particular, we can easily compute h_2 using the definition in Eq. (12.101). We always remain under the assumption of stationary limit distribution and translation-invariant correlation functions. The probability of having event i is, therefore, independent of time and is $u(i) = s_i(1)$, the ith element of the left eigenvector $\langle 1|$ of spectral radius $\lambda(1) = 1$ of the stochastic matrix P of the transition probabilities along the chain. The joint probability of having event i at time n_0 and event k at time $n + n_0$ depends only on the difference of the time steps and is given by

$$P(i(n_0), k(n_0 + n)) = p(i(n_0))P(k(n + n_0)\,|\,i(n_0)) = u(i)P(k(n)|i(0)) = s_i(1)P^{(n)}_{i,k}\,.$$

$$(12.109)$$

The average value of $A(n) = \sum_{t=1}^{n} a(k_t)$ is given by

$$\langle A(n)\rangle = nh_1 = n\langle 1|\hat{a}|1\rangle = n\langle a\rangle\,.$$

$$(12.110)$$

The variance $\langle A^2(n)\rangle_c$ is given by nh_2, where

$$h_2 = \sum_{j=-\infty}^{\infty} \langle a(k_0)a(k_j)\rangle_c = \langle a(k_0)^2\rangle_c + 2\sum_{t=1}^{\infty}\langle a(k_0)a(k_t)\rangle_c$$

$$= \langle a(k_0)^2\rangle - \langle a(k_0)\rangle^2 + 2\sum_{t=1}^{\infty}[\langle a(k_0)a(k_t)\rangle - \langle a(k_0)\rangle\langle a(k_t)\rangle]\,.$$

$$(12.111)$$

Using Dirac notation as in Eq. (12.12), the disconnected correlation functions of the previous expression can be written as

$$\langle a(k_0)\rangle = \sum_{k_0} u_{k_0} a(k_0) = \langle 1|\hat{a}|1\rangle\,,$$

$$(12.112)$$

$$\langle a(k_0)^2\rangle = \sum_{k_0} u_{k_0} a(k_0)^2 = \langle 1|\widehat{a^2}|1\rangle\,,$$

$$(12.113)$$

at $t = 0$, and for $t > 0$ as

$$\langle a(k_0)a(k_t)\rangle = \sum_{k_0,k_t} u_{k_0} P^{(t)}_{k_0,k_t} a(k_0)a(k_t) = \sum_{\ell=1}^{n} \lambda^t(\ell)\langle 1|\hat{a}|\ell\rangle\langle \ell|\hat{a}|1\rangle\,.$$

$$(12.114)$$

In the last step, we have used the spectral decomposition (10.69) of the non-Hermitian stochastic matrix P. With this expression, recalling that for an irreducible ergodic Markov chain $\lambda(1) = 1$ is the only eigenvalue of modulus 1 and taking for simplicity a reversible chain ($\lambda(\ell) \in \mathbb{R}$) with non-degenerate eigenvalues, ordered like $\lambda(1) = 1 > \lambda(2) > \lambda(3) > \cdots > \lambda(n)$ (see Section 11.2), Eq. (12.111) for the variance becomes

$$h_2 = \langle 1|\widehat{a^2}|1\rangle - \langle 1|\hat{a}|1\rangle^2 + 2\sum_{t=1}^{\infty}\left[\sum_{\ell=1}^{n} \lambda^t(\ell)\langle 1|\hat{a}|\ell\rangle\langle \ell|\hat{a}|1\rangle - \langle 1|\hat{a}|1\rangle^2\right]$$

$$= \langle 1|\widehat{a^2}|1\rangle - \langle 1|\hat{a}|1\rangle^2 + 2\sum_{\ell=2}^{n} \frac{\lambda(\ell)}{1 - \lambda(\ell)}\langle 1|\hat{a}|\ell\rangle\langle \ell|\hat{a}|1\rangle\,,$$

$$(12.115)$$

which coincides with the expression (12.15) that was obtained by disturbing the formal parallel with perturbation theory in quantum mechanics. The first term comes from the contribution at equal times and the second term, containing the correlations contribution, from that at different times.

12.5 Strongly Correlated Events and Phase Transitions

We have seen in formulas (12.97) and (12.98) that, in order to have a multivariate Gaussian central limit, the cluster decomposition of the connected correlation functions $c(d)$ must occur faster than d^{-2} for large d. We now want to study what happens when the correlations go to zero so slowly that the series $\sum_{d=0}^{\infty} c(d)$ is divergent. In general, we want to understand the limits of validity of the central limit theorem.

In the case of recurrent events, if the second moment of the first return probability distribution

$$\sum_{n=1}^{\infty} n^2 f(n) \tag{12.116}$$

is divergent, the variance of the first return time with respect to the mean recurrence time $\mu = \sum_n n f(n)$ diverges and the central limit theorem does not hold any more. This happens when, for large n, $f(n) \propto n^{-\gamma}$ with $2 < \gamma \leq 3$. The second derivative of the generating function $F(s)$ has a singular term at $s = 1$ of the type $(1 - s)^{-(3-\gamma)}$ and generalized central limit theorems are valid, with asymptotic distributions similar to those we saw in Chapter 3, Section 3.3.

The validity or the invalidity of the central limit theorem is crucial in statistical mechanics for critical phenomena. Let us consider for simplicity the Ising model, introduced in Section 11.3.4. The spin variables $\sigma(i)$ can take only values ± 1 and the stationary probability distribution for configurations $\sigma(i)$, with $i \in [1, L]$ on a one-dimensional lattice, is

$$P[\sigma] = \frac{\exp(-\beta H[\sigma])}{\sum_{\{\sigma\}} \exp(-\beta H[\sigma])},$$

with

$$H[\sigma] = -\sum_{i \neq k}^{1,L} J(|i - k|)\sigma(i)\sigma(k), \tag{12.117}$$

where we have used square brackets to emphasize that the functions P and H depend on the set of all variables $\{\sigma_1, \ldots, \sigma_{L-1}, \sigma_L\}$, and where $\beta = 1/(k_B T)$ is the inverse temperature (k_B is the Boltzmann constant). For simplicity we assume that both J and β are positive.

Suppose the function J is such that, in the limit $L \to \infty$, the probability distribution of $\sigma(i)$ has a limit. The necessary and sufficient conditions that the function J must satisfy for this to happen are well known in the case of positive J, and are essentially that

$$\sum_{d=1}^{\infty} J(d) < \infty. \tag{12.118}$$

The system we have described corresponds, in statistical mechanics, to the canonical ensemble of an Ising spin system in equilibrium at the thermal bath temperature T. In this context, the limit $N \to \infty$ is often referred to as the *infinite volume* limit or *thermodynamic limit*.

Let us introduce the total magnetization variable,

$$M(L) \equiv \sum_{i=1}^{L} \sigma(i).$$

This is a sum of L random variables of equiprobable value ± 1. The mean is, therefore, zero. Studying the central limit in this case means knowing what is the asymptotic behavior in the limit of large L of the variable

$$\mathcal{N}(L) \equiv [\langle M^2(L) \rangle]^{1/2}, \tag{12.119}$$

i.e., the mean square deviation of the total magnetization $M(L)$. Furthermore, we want to know the probability distribution of the rescaled variable:

$$m(L) \equiv \frac{M(L)}{\mathcal{N}(L)}.$$

Note that, for reasons of symmetry, $\langle m \rangle = \langle M \rangle = 0$ and that, by construction, $\langle m^2 \rangle = 1$. Hence, $\forall L$ the variance of $m(L)$ is always 1. For physicists, we underline that $m(L)$ does not coincide with the mean magnetization per spin $M(L)/L$.

If the behavior of $J(d)$ is such that it satisfies the hypothesis that the clustering decomposition of $\langle \sigma(0)\sigma(d) \rangle_c$ occurs faster than d^{-2}, the central limit distribution is Gaussian and we have that

$$\mathcal{N}(L) \underset{L \to \infty}{\sim} A L^{1/2}. \tag{12.120}$$

For physicists, it is interesting to note that the magnetic susceptibility χ in one dimension, for an L spin system, is given by the relation

$$\chi = \beta L \left\langle \left(\frac{M(L)}{L} \right)^2 \right\rangle \underset{L \to \infty}{\sim} \beta A^2.$$

A detailed analysis of the behavior of the system in the case where $J(i-k)$ still satisfies (12.118) for large $|i-k|$ but the clustering decomposition is too slow to guarantee a Gaussian central limit, goes beyond the scope of this book. In order to show the complexity of such an analysis, however, we report, without even trying to prove anything, a series of results fundamental in the theory of critical phenomena.

12.5.1 Gaussian Central Limit and the Absence of a Phase Transition

This is the case when the magnetic exchange interaction $J(k)$ goes to zero faster than k^{-2}, as the distance k between spins goes to infinity. In this case, the standard central limit theorem holds for all values of β, yielding a normal Gaussian distribution of $m(L)$ (unit variance and zero average). The mean square deviation $\mathcal{N}(L)$ of $M(L)$ is always proportional to $L^{1/2}$ at any temperature.

12.5.2 Generalized Central Limit and Critical Phenomena

In this case, when k grows to infinity, the function $J(k)$ goes to zero as $k^{-\omega}$ with $1 < \omega < 2$. A phase transition occurs for some value β_c. Three distinct regimes can be observed depending on the temperature.

- For $\beta < \beta_c$ the system is in a high-temperature phase $T > T_c = 1/(k_B\beta_c)$, the distribution of $m(L)$ is Gaussian in the thermodynamic limit

$$P(m) \to \frac{e^{-m^2/2}}{\sqrt{2\pi}},$$

and $\mathcal{N}^2/L \to 1$ as $L \to \infty$.

- For $\beta > \beta_c$ we are in a low-temperature phase, the distribution of the variable m is a bimodal or Rademacher distribution

$$P(m) = \tfrac{1}{2}\delta(m - 1) + \tfrac{1}{2}\delta(m + 1) = \delta(m^2 - 1), \tag{12.121}$$

and $\mathcal{N}(L)$ is now proportional to L, with a proportionality factor

$$\lim_{L\to\infty} \frac{\mathcal{N}(L)}{L} = |\bar{m}|.$$

In this case, a detailed computation shows that the connected correlation function does not go to zero for large distances; rather

$$C(d) = \langle \sigma(0)\sigma(d)\rangle_c \xrightarrow[d\to\infty]{} \bar{m}^2. \tag{12.122}$$

The *cluster decomposition* is not valid any more. Physically, this situation corresponds to the presence of a spontaneous magnetization of the spins, the value of which is precisely the thermodynamic limit of the average magnetization per spin, $\bar{m} = \lim_{L\to\infty} M(L)/L$.

- The behavior at $\beta = \beta_c$, i.e., at the critical point, is very interesting. In this case the cluster decomposition is still valid, but too slow for the Gaussian central limit to hold. Indeed,

$$\mathcal{N}(L) \sim L^{\alpha} \quad \text{with} \quad \alpha = (3 - \omega)/2 \in (1/2, 1), \quad \omega \in (1, 2).$$

Regarding the asymptotic probability distribution, we must make a further distinction at the critical point. Let us define $\omega_c = 3/2$.

- For $\omega_c < \omega < 2$, the limit distribution $P(m)$ is Gaussian.
- For $1 < \omega < \omega_c$, the central limit distribution $P(m)$ is not Gaussian. However, it has nothing to do with the distributions found in the case of the generalized central limit theorem in Section 3.3. Here, $P(m)$ in the thermodynamic limit has, rather, a behavior of the type $P(m) \sim \exp(-am^2 - bm^4)$, with appropriate a and b.

The distribution $P(m)$ at the critical point is a continuous function of the parameter ω. Furthermore, in the limit $\omega \to \omega_c^-$ we have that

$$P(m) \to \frac{e^{-m^2/2}}{\sqrt{2\pi}}, \tag{12.123}$$

whereas in the limit $\omega \to 1$ we have

$$P(m) \to \delta(m^2 - 1). \tag{12.124}$$

In other words, when ω approaches ω_c, the behavior at β_c becomes similar to that of the high-temperature phase, whereas when ω approaches 1, the critical behavior matches the low-temperature phase behavior.

As can be observed from this concise discussion, the problem of the validity of the central limit theorem for strongly correlated events is deeply connected to the physics of phase transitions.

12.5.3 Asymptotic Probability in the Correlated Case

It is interesting to introduce the variables[5]

$$x^{(n)}(i) = \frac{\sum_{j=1}^{L} \sigma(Li + j)}{\mathcal{N}(L)}, \tag{12.125}$$

where $L = 2^n$ (also $\langle x^{(n)}(i) \rangle = 0$ and $\langle (x^{(n)}(i))^2 \rangle = 1$). The variables x will satisfy the recursion formula (3.63) that we introduced in the study of stable distributions and which we rewrite here for convenience:

$$x^{(n+1)}(i) = 2^{-\rho}[x^{(n)}(2i) + x^{(n)}(2i + 1)]. \tag{12.126}$$

The previous equation allows us to compute the probability distribution $P_{n+1}[x]$ of $x^{(n+1)}$ in terms of the probability distribution $P_n[x]$ of $x^{(n)}$, a relationship that we can write in symbolic form as

$$P_{n+1} = \mathcal{R}[P_n]. \tag{12.127}$$

In the case where the variables are uncorrelated, we have seen in Section 3.4 that the recursion equation takes a simple form (3.64). In the general correlated case, however, the $P_n[x]$ function depends on an infinite number of variables, and it is not possible

---

[5] Imagine that the spins are on a lattice much longer than L, or that there are periodic conditions at the boundaries for which $\sigma(Li + \ell) = \sigma(\ell), \forall i$.

to put the previous transformation into an explicit form from which the asymptotic behavior is easy to extract.

We can expect that in the limit in which the iteration step n tends to infinity, the probability distribution of the variables $x^{(n)}$ tends to a well-defined limit $P_\infty[x]$. This will necessarily satisfy conditions that generalize to the case of correlated events those conditions that we have seen in the case of uncorrelated events in Section 3.4. Indeed, we expect that the asymptotic probability distribution satisfies an equation that generalizes (3.65):

$$P_\infty = \mathcal{R}[P_\infty]. \tag{12.128}$$

Unfortunately, despite a great deal of effort over the past thirty years, we are a long way from being able to classify all the solutions of the fixed point equation (12.128), although by various methods we are able to check some of these solutions.

Mathematical Appendices

12.A Asymptotic Behavior of Series and Complex Singularities

As we have repeatedly seen in Chapter 8, the radius of convergence r of the generating function of a series $\{b(n)\}$,

$$B(s) \equiv \sum_{n=0}^{\infty} b(n)s^n , \tag{12.129}$$

is given by the modulus $r = |s_0|$ of the closest complex singularity to the origin. At first sight, this means that for large n the various terms of the series must go to zero as

$$b(n) \sim \frac{1}{r^n} . \tag{12.130}$$

We can be more precise in establishing the relationship between the singularities closest to the origin of the generating function $B(s)$ and the asymptotic behavior of the series $b(n)$. To this end, we report in a simplified form the statement of Appell's theorem.

12.A.1 Appell's Theorem for Series Convergence

To avoid a too detailed discussion, let us assume that there exists a circle of radius R around the origin and that within this disk the function $B(s)$ has a finite number M of non-zero singularities in the complex plane: $s_k = r_k e^{i\phi_k}$, with $k = 1, \ldots, M$. We further assume that all singularities are poles or power cuts, i.e., they are of the form

$$\frac{1}{(s - s_k)^{\beta_k}} \sim \alpha_k \left(1 - \frac{s}{s_k}\right)^{-\beta_k} . \tag{12.131}$$

Using Newton's binomial for a negative power,

$$\frac{1}{(1-x)^{\beta}} = \sum_{n=0}^{\infty} \binom{\beta+n-1}{n} x^{\ell} = \sum_{n=0}^{\infty} \frac{\Gamma(\beta+n)}{\Gamma(n+1)\Gamma(\beta)} x^{\ell}, \quad \beta > 0,$$

we find that the behavior of the generating function in the neighborhood of singularities is given by

$$B(s) \simeq \sum_{k=1}^{M} \frac{\alpha_k}{(1-s/s_k)_k^{\beta}} = \sum_{k=1}^{M} \alpha_k \sum_{n=0}^{\infty} \frac{\Gamma(\beta_k+n)}{\Gamma(\beta_k)\Gamma(n+1)} \left(\frac{s}{s_k}\right)^{n}, \tag{12.132}$$

so that the series element $b(n)$ is

$$b(n) = \sum_{k=1}^{M} \frac{\alpha_k}{\Gamma(\beta_k)} \frac{\Gamma(n+\beta_k)}{\Gamma(n+1)} \frac{1}{s_k^n} + O\left(\frac{1}{R^n}\right). \tag{12.133}$$

The terms with the smallest modulus $|s_k|$ are dominant. If there are more terms with the same value of $|s_k| = r$ but with a different phase ϕ_k, the dominant term is given by the higher-order poles, i.e., those with the largest value of β_k, $\beta = \sup_k \beta_k$ (e.g., of a subset of singularities $m \le M$).

Considering the behavior for large n of the ratio

$$\frac{\Gamma(n+\beta)}{\Gamma(n+1)} = \frac{(n+\beta)\cdots(n+2)}{\Gamma(n+1)} \simeq n^{\beta-1},$$

for $\beta \ge 1$, we obtain that the elements of the series behave according to Appell's formula,

$$b(n) \underset{n\to\infty}{\sim} \frac{n^{\beta-1}}{r^n} \frac{1}{\Gamma(\beta)} \sum_{k=1}^{m} \alpha_k e^{\imath n\phi_k}, \tag{12.134}$$

where the sum may often contain just one term.

Conversely, if the formula (12.134) is true, we can say that the m complex singularities closest to the origin and of higher order have the form described in Eq. (12.131) and are located on the circle $|s| = r$, at the points $s_k = r e^{\imath \phi_k}$, all with the same value $\beta_k = \beta$.

13

Continuous-Time Markov Processes

In this chapter we will deal with the study of stochastic Markov processes, i.e., random sequences of correlated but memory-less events that depend on continuous time. It is, essentially, the continuous-time generalization of *Markov* chains with integer time steps. Indeed, the name *chain* is traditionally reserved for processes for which time can only take discrete values. We will start from the simplest case and gradually consider processes with more and more diversified properties.

13.1 Poisson Processes

The state of a system realizing a Poisson process is characterized by an integer function $n(t)$ of time $t \in \mathbb{R}$, which is non-negative and non-decreasing. Conventionally, $n(0) = 0$ is assumed. Very often the function $n(t)$ is interpreted as the number of counts of an event that have occurred up to the time t.

Unless it is constant, the function $n(t)$ is not continuous. The Poisson process is characterized by the fact that the jump of the function n at the discontinuity is always 1 and that the discontinuities are uniformly distributed with a probability per unit time equal to λ. More precisely, if we denote by $o(t)$ something that goes to zero faster than t ($\lim_{t \to 0} o(t)/t = 0$), we have that the probability that $n(t + \Delta t) - n(t) = 1$ is given by $\lambda \Delta t + o(\Delta t)$, for small Δt, while the probability that $n(t + \Delta t) - n(t) = 2$, or more, is given by $o(\Delta t)$. With this construction, the state evolution dynamics of a system exclusively depends on the preceding state in the time interval Δt, so the evolution is memory-less of all previous dynamics: it is Markovian.

We propose to compute the probability $P_n(t)$ that $n(t)$ is equal to n (assuming $n(0) = 0$). The computation can be carried out with the elementary methods of Chapter 2. However, we want to set up the problem so that we can extend the same approach to more complicated cases. We will, therefore, proceed by deriving a system of differential equations for $P_n(t)$.

For this purpose, we observe that if Δt is small and we neglect terms of order $o(\Delta t)$, and if $n(t + \Delta t)$ is equal to n, only two things can have happened in the recent past:

- $n(t) = n$ and no transition in time Δt occurred (this case has probability $1 - \lambda \Delta t$); or

- $n(t) = n - 1$ and a transition occurred in the interval Δt, with probability $\lambda \Delta t$.

At this point, we can write that the probability of realizing the value n at time $t + \Delta t$ is

$$P_n(t + \Delta t) = P_n(t)(1 - \lambda \Delta t) + P_{n-1}(t)\lambda \Delta t + o(\Delta t). \tag{13.1}$$

This equation can be rewritten as

$$\frac{P_n(t + \Delta t) - P_n(t)}{\Delta t} = -\lambda P_n(t) + \lambda P_{n-1}(t) + \frac{o(\Delta t)}{\Delta t}. \tag{13.2}$$

In the limit $\Delta t \to 0$, we obtain the so-called *master equation* for Poisson processes:

$$\frac{dP_n}{dt} = -\lambda P_n(t) + \lambda P_{n-1}(t). \tag{13.3}$$

Obviously, the previous equation is only valid for $n > 0$ unless we conventionally define $P_{-1}(t) = 0$. In fact,

$$\frac{dP_0}{dt} = -\lambda P_0(t). \tag{13.4}$$

Since Eq. (13.3) contains only P_n and P_{n-1}, the equations at all n can be solved in cascade from (13.4), first for $P_0(t)$, then for $P_1(t)$, and so forth.[1]

The previous equations have one and only one solution, which is easy to guess and to verify, namely, the Poisson distribution with average λt:

$$P_n(t) = \frac{(\lambda t)^n}{n!} e^{-\lambda t}. \tag{13.5}$$

The mean value of the population, hence, is the average value of the Poisson distribution (13.5) and grows linearly over time. So does its variance $\langle n(t) \rangle = \langle (n(t) - \langle n(t) \rangle)^2 \rangle = \lambda t$.

To derive this, we can use the generating function, from the definition (8.1),

$$P(s,t) = \sum_{n=0}^{\infty} P_n(t)s^n, \tag{13.6}$$

for which the master equation (13.3) becomes

$$\frac{dP(s,t)}{dt} = \lambda(s - 1)P(s,t), \tag{13.7}$$

whose solution is

$$P(s,t) = P(s,0)e^{(s-1)\lambda t}. \tag{13.8}$$

If we take as an initial condition the state of zero population, $P_0(0) = 1$, then $P(s,0) = 1$. With a Taylor expansion of $e^{s\lambda t}$ we obtain

$$P(s,t) = e^{-\lambda t} \sum_{n=0}^{\infty} \frac{(\lambda t)^n}{n!} s^n, \tag{13.9}$$

and by the definition of (13.6) $P_n(t)$ is the distribution (13.5).

---

[1] This property is also true for processes of pure birth, which we will discuss in the next section.

13.2 Pure Birth Processes and Feller's Theorem

We will deal in the following with birth processes and with the case in which such processes may diverge at finite times. We will consider a system that at time t may be in one of the states $0, 1, \ldots, n$. The probability rate coefficients λ_n are defined in such a way that the probability of the system passing from state n to state $n + 1$ in a small enough time interval Δt is

$$P(n \to n + 1) = \lambda_n \Delta t + o(\Delta t). \tag{13.10}$$

For $\lambda_n = \lambda$ we are back to the Poisson case. A possible application of this formalism, as we said, is useful to study birth processes. We can in this case consider as state n the size of the population of a certain species. The transition to the $n + 1$ state is caused by the birth of a new specimen of the species. The λ_n coefficients are inversely proportional to the time required for the group of individuals to give birth to a new specimen, when there are already n. When λ_n grows with n, the time that passes on average between one birth and the next decreases. The system, hence, tends to spend less and less time in a given n state as n increases.

The quantity λ_n is, in a slightly different language, proportional to the inverse of the average life of the state n. The fact that the value of n can only grow means that we are considering a process of *pure birth*, and we are ignoring unpleasant eventualities such as the death of a group member. As we will see in Section 13.3, a birth *and death* process also includes a probability (which we will call $\mu_n \Delta t$) that the system will go from the n state to the $n - 1$ state. Finally, we note that the fact that the coefficients λ_n depend only on the starting n state (and not on the system's past history) implies that the stochastic process we are considering is of the Markov type. Furthermore, the fact that λ_n does not depend on the absolute time in which the state n is realized, but only depends on the value of n, implies that the process is stationary.

Proceeding exactly as for Poisson processes, we can derive the master differential equation of probability evolution:

$$\frac{dP_n}{dt} = -\lambda_n P_n(t) + \lambda_{n-1} P_{n-1}(t). \tag{13.11}$$

Exactly as in the case of Poisson processes, the solution of the system of differential equations exists and is unique. In the generic case (all different λ values), it is easy to verify that it can be written in the form

$$P_n(t) = \sum_{k=0}^{n} C_{k,n} \exp(-\lambda_k t), \tag{13.12}$$

where each λ_k contributes exponentially and where the constants $C_{k,n}$ depend on the initial condition. These constants can be obtained recursively. Indeed, the initial condition implies that

$$\sum_{k=0}^{n} C_{k,n} = P_n(0). \tag{13.13}$$

Furthermore, the master equations imply that

$$- \lambda_k C_{k,n} = -\lambda_n C_{k,n} + \lambda_{n-1} C_{k,n-1} \quad \text{for } k < n. \tag{13.14}$$

We thus have the relation, valid when λ_k are all distinct,

$$C_{k,n} = \frac{\lambda_{n-1}}{\lambda_n - \lambda_k} C_{k,n-1}, \tag{13.15}$$

which (combined with the initial condition) allows us to compute recursively all $C_{k,n}$.

In the case of the usual initial condition $n(0) = 0$, it is also possible to prove that the constants $C_{k,n}$ can be found by solving the system of linear equations:

$$\sum_{k=0}^{n} C_{k,n}(-\lambda_k)^s = 0 \quad \text{for } s < n, \tag{13.16}$$

$$\sum_{k=0}^{n} C_{k,n}(-\lambda_k)^n = \prod_{k=0}^{n-1} \lambda_k, \tag{13.17}$$

where $s = 0$ is obtained from (13.13). This formulation allows us to obtain $C_{k,n}$ for given n, without having to compute the $C_{k,n}$ for lower n values. This can be demonstrated from the condition $n(0) = 0$, from which it follows that $P_0(0) = 1$ and $P_{n>0}(0) = 0$. As a consequence, from the master equation for $P_1(t)$, we have

$$\dot{P}_1(0) = \lambda_0,$$

while for $n > 1$ we have $\dot{P}_n(0) = 0$. Carrying out higher-order derivatives, it is easy to see that no term survives at $t = 0$ unless it is of order equal to that of the derivative:

$$\left. \frac{d^s P_n(t)}{dt^s} \right|_{t=0} = 0 \quad \text{if } s < n.$$

By inserting the general form (13.12) of the solution, this leads to (13.16). In the case $s = n$ we obtain the derivative of order n from the master equation (13.11),

$$\left. \frac{d^n P_n(t)}{dt^n} \right|_{t=0} = -\lambda_n \left. \frac{d^{n-1} P_n(t)}{dt^{n-1}} \right|_{t=0} + \lambda_{n-1} \left. \frac{d^{n-1} P_{n-1}(t)}{dt^{n-1}} \right|_{t=0} = \lambda_{n-1} \left. \frac{d^{n-1} P_{n-1}(t)}{dt^{n-1}} \right|_{t=0},$$

which is a recursion formula from $\dot{P}_{n-1}(0)$ to $\dot{P}_n(0)$. Starting from $\dot{P}_1(0) = \lambda_0$, we then arrive at the formula (13.17).

Note that the equations (13.16) and (13.17) have one and only one solution only if all the λ_n are different from each other. Only in this case does the associated linear homogeneous equation have the null solution. In fact, the determinant of the matrix, called the Vandermonde matrix,

$$M_{k,s} \equiv (\lambda_k)^s \tag{13.18}$$

is zero only if two λ values are equal.

13.2.1 Poisson Process Limit

To see how we get back to the Poisson case, we can start with the explicit solution. Let us assume $P_0(0) = 1$ and take the first two probabilities $P_0(t) = e^{-\lambda_0 t}$ and

$$P_1(t) = C_{0,1}(t)e^{-\lambda_0 t} + C_{1,1}(t)e^{-\lambda_1 t}, \tag{13.19}$$

where we have left open the possibility that the coefficients depend on time. If we insert the expression of $P_1(t)$ on both sides of its master equation, we obtain the condition

$$e^{-\lambda_0 t}[\dot{C}_{0,1} - \lambda_0 C_{0,1} + \lambda_1 C_{0,1} - \lambda_0] + e^{-\lambda_1 t}\dot{C}_{1,1} = 0, \tag{13.20}$$

from which $C_{1,1} = C_{0,1} = $ constant with

$$C_{0,1} = \frac{\lambda_0}{\lambda_1 - \lambda_0},$$

hence (13.15) for $n = 1$. If, however, $\lambda_1 = \lambda_0 = \lambda$, then (13.20) becomes

$$e^{-\lambda t}[\dot{C}_{0,1} - \lambda + \dot{C}_{1,1}] = 0, \tag{13.21}$$

from which $C_{0,1} + C_{1,1} = \lambda t$ and the solution (13.19) reads

$$P_1(t) = [C_{0,1}(t) + C_{1,1}(t)]e^{-\lambda t} = \lambda t e^{-\lambda t}, \tag{13.22}$$

which is the probability of a Poisson event. For each generic term n, this implies that the solution tends to (13.5) as $\lambda_k \to \lambda$, $\forall k$.

13.2.1.1 Long-Time Birth Processes

The qualitative behavior of the solution of the system of differential equations at long times depends on the form of the λ_n. For example, the qualitative behavior of the expected value $\langle n(t) \rangle = \sum_{n=0}^{\infty} n P_n(t)$ of the population over time changes.

We can have three behaviors that in the biological metaphor correspond to different limiting factors of the growth of the population.

1. $\lambda_n \to$ constant. In this case, the limiting factors do not depend on the size of the population and we have that $\langle n(t) \rangle$ increases linearly with time, as seen at the end of Section 13.1. The asymptotic behavior is the Poisson distribution.
2. $\lambda_n \propto n$. In this case, the main limiting factor is the size of the population. It can be realized in asexual reproduction, for example, in colonies of bacteria, or also in sexual reproduction where the limiting factor is, for example, the length of gestation, or the appearance of a new gene mutation. In the latter case, the process is also called the Yule process, after Yule, who was the first to study it in the context of the mathematical theory of evolution [81]. In this regime, the expected value of the population $\langle n(t) \rangle$ increases exponentially with time. This can be simply seen by means of the generating function of the population distribution over time

$$P(s, t) = \sum_{n=0}^{\infty} P_n(t)s^n.$$

In terms of the generator, the master equation (13.11) is rewritten as

$$\frac{\partial P(s,t)}{\partial t} \propto -(1-s)\frac{\partial P(s,t)}{\partial s} . \tag{13.23}$$

Differentiating both members with respect to s and setting $s = 1$, we obtain the equation for the mean value $\langle n \rangle = (\partial P/\partial s)|_{s=1}$:

$$\frac{d\langle n(t)\rangle}{dt} \propto \langle n(t)\rangle , \tag{13.24}$$

the solution of which is exponential in t.

3. $\lambda_n \propto n^2$. This limiting case can be realized for the population of a species with sexual reproduction that lives in a low-density environment (e.g., hamsters in the Syrian desert), where the main limiting factor at low intensity is the meeting of adult specimens of different sexes, the probability of which is in fact proportional to n^2. In this case, as we shall see in the next section, $\langle n(t)\rangle$ becomes infinite at finite times. Obviously, before this could happen in nature, other limiting factors become dominant.

13.2.2 Divergent Birth Processes: Feller's Theorem

In the following, we will be interested in the case where the coefficients λ_n diverge as the population diverges, i.e., the case where the time it takes for the population to generate a new specimen ($\tau_n \equiv \lambda_n^{-1}$) tends to zero when the number of members of the population grows. This is reasonable (under our assumption of unlimited resources), since a large population will tend to have children more frequently than a small population. In this case *Feller's theorem* [28] holds.

Theorem. (Feller's theorem) *The convergence of the series*

$$T \equiv \sum_{n=0}^{\infty} \frac{1}{\lambda_n} = \sum_{n=0}^{\infty} \tau_n \tag{13.25}$$

is a necessary and sufficient condition for the variable n to assume an infinite value in a finite time.

Note also that the quantity

$$\tau_n \equiv \lambda_n^{-1} \tag{13.26}$$

is the mean lifetime of the state n, i.e., the time the system spends before jumping to $n+1$ elements. We can, therefore, expect that the average time needed for the population to diverge, $n \to \infty$, to be precisely given by $T = \sum_n \tau_n$.

This time T is finite, i.e., Eq. (13.25) converges, if the coefficients λ_n diverge fast enough with n, i.e., if the procreation rate increases strongly enough with the number of individuals. The fact that the system reaches an arbitrarily large state in finite time also implies that, in a finite time, an infinite number of transitions occur. The formalism we are developing serves, for example, to explain the onset of chain reactions, studied as

Markov chains in discrete times in Chapter 8, Section 8.2. The possibility of reaching infinity in finite time is a new feature that can occur in processes but not in chains, where for finite time the number of transitions is necessarily finite.

Intuitively, T is the average time the system takes to arrive at infinity. Hence, it will be quite straightforward to demonstrate that the condition of convergence of the series in (13.25) is sufficient. It will be only slightly more complicated to prove that the condition is also necessary.

We will study this problem in detail, arriving at the same conclusion using quite different arguments. The reason why we will present four different ways of proving the same result is to illustrate different techniques, so that the reader can see the advantages and disadvantages of each approach, one heuristic, the other three rigorous. Also note that in this simple case (pure birth), the proofs are very simple, whereas an equivalent demonstration in the case of birth and death (which we will study in the next section) would be much more difficult and even infeasible with some of the techniques that we will use.

13.2.3 An Approximate Analysis of Diverging Birth Processes

The basic idea is to reduce the infinite system of linear equations for the probability distribution to a small number of nonlinear equations for a deterministic quantity. In the simplest case that we are going to illustrate, we will reduce everything to a single equation for the mean value $\langle n(t) \rangle$ of the population. Let us introduce the notation:

$$\bar{n}(t) \equiv \langle n(t) \rangle = \sum_{n=0}^{\infty} n P_n(t). \tag{13.27}$$

We first prove that

$$\frac{d\bar{n}(t)}{dt} = \sum_{n=0}^{\infty} \lambda_n P_n(t) \equiv \langle \lambda_n \rangle_t. \tag{13.28}$$

Indeed, we have

$$\frac{d\bar{n}(t)}{dt} = \sum_{n=0}^{\infty} n[-P_n(t)\lambda_n + P_{n-1}(t)\lambda_{n-1}]$$

$$= -\sum_{n=0}^{\infty} \lambda_n n P_n(t) + \sum_{n=0}^{\infty} \lambda_n(n+1)P_n(t) = \sum_{n=0}^{\infty} \lambda_n P_n(t) = \langle \lambda_n \rangle_t, \tag{13.29}$$

where we have made the assumption that the sum $\sum_{n=0}^{\infty} \lambda_n n P_n(t)$ is convergent.

If, in a first approximation, we assume that the function $P_n(t)$ has a narrow peak for $n \approx \bar{n}(t)$ (i.e., a peak whose width relative to $n(t)$ is small, exactly as happens for the Poisson distribution for large times), we can write the mean value of λ_n as the peak value of the birth rate:

$$\langle \lambda_n \rangle_t \approx \lambda_{\bar{n}(t)}. \tag{13.30}$$

This hypothesis, which will not be exactly true, allows us to obtain an initial qualitative result, to be later more rigorously confirmed. Now $\bar{n}(t)$ is not necessarily an integer and, therefore, it is not a good index for λ_n. However, we can introduce a smooth $\lambda(n)$ function that analytically continues λ_n to non-integer values, such as $\lambda(\bar{n}(t))$. With this interpolation we can integrate Eq. (13.28):

$$t = \int_{\bar{n}(0)}^{\bar{n}(t)} \frac{dn}{\lambda(n)}. \tag{13.31}$$

The divergence time T of the population, such that $\bar{n}(T) = \infty$, is therefore defined by the condition

$$T = \int_{\bar{n}(0)}^{\infty} \frac{dn}{\lambda(n)}. \tag{13.32}$$

If the previous integral is convergent, the population diverges in a finite time. If we compare it with the convergence condition (13.25) of Feller's theorem, we realize that it is simply its continuous proxy.

It is evident that we have made an approximation that we do not control well. We can improve things by shifting from the study of the mean value of the population over time to the study of a Gaussian distribution of n at a given time. This provides a more refined approximation, which can be improved in a systematic manner. If we look at the variance of the distribution,

$$\sigma^2(t) = \langle n(t)^2 \rangle - \langle n(t) \rangle^2, \tag{13.33}$$

we obtain the differential equation

$$\frac{d\sigma^2(t)}{dt} = \langle \lambda \rangle_t + 2(\langle \lambda_n n \rangle_t - \langle \lambda_n \rangle_t \langle n \rangle_t). \tag{13.34}$$

If we assume that the distribution of n is Gaussian, the right-hand sides of Eqns. (13.33) and (13.34) can be computed as functions of $\bar{n}(t)$ and $\sigma^2(t)$ and thus the previous equations provide nonlinear evolution equations that generalize Eq. 13.28.

We will now present three different proofs of the result just obtained with this approximated method. The first, in Section 13.2.4, is Feller's original proof. The second, in Section 13.2.6, is a constructive proof, in the sense that we can explicitly control the form of some of the trajectories in n that arrive at infinity. Finally, in a third proof reported in Section 13.2.7, we can explicitly control the distribution of the divergence time of the birth process.

13.2.4 Feller's Theorem Proof

In the original formulation, Feller looks at a cumulative distribution at time t,

$$S(t) \equiv \sum_{n=0}^{n_{\max}(t)} P_n(t), \tag{13.35}$$

where the sum runs from $n = 0$ to the maximum population $n_{\max}(t)$ achieved in a time t. In this equation $S(t)$ is the probability that in a finite time t a finite number of state transitions of the birth type, $n \rightarrow n+1$, occur. If this probability is 1, the maximum population reached at finite time t is a finite number. The population will tend to infinity only if the time interval over which the process occurs is infinite: $S(t) = 1$ corresponds to an infinite divergence time.

On the other hand, if, from a certain point onwards, $t \geq T$, we have that $S(t) < 1$, it means that there is a non-zero probability $1 - S(t)$ that at time t an infinite number of transitions will occur, reaching $n = \infty$ in a finite time. To prove Feller's theorem, we are, therefore, interested in seeing under what conditions $S(t) < 1$, for t sufficiently large.

Feller's original proof   To this end, it is convenient to define the cumulative population distribution at time t of having a population less than or equal to N elements,

$$S_N(t) \equiv \sum_{n=0}^{N} P_n(t), \tag{13.36}$$

which will be a monotonic function of N, increasing towards the limit $S(t)$ as N grows to its superior limit $n_{\max}(t)$. We can directly set $n_{\max}(t) = \infty$ in the definition (13.35) of $S(t)$, since, if the population is not divergent at time t, all $P_{n > n_{\max}}(t) = 0$.

Using the master equation (13.11) for the probability, we immediately derive the relationship

$$\frac{dS_N(t)}{dt} = \sum_{n=0}^{N} [- \lambda_n P_n(t) + \lambda_{n-1} P_{n-1}(t)] = -P_N(t)\lambda_N . \tag{13.37}$$

This equation has a very clear meaning: $S_N(t)$ is the probability that $n(t) \leq N$ and this probability is influenced only by the transition $N \rightarrow N+1$ occurring with a probability rate λ_N. If we take as an initial condition $n(0) = 0$ we have $P_0(0) = 1$ and $P_{n>0}(0) = 0$. For the cumulant, it implies $S_N(0) = 1$. Integrating the previous equation we therefore find

$$1 - S_N(t) = \lambda_N \int_0^t d\tau \, P_N(\tau). \tag{13.38}$$

Since $0 \leq S_N(t) \leq S(t)$ we can write the inequalities

$$\frac{1 - S(t)}{\lambda_N} \leq \int_0^t d\tau \, P_N(\tau) \leq \frac{1}{\lambda_N} . \tag{13.39}$$

Summing over all states N, up to ∞, and using the definition (13.25) for the parameter T, we obtain

$$[1 - S(t)]T \leq \int_0^t d\tau \, S(\tau) \leq T , \quad \forall t . \tag{13.40}$$

We note that, for any finite time t, the time integral of $S(t) \leq 1$ is finite and can – eventually – tend to infinity only for $t \rightarrow \infty$.

If $T < \infty$ the integrand function $S(\tau)$ cannot be equal to 1 for all times. At most, it can be 1 only for times up to T. For large times it must decay, otherwise the integral cannot converge to something not larger than T. So $T < \infty$ implies that $S(t) < 1$, at least for $t \geq T$. This implies that the probability $1 - S(t)$ of reaching an infinite population in a finite interval of time t must be non-zero.

Conversely, $1 - S(t) > 0$ implies that $T < \infty$. Indeed, let us look at the inequality on the left-hand side of (13.40), which we rewrite as

$$1 - S(t) \leq \frac{\int_0^t d\tau\, S(\tau)}{T}.$$

Since the integral in the numerator is finite for each $t < \infty$, only if T is also finite can one have $S(t) < 1$. If, instead, $T = \infty$, the inequality results in $1 - S(t) \leq 0$, whence $S(t) = 1$.

A finite T is a necessary and sufficient condition for a divergent pure birth process. $\blacksquare$

13.2.5 Average Permanence Time of an *N*-Element Population

We observe that a system starting from a population 0 will arrive at a population $N+1$ in a certain time t_N, which is the time the system remains in a state of population $n \leq N$. We also observe that the cumulant $S_N(\tau)$ is the probability that the system remains in the region $n \leq N$ for an interval of time τ. If we integrate over all possible times, therefore, $\int_0^\infty d\tau\, S_N(\tau)$ is precisely the (average) time in which the system remains in the region $n \leq N$, as any history of the system up to the transition $N \to N+1$ gives a contribution τ to the previous integral. Hence, we can write

$$\langle t_N \rangle = \int_0^\infty d\tau\, S_N(\tau) = \lim_{t \to \infty} \sum_{k=0}^{N} \frac{1 - S_k(t)}{\lambda_k} = \sum_{k=0}^{N} \frac{1}{\lambda_k}, \tag{13.41}$$

where in the first step we used the formula (13.38) for which

$$\sum_{k=0}^{N} \frac{1 - S_k(t)}{\lambda_k} = \int_0^t d\tau \sum_{k=0}^{N} P_k(\tau) = \int_0^t d\tau\, S_N(\tau),$$

and in the second step we used the fact that, for a birth process,

$$\lim_{t \to \infty} S_k(t) = 0$$

for $k < \infty$ (unless one of the λ_k is zero). The equation

$$\langle t_N \rangle = \sum_{k=1}^{N} \tau_n$$

states the obvious fact that the average time the system remains in the $n \leq N$ region is equal to the sum of the average lifetimes of each one of the states of population $n \leq N$ during the birth process. Moving on to the limit $N \to \infty$, we find that the average divergence time of the population is precisely T.

13.2.6 A Constructive Proof of Feller's theorem

In this case, we give a proof under stricter hypotheses. That is, we will assume that the

$$\sum_{n=0}^{\infty} \frac{\log \lambda_n}{\lambda_n} \equiv \tilde{T} \tag{13.42}$$

is convergent. The convergence of (13.42) obviously implies the convergence of (13.25). Put simply, we are making a more stringent request than would be necessary on the sequence of λ_n, and we are asking that the typical birth time decreases with n faster than would really be necessary to have a divergent population. We denote the procreation time, i.e., the time it takes the system to go from $n \to n+1$, which is also the average lifetime of the state n, by

$$t_n \equiv \frac{\log \lambda_n}{\lambda_n} . \tag{13.43}$$

This is a frequently used procedure in physics: in order to gain the simplicity of the proof (either the transparency of the procedure or the very fact that a proof exists), we pay the price of a somewhat lower generality of the result obtained.

We assume that Eq. (13.42) is convergent, and show that the system reaches an infinite population state in finite time. To prove the result, we will show that the system reaches the state $n = \infty$ in finite time by following a specific chosen roadmap: the population will diverge if the system remains in each state n for a time less than or equal to t_n, that is preassigned. The system will of course also diverge if it remains in each state for a time larger than t_n, but this additional possibility will only make divergence at finite time more likely. Our ability to exhibit one sequence of times $\{t_n\}$ that guarantees the divergence of n in finite time proves the theorem.

Constructive proof of Feller's theorem.   Let us denote by $P_\infty(t)$ the probability that the system reaches $n = \infty$ in finite time t. Our method will exhibit a value of $\tilde{T}$ for which $P_\infty(\tilde{T})$ is certainly non-zero. This is a lower bound: it is a sufficient condition for a divergent birth process. Nothing precludes that for another roadmap one has $P_\infty(t < \tilde{T}) > 0$, or that there are more sequences of average lifetimes that contribute to the divergence of the population at time $\tilde{T}$, so that the true probability of divergence at $\tilde{T}$ is larger than our estimate. We show that the divergence probability at finite time $\tilde{T}$ (13.42) is non-zero, and it is lower-bounded by $P_\infty(\tilde{T}) > 0$.

Let us consider the preassigned sequence $\{t_n\}$ (the roadmap), which we choose to be composed of times (13.43), and let us write the time $\tilde{T}$ (13.42) as

$$\tilde{T} \equiv \sum_{n=0}^{\infty} t_n .$$

We denote by $p_n(\tau)$ the probability that the system moves from n to $n+1$ elements in a time interval τ. We then define the probability $\tilde{P}$ that the population diverges following the convergent time series t_n:

$$\tilde{P} \equiv \lim_{N \to \infty} \prod_{n=0}^{N} p_n(t_n). \tag{13.44}$$

As previously discussed, this is a lower bound on the probability that $n \to \infty$ in finite time: for times $t \geq \tilde{T}$ the probability that the population of the system will diverge will be $P_\infty(t) \geq \tilde{P}$. In fact, the sequence $\{t_n\}$ is only one of the possible sequences of times that cause the population to diverge in finite time.

We only have to show that, if the system obeys our time roadmap (13.43), then $\tilde{P} > 0$ strictly. To do this we need the explicit form of the transition probabilities $n \to n+1$, i.e., $p_n(t_n)$.

Pure birth process transition probability.　Let us consider a system that can make a transition from the state A to the state B in the infinitesimal time interval Δt with probability $p_A(\Delta t) = \lambda \Delta t$. We want to obtain the probability $p_A(t)$ that the system will leave the A state to go to the B state in the finite time t. Let $q_A(t) \equiv 1 - p_A(t)$ be the probability that the system has not made the transition in time t. We will have that $q_A(\Delta t) = 1 - \lambda \Delta t$. We use the multiplication property for independent event probabilities to find that the probability that the system had no transitions up to time $t + \Delta t$ is given by

$$q_A(t + \Delta t) = q_A(t) q_A(\Delta t) = q_A(t)(1 - \lambda \Delta t),$$

that is,

$$\frac{q_A(t + \Delta t) - q_A(t)}{\Delta t} = -\lambda q_A(t).$$

Taking the limit $\Delta t \to 0$ we find that

$$\dot{q}_A(t) = -\lambda q_A(t),$$

whose solution is

$$q_A(t) = q_0 e^{-\lambda t}, \tag{13.45}$$

with $q_0 = 1$, since the system is initially in the state A. We thus obtain, for the transition probability from the state A in a time t,

$$p_A(t) = 1 - e^{-\lambda t}. \tag{13.46}$$

We use the result (13.46) to compute the value that

$$\tilde{P} = \prod_{n=0}^{\infty} p_n(t_n)$$

takes on our roadmap (13.43). Given (13.46) the probability of leaving the state of population n in time t_n will be

$$p_n(t_n) = 1 - \exp(-\lambda_n t_n). \tag{13.47}$$

For an infinite product to be different from zero, the sum of the logarithms of the multiplicands must be finite. With the result (13.46), we obtain, due to the dominant behavior at large n,

$$\ln \tilde{P} = \sum_{n=0}^{\infty} \ln p_n(t_n) = \sum_{n=0}^{\infty} \ln \left[1 - e^{-t_n \lambda_n} \right],$$

which on our roadmap (13.43) yields

$$\ln \tilde{P} = \sum_{n=0}^{\infty} \ln \left(1 - \frac{1}{\lambda_n} \right).$$

For large n (hence large λ_n), we have

$$\ln \tilde{P} \simeq - \sum_{n=0}^{\infty} \frac{1}{\lambda_n},$$

and this sum, as we have already discussed, is all the more convergent since, by hypothesis, (13.42) holds. Thus

$$\tilde{P} \simeq \exp \left(- \sum_n \frac{1}{\lambda_n} \right) > \exp \left(- \sum_n \frac{\ln \lambda_n}{\lambda_n} \right) = e^{-\tilde{T}} > 0,$$

proving the intermediate result needed. ∎

The theorem is proved, and it is true that, under our assumptions, the birth process leads to a divergent population in finite time. ∎

For our proof to be valid, it is necessary that $\tilde{T}$, i.e., the formula (13.42), converges, i.e., that the coefficients λ_n diverge with n faster than $n \log(n)$. Recall, however, that this is only a technical limitation of our proof, and that the real limit is given by the condition that the λ_n diverge faster than n.

13.2.7 Explicit Computation of Divergence Time Distribution

We want to explicitly compute the probability distribution $\mathcal{P}(T)$ of the time T it takes for the system to reach a divergent population.

Divergence time distribution proof of Feller's theorem.   If we denote by t_n the average lifetime of the nth state, by $q_n(t_n)$ the probability of remaining in state n for a time t_n (i.e., the complement of $p_n(t_n)$), and by

$$T = \sum_{n=0}^{\infty} t_n \tag{13.48}$$

the divergence time, its probability distribution $\mathcal{P}(T)$ will be given by the convolution product of the permanence probabilities of remaining in the state n for a time t_n, from $n = 0$ to $n \to \infty$.

Consequently, the Laplace transform of $\mathcal{P}(T)$ is given by the product of the Laplace transforms of the permanence probabilities $\hat{q}_n(s)$:

$$\widehat{\mathcal{P}}(s) = \prod_{n=0}^{\infty} \hat{q}_n(s).$$ (13.49)

We have already seen from (13.45) that

$$q_n(t_n) = \lambda_n e^{-\lambda_n t_n},$$ (13.50)

where the prefactor λ_n is now obtained from the normalization condition: at time t the system must be with certainty in any one of the states $n \in [0, \infty)$. Thus, we have

$$\hat{q}_n(s) = (1 + \tau_n s)^{-1},$$ (13.51)

with $\tau_n = \lambda_n^{-1}$. Eventually, we obtain the Laplace transform of the divergence time distribution:

$$\widehat{\mathcal{P}}(s) = \prod_{n=0}^{\infty} (1 + \tau_n s)^{-1} = \exp\left(-\sum_{n=0}^{\infty} \log\left(1 + \tau_n s\right)\right).$$ (13.52)

A detailed analysis shows that the condition for the convergence of the infinite product to a non-zero value is that the series τ_n is summable, and thus we find once again the condition (13.25), provided by Feller's theorem. Conversely, if the sum of the series τ_n is divergent, we find that

$$\lim_{N \to \infty} \prod_{n=0}^{N} \hat{q}_n(s) = 0$$ (13.53)

as long as $s > 0$, and thus the probability that the variable

$$T_N = \sum_{n=0}^{N} t_n$$ (13.54)

has a finite value that tends to zero when $N \to \infty$.

If the series τ_n is summable, $\widehat{\mathcal{P}}(s)^{-1}$ is an integer analytic function whose asymptotic behavior is connected to the distribution of its zeros $s = -\lambda_n$.

If we assume that

$$\lambda_n \propto n^{\alpha},$$ (13.55)

the interesting case, where divergent processes are possible, corresponds to the region $\alpha > 1$, where we find that, for positive and sufficiently large s, the $\widehat{\mathcal{P}}(s)$ given by (13.52) goes to zero as[2]

$$\widehat{\mathcal{P}}(s) \sim \exp\left(-As^{1/\alpha}\right), \quad s \ll 1.$$

---

[2] This can be seen heuristically by doing the analytic continuation of the function in the exponent and going from the discrete sum to the integral on the continuum:

$$\sum_n \ln\left(1 + \frac{s}{n^{\alpha}}\right) \approx \int_0^{\infty} dn \ln\left(1 + \frac{s}{n^{\alpha}}\right) = \frac{s^{1/\alpha}}{\alpha} \int_0^{\infty} dy\, y^{1/\alpha - 1} \ln\left(1 + \frac{1}{y}\right) \propto s^{1/\alpha},$$

where we have made the change of integration variable $y = s^{-1} n^{\alpha}$.

At this point, it is sufficient to note that the behavior for large s of the Laplace transform of a function is closely related to the behavior of the function at the origin. If, for example, $\mathcal{P}(T)$ goes to zero as

$$\mathcal{P}(T) \simeq \exp(-BT^{-\gamma}), \quad \text{for } T \to 0, \tag{13.56}$$

using, for example, the Laplace maximum method, for large s we find that

$$\widehat{\mathcal{P}}(s) \propto \exp(-Cs^{\gamma/(\gamma+1)}), \quad \text{for } s \ll 1.$$

Putting all this together, we find the relation

$$\gamma = \frac{1}{\alpha - 1}.$$

The distribution of times for small T is, therefore,

$$\mathcal{P}(T) \propto \exp(-BT^{-1/(\alpha-1)}), \tag{13.57}$$

$$\alpha \equiv \lim_{n \to \infty} \frac{\ln(\lambda_n/\lambda_1)}{\ln n},$$

and the probability of the divergence time T, therefore, tends to zero for small T with a rate that depends on the asymptotic behavior of λ_n for large n. The faster λ_n grows, the more likely it is that a small divergence time will be realized. However, we note that, even if exponentially small, $\mathcal{P}(T)$ is always strictly positive as long as $T > 0$. This means that, if the process can diverge, it can diverge at any, even arbitrarily small, time.   ■

13.3 Birth and Death Processes

Birth and death processes are characterized by the fact that the quantity n, which is always assumed to be non-negative, can both increase and decrease by one unit. The conditional probability that the system in the state n passes to the state $n+1$ in a small enough time interval Δt is

$$P(n \to n+1) = \lambda_n \Delta t, \tag{13.58}$$

whereas the probability of the transition in the reverse direction is given by

$$P(n \to n-1) = \mu_n \Delta t. \tag{13.59}$$

Again, for the evolution of the probability, we derive a master differential equation

$$\frac{dP_n}{dt} = -\lambda_n P_n(t) + \lambda_{n-1} P_{n-1}(t) - \mu_n P_n(t) + \mu_{n+1} P_{n+1}(t). \tag{13.60}$$

To study the behavior of the solution to this equation, we can introduce the net probability flow between the population states n and $n+1$, which we will denote as the flow through the state "$n+1/2$",

$$\Phi\left(n + \frac{1}{2}, t\right) = \lambda_n P_n(t) - \mu_{n+1} P_{n+1}(t). \tag{13.61}$$

This is the probability that the system goes from n to $n + 1$ due to a birth, minus the probability of the transition in the opposite direction due to a death. In this notation, the master equation is rewritten as

$$\frac{dP_n}{dt} = -\Phi\left(n + \frac{1}{2}, t\right) + \Phi\left(n - \frac{1}{2}, t\right). \tag{13.62}$$

In the case $n = 0$, the previous equation is simplified, $\mu_0 = 0$, and becomes

$$\frac{dP_0}{dt} = -\Phi\left(\frac{1}{2}, t\right). \tag{13.63}$$

A necessary condition for the existence and uniqueness of the solution of the system of differential equations (13.62) and (13.63) is more complicated than in the case of pure birth in Section 13.2, as the equations are all coupled and we cannot proceed by solving the equations in series from the one at $n = 0$.

It is possible to prove the existence and uniqueness of the solution, although in the general case it is quite complicated and beyond the scope of the present description. For example, if there exists a constant A, such that for n large enough

$$\lambda_n < A \quad \text{and} \quad \mu_n > A, \tag{13.64}$$

this is a sufficient condition for the solution to exist and be unique [28]. The solution will still be unique even if the birth and death rates λ_n and μ_n grow slowly enough with n, but for too fast growths of one or the other the solution may not be unique and it may also be the case that $\sum_{n=0}^{\infty} P_n(t) < 1$.

Let us restrict ourselves to stationary solutions, assuming such a solution exists. We will first construct it and then reason about the conditions for the existence of this solution. In certain birth–death processes, it is, indeed, possible to reach a stationary state, i.e., it may be that

$$\lim_{t \to \infty} P_n(t) = u_n > 0. \tag{13.65}$$

We propose to derive this solution and to give a necessary condition for it to exist. First, we observe that, since the stationary distribution, the solution of

$$\frac{d P_n(t)}{dt} = 0,$$

does not depend on time, the master equation reduces to

$$-\Phi\left(n + \frac{1}{2}\right) = \Phi\left(n - \frac{1}{2}\right), \tag{13.66}$$

where (13.61) is computed using u and no longer depends on time:

$$\Phi\left(n + \frac{1}{2}\right) = \lambda_n u_n - \mu_{n+1} u_{n+1} \quad \text{if } n > 0 \tag{13.67}$$

and

$$\Phi\left(\frac{1}{2}\right) = 0 \quad \text{if } n = 0. \tag{13.68}$$

Equations (13.66) and (13.68) imply that, in the steady state, all Φ are null. We will therefore have

$$u_{n+1} = \frac{\lambda_n}{\mu_{n+1}}\, u_n \equiv v_n u_n \,, \tag{13.69}$$

where we have defined the ratio between the transition rates $n \to n+1$ and $n+1 \to n$ as $v_n \equiv \lambda_n/\mu_{n+1}$. We then obtain, by recursion, that

$$u_n = \prod_{k=0}^{n-1} v_k u_0 \,, \tag{13.70}$$

which we can shorten to $u_n = R_n u_0$, by defining $R_n \equiv \prod_{k=0}^{n-1} v_k$. The previous relationship allows us to compute all u_n in terms of the u_0. Under the assumption that $\{u\}$ is a well-defined stationary probability, imposing the normalization condition $\sum_n u_n = 1$ gives us the expression for the stationary probability of having a population of zero elements:

$$u_0 = \frac{1}{\sum_{n=0}^{\infty} R_n} \,. \tag{13.71}$$

We note that this solution is possible only if

$$\sum_{n=0}^{\infty} R_n < \infty \,, \tag{13.72}$$

and, therefore, this is a necessary condition for achieving a steady stationary state. A sufficient condition for the above series to be convergent is that, for k large enough ($k \geq k^*$), we have

$$v_k = \frac{\lambda_k}{\mu_{k+1}} < \frac{1}{1+ck} \,, \quad \text{for } k \geq k^* \,, \tag{13.73}$$

where c is a positive constant. Under this condition one has

$$R_n = \prod_{k=0}^{n-1} e^{\ln v_k} = \exp\left(\sum_{k=0}^{k^*-1} \ln v_k + \sum_{k=k^*}^{n-1} \ln v_k \right) \leq K \exp\left(-\sum_{k=k^*}^{n-1} \ln(1+ck) \right)$$

$$\leq K e^{-(n-k^*)\ln(1+ck^*)} \leq K'(1+ck^*)^{-n}. \tag{13.74}$$

Thus, $R_n \to 0$ exponentially when $n \to \infty$ and the condition (13.72) for the stationary solution (13.70) and (13.71) is satisfied.

13.4 Markov Processes

Markov processes can be considered as the generalization of birth–death processes in the presence of multiple births and simultaneous causes of death such as traffic accidents. In a birth–death process, the only allowed transitions are $n \to n+1$, with probability $P_{n,n+1} = \lambda_n \Delta t$, and $n \to n-1$, with probability $P_{n,n-1} = \mu_n \Delta t$. We can describe these as transitions between states of a Markov chain for a time step Δt, for which the stochasticity property of transition matrices applies:

$$\sum_j P_{i,j} = \lambda_i \Delta t + \mu_i \Delta t + P_{i,i} = 1 ,$$

from which the probability of remaining in the state i in the interval Δt is $P_{i,i} = 1 - (\lambda_i + \mu_i)\Delta t$ as in Eq. (13.60).

As in a Markov chain, also in a Markov process there will be, in general, transition probabilities between all accessible states. In this case the probability of transition from the state i to the state j in the time interval Δt is given by

$$P_{i,j} = K_{i,j} \Delta t , \tag{13.75}$$

where the diagonal elements of the matrix K are conventionally equal to zero, and the diagonal elements of the stochastic matrix are given by

$$P_{m,m} = 1 - \sum_{j \neq m} P_{m,j} = 1 - \sum_j K_{m,j} \Delta t .$$

We can denote by $P_{i,k}(t',t)$ the probability of going from state $i(t')$ at time t' to the state $k(t)$ at time t. Since in the stationary limit the transition probabilities are time-translation-invariant and depend only on the time interval, we write[3]

$$P_{i,k}(t',t) = P_{i,k}(t-t') . \tag{13.76}$$

13.4.1 The Forward Kolmogorov Master Equation

Using the same arguments as in the previous sections, computing the probability of a transition as time-of-arrival changes

$$P_{i,m}(t',t+\Delta t) = P_{i,m}(t',t)\left[1 - \sum_j K_{m,j}\Delta t\right] + \sum_k P_{i,k}(t',t)K_{k,m}\Delta t ,$$

we obtain the master equation for the generic Markov process:

$$\frac{dP_{i,m}(t',t)}{dt} = -P_{i,m}(t',t)\sum_j K_{m,j} + \sum_k P_{i,k}(t',t)K_{k,m} = -\sum_k P_{i,k}(t',t)T_{k,m} , \tag{13.77}$$

where we have defined the matrix

$$T_{k,m} \equiv \delta_{k,m} \sum_j K_{m,j} - K_{k,m} . \tag{13.78}$$

This master equation for discrete stochastic processes is also called the forward Kolmogorov equation, because we are looking at the variation of the transition

---

[3] The reader should pay attention to the notation: the first time corresponds to the initial state, the second time to the final state of the transition.

probability as the arrival time varies. Using time translation invariance (13.76), we can rewrite it as

$$\frac{dP_{i,m}(t)}{dt} = -P_{i,m}(t)\sum_j K_{m,j} + \sum_k P_{i,k}(t)K_{k,m} = -\sum_k P_{i,k}(t)T_{k,m}. \tag{13.79}$$

Obviously, information about the initial conditions must be added. If at time zero the system starts with certainty from the state i, we will have that

$$P_{i,j}(0) = \delta_{i,j}.$$

The previous equation is not immediately derivable, if we want to carry out the steps carefully. For example, it is evident that the equation can only be written if

$$\sum_j K_{i,j} < \infty, \tag{13.80}$$

otherwise some diagonal elements of the matrix T would be infinite.

13.4.2 The Backward Kolmogorov Master Equation

We can also look at the variations of the transition probability when the starting time varies, rather than the arrival time. If we start from the state i at time t' and arrive in the state m at time t, we can ask ourselves what happens if the starting time varies by $\Delta t'$. It may happen that in the interval $\Delta t'$ the system remains in the state i (with probability $1 - \sum_j K_{i,j}\Delta t$) and then arrives in m at time t. This option has a probability

$$\left(1 - \sum_j K_{i,j}\Delta t\right)P_{i,m}(t' + \Delta t', t).$$

Alternatively, the system may jump from i into another state k at time $\Delta t'$ (with probability $K_{i,k}\Delta t'$) before arriving in the state m at time t. The probability contribution of the set of all possible paths of this type will be

$$\sum_k K_{i,k}\Delta t' P_{k,m}(t' + \Delta t', t).$$

Putting all the contributions together, we can write the probability of going from $i(t')$ to $m(t)$ in terms of what happens by varying the initial time $\Delta t'$,

$$P_{i,m}(t', t) = \left(1 - \sum_j K_{i,j}\Delta t'\right)P_{i,m}(t' + \Delta t', t) + \sum_k K_{i,k}\Delta t' P_{k,m}(t' + \Delta t', t),$$

from which, in the limit $\Delta t' \to 0$, we obtain

$$-\frac{dP_{i,m}(t', t)}{dt'} = \sum_j [K_{i,j}P_{j,m}(t', t) - K_{i,j}P_{i,m}(t', t)] = -\sum_j T_{i,j}P_{j,m}(t', t), \tag{13.81}$$

where the matrix T is defined in (13.78).

Exploiting time translation invariance and renaming t by the difference $t - t'$, we finally have the master equation, called the *backward* Kolmogorov equation

$$\frac{dP_{i,m}(t)}{dt} = -\sum_j T_{i,j} P_{j,m}(t).$$

(13.82)

The existence of two equations, forward and backward, for the same process should not come as a surprise. It is a common fact that we have already encountered in similar contexts: for example, for Markov chains, we have that

$$P_{i,k}^{(n+1)} = \sum_j P_{i,j}^{(n)} P_{j,k} = \sum_j P_{i,j} P_{j,k}^{(n)}.$$

(13.83)

In general, for an arbitrary $\mathbf{A}$ matrix, we have that

$$\mathbf{A}^{n+1} = \mathbf{A}^n \mathbf{A} = \mathbf{A}\mathbf{A}^n.$$

(13.84)

If we do not miscompute, the matrices $\mathbf{A}$ and $\mathbf{A}^n$ commute and so the information contained in the two equations is the same.

If the process is reversible, i.e., invariant under time reversal, then, as in the case of the Markov chains of Chapter 11, the relation of detailed balance applies. For continuous times, it reads

$$u_i P_{i,m}(t', t) = u_m P_{m,i}(t', t).$$

(13.85)

In this case it can be shown that the solutions of the forward and backward Kolmogorov equations are linked by (13.85): if $P_{m,i}(t', t)$ is the solution of the backward equation (13.81), the transition probability

$$\hat{P}_{i,m}(t', t) \equiv \frac{u_m}{u_i} P_{m,i}(t', t)$$

will be a solution of the forward equation (13.77).

In the case of infinite states, it is non-trivial to find necessary and sufficient conditions that ensure us of the existence of one and only one solution of the Kolmogorov equation. A sufficient (but overly restrictive) condition is that there exists a constant C, independent of i, such that

$$T_{i,i} = \sum_j K_{i,j} < C.$$

(13.86)

Obviously, if the previous sum were divergent, we would not be able to define the matrix T and, therefore, our poor differential equations for the probabilities would be meaningless. In the case where the previous sum is convergent and uniformly bounded, the transition probabilities are also uniformly bounded. In this case, it is evident that the condition

$$\sum_i P_{m,i}(t) = 1$$

(13.87)

implies that

$$\left| \frac{dP_{m,i}(t)}{dt} \right| < C.$$

(13.88)

The derivative with respect to time of the transition probabilities can never become singular, and by exploiting theorems on systems of differential equations one can prove the existence and uniqueness of the solution.

Obviously, in the case of a finite number of states, things are much simpler. Formally, the solution of the forward Kolmogorov equation (13.77) can be written as

$$P_{i,m}(t) = \sum_j P_{i,j}(0)(e^{-tT})_{j,m} \tag{13.89}$$

and the exponential of a finite-dimensional matrix, in the case of non-degenerate eigenvalues, can easily be written in the form

$$P_{i,m}(t) = \sum_\ell d_i(\ell)e^{-t\lambda(\ell)}s_m(\ell), \tag{13.90}$$

where $\lambda(\ell)$ are the eigenvalues, and $\vec{s}$ and $\vec{d}$ are the left and right eigenvectors forming the orthonormal skew basis of the non-Hermitian matrix T. We saw this in Appendix 10.B where we studied the spectral decomposition of non-Hermitian matrices.

The classification of Markov processes can be done along the lines of the one we performed for Markov chains (reducible–irreducible processes, decomposable–indecomposable, ergodic, ...), with the simplification of the absence of periodic processes.

We do not report these results but merely point out that there is a very close relationship between Markov chains and Markov processes. If we take $t = n\Delta t$ with an arbitrary Δt, the operator of Eq. (13.89) can be written as

$$e^{-n\Delta t T} = e^{-\Delta t T}e^{-\Delta t T}\cdots e^{-\Delta t T} = (e^{-\Delta t T})^n, \tag{13.91}$$

and if we define $P(\Delta t) \equiv e^{-\Delta t T}$, Eq. (13.89) becomes

$$P_{i,m}(n\Delta t) = \sum_j P_{i,j}(0)(e^{-\Delta t T})^n_{j,m} = \sum_j P_{i,j}(0)(P(\Delta t)^n)_{j,m}.$$

If we take $P_{i,j}(0) = \delta_{i,j}$, we have $P_{i,m}(n\Delta t) = (P(\Delta t))^n_{i,m}$ and the evolution of the absolute probability is

$$p_m(n\Delta t) = \sum_i p_i(0)P_{i,m}(n\Delta t) = \sum_i p_i(0)(P^n(\Delta t))_{i,m}, \tag{13.92}$$

which in time step units Δt leads to

$$p_m(n) = \sum_i p_i(0)(P^n)_{i,m}. \tag{13.93}$$

Equation (13.93) coincides with the characteristic equation (10.10) of a Markov chain. In other words, if we observe a Markov process at multiple time intervals of Δt (stroboscopic vision), the Markov process is reduced to a Markov chain. Furthermore, it is possible to use the expansion in powers of Δt (which is certainly convergent for small Δt if the condition (13.86) is satisfied) to compute the P matrix, always in the region of small Δt. In this way, it is possible to derive the properties of the P matrix from those of the T matrix.

Obviously, we can extend both the concept of the Markov chain and the Markov process to the case where the set of states is not countable, e.g., it is labeled by a real number. An explicit example of a Markov chain defined on a continuous space is given by random walks on the continuum, see Sections 7.3–7.6. The processes described by the Langevin equation are Markov processes defined on the continuum. The behavior of Markov processes with variables with values both on the continuum and with discrete jumps can be studied in general. More general equations can be written that include both the master equations studied here and the Fokker–Planck equations [56] that we encountered in Chapter 7, sometimes called Chapman–Kolmogorov integro-differential equations [30]. These are a generalization to continuous times and both discrete and continuous variables of Eqns. (7.37), (10.11), (13.77), and (13.81). A study of this theory of stochastic processes is beyond the scope of this book.

14

Entropy, Probability, and Statistical Mechanics

The concept of entropy[1] is one of the most interesting and ubiquitous concepts since the advent of commercial steam engines in the Industrial Revolution. Although a physical quantity by birth, after Sadi Carnot's studies of thermodynamics and the efficiency of engines [82], entropy is a quantity now found in many different areas, with different meanings. The quantity "entropy" introduced by Rudolf Clausius [83] in thermodynamics to define the relationship between the heat exchanged by a system and its temperature, and to state the second law, was defined at the microscopic level in statistical mechanics by Ludwig Boltzmann and Willard Gibbs [57]. But entropy is also a stochastic concept, a founding quantity of information theory, where it was reintroduced by Claude Shannon [84], as we shall see at length in Section 14.1.1, and, as such, is the basis of a probabilistic formulation of statistical mechanics, according to the variational principle of maximum entropy [85] described in Section 14.4. Entropy, or rather entropy production, is a quantity related to chaos in dynamical systems, where it takes the name of Kolmogorov–Sinai entropy, which we will mention in Section 14.2. In complex systems, it is also the quantity that describes the multiplicity of out-of-equilibrium states, of possible histories that have not (yet) occurred, of *hidden* states in the space of multi-equilibria. The field of complex disordered systems is very broad and far beyond the limits of this book. The interested reader might look at ref. [86]. We will try to sketch an idea for entropy in glassy systems in Section 14.6.

As we shall see in various cases that we shall illustrate below, depending on the field, entropy quantifies irreversible changes (dissipations) in a system, or it measures our ignorance of a system or, equivalently, the information we need to understand it, or, again, it estimates a system's disorder, which may also take the meaning of freedom of choice, or of uncertainty, depending on the point of view we shall adopt along the way. Before we weigh anchor and sail into the archipelago of entropies, we can anticipate a common trait in the various representations and interpretations of entropy: *exponential proliferation* [87]. Readers who have studied statistical mechanics will know that the volume Ω of the phase space at constant energy E of a system of N degrees of freedom (thus, in a *microcanonical ensemble*) is $\Omega(E) \sim e^{NS(E)}$, where S is the universally preferred letter for entropy, or nearly so. In information theory, we will see that the number of messages of a certain length L that can be encoded and transmitted by S bits is $\sim 2^{LS}$. In chaotic dynamic systems, the uncertainty of position after a sufficiently large time n grows as e^{Kn}. (Here the symbol for entropy is K in honour of

---

[1] From the ancient Greek ἐν (en), "inside", and τροπή (tropé), "transformation", in parallel with the etymology of energy, from the ancient Greek ἐν and ἔργον (ergon), "work".

Kolmogorov! But it is an entropy with the unit of time.) In glassy systems, and more generally in complex systems with a high multiplicity of metastable states, the number of hidden glassy states, alternative to the glassy state actually realized in a single history of the material, for a system of N degrees of freedom grows as e^{NS_c}, where the subscript c on S can mean *configurational* or *concealed states* or even *complexity function*, depending on taste and context.

Let us, therefore, begin our journey in a strictly entropic manner, in neither chronological nor alphabetical order.

14.1 Microscopic Entropy and Information Theory

14.1.1 Shannon's Entropy Definition

The definition of the entropy of a discrete probability distribution $\{P\}$ introduced by Shannon is

$$S[P] = -\langle \log P \rangle_P = -\sum_{n=1}^{N} P_n \log(P_n),\tag{14.1}$$

where N can also be infinite. It is immediate to show that entropy cannot be negative, since $P_n \in [0, 1]$ is a probability. For Shannon, who laid the foundations of information theory [84], this is the mean value of *information*, $-\ln P$, which is transmitted between two agents (transmitter and receiver) who know the distribution $\{P\}$ with which the transmitted events are generated. How information and entropy of a certain distribution turn out to have precisely these forms and not other functions of $\{P\}$ is explained in Appendix 14.A.

14.1.1.1 Shannon Entropy in the Continuum and the Kullback–Leibler Divergence

Analogous to (14.1), which is valid for discrete probabilities, we might define the entropy of a continuous probability distribution as

$$S[P] = -\int dx\, P(x) \log P(x) \equiv -\langle \log(P) \rangle_P.\tag{14.2}$$

In the case of a continuous distribution, however, the entropy is no longer necessarily positive. For example, if $P(x) = h^{-1}x$ for $0 \leq x \leq h$, the entropy is precisely $\log(h)$, a function that can have an arbitrary sign. The continuous expression also has another problem: if we make a change of variables $y = y(x)$ the entropy is not invariant.

This can be solved in the continuum by defining an entropy relative to a reference probability density $R(x)$ as

$$S[P|R] = \int dx\, P(x) \log \frac{P(x)}{R(x)} = -\langle \log(R) \rangle_P + \langle \log(P) \rangle_P.\tag{14.3}$$

This regularized entropy is called the Kullback–Leibler divergence. "Divergence" because it tells us how different two probability distributions are.

Proof of non-negativity of the Kullback–Leibler divergence   It can be shown that this expression is always positively defined, for any form of $R(x)$ (as long as it is a well-normalized distribution), and is invariant by reparameterization of the variables. We start with the inequality $\log y \leq y - 1$, where the equality holds only when $y = 1$. Indeed, recalling that $d \log y / dy = 1/y$, it is immediate to see that for $y > 1$ the logarithm grows less than linearly, whereas for $y < 1$ it has a larger slope and always remains below the straight line $y - 1$. If we chose $y = R(x)/P(x)$ this implies

$$-\int dx\, P(x) \log \frac{R(x)}{P(x)} \geq -\int dx\, P(x) \left[\frac{R(x)}{P(x)} - 1\right] = -\int dx\, R(x) + \int dx\, P(x) = 0.$$

We obtain that $S[P|R] \geq 0$ and this is called the Gibbs inequality.   ■

14.1.1.2 The Limiting Cases of Shannon's Maximum Entropy

Let us focus here on the discrete case and analyze the two limiting cases.

1. In the first case there exists a k such that

$$P_k = 1, \quad \text{while} \quad P_n = 0,\ \forall n \neq k. \tag{14.4}$$

In this case, there is only one state, the system is deterministic, the information transmitted is null (the receiver already knows with certainty the content of the message transmitted), and the entropy of this distribution is $S[P] = 0$.

2. In the second case,

$$\begin{aligned} P_k &= 1/M, \quad \text{for } k \leq M, \\ P_k &= 0, \quad\quad\ \text{for } k > M. \end{aligned} \tag{14.5}$$

All states with non-zero probability have uniform probability. This is the most disordered case and the entropy of this distribution,

$$S[P] = \log(M), \tag{14.6}$$

is the maximum value for a system of M states.

We emphasize that this second case is an important result. When all the states of a stochastic source of events are equiprobable, the choice is the maximum possible and is represented by the logarithm of the number of states, which is the maximum entropy of the source. We note that having a lot of choice means that we are very ignorant about the next event that may occur. This link between maximum ignorance and the logarithm function is not entirely new to us. We got a taste of it when we studied the choice of the prior distribution in statistical inference in Section 5.3.

14.1.1.3 Capacity of the Transmission Source

We anticipate that, in information theory, the maximum entropy of signals generated by a source with a distribution $\{P\}$ and transmitted through a given communication channel is also called the *capacity* of the source. This capacity is nothing more than the increase rate of the number of signals transmitted in a unit of time or length. Usually, capacity is measured in the number of bits transmitted per second, or in the number of bits transmitted per unit of the characters of the alphabet that make up the messages (bits per symbol). We will in fact see that $S[P]$, or rather,

$$S_2[P] = \frac{S[P]}{\ln 2},$$

represents a number of bits.

At this point we have realized that, in information theory, the source of information is represented by a stochastic process, such as those we have encountered in previous chapters. If the distribution of the process admits only one value, the process is no longer stochastic, there is no ignorance, there is no information to be transmitted, and the entropy of the source is zero. If all the values of the distribution are equiprobable its entropy is maximum.

14.1.1.4 Properties of Entropies

Let us now recall other properties of entropy, or rather of *entropies*.

Entropy of joint events. Let us take two types of events (two sources, two stochastic processes):

- X, with distribution $P^{(X)}(n)$, $n = 1, \ldots, N$, and
- Y, with distribution $P^{(Y)}(r)$, $r = 1, \ldots, R$.

If $P^{(X,Y)}(n, r)$ is the joint probability that $X = n$ and $Y = r$, the joint entropy is defined as

$$S[P^{(X,Y)}] = -\sum_{r=1}^{R} \sum_{n=1}^{N} P^{(X,Y)}(n, r) \log P^{(X,Y)}(n, r). \tag{14.7}$$

Marginalizing with respect to one or the other event, e.g.,

$$P^{(X)}(n) = \sum_{r=1}^{R} P^{(X,Y)}(n, r), \tag{14.8}$$

one can rewrite the entropies of the process X and the process Y, see (14.1), as

$$S[P^{(X)}] = -\sum_{n=1}^{N} \sum_{r=1}^{R} P^{(X,Y)}(n, r) \log \sum_{r=1}^{R} P^{(X,Y)}(n, r), \tag{14.9}$$

$$S[P^{(Y)}] = -\sum_{r=1}^{R} \sum_{n=1}^{N} P^{(X,Y)}(n, r) \log \sum_{n=1}^{N} P^{(X,Y)}(n, r). \tag{14.10}$$

Relationship between joint entropy and entropies of individual variables. Proceeding as in the case of the Kullback–Leibler divergence, using the Gibbs inequality, from

$$\log \frac{P^{(X)} P^{(Y)}}{P^{(X,Y)}} \leq \frac{P^{(X)} P^{(Y)}}{P^{(X,Y)}} - 1 \,,$$

we have that

$$S[P^{(X)}] + S[P^{(Y)}] - S[P^{(X,Y)}] \geq 0 \,,$$

i.e.,

$$S[P^{(X,Y)}] \leq S[P^{(X)}] + S[P^{(Y)}] \,. \tag{14.11}$$

The entropy of a joint process is not greater than the sum of the entropies of the two individual processes. The equality holds only if X and Y are independent.

Conditional entropy. Given two processes X and Y, the entropy of Y conditioned on X is defined as

$$S[P^{(Y|X)}] = - \sum_{n,r} P^{(X,Y)}(n,r) \log P^{(Y|X)}(r|n) \,, \tag{14.12}$$

where

$$P^{(Y|X)}(r|n) - \frac{P^{(X,Y)}(n,r)}{P^{(X)}(n)}$$

is the conditional probability defined in (1.30). The entropy $S[P^{(Y|X)}]$ measures the mean uncertainty about the knowledge of Y, if X is known. Substituting the definition of conditional probability into the formula (14.12) and using the expression (14.8) of the marginalization on Y, we obtain

$$S[P^{(X,Y)}] = S[P^{(X)}] + S[P^{(Y|X)}] \,. \tag{14.13}$$

The entropy of the joint processes X and Y is equal to the entropy of X plus the entropy of Y conditional on the knowledge of X. Obviously, the mirror relation is also valid:

$$S[P^{(X,Y)}] = S[P^{(Y)}] + S[P^{(X|Y)}] \,.$$

Using the relationship (14.11) one can easily prove the important property that

$$S[P^{(Y)}] \geq S[P^{(Y|X)}] \,. \tag{14.14}$$

The entropy of Y never increases due to the knowledge of X. If X and Y are related in some way, knowing X decreases our uncertainty about Y. This last property is useful to understand why entropy is a good representation of uncertainty.

Entropy increase at the balance of probabilities. Any change in the probabilities of the P_n towards their equalization increases the entropy $S[P]$. If we have $P(n) < P(n')$ and we increase $P(n)$ or decrease $P(n')$ so that they are more similar to each other,[2]

---

[2] Obviously taking into account the normalization of $\{P\}$, we will either have to do both or vary some other probability. In the latter case, we will assume that such variations are small, well distributed, and do not cause large variations in entropy.

entropy increases. More precisely, if we introduce weights $w_{r,n} \geq 0$ such that $\sum_n w_{r,n} = \sum_r w_{n,r} = 1$, and by means of a linear transformation we define new probabilities $\{Q\}$ that are each an average of $\{P\}$, i.e.,

$$Q(r) = \sum_n w_{r,n} P(n), \tag{14.15}$$

then $S[Q] \geq S[P]$.

It is easy to see that if the entropy is zero or $\log(M)$, the probability distribution must be that described in one of the two examples (14.4) or (14.5).

In statistical mechanics it is often stated that the quantity $M_{\mathrm{eff}} = \exp(S[P])$ has the meaning of the volume of the phase space actually occupied by the system, i.e., the number of *actual* or *typical* states. It is a formulation of Boltzmann's principle according to which entropy is defined, barring a multiplicative constant, as the logarithm of the number of typical microscopic configurations. This apparently vague statement, which is clear only in the two extreme cases we have illustrated, can be made more precise by formulating the source coding theorem, also known as Shannon's first theorem [84, 85, 88, 89].

14.1.2 Coding of a Message

In order to state Shannon's first theorem, we first introduce the concept of *coding*. Shannon's theorem can be stated as follows. What is the purpose of encoding a message? It serves to reduce the number of bits per second required to transmit the message, i.e., to *compress* it. Compression can be carried out by exploiting knowledge of the statistical properties of the source. In fact, as a rule, the message that is transmitted is not a sequence of random numbers. The message has a meaning, it is placed in a context, it is written in a language that has a vocabulary and a grammar. Different messages of equal length, therefore, have different probabilities of being transmitted. If both the person who transmits and the person who receives the message are aware of the probability distribution of the message source, they can agree to use a coding. That is, they agree to associate a word with a number, in a unique way, so that the message, once encoded by the sender, is, then, unencoded by the receiver.[3]

In other words, an encoding consists of assigning an $\mathcal{M}(\mathbf{c})$ function on the set of all $\{\mathbf{c}\}$ messages generated by a source, which is invertible, i.e., given the value of $\mathcal{M}(\mathbf{c})$, we can uniquely trace back to $\mathbf{c}$. The encoding is not unique. There will, in general, be several possible bi-univocal encodings of the same messages. Some of these encodings will be more efficient, and others less so, in compressing a message to be sent along the transmission channel.

Associating the most likely words with smaller numbers is convenient to reduce the number of bits needed to transmit the true message, without losing information. In general, we can ask ourselves how much we can gain if we follow this strategy. What is the minimum length of the bit string if we use an optimal compression coding and

---

[3] In information theory, the object that performs the encoding is called a transducer. If the messages encoded by a transducer are invertible, this is called a non-singular transducer.

what is this coding? The answers can be found in Shannon's theorem of source coding, which we will state in a moment.

Let us first see the simplest example of an encoding. Suppose we have an alphabet composed of N characters. Consider a sequence (a possibly long word) that is made up of L characters c_i, $i = 1, \ldots, L$, where the c_i characters take values between 0 and $N - 1$, extremes inclusive. This sequence, which we will denote by $\mathbf{c}$, is the message we wish to transmit. It is evident that there is a bi-univocal correspondence, which we will call a *coding*, between the words of length L and the numbers, which we will call K, between 0 and $N^L - 1$. This correspondence is given by the *lexicographic* function

$$K(\mathbf{c}) \equiv \sum_{i=1}^{L} c_i N^{i-1} . \tag{14.16}$$

In this way, to transmit a word $\mathbf{c}$, we can simply transmit the integer number $K(\mathbf{c})$, or, better still, the binary digits that identify it. To transmit a number M of binary digits, we need to transmit M bits of information.[4] The words $\mathbf{c}$ that correspond to low values of $K(\mathbf{c})$ can be transmitted using a low number of bits. Words corresponding to high K will require a higher number of bits. How many? Given a large K, the number of bits required to transmit the corresponding words is equal to the logarithm in base 2 of K. If in this instance we assume that all words are equiprobable, then, all numbers K between 0 and $N^L - 1$ are equiprobable, and the average number of bits (per character) that need to be transmitted will be given, in the limit of large L, see Eq. (14.6), by

$$C = \lim_{L \to \infty} \frac{1}{L} \log_2 (N^L) = \log_2 N , \tag{14.17}$$

where $\log_2 (N) \equiv \ln (N)/\ln(2)$ is the logarithm in base 2. This limit of maximum ignorance (and maximum information to be transmitted in order to understand something) is the capacity of the source, which we mentioned in the previous section. The capacity is the maximum entropy (14.6).

Let us consider the simplest case where each letter c_i has a probability p_i of being drawn and the message has been constructed with letters randomly distributed without any correlation (this is certainly not the case for natural languages).

We recall the definition of entropy in bits per character $S_2[P] = S[P]/\log 2$ of the message source characterized by the stochastic process of the distribution $\{P\}$ of characters, see formula (14.1). As we said, its maximum over all character distributions is the capacity C in bits per character (14.17) and we can move on to state the data compression theorem known as the Shannon source coding theorem.

14.1.3 Shannon's Source Coding Theorem (First Theorem)

Theorem. (Shannon's source coding theorem) *Let us take a message of L characters, with large L, issued by the distribution source $\{P\}$, then*

---

[4] We actually also need to transmit a stop signal, i.e., an end of message. How such a signal can be transmitted is not crucial to the considerations we want to make. The curious reader may find a brief discussion in Appendix 14.B.

1. *there exists a coding of the message such that the probability that more than $LS_2[P]$ bits are required for its transmission tends to zero for $L \to \infty$, and*
2. *there is no coding such that the probability that the message can be transmitted with less than $LS_2[P]$ bits tends to be larger than 0 as $L \to \infty$.*

In concrete terms, the message of L characters (if $L \gg 1$) can be encoded so that it is transmitted without loss of information using only, but not less than, $LS_2[P]$ bits. You do not need more and you cannot use less (the probabilities of both events tend to zero for long messages). This coding is a *compression* of the data to be transmitted.

An extreme case of Shannon's theorem is given by $S = 0$. If the distribution of letters is known to both sender and receiver, and consists of a single letter, then no message is necessary. There is no uncertainty. The information to be exchanged is null.

We stress that it is necessary that both the sender and the receiver know the probability distribution P of the source. However, this is not a requirement that changes the situation much, because the transmission of the probability distribution itself takes a small number of bits, and we might safely put it beforehand on the message without losing terms proportional to L.

The capacity of the source is given by the maximum entropy, so every coding is equiprobable; see (14.17). According to Shannon's theorem, the knowledge of the probability distribution makes it possible to devise a coding that saves the transmission of $\hat{I} \equiv L(\log_2 N - S_2[P])$ bits, i.e., it allows the message to be compressed by $\hat{I}$ bits. We can call this $\hat{I}$ the information contained in the $\{P\}$ (concerning the transmission of a message of L letters). This should not be confused with the information to be transmitted to the receiver in order to decode the message sent by the source, which is precisely the entropy $LS_2[P]$, i.e., the average of the information (14.97) (see Appendix 14.A) of the symbols composing the messages.

14.1.4 Proof of Shannon's First Theorem

In the special case (14.5) in which all sequences are equiprobable, the proof of the theorem is immediate. The function $K(\mathbf{c})$ that we have previously defined, or any other coding algorithm $\mathcal{M}(\mathbf{c})$, satisfies the conditions of the theorem, and it is evident that it cannot be improved.

In the general case, the proof can be carried out in three steps.

1. Construct a coding following an appropriate algorithm $\mathcal{M}[\mathbf{c}]$.
2. Prove that this coding yields an information compression per character down to $S_2[P]$ (first thesis of the theorem).
3. Prove that such compression cannot be improved (second thesis of the theorem).

Proof

1. Construction of a coding.

The first step is very easy. We first note that each message $\mathbf{c}$ of L characters c_i is associated with a probability

$$\mathcal{P}_L(\mathbf{c}) = \prod_{i=1}^{L} P_{c_i}, \tag{14.18}$$

where P_{c_i} is the probability of the ith symbol in the message. Indeed, if the letters are chosen at random with the distribution $\{P\}$, the equation above gives precisely the probability of the message $\mathbf{c}$.

We can sort messages of length L in order of decreasing probability. In other words, if we denote by $O(\mathbf{c})$ the ranking number of the message in the ordered list (the *order number*), the chosen ordering implies that

$$\mathcal{P}_L(\mathbf{c}^{(1)}) > \mathcal{P}_L(\mathbf{c}^{(2)}) \quad \Longrightarrow \quad O(\mathbf{c}^{(1)}) < O(\mathbf{c}^{(2)}). \tag{14.19}$$

With this criterion, we can, for instance, also define the order number $O(\mathbf{c})$ as the integer between $1/\mathcal{P}_L(\mathbf{c})$ (inclusive) and $1 + 1/\mathcal{P}_L(\mathbf{c})$ (exclusive),

$$O(\mathbf{c}) = \left[\frac{1}{\mathcal{P}_L(\mathbf{c})} \right].$$

Finally, when the two order numbers of different messages are equal, we can resolve this degeneracy with some further criterion. For example, we can impose that these messages are ordered in lexicographic order, i.e., that

$$K(\mathbf{c}^{(1)}) < K(\mathbf{c}^{(2)}) \quad \Longrightarrow \quad O(\mathbf{c}^{(1)}) < O(\mathbf{c}^{(2)}). \tag{14.20}$$

Eventually, we have uniquely constructed our coding. We can, of course, also directly define the encoding $K(\mathbf{c})$ as an order number, though this is less effective in compressing.

2. Computation of the encoding information: $S_2[P]$.
Let us move on to the second part of the proof. We consider in general a function f_c of the letters. Since the letters are random variables, also f will be a random variable. As usual, we denote the expectation value by

$$\langle f \rangle = \sum_{c=1}^{N} f_c P_c. \tag{14.21}$$

With each specific message $\mathbf{c}$ of L letters, we can then associate the quantity $F(\mathbf{c})$, defined by

$$F(\mathbf{c}) = \frac{1}{L} \sum_{i=1}^{L} f_{c_i}. \tag{14.22}$$

This is nothing other than an empirical average of the f of a particular realization of a string of L characters. The law of large numbers (3.6) implies that the probability that the message $\mathbf{c}$ belongs to the set for which $|F(\mathbf{c}) - \langle f \rangle| > \epsilon$, with arbitrarily small ϵ, tends to zero when L tends to infinity. We apply this result to the case

$$f_c = \log P_c, \tag{14.23}$$

hence

$$\langle f \rangle = \langle \log P \rangle = -S[P], \tag{14.24}$$

$$F(\mathbf{c}) = \frac{1}{L} \sum_{i=1}^{L} \log P_{c_i} = \frac{1}{L} \log \mathcal{P}_L(\mathbf{c}). \tag{14.25}$$

The law of large numbers thus reads

$$\lim_{L \to \infty} P\left(\left| S[P] + \frac{1}{L} \log \mathcal{P}_L(\mathbf{c}) \right| > \epsilon \right) = 0$$

for arbitrary ϵ. Neglecting messages belonging to a subset whose total probability tends to zero for large L, we have

$$\exp\{-L(S[P] + \epsilon)\} < \mathcal{P}_L(\mathbf{c}) < \exp\{-L(S[P] - \epsilon)\}. \tag{14.26}$$

We will denote messages with this probability as *typical messages*. Under the same assumptions, this result implies that the order numbers for typical messages satisfy the constraints

$$\exp\{L(S[P] - \epsilon)\} < O(\mathbf{c}) < \exp\{L(S[P] + \epsilon)\}. \tag{14.27}$$

The typical message therefore belongs to a set of messages where the relation (14.26) is satisfied and whose order numbers scale with L as $\exp\{LS[P]\}$. In base 2 it reads $O(\mathbf{c}) \sim 2^{L\,S_2[P]}$.

We note that, when L is large, this order number of the coding is of the order of magnitude of the total number of messages belonging to the set for which the probability scales like $\sim e^{-LS_2[P]}$; see (14.26). Indeed, the latter must be of the order of $\exp\{LS[P]\}$, as each message has a probability of the order of $\exp\{-LS[P]\}$ and the total probability of all typical messages, the sum of the individual probabilities of each one of these messages, is equal to a finite number (close to 1):

$$\lim_{L \to \infty} \sum_{\mathbf{c} \in \text{typical}} \mathcal{P}_L(\mathbf{c}) = 1. \tag{14.28}$$

Since the number of bits required to transmit a message is the binary base logarithm of the coding, from (14.27) we find an information equal to

$$\log_2 O(\mathbf{c}) \simeq LS_2[P]$$

and we have proved the second part of the theorem.

We note that we have only used the law of large numbers here. We could have used the central limit theorem, as well, as the variance is finite (it is the sum of a finite number of terms), to obtain more precise estimates, but these are not difficult to derive.

3. $S_2[P]$ is the best possible compression.

In addition to this type of typical (or *central*) message, there may also be messages of length L with probability larger than $\exp\{-LS[P]\}$. Although a single message of this type is more likely than a single typical message, by the law of large numbers the total probability of atypical messages of finite length L is strictly smaller than 1 and tends to zero as $L \to \infty$:

$$\lim_{L \to \infty} \sum_{\mathbf{c}\, \in\, \text{atypical}} \mathcal{P}_L(\mathbf{c}) = 0. \tag{14.29}$$

Their number must, therefore, be very small and grow much more slowly than $\exp\{LS[P]\}$ as L grows.

Furthermore, there are other atypical messages that have an extremely large order number, greater than $\exp\{LS[P]\}$: their total probability is exponentially negligible when L tends to infinity, and they fall, as well, within a set of null measure of the law of large numbers, cf. Eq. (14.29).

To recapitulate, we have divided all possible transmitted sequences of length L into three sets, which we report in order of decreasing sequence probability.

(A) High-probability atypical messages. In this first set, the probability $\mathcal{P}_L(\mathbf{c})$ of transmitting a message $\mathbf{c}$ of length L is significantly larger than $2^{-LS_2[P]}$. The number of elements in this set is so much smaller than $2^{LS_2[P]}$, though, that they are irrelevant for the transmission of information.

(B) Typical messages. In this second set,

$$\left| \log_2 \frac{1}{\mathcal{P}_L(\mathbf{c})} - LS_2[P] \right| < L\epsilon, \tag{14.30}$$

for arbitrarily small ϵ, i.e., the probability $\mathcal{P}_L(\mathbf{c}) \sim 2^{-LS_2[P]}$. The number of elements in this set is approximately $2^{LS_2[P]}$ and their overall probability is $O(1)$, tending to 1 as $L \to \infty$.

(C) Extremely rare atypical messages. In this third set, the probability $\mathcal{P}_L(\mathbf{c})$ is much smaller than $2^{-LS_2[P]}$. These messages are potentially numerous; their order number might even increase more than $\exp\{LS[P]\}$, but they occur so rarely that their total probability is very small, vanishing in the limit of large L.

We have derived the typical information $S_2[P]$ (and proved the second part of the theorem) by exploiting the fact that the total probability of messages belonging to sets A and C above for large L is arbitrarily small, while the probability of set B of typical messages is arbitrarily close to 1.

To prove the third part of the theorem, it suffices to observe that the candidate messages with a number of bits per symbol less than $S_2[P]$ are the high-probability atypical messages (set A). Although single messages are more probable, for large L the cumulative probability of sequences in the A group tends to zero, and it is not possible to find a coding that allows the identification of a sequence with a number of bits per character less than $S_2[P]$. Hence, it is not possible to find an asymptotically better coding to compress the message without losing information. ∎

Very often the method outlined here is not practicable for large L, as one must examine all N^L sequences to determine the coding. There are methods that allow one to compress the message in simpler ways [88, 89]. However, the discussion of that issue, of great interest in communication and computer data compression, would take us too far off track and will not be pursued further.

14.1.5 Transmission of Information in the Presence of Noise

Until now, we have dealt with the problem of data compression without any loss of information, in the absence of transmission errors. A transmission channel, however, is generally not free of errors. If the errors are different with each communication of the same message, then we are in the presence of *noise*, which, as we know, is a random process. If, on the other hand, the errors in the message are always the same, we are in the presence of *distortion*. Distortion is a systematic, deterministic error that can be corrected by knowing the original message, and is of no interest to us here. Because of the noise, some bits will be wrong. To correct them, we should know which ones. We can quantify the information we need to make corrections as the information that was lost in transmission or, equivalently, as the uncertainty about the original message. Our favourite measure of uncertainty is, of course, entropy.

In this case, which we represent with the drawing in Figure 14.1, transmission involves two stochastic processes: the source generating the messages, with its distribution $\{P\}$, and the noise in their transmission. We will, therefore, have several entropies to consider in order to construct the significant quantities for describing the transmission as a whole:

- the entropy of the source $S[P^{(X)}]$, which is equal to the entropy of the input to the transmission channel, since the act of (reversible) coding does not increase the entropy;
- the entropy of the transmission channel output, $S[P^{(Y)}]$;
- the conditional entropies $S[P^{(X|Y)}]$ and $S[P^{(Y|X)}]$; and
- the joint entropy $S[P^{(X,Y)}]$, which, see (14.13), reads

$$S[P^{(X,Y)}] = S[P^{(X)}] + S[P^{(Y|X)}] = S[P^{(Y)}] + S[P^{(X|Y)}] \qquad (14.31)$$

as a function of the preceding ones.

Figure 14.1 Illustration of the process of transmitting a message through a noisy communication channel. The source is characterized by an alphabet of symbols and generates a message $\mathbf{c}$ composed of characters c_i that take values on the symbols. Before being transmitted, the message $\mathbf{c}$ is encoded into an input signal $\mathbf{X} = \mathcal{M}(\mathbf{c})$ by means of a function $\mathcal{M}$, see (14.16) for example. This is bi-univocal and, therefore, invertible. The input signal is transmitted through the channel, from which it exits, with some error due to noise, as the output $\mathbf{Y}$ signal. This is decoded using the inverse function of $\mathcal{M}$: $\hat{\mathbf{c}} = \mathcal{M}^{-1}(\mathbf{Y})$. Because of the noise, $\hat{\mathbf{c}} \neq \mathbf{c}$ and the difference between the two is given by the "equivocation" (see Section 14.1.5.1) $S(\mathbf{c}|\hat{\mathbf{c}}) = S_{X|Y}$.

14.1.5.1 Real Rate of Information Production and Transmission Equivocation

What is the measure of information lost in the transmission? How much of the input signal X is lost once the output Y is received? We know the output; all that remains is to quantify the entropy of the source conditioned on the received signal: $S[P^{(X|Y)}]$. This quantity was called *equivocation* by Shannon [84].

In the presence of noise, the real rate R of information actually transmitted (the number of correct bits per symbol) can then be expressed as the *mutual entropy*, defined as

$$R \equiv S^{(\mathrm{m})}[P^{(X)}, P^{(Y)}] = S[P^{(X)}] - S[P^{(X|Y)}], \tag{14.32}$$

i.e., the entropy in the noise-free transmission minus the equivocation caused by the noise. It is also called *mutual information*. Using the formula (14.31) we can rewrite it also as

$$S^{(\mathrm{m})}[P^{(X)}, P^{(Y)}] = S[P^{(Y)}] - S[P^{(Y|X)}] = S[P^{(X)}] + S[P^{(Y)}] - S[P^{(X,Y)}]. \tag{14.33}$$

The second expression represents the measure of information received minus the portion of information corrupted by the noise. The third expression makes explicit the symmetry of the mutual information under exchange of input and output, and corresponds to the average number of bits of the input and output signal in the channel minus their joint entropy. In practice, it indicates the number of bits common to input and output.

An extreme example is the case where input and output have nothing to do with each other. In (14.11) the equals sign applies, then, for independent X and Y and the mutual information is zero – zero bits in common.

If noise is present, the capacity (14.17) of the channel will be reduced compared to the information loss-less case. However, it is always defined as a maximum entropy over all possible source distributions: the maximum mutual entropy,

$$C = \max_{P^{(X)}}\{S^{(\mathrm{m})}[P^{(X)}, P^{(Y)}]\}. \tag{14.34}$$

The distribution $P_{\mathrm{opt}}^{(X)}$ maximizing the above expression is called the optimal input distribution. If the channel is noise-less, the equivocation $S[P^{(X)}|P^{(Y)}]$ is zero and the channel capacity reverts to the source capacity (14.17).

For convenience, we will use the abbreviations below:

$$S_X \equiv S_2[P^{(X)}], \quad S_Y \equiv S_2[P^{(Y)}], \quad S_{X,Y} \equiv S_2[P^{(X,Y)}],$$

$$S_{X|Y} \equiv S_2[P^{(X)}|P^{(Y)}], \quad \text{and} \quad S_{XY}^{(\mathrm{m})} \equiv S_2^{(\mathrm{m})}[P^{(X)}, P^{(Y)}].$$

In the presence of a transmission channel, we can also generalize the concept of *typicity*, cf. Eq. (14.28) or (14.30), of the L-characters-long messages generated by the source. The conditions of typical input and output signals will, therefore, be

$$\left|\log_2 \frac{1}{P^{(X)}} - LS_X\right| < L\epsilon, \tag{14.35}$$

$$\left| \log_2 \frac{1}{P(Y)} - L S_Y \right| < L\epsilon . \tag{14.36}$$

Furthermore, we can define the set of input/output *mutually typical* messages transmitted through the channel as those for which their joint probability distribution and their joint entropy for large L satisfy the constraint

$$\left| \log_2 \frac{1}{P(X,Y)} - L\, S_{X,Y} \right| < L\epsilon ,$$

with ϵ as small as desired.

14.1.6 Shannon's Noisy-Channel Coding Theorem (Second Theorem)

In the presence of noise in the transmission of a message, Shannon stated a second fundamental theorem [84].

Theorem. (Shannon's noisy-channel coding theorem) *Given a transmission channel of capacity C and a source of entropy S_X, then*

1. *if $S_X \le C$ there exists a coding such that the signal X from the source can be transmitted along the channel with an arbitrarily small equivocation,*
2. *if $S_X > C$ there exists a coding of the signal such that the equivocation $S_{X|Y}$ of the transmission is smaller than $S_X - C + \epsilon$, with $\epsilon > 0$ as small as you like,*

   but

3. *there exists no coding yielding an equivocation $S_{X|Y}$ strictly smaller than $S_X - C$.*

Figure 14.2 summarizes the contents of the three parts of the theorem.

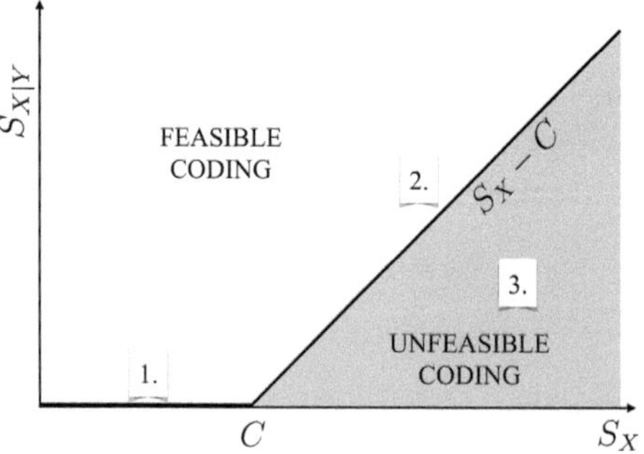

Figure 14.2 Behavior of the equivocation $S_{X|Y}$ in the transmission of a signal X through a noisy channel of capacity C (with known output Y) as a function of the original signal information S_X in the absence of noise. The three marked zones represent the cases predicted by Shannon's second theorem.

14.1.7 Proof of Shannon's Second Theorem

The three parts of the theorem are proved separately as follows.

Proof

1. If the information S_X is not larger than the channel capacity C, the equivocation in transmission can be arbitrarily small, $S_{X|Y} < \epsilon$.

Shannon's proof of the first part of the theorem is interesting in itself. Indeed, Shannon does not explicitly construct a coding with equivocation tending to zero, rather he demonstrates that at least one such coding exists in a certain set of codings. The stratagem he uses is simple: if the average of a set of numbers is less than a certain value ϵ, then at least one of the numbers must obviously be less than ϵ. This is not a necessary condition, but it is certainly sufficient. By the same principle, if we can somehow compute the expected value of the number of wrong bits by averaging over many codings and find that it is less than a certain value, then at least one coding carries an uncertainty less than that value. If, then, that value ϵ is arbitrarily small, we have proved the first part of the theorem.

We thus want to compute the average coding error on all typical codes, neglecting the atypical ones whose cumulative probability is very small, tending to zero for very long messages, cf. Eq. (14.29), and, thus, irrelevant in the average.

Let us first take as reference a source that has mutual entropy almost equal to the capacity C of the transmission channel. Here "almost" stands for "less than any small δ in the limit of long signals". The distribution of input signals is, therefore, $P_{\text{opt}}^{(X)}$; see (14.34).

Let us, then, consider sequences of length L transmitted and received. According to the source coding theorem, cf. Section 14.1.3, the average information in number of bits transmitted[5] as input is equal to LS_X.

Now consider:

- the set of typical input sequences of length L, according to (14.35), consisting of approximately 2^{LS_X} elements; and
- the set of typical output sequences of length L, according to (14.36), consisting of approximately 2^{LS_Y} elements.

Each typical output sequence can be produced at the source by about $2^{LS_{X|Y}}$ different inputs (corrupted by noise at different points). The cumulative probability of all other sequences is too small to be taken into account.

We then take a second generic source X', whose average rate of information R is less than the channel capacity: $R < C$. In other words, $P^{(X')}$ is not the optimal distribution defining the channel capacity (14.34) (indeed, we did not subscript it "opt").

---

[5] If we associate each character with a transmission time, instead of the length L of the messages (in characters), we can equivalently use the duration T of the transmitted sequence. In this case the entropy indicates the rate of information transmitted in number of bits per second, and the average information in the period T is TS_X [84].

To prove the theorem, we want to combine typical outputs with a selection of typical inputs, so that the frequency of errors on the transmitted bits is practically zero. Using exclusively typical inputs (defined according to our optimal reference case X), we establish all possible associations between inputs and outputs, generated by all possible X' coding, and we average the frequency of errors on the outputs over this set. This is equivalent to computing the average frequency of errors for a random association of input and output of length L.

Let us take a given received output, y^*. It certainly corresponds to an input transmitted through the channel. But it is corrupted, and several other inputs could have generated it. What is the probability that more than one input generates y^*?

Of the 2^{LS_X} typical input messages of length L, there are 2^{LR} transmitted through the corrupted channel. In Figure 14.3 every possible typical input is represented by a black dot. The transmitted messages are 2^{LR} and the probability that a black dot is a transmitted message is $2^{L(R-S_X)}$. Thus the probability that a black dot is not a message is $1 - 2^{L(R-S_X)}$.

The number of corrupted messages due to transmission channel noise is $2^{LS_{X|Y}}$. These are the ambiguous messages, which resemble the correct input message but are not. In Figure 14.3 they are represented by those black dots connected to the red dot y^*. The probability that none of these correspond to a real input message is

$$p_0 \equiv [1 - 2^{L(R-S_X)}]^{2^{LS_{X|Y}}} . \tag{14.37}$$

Now $R < C \equiv S_X - S_{X|Y}$ (cf. (14.32) and (14.34), we are omitting the subscripts "opt" to X). Given any $\eta > 0$ we can write

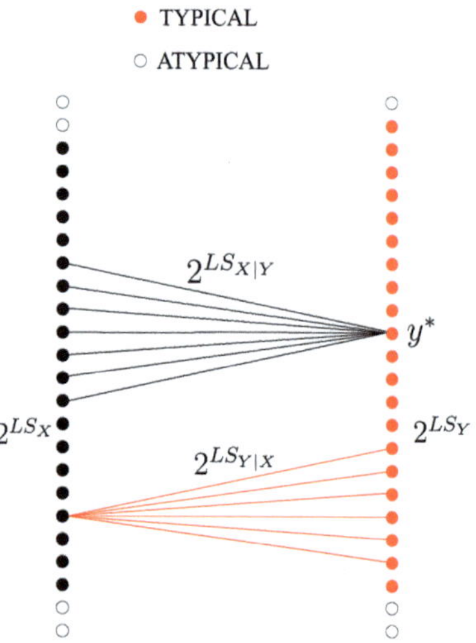

Figure 14.3 Relationship between input and output signals across a noisy transmission. Typical signals are represented by full points, atypical ones by open points. Input signals are black (open or full), output signals are red. Each of the 2^{LS_Y} noise-corrupted typical output signals y^* (full red points) can, in principle, be generated by any one among the $2^{LS_{X|Y}}$ input signals. Conversely, each of the 2^{LS_X} typical input signals (full black points) might correspond to $2^{LS_{Y|X}}$ different outputs due to the number of wrong bits in the transmitted signal.

$$R - C = R - S_X + S_{X|Y} = -\eta.$$

Substituting in (14.37) we arrive at

$$p_0 \equiv \left[1 - \frac{1}{2^{L\eta + LS_{X|Y}}} \right]^{2^{LS_{X|Y}}} \simeq 1 - \frac{1}{2^{L\eta + LS_{X|Y}}} 2^{LS_{X|Y}} = 1 - 2^{-L\eta} \to 1, \quad \text{for } L \to \infty.$$

$$(14.38)$$

For large signals the probability that there are no other inputs responsible for a given output tends to 1 and equivocation completely dissipates. Thus, on a random set of codings $X' = \mathcal{M}(\mathbf{c})$, the *average* probability of transmission errors tends to zero. This implies that among all codings for which $S_{X'} < C$, there necessarily exists at least one whose probability of error tends to zero, and the first part of the theorem is proved.

2. If the information S_X overcomes the channel capacity C, the equivocation can be smaller than $S_X - C + \epsilon$.

Let us come to the second part. If we had a source of entropy $S_{X'} > C$, the most we could do is transmit C bits per symbol, compulsorily losing the information $S_{X'} - C$ contained in the bits that are not transmitted, that will arrive nonsensically at the receiver. The equivocation in reception in this case reduces to

$$S_{X'|Y} = S_{X'} - C + \epsilon,$$

with $\epsilon > 0$ arbitrarily small. The second part of the theorem is proved.

3. No coding exists yielding equivocation smaller than $S_X - C$.

The third part of the theorem can be proved by contradiction. Let us assume that there exists a source coding $\tilde{X} = \mathcal{M}(\mathbf{c})$, whose entropy is still $S_{\tilde{X}} > C$, but generates an equivocation lower than $S_{\tilde{X}} - C$:

$$S_{\tilde{X}|Y} = S_{\tilde{X}} - C - \Delta, \quad \Delta > 0.$$

In this case the information production rate would be

$$R \equiv S_{\tilde{X}} - S_{\tilde{X}|Y} = C + \Delta,$$

higher than the channel capacity: $R > C$! This thus contradicts the definition of capacity (14.34) as the maximum mutual entropy. There is, hence, no coding $\tilde{X}$ with equivocation less than $S_{\tilde{X}} - C$, and the third part of the theorem is proved. ■

14.2 Entropy in Dynamical Systems

14.2.1 Lyapunov Principal Exponent

Shannon's microscopic entropy finds its generalization in the application to dynamical systems. This is so, even in the case of completely deterministic systems, where, however, the dynamics rules are such that an arbitrarily large departure of trajectories can develop starting from very similar initial conditions. This is the phenomenon of chaos, with which one can associate *entropy production* describing the loss of information about the position of the dynamical system as time goes by. This entropy production is called Kolmogorov–Sinai entropy [29, 90, 91]. In the following, we shall limit ourselves to making a few concise remarks on its definition and its relation to Lyapunov exponents.

As an example of a dynamical system, we consider the sequences

$$\vec{x}(n + 1) = \vec{f}(\vec{x}(n)), \tag{14.39}$$

where n is the step of the sequence (discrete time), $\vec{x}$ is an N-dimensional vector, and $\vec{f}$ is a smooth function with vector values that carries a finite set $\mathcal{D}$ (belonging to the N-dimensional space) into itself. The function $\vec{f}$ is also termed a *map*.

The previous expression can also be considered as the discretization of a differential evolution equation, a first-order, or even higher-order, one. Indeed, if we make the definitions $\vec{x} \equiv (x, v)$ and $\vec{f} \equiv (v, f)$, we can represent the second-order equation $\ddot{x} = f$ by means of the two first-order equations $\dot{v} = f$ and $\dot{x} = v$. In reality, all the considerations we are making here can be applied (using due precautions) to systems of differential equations of the type

$$\frac{d\vec{x}}{dt} = \vec{F}(x).$$

If $\vec{x}(0)$ belongs to the domain $\mathcal{D}$, the sequence of $\vec{x}$ is well defined and computable with certainty (there are no probabilistic elements): $\vec{x}(n)$ is deterministically given by

$$\vec{x}(n) = \vec{f}^{(n)}(\vec{x}(0)), \tag{14.40}$$

where $\vec{f}^{(n)}(x)$ indicates the application of the map function $\vec{f}$ for n times.

Let us assume that we know the initial condition with some precision. This uncertainty is reflected in the value of $\vec{x}$ at time n and may either increase or decrease with n. If the uncertainty decreases with n, the behavior at large times is not affected by small errors in the initial data. The interesting case is, instead, when the uncertainty becomes larger and larger with n. We can be more quantitative if we introduce the *Lyapunov principal exponent* Λ, which tells us how fast the dynamical system moves away from or closer to its initial conditions $\vec{x}$ given a certain initial uncertainty $\Delta\vec{x}$:

$$|\Delta\vec{x}(n)| \equiv \left|\vec{f}^{(n)}(\vec{x} + \Delta\vec{x}) - \vec{f}^{(n)}(x)\right| \sim |\Delta\vec{x}|e^{n\Lambda}. \tag{14.41}$$

The above behavior is valid in the region of small initial $|\Delta\vec{x}|$, large n, but small $|\Delta\vec{x}(0)| \exp(n\Lambda)$. In stricter terms, the Lyapunov principal exponent is defined as

$$\Lambda \equiv \lim_{n \to \infty} \lim_{|\Delta\vec{x}| \to 0} \frac{1}{n} \ln \frac{|\Delta\vec{x}(n)|}{|\Delta\vec{x}|}. \tag{14.42}$$

A positive Lyapunov principal exponent implies that two trajectories that differ by a small amount in the initial condition will differ by a large quantity as soon as

$$n\Lambda > -\ln(|\Delta\vec{x}|),$$

i.e., as soon as the quantity $|\Delta\vec{x}| \exp(n\Lambda)$ ceases to be small.

The previous formulation can be put into a more precise form by using derivatives and limits, but, for simplicity, here we will just try to give the main ideas. By introducing the various $\alpha = 1, \ldots, N$ components of the N-dimensional phase space, we can write

$$f_\alpha^{(n)}(\vec{x} + \Delta\vec{x}) - f_\alpha^{(n)}(\vec{x}) \approx \sum_{\gamma=1}^{N} T_{\alpha\gamma}^{(n)}(\vec{x})\Delta x_\gamma\,. \tag{14.43}$$

In other words, a sphere of small radius centered at $\vec{x}(0)$ is transformed into an ellipsoid at time n, unless the matrix $\mathbb{T}^{(n)}$ is proportional to the identity matrix for each n, in which case we always have a sphere. The eigenvalues of the matrix $\mathbb{T}^{(n)}$, for large n, are proportional to $\exp(n\Lambda(\alpha))$, where the $\Lambda(\alpha)$ are sorted in decreasing order. The exponent of the largest eigenvalue, $\Lambda(1)$, coincides with Lyapunov's principal exponent Λ.

In the good cases, in which the functions $\vec{f}$ satisfy a number of conditions that we are not going to list, the Lyapunov exponents do not depend on the starting point $\vec{x}(0)$ (apart from exceptional points of zero size in the domain $\mathcal{D}$). The list of Lyapunov exponents gives much more detailed information on the behavior of the system of the main exponent alone. It is, in fact, possible that the uncertainty in one direction increases with n while it decreases in other directions.

As as example, we can take the evolution of Hamiltonian systems (or a suitable discretization of them) for which the Liouville theorem implies that the volume of the phase space remains invariant [92]: our initial sphere of radius $|\Delta\vec{x}|$, centered at $\vec{x}$, evolves with time into an ellipsoid of equal volume. This is proportional to

$$V(\mathcal{D}) \propto \prod_{\alpha=1}^{N} \exp(n\Lambda(\alpha))\,, \tag{14.44}$$

and since it is invariant in time n, we have that in these systems the Lyapunov exponents must satisfy the relation

$$\sum_{\alpha=1}^{N} \Lambda(\alpha) = 0\,. \tag{14.45}$$

Therefore, if there are positive Lyapunov exponents, there must also be negative ones. Consider, for example, the function f defined on a one-dimensional space:

$$f(x) = ax \bmod(1)\,, \quad x \in [0, 1)\,. \tag{14.46}$$

This function brings the interval $[0, 1)$ into itself. We can distinguish three cases:

- $a < 1$. In this case $x(n)$, the iteration of (14.46) at the nth step, is given by $f^{(n)}(x) = a^n x(0) = x(0)^{n\ln a}$ and the exponent Λ, cf. (14.42), is given by $\ln a < 0$.
- $a = 1$. The transformation is trivial and $\Lambda = 0$.
- $a > 1$. Also in this case, at the nth iteration, $x(n) = a^n x(0) \bmod(1)$ and the exponent Λ is given by $\ln a$, but this time it is positive.

Generalizing the previous example, we can consider the function $\vec{f}$ defined on a two-dimensional space:

$$f_1(x, y) = ax \bmod(1), \quad f_2(x, y) = by \bmod(1). \tag{14.47}$$

It is easy to see that the function $\vec{f}$ carries a point of the square $x \in [0, 1)$, $y \in [0, 1)$ inside the same square, and that the two Lyapunov exponents are $\log(a)$ and $\log(b)$. The case where $a > 1$ and $b < 1$ corresponds to an increase in uncertainty on x and a decrease in uncertainty on y as n increases.

14.2.2 Entropy Production

The basic idea of Kolmogorov and Sinai is to characterize the loss of information on the point position $\vec{x}(n)$ as n increases by an increase in the entropy of the probability of the point $\vec{x}$. Proceeding in a naive manner, we could suppose that, at the initial time, the point $\vec{x}(0)$ is known with very good precision and is characterized by a probability $P^{(0)}$ concentrated in a small region of the phase space and, therefore, by a very low entropy

$$S(0) \equiv S[P^{(0)}]. \tag{14.48}$$

As n increases, if the trajectory moves away from the initial point and can occupy an increasing region of space, the entropy will increase. Entropy is *produced* in the dynamics. For intermediate values of n, we can have a linear growth regime of the type

$$S(n) \equiv S[P^{(n)}] = S(0) + nI, \tag{14.49}$$

where I is, by definition, the loss of information per unit of time. The above formula, especially in the case of positive I, cannot be valid for n that are too large, as the entropy cannot increase disproportionately. We know, indeed, that it is certainly always smaller than the maximum entropy of the volume $V(\mathcal{D})$ of the phase space domain $\mathcal{D}$ of definition of the system: $S(n) \leq \log V(\mathcal{D})$. Taking the initial entropy $S(0)$ sufficiently small, however, the range in n where a linear increase is observed becomes arbitrarily large.

The definition (14.49) of I is perfectly legitimate, but has the disadvantage of giving, in the example we defined above, formula (14.47),

$$I = \log(ab) = \log(a) + \log(b). \tag{14.50}$$

In this expression, any increase in uncertainty in x if $a > 1$ can be compensated by a decrease of uncertainty in y if $b < 1$ and $ab < 1$. Even if a is larger than 1, we may thus have a negative information loss I, despite the fact that the position of the system for large n is completely random in the coordinate x (albeit very close to the line $y = \text{constant}$).

More generally, it is possible to prove that

$$I = \sum_{\alpha=1}^{N} \Lambda(\alpha). \tag{14.51}$$

The compensation between positive and negative exponents that we saw in the previous example is, in particular, always present for Hamiltonian systems, where the volume of the phase space is conserved, due to Liouville's theorem. The entropy (14.49) $S(n)$ does not depend, therefore, on n and we obtain the trivial result $I = 0$. The definition of I does not give us a good parameter to quantify the loss of information in the Hamiltonian case.

14.2.3 The Kolmogorov–Sinai Entropy

Kolmogorov and Sinai gave a definition of the loss of information or, equivalently, of entropy production, associated with dynamic processes that well captures the increase in uncertainty even along one direction of the space $\mathcal{D}$. In the example (14.47) this occurred in the region $a > 1$ and $b < 1$. In general, in other systems with Lyapunov exponents of different signs, the Kolmogorov–Sinai entropy does not have the disadvantages of the previous definition, but this is slightly more complicated. Here we present a version of it that is as simplified as possible.

We start as in the previous case from a $\vec{f}(\vec{x})$ function that brings a limited region of the N-dimensional space into itself. We take a small number ϵ and define integer coordinates

$$i_\alpha = \text{int}(x_\alpha \epsilon^{-1}), \quad \alpha = 1, \ldots, N, \tag{14.52}$$

where $\text{int}(z)$ is the integer part of z. The space is then divided into many small boxes of volume ϵ^N consisting of all points having the same integer part coordinates $\vec{i} = \{i_1, i_2, \ldots, i_N\}$. We denote each of these regions by $B(\vec{i})$. Given a probability density $p(\vec{x})$ we can define a probability $P(\vec{i})$ as the probability of the box labeled by $\vec{i}$, i.e.,

$$P(\vec{i}) = \int_{\vec{x}\,\in\,B(\vec{i})} \mathcal{D}x\; p(\vec{x}). \tag{14.53}$$

For convenience we will identify the little box $\vec{i}$ in space with its lexicographic coordinate[6] ℓ. We can then define an ϵ-regularized entropy in the discrete space, i.e., the entropy of the distribution P_ℓ:

$$S_\epsilon[P] = -\sum_\ell P_\ell \log P_\ell. \tag{14.54}$$

In the case in which the probability density $p(\vec{x})$, whose microscopic entropy is given by (14.2), is a sufficiently regular function, it is easy to see, for small ϵ, that $p(\vec{x})$ is approximately constant within each box,

$$P_\ell \simeq \epsilon^N p(\vec{x} \in B(\ell)),$$

---

[6] Just as we have seen for the encoding (14.16) of a word into a number, we can encode a coordinate vector into a number that bijectively corresponds to the vector. For example, in two dimensions ($N = 2$), if each discrete coordinate takes values $i_\alpha = 0, 1, \ldots, L - 1$, we can define the lexicographic coordinate $\ell = i_1 + L i_2$, where $\ell \in [0, L^2 - 1]$. In N dimensions, one has $\ell = \sum_{k=1}^{N} i_k L^{k-1}$.

and thus we have the following relationship between discrete entropy and exact entropy:

$$S_\epsilon[P] = S[P] - N \log(\epsilon). \tag{14.55}$$

Obviously, the previous relation has no possibility of being true if the function $p(\vec{x})$ has variations on the scale ϵ, because the regularized entropy is absolutely blind to these variations.

The definition of the ϵ-regularized entropy is natural from the physical point of view. In fact, if we can measure the position of a system with a certain sensitivity ϵ, we are not able to distinguish systems whose coordinates differ by less than ϵ. If we perform measurements on a set of systems, we can, by computing a histogram, reconstruct the probabilities P_ℓ and from these reconstruct the corresponding ϵ-adjusted entropy, but we can never compute the true entropy.

Let us now take an initial distribution that is constant inside one of the small boxes and zero elsewhere. In this case the initial entropy is zero. If we evolve such a distribution and compute its regularized entropy, this entropy will be sensitive to the fact that the probability distribution widens in some directions, but will not be sensitive to a narrowing of the probability distribution in other directions, since this narrowing produces variations on a scale smaller than ϵ.

For values of n that are not too large, we will have a linear increase of the entropy with time,

$$S_\epsilon(n) = S_\epsilon[P^{(n)}] = \text{constant} + Kn, \tag{14.56}$$

and, in this case, we can take the rate of change of the entropy, K, as a measure of the increase of entropy and, hence, of the decrease of information. The quantity K is called the Kolmogorov–Sinai entropy.

In the case of the transformation (14.47), it is easy to see that $K = \ln a$ in the case $b < 1$. In the case $a > 1$ and $b > 1$, we have $K = \ln a + \ln b$; while in the case $a < 1$ and $b < 1$, we have $K = 0$: there is no entropy production as time goes by.

More generally, in the case of N dimensions, the following Pesin's formula is valid, stating that the Kolmogorov–Sinai entropy is the sum of all the positive Lyapunov exponents:

$$K = \sum_{\alpha=1}^{N} \Lambda(\alpha)\theta(\Lambda(\alpha)). \tag{14.57}$$

Pesin's formula reproduces the results for K in the case of the transformation (14.47).

The Kolmogorov–Sinai entropy is positive as long as there is one positive Lyapunov exponent and, therefore, can have positive values even in Hamiltonian systems and, more generally, for transformations that leave the volume unchanged. Indeed, it was demonstrated by Sinai that, for simple Hamiltonian systems, the Kolmogorov–Sinai entropy is positive [91]. The simplest example is a ball moving on a billiard table with "walls" of an appropriate shape.

14.3 Intermezzo: Fundamentals of Statistical Mechanics

Since, in the following, we will often refer to the properties of statistical mechanics when discussing entropy, we want to give a quick reminder of some basic concepts for those who may need them. For a more complete study, we refer to any textbook of basic statistical mechanics, for example, Huang's text [57].

14.3.1 Boltzmann–Gibbs Distribution

In this section, we consider only classical systems. We denote by $\mathcal{H}[\mathbf{x}]$ the Hamiltonian function of the configurations $\mathbf{x}$ of the system. We define the partition function

$$Z(\beta) = \int \mathcal{D}\mathbf{x} \, \exp\{-\beta \mathcal{H}[\mathbf{x}]\}, \tag{14.58}$$

where $\beta = (k_B T)^{-1}$ is the inverse temperature, k_B is the Boltzmann constant, and T is the absolute temperature. In the case of a system with a discrete number of configurations, the integral is replaced by a sum.

Standard arguments [57] imply that for a system in contact with a thermal bath, or for a large isolated system of which we observe only a small part, after a sufficiently long time the probability distribution is the Boltzmann–Gibbs distribution:

$$P_{\mathrm{eq}}[\mathbf{x}] = \frac{\exp\{-\beta \mathcal{H}[\mathbf{x}]\}}{Z(\beta)}. \tag{14.59}$$

This distribution characterizes thermal equilibrium.

All systems in contact with a thermal bath reach thermal equilibrium sooner or later, in theory. One must, however, take into account that in the real world there are physical systems that reach equilibrium after much longer times than those typical of human experiments: such systems find themselves out of equilibrium, sometimes in states that are metastable. The study of systems with metastable states is much more complicated than that of thermal equilibrium, as the system's behavior depends on the system's history. A typical example is metal alloys, where the physical properties (stress resistance, brittleness, response to changes in magnetic fields) depend on how the alloy has been cooled: quickly (think of the typical scene in a movie about knights, where the glowing sword is quickly immersed in water), slowly, or by undergoing certain thermal cycles. Metallic alloys with magnetic impurities are called *spin-glasses*. Another example is proper window glass, for which the same transition temperature from liquid to glassy amorphous solid depends on the cooling rate.

14.3.2 Linear Response Theorem

In general, given a function of microscopic configurations $A[\mathbf{x}]$, its mean expected value at equilibrium at a given value of β is defined as

$$\langle A \rangle \equiv \int \mathcal{D}\mathbf{x}\, P_{\mathrm{eq}}[\mathbf{x}] A[\mathbf{x}] = \frac{\displaystyle\int \mathcal{D}\mathbf{x}\, A[\mathbf{x}] \exp\{-\beta \mathcal{H}[\mathbf{x}]\}}{Z(\beta)}, \tag{14.60}$$

where, to lighten the notation, we have not indicated the dependence on β of $\langle A \rangle$. Let us consider a Hamiltonian linearly dependent on a small parameter ϵ, for example,

$$\mathcal{H}_\epsilon[\mathbf{x}] = \mathcal{H}_0[\mathbf{x}] + \epsilon B[\mathbf{x}]. \tag{14.61}$$

The linear response theorem [56, 59, 93] states that

$$\frac{d}{d\epsilon}\langle A \rangle_\epsilon \bigg|_{\epsilon=0} = \langle AB \rangle_c \equiv \langle AB \rangle - \langle A \rangle\langle B \rangle, \tag{14.62}$$

where $\langle \cdot \rangle_\epsilon$ denotes the expectation value computed with the Hamiltonian $\mathcal{H}_\epsilon[\mathbf{x}]$, while the expectation values on the right-hand side of the equation were computed with the Hamiltonian $\mathcal{H}_0[\mathbf{x}]$. The quantity $\langle AB \rangle_c$ is the connected correlation function, see Section 12.3. The proof is simple: just write $\langle A \rangle_\epsilon$ explicitly as in formula (14.60) with $\mathcal{H}_\epsilon[\mathbf{x}]$ in place of $\mathcal{H}[\mathbf{x}]$, perform the derivative with respect to ϵ, and at the end set $\epsilon = 0$.

The formula (14.62) is the static version of the fluctuation–dissipation theorem. Normally, the fluctuation–dissipation theorem in its complete form involves time-dependent correlation and response functions [56, 59].

14.3.3 Thermodynamic Potentials

From the partition function, one can reconstruct the various thermodynamic quantities of the system. For example, the free energy $F(\beta)$ and the internal energy $E(\beta) \equiv \langle H \rangle$ are given, respectively, by

$$F(\beta) = -\frac{1}{\beta} \ln Z(\beta), \tag{14.63}$$

$$E(\beta) = \langle H \rangle = \frac{\partial}{\partial \beta} \ln Z(\beta). \tag{14.64}$$

Using Shannon's definition of microscopic entropy, cf. Eq. (14.1), the free energy can also be written as

$$F(\beta) = E(\beta) - TS(\beta), \tag{14.65}$$

where we use the units with $k_B = 1$, and for the computation of S we have taken $P_{eq}[\mathbf{x}]$ as the probability distribution. Indeed

$$\log P_{\mathrm{eq}}[\mathbf{x}] = -\beta \mathcal{H}[\mathbf{x}] - \ln Z(\beta),$$

and thus

$$S[P_{\mathrm{eq}}] = -\langle \log P_{\mathrm{eq}} \rangle = \beta E(\beta) - \beta F(\beta), \tag{14.66}$$

whence the equation (14.65). A detailed computation shows that the thus-defined thermodynamic function S is the macroscopic entropy, pertaining to a macroscopic thermodynamic state, to which various microscopic configurations correspond, distributed according to the Boltzmann–Gibbs distribution (14.59). Macroscopic entropy is the heat integrating factor, defined in the second law of thermodynamics, from the relation between quasi-static adiabatic variation of heat and entropy [57, 83, 94],

$$\frac{\delta Q}{T} = dS \, .$$

Here S is the macroscopic entropy, introduced by Clausius [83] in thermodynamics as a measure of the irreversibility of the process in which heat exchange takes place.

Finally, we may note that if we consider the thermodynamic function

$$\Phi(\beta) = \beta F(\beta), \tag{14.67}$$

its Legendre transform (see Section 4.5) is precisely the entropy $S(E)$, with conjugated variables E and β. Indeed, Eq. (14.65) is nothing other than $\Phi(\beta) = \beta E - S(E)$ and Eq. (14.64) is the maximum equation

$$E = -\frac{\partial \Phi(\beta)}{\partial \beta}$$

relating E and β. We will return to this relationship in more detail in Section 14.5.

14.3.4 Variational Principles

It is commonly said that the free energy of an isolated system cannot increase. This statement can be formalized by introducing an ensemble of systems and a probability distribution $P[\mathbf{x}]$. For each possible probability distribution P, we associate a free energy functional $F[P]$. We wish to demonstrate that, as the distribution $P[\mathbf{x}]$ of the configurations varies, the functional $F[P]$ assumes its minimum value when $P[\mathbf{x}] = P_{eq}[\mathbf{x}]$, given by Eq. (14.59). The statement that $F[P]$ is a monotonic non-increasing function of time implies that systems cannot move away from the equilibrium state once it has been reached.

We start with the definition of the entropy of a probability distribution $S[P]$, cf. formulas (14.1) and (14.2). We then define the variational free energy as

$$F[P] = U[P] - TS[P], \tag{14.68}$$

where

$$U[P] = \langle H \rangle_P \equiv \int \mathcal{D}\mathbf{x}\, P[\mathbf{x}]\mathcal{H}[\mathbf{x}] \, . \tag{14.69}$$

A simple computation based on Lagrange multipliers [85] shows that the minimum of $F[P]$ is given by the equilibrium Boltzmann–Gibbs distribution (14.59).

The reader interested in learning more about entropy in statistical mechanics and the use of the variational principle for the approximate computation of free energy, known as the mean-field approximation, can consult a book on statistical mechanics, for example, ref. [94].

Let us now look at a very important example of a variational principle for the quantity to which this chapter is devoted: the maximum entropy principle.

14.4 Maximum Entropy Principle

Given a multi-dimensional system $\mathbf{x}$, which we know to be stochastic, we would like to estimate its probability distribution $\{P\}$ in the least conditional way possible, accepting the maximum possible amount of ignorance compatible with the phenomenon. This is a bit like we did in Chapters 5 and 6 when we knew nothing about the distribution of the data except that it was a distribution, and that the values $\mathbf{x}_i$ of the data measured at the ith measurement were in its event space. We now denote by $\mathbf{x}_i$ one of the $\mathcal{N}$ possible configurations of the process $\mathbf{x}$. The first condition will be the normalization condition:

$$\sum_{i=1}^{\mathcal{N}} P[\mathbf{x}_i] = 1 \,. \tag{14.70}$$

We add that we want to represent a process on which a generic function $A[\mathbf{x}]$ has a known mean value $\langle A \rangle$. Hence the second condition is:

$$\sum_{i=1}^{\mathcal{N}} P[\mathbf{x}_i] A[\mathbf{x}_i] = \langle A \rangle \,. \tag{14.71}$$

As we have seen in Section 14.1.1, information theory unambiguously defines the amount of ignorance of a stochastic process characterized by a certain probability distribution: the entropy (14.1). The minimum conditioning on $\{P\}$ therefore corresponds to the maximum entropy principle [85]. To determine the least distorted $\{P\}$ by any spurious conditions, what we need to do is to look for the maximum Shannon entropy on all distributions constrained only by the conditions (14.70) and (14.71):

$$\max_{\{P\}} S[P] \bigg|_{\substack{\sum_i P[\mathbf{x}_i]=1, \\ \sum_i P[\mathbf{x}_i]A[\mathbf{x}_i]=\langle A \rangle}} = \max_{\{P\}} \{\tilde{S}[P|\mu,\lambda]\},$$

where, in

$$\tilde{S}[P|\mu,\lambda] \equiv - \sum_{i=1}^{\mathcal{N}} P[\mathbf{x}_i] \ln P[\mathbf{x}_i] - \lambda \sum_{i=1}^{\mathcal{N}} P[\mathbf{x}_i] - \mu \sum_{i=1}^{\mathcal{N}} P[\mathbf{x}_i] A[\mathbf{x}_i], \tag{14.72}$$

we have introduced the Lagrange multipliers λ and μ to implement the constraints.

By imposing the maximum conditions

$$\frac{\partial \tilde{S}}{\partial P[\mathbf{x}_i]} = 0 \quad \Longrightarrow \quad -\ln P[\mathbf{x}_i] - 1 - \lambda - \mu A[\mathbf{x}_i] = 0 \,,$$

we obtain the form

$$P_{\max}[\mathbf{x}_i] = \frac{e^{-\mu A[\mathbf{x}_i]}}{e^{\lambda+1}} \,.$$

We now have to find the optimal values of the two multipliers. We obtain a first equation from the constraint 1 on $P_{\max}$:

$$e^{\lambda+1} = \sum_{i=1}^{\mathcal{N}} e^{-\mu A[\mathbf{x}_i]} \equiv Z(\mu), \tag{14.73}$$

where the last abbreviation will be called the *partition function*. We obtain the other equation by observing that

$$\frac{\partial \ln Z(\mu)}{\partial \mu} = \frac{1}{Z(\mu)} \sum_{i=1}^{\mathcal{N}} e^{-\mu A[\mathbf{x}_i]}(-A[\mathbf{x}_i]) = -\langle A \rangle, \tag{14.74}$$

a value set by the second constraint. In the end, using the substitution (14.73), we have that the distribution obtained according to the maximum entropy principle is

$$P_{\max}[\mathbf{x}] = \frac{e^{-\mu A[\mathbf{x}]}}{Z(\mu)}, \tag{14.75}$$

and the maximum entropy is written as

$$S[P_{\max}] = S(\langle A \rangle) = \mu \langle A \rangle + \ln Z(\mu), \tag{14.76}$$

where μ is determined by Eq. (14.74).

As we shall see shortly, the maximum entropy principle naturally leads to the Boltzmann–Gibbs distribution. If, instead of an average, we can set two known averages, the whole procedure generalizes with the use of two Lagrange multipliers μ_1 and μ_2.

14.4.1 A Probabilistic Derivation of Statistical Mechanics

Let us make an observation. If we just rename $\langle A \rangle = E$ and $\ln Z(\mu) = \Phi(\mu)$, Eqs. (14.74) and (14.76) can be rewritten as

$$S(E) = \mu E + \Phi(\mu), \tag{14.77}$$

$$E = -\frac{\partial \Phi}{\partial \mu}. \tag{14.78}$$

As we have already seen in Section 4.5, the entropy and the logarithm of the partition function are Legendre transforms of each other with conjugate variables μ and E. For the involutive property of the Legendre transform, it also holds that

$$\Phi(\mu) = -\mu E + S(E), \tag{14.79}$$

$$\text{with} \quad \mu = \frac{\partial S}{\partial E}. \tag{14.80}$$

A special and particularly significant case is when $\mathbf{x}$ are the configurations of the degrees of freedom of a Hamiltonian system and $A[\mathbf{x}] = \mathcal{H}[\mathbf{x}]$ is the Hamiltonian function representing the energy of the configuration $\mathbf{x}$, whose mean value $\langle \mathcal{H} \rangle = E$ is the internal energy. In this case, Eq. (14.75) becomes

$$P_{\max}[\mathbf{x}] = \frac{e^{-\mu \mathcal{H}[\mathbf{x}]}}{Z(\mu)}, \qquad (14.81)$$

and Eq. (14.80) is Maxwell's relation, which in statistical mechanics defines the inverse temperature $\beta = 1/T$ (barring multiplicative constants, like the Boltzmann constant). Therefore, $\mu = \beta$ and consequently $P_{\max}[\mathbf{x}] = P_{\mathrm{eq}}[\mathbf{x}]$ is the equilibrium Boltzmann–Gibbs distribution (14.59), and

$$F(\beta) = -\frac{\Phi(\beta)}{\beta}$$

is the Helmholtz free energy of the system.

Hence, we obtain the well-known formulas (14.65) and (14.66) relating the thermodynamic potentials of the microcanonical and the canonical ensembles, i.e., the entropy

$$S(E) = \beta E - \beta F(\beta),$$

and the free energy

$$F(\beta) = E - T S(E).$$

As is well known, in statistical mechanics the state of maximum entropy denotes the thermodynamic state in the microcanonical ensemble, while the minimum free energy denotes the thermodynamic state in the canonical ensemble. Just applying the maximum entropy principle to a Hamiltonian system of statistical mechanics, we find the distribution (14.59) of the equilibrium configurations in the canonical ensemble, the Boltzmann–Gibbs distribution. We did it all starting from the probabilistic definition of entropy derived by Shannon.

We also derived that entropy and free energy are the Legendre transforms of one another. This is a general property that holds as long as the ensembles are equivalent. We have studied Legendre transforms in the context of the theory of large deviations in Chapter 4, Section 4.5, where we briefly mentioned the example of statistical mechanics in the formula (4.60). The time has finally come to take up and explore this relationship between large deviations and thermodynamics.

14.5 Large Deviations and Thermodynamics

In equilibrium statistical mechanics, many problems can be formulated in variational form. For example, we can consider a system composed of N variables x_i (discrete or continuous), on which a Hamiltonian function $\mathcal{H}[\mathbf{x}]$ is defined and which are at equilibrium, distributed according to (14.59). The partition function $Z_N(\beta)$ can be written with the formula (14.58). If the Hamiltonian is not pathological, the limit for N tending to infinity of the free energy density (14.63) exists and is finite:

$$f(\beta) = \lim_{N\to\infty} -\frac{1}{\beta N} \log Z_N(\beta). \tag{14.82}$$

For the study of large deviations, we define the random variable sum of N random variables,

$$A_N[\mathbf{x}] \equiv \sum_{i=1}^{N} x_i.$$

Generalizing what was done in Chapter 4, we consider the $\mathbf{x}$ variables to be correlated (see Section 12.4.1), rather than independent, and study the behavior of the large deviations: $A_N = \lambda N$, with finite λ. To do so, we define the generating function of the moments of A_N at equilibrium, a generalization of the function $e^{R(h)}$ in (4.14), i.e., the mean value over the Boltzmann probability distribution of e^{hA_N}:

$$e^{R_N(h)} \equiv \langle e^{hA_N} \rangle = \frac{\displaystyle\int \mathcal{D}\mathbf{x}\, e^{-\beta\mathcal{H}[\mathbf{x}]+hA_N[\mathbf{x}]}}{Z_N(\beta)} = \frac{Z_N(\beta, h)}{Z_N(\beta)}. \tag{14.83}$$

Here in the last step we defined the partition function and the Hamiltonian perturbed by a field h/β:

$$Z_N(\beta, h) \equiv \sum_{\{\mathbf{x}\}} e^{-\beta\mathcal{H}_h[\mathbf{x}]}, \tag{14.84}$$

$$\mathcal{H}_h[\mathbf{x}] \equiv \mathcal{H}[\mathbf{x}] - \frac{h}{\beta} A_N[\mathbf{x}]. \tag{14.85}$$

The corresponding free energy density $f(\beta, h)$ is defined as

$$f(\beta, h) = \lim_{N\to\infty} -\frac{1}{\beta N} \log Z_N(\beta, h). \tag{14.86}$$

From (14.83) it is easy to see that

$$\mathcal{R}(h) \equiv \lim_{N\to\infty} \frac{R_N(h)}{N} = \lim_{N\to\infty} \frac{\log\langle e^{hA_N[\mathbf{x}]}\rangle}{N} = -\beta[f(\beta, h) - f(\beta, 0)]. \tag{14.87}$$

In other words, the problem of computing the cumulant generating function of the random variable $A_N[\mathbf{x}]$ coincides with the problem of computing the variation of free energy density when adding a perturbation proportional to $A_N[\mathbf{x}]$ in the Hamiltonian. Knowing the probability distribution of the quantity A_N in the region of large deviations, $A_N = \lambda N$, is equivalent to computing the partition function of a statistical mechanics problem for arbitrary values of the temperature. Let us call $S_N(\lambda)$ the decrease function in extended form: $S_N(\lambda) = N S(\lambda)$. If the function $\mathcal{R}(h)$ (14.87) is differentiable for every h, then by the Gärtner–Ellis theorem, see Section 4.6.2, the distribution $d_N(\lambda)$ satisfies the large deviations theorem,

$$d_N(\lambda) \sim e^{S_N(\lambda)} = \exp\{-N[\bar{h}(\lambda)\lambda - \mathcal{R}(\bar{h}(\lambda))]\} \propto \exp\{-N[\bar{h}(\lambda)\lambda + \beta f(\beta, \bar{h}(\lambda))]\}, \tag{14.88}$$

Table 14.1 Extension of the definitions of large deviations functions to the multi-dimensional correlated case.

Formula	I.i.d. variables		Correlated variables	Expression name
(4.14)	$e^{R(h)} = \int dx\, e^{hx} p(x)$	$\implies$	$e^{R_N(h)} = \int \mathcal{D}\mathbf{x}\, e^{hA_N[\mathbf{x}]} P[\mathbf{x}]$	cumulants generator
(4.16)	$S(\lambda) = -h\lambda + R(h)$	$\implies$	$S_N(\lambda) = -h\lambda N + R_N(h)$	LD decrease function
(4.17)	$p_h(x) = p(x)\, e^{hx - R(h)}$	$\implies$	$P_h[\mathbf{x}] = P[\mathbf{x}]\, e^{hA_N[\mathbf{x}] - R_N(h)}$	perturbed distribution
(4.18)	$dR/dh = \langle x \rangle_h = \lambda$	$\implies$	$dR_N/dh = \langle A_N[\mathbf{x}] \rangle_h = \lambda N$	saddle point equation

with $\bar{h}(\lambda)$ determined by the saddle point equation (4.18), which can be rewritten in this case as

$$\frac{d\mathcal{R}(h)}{dh} = -\frac{\partial \beta f(\beta, h)}{\partial h} = \lambda. \tag{14.89}$$

One could validly support the view (perhaps a little too one-sided) that all statistical mechanics is a special case of the theory of large deviations.

If we review the theory of large deviations in Chapter 4, Sections 4.2–4.4, it is perhaps useful to schematically report in Table 14.1 the relationship between the functions $R(h)$ and $S(\lambda)$ for independent variables, and the related functions $R_N(h)$ and $S_N(\lambda)$ in the case of correlated variables, as, for example, the dynamic variables of an interacting system of statistical mechanics.

In Section 4.2 we considered the sum of independent and identically distributed variables, whose distribution was therefore factorized as

$$P[\mathbf{x}] = \prod_{i=1}^{N} p_i(x_i) = \prod_{i=1}^{N} p(x_i),$$

and we introduced the function $R(h)$, the logarithm of the one-dimensional Laplace transform of $p(x)$. We are now considering the case of N variables whose distribution (14.59) depends on the function $\mathcal{H}[\mathbf{x}]$, which may be the Hamiltonian of a system of interacting variables, such as, for example, in the case of the Ising model studied in Section 11.3.4. Generalizing the formulas in Section 4.2 to the case of the sum $A_N[\mathbf{x}] = \sum_{i=1}^{N} x_i$ of N variables that are no longer independent, we obtain the relations in the third column in Table 14.1. From these, we obtain the generalization of the decrease function in λ, $S_N(\lambda)$, as the Legendre transform of $R_N(h)$ in the conjugate variables h and λ.

14.5.1 Helmholtz and Gibbs Free Energies

What is the physical meaning of the Legendre transform of $R_N(h)$ with respect to h? The large deviations function, cf. Eq. (14.88),

$$S_N(\lambda) = -h\lambda N + R_N(h),$$

depends on the mean,

$$\lambda = \frac{1}{N} \sum_{i=1}^{N} \langle x_i \rangle_h \,,$$

rather than on the external parameter h. To cite an example known to us, in the case of the Ising model, see Section 11.3.4,

$$\mathcal{H}_h[\sigma] = \mathcal{H}[\sigma] - \frac{h}{\beta} \sum_i \sigma_i \,,$$

where h/β represents an external magnetic field,[7] $\langle \sigma_i \rangle_h = m_i$ is the average magnetization of the spin i, and

$$\frac{dR_N}{dh} = \lambda N = \sum_i m_i = M$$

is the total average magnetization $M = Nm$. The large deviations decrease function $S_N(m)$ is, therefore, proportional to a thermodynamic potential that depends on the mean magnetization, rather than on the external field conjugated to it:

$$S_N(m) = -hM + R_N(h) \quad \Longrightarrow \quad \mathcal{S}(m) = -hm + \mathcal{R}(h) = -hm - \beta f(h) + \beta f(0),$$

$$(14.90)$$

with

$$m = \frac{d\beta f}{dh} = \frac{d\mathcal{R}}{dh} \,.$$

Here we have defined

$$\mathcal{S} \equiv \lim_{N \to \infty} \frac{S_N}{N}$$

and used the relation (14.87) to express the generating function of the cumulants as a function of the Helmholtz free energy of the field h (we have omitted to write the dependence of β, not involved in the transform). From (14.90) we have that the Legendre transform of $\beta f(h)$ with respect to the magnetic field is

$$\beta f(0) - \mathcal{S}(m) = hm + \beta f(h).$$

If we divide by β and appropriately rescale the magnetic field, $h/\beta \to h$, we have that the Legendre transform of the Helmholtz free energy density $f(h)$, a function of the field, is the Gibbs free energy, a function of the average magnetization:

$$g(m) = mh + f(h).$$

One can generalize the relationship between free energies to heterogeneous magnetic fields, with values depending on the spin variable on which they act, obtaining the multivariate transform

---

[7] A warning for experts: in the statistical mechanics literature, the magnetic field is usually denoted by h. We will redefine it accordingly in a few lines, in case you might be annoyed.

$$g(\mathbf{m}) = \sum_{i=1}^{N} m_i h_i + f(\mathbf{h}) \,. \tag{14.91}$$

A very interesting case is where $P[\sigma]$ is approximated by a factorizable distribution. It is called the *mean-field approximation*: each variable feels the interaction with the other variables only through a mean field $\tilde{h}_i$, so that the Hamiltonian is approximated by

$$\mathcal{H}[\sigma] = -\sum_{i=1}^{N} \sigma_i \tilde{h}_i \,.$$

Adding the perturbation gives

$$\mathcal{H}[\sigma, h] = \mathcal{H}[\sigma] - \sum_{i=1}^{N} h_i \sigma_i = -\sum_{i=1}^{N} \sigma_i (\tilde{h}_i + h_i) \,, \tag{14.92}$$

and the equilibrium distribution is factorized.

14.5.2 Entropy and Large Deviations Function

So far we have looked at the conjugate variables A_N and h as they appear in (14.83)–(14.87), at the distribution of large deviations $d_N(\lambda) \sim e^{S_N(\lambda)}$ of $A_N = \lambda N$, and at the generator $R_N(h)$ of the cumulants of A_N. We can also consider the conjugate variables $\mathcal{H}$ and β as they appear in the partition function (14.58). The generator of the cumulants of $\mathcal{H}$ is, therefore, $-\ln Z(\beta) = \beta F(\beta)$, and the function S_N of the large deviations of $\mathcal{H}$ is the Legendre transform of $\ln Z(\beta)$ with respect to β, i.e., the entropy

$$S(E) = \beta E + \ln Z(\beta) = \beta(E - F(\beta)) \,, \tag{14.93}$$

with

$$E = \langle \mathcal{H}[\mathbf{x}] \rangle$$

and $\beta = \bar{\beta}(E)$ determined by the saddle point equation of the large deviations, which is now written

$$\frac{\ln Z(\beta)}{d\beta} = \frac{dF(\beta)}{d\beta} = -E \,. \tag{14.94}$$

Let us add an observation for the more interested reader. Entropy, if the system is well defined, is always non-negative, by Nernst's principle. The decrease function, on the other hand, is by definition negative, otherwise the distribution of large deviations would explode as N increases, instead of being very small. The difference lies in the proportionality factors that do not depend on the variables and are, hence, part of the normalization of the distribution $d_N(\lambda)$.

Let us give an explicit example. In the case of the decrease function of the large deviations of the sum of Rademacher variables (or Ising spins) $x_i = \pm 1$ in Section 4.3.2, we found the formula (4.28),

$$S_{\mathrm{ld}}(\lambda) = -\frac{1+\lambda}{2}\ln(1+\lambda) - \frac{1-\lambda}{2}\ln(1-\lambda),$$

with $\lambda \in [-1, 1]$, which is always negative and acquires its minimum value when all $x_i = 1$ (or all $x_i = -1$): $S_{\mathrm{ld}}(\pm 1) = -\ln 2$.

In Chapter 11, Section 11.3.4, we studied the Ising model, a model of statistical mechanics whose variables, the spins, take on values $\sigma_i = \pm 1$. In this case, $\lambda = \sum_i \sigma_i/N = M/N = m$ is the average magnetization per spin. Although we will not show it here, in the mean-field approximation (14.92) it is possible to compute the entropy of the Ising model as a function of the average magnetization per spin, $m \in [-1, 1]$. The result is [94]

$$S(m) = -\frac{1+m}{2}\ln\frac{1+m}{2} - \frac{1-m}{2}\ln\frac{1-m}{2} = \ln 2 + S_{\mathrm{ld}}(m), \tag{14.95}$$

which is always non-negative and only cancels in the two configurations of perfect spin alignment.

14.6 Configurational Entropy of Glassy Systems

A glass is an amorphous solid. This means that it does not have a shape predetermined by the microscopic ordering of its atomic components, as occurs, for example, in crystals when the temperature is lowered. A glass takes the shape of the mold in which it is cooled or that the master glassmaker induces by, for example, blowing into a blowpipe.

If we use a microscope to look at an image of a configuration of the molecules of a liquid able to vitrify (with a suitable procedure), and at an image of a configuration of the same substance already vitrified (at a lower temperature), there is nothing to help us discriminate between the two cases by eye. The situation is different if we watch videos. In the liquid, the molecules rapidly diffuse, whereas in the glass, they oscillate a little around a central position, but do not move much. In the latter case, if the video is long enough and we are very lucky, at most we can see the creation of enough space between the various molecules to allow two of them to swap places.

The transition from liquid to glass is not an equilibrium transition, like the transition from a paramagnet to a ferromagnet, e.g., in the Ising model, or like the transition from water to ice or, in general, from liquid to crystal. The glass transition is dynamic in nature and occurs because the system remains out of equilibrium.

Let us understand this statement better. In a glass-forming liquid, there are faster processes and slower processes. For example, near the vitrification temperature, the thermal fluctuations of molecules are faster than the diffusion of molecules or the rotation of large or particularly long molecules, such as polymers. Each of these processes will be characterized by its own time scale. For fast processes, the time scale is short; for slow processes, it is long. More or less all processes slow down on lowering the temperature, but some, those we call *slow*, slow down much more, and others, termed *fast*, much less. By cooling a glass-former, at a certain temperature the slow processes become so slow that their time scale exceeds the observation time scale of the experiment. For experiments, we can consider observation times of hours, days, or even years.

Below the glass transition temperature, we see a solid, even though no new bonds have been created between the molecules and no new optimal ordering of the molecules' positions has been enucleated corresponding to a global minimum of the system's energy. The true global stable minimum usually remains the liquid state, but it takes too long to reach it, longer than the experimental observation time, and the system settles into a *metastable* state. That is, a state that is stable, i.e., corresponding to a minimum of some thermodynamic functional, but at an excited level, higher than the lowest minimum.

The fact is that there are very many glassy metastable states exhibiting the same macroscopic physical properties at given conditions of temperature, pressure, or other parameters. There are "exponentially" many, in the sense that their number grows as

$$\mathcal{N} \propto e^{N s_c}$$

with the number N of degrees of freedom of the molecules. Here, by s_c, we denote the entropy density relative to all the possible glassy states in which a liquid may in principle vitrify (e.g., by suddenly lowering the temperature). When the liquid vitrifies in one of these states, all the others become inaccessible, except on very, very long time scales. If one waits a long time, the system may exit one glassy state and re-accommodate itself into another one (degenerate, physically equivalent), but it will not be able to explore the whole set of metastable states and find, at the bottom of the exponential sea of all glassy states, the liquid state represented by the global minimum.

Let us try to clarify the concept of separating the time scales of slow and fast processes in a glass and glass-forming liquids a little more formally and tie it to a static representation of a thermodynamic potential that has many local minima and in which we can define the entropy s_c of alternative histories.

We take the set of degrees of freedom $\mathbf{x}$ that characterizes a configuration of the system and give ourselves license to classify these degrees of freedom into fast and slow: $\mathbf{x} = \{\mathbf{x}_f, \mathbf{x}_s\}$. Configurations are configurations, sets of variable values, but we now group them according to the ease with which they can update their values, based on their dynamic properties. From the point of view of fast variables, the slow ones are stationary, locked, *quenched* to use the metallurgical jargon. Imposing the separation of time scales, we can think of writing the probability distribution of configurations $\mathbf{x}$ as a conditional probability:

$$P(\mathbf{x}) = P(\mathbf{x}_f | \mathbf{x}_s) P_s(\mathbf{x}_s).$$

In particular, conjecturing for the distribution (at least for the dependence on the fast variables) the form of the Boltzmann–Gibbs distribution, we can write

$$P(\mathbf{x}) = P(\mathbf{x}_f | \mathbf{x}_s) P_s(\mathbf{x}_s) \propto e^{-\beta \mathcal{H}[\mathbf{x}_f | \mathbf{x}_s]}.$$

Consequently, we can define the free energy density of the subset of fast variables as

$$f_f(\beta | \mathbf{x}_s) = \lim_{N \to \infty} -\frac{1}{\beta N} \ln Z[\beta | \mathbf{x}_s] = \lim_{N \to \infty} -\frac{1}{\beta N} \int \mathcal{D}\mathbf{x}_f \, e^{-\beta \mathcal{H}[\mathbf{x}_f | \mathbf{x}_s]}. \quad (14.96)$$

This formalization is legitimate because even in the worst case (that of the solid glass) at least the fast variables are at equilibrium, because they evolve on time scales much shorter than the observation times.

From the point of view of the slow variables, which remain out of equilibrium when the system vitrifies, the fast ones are noise, to be marginalized. From this point of view, Eq. (14.96) is nothing other than an effective Hamiltonian of the slow variables:

$$\mathcal{H}_{\text{eff}}[\mathbf{x}_s] = N f_f(\beta | \mathbf{x}_s).$$

This last formalization must be considered with several grains of salt, since, below the glass transition temperature, slow processes – by definition – are not at equilibrium with the thermal bath. The description is more sophisticated than that, and requires the definition of a good order parameter for the glass transition: the similarity between states. This is not the place to go into this in-depth, but interested readers may consult, for example, ref. [95].

The glassy metastable states, mentioned earlier, can be represented as local excited minima of the thermodynamic potential $f_f(\beta | \mathbf{x}_s)$ in the landscape of the configurations of the slow variables $\mathbf{x}_s$. Each minimum (corresponding to a given $\mathbf{x}_s$) has a free energy $f_f(\beta | \mathbf{x}_s) = e(\beta | \mathbf{x}_s) - T s(\beta | \mathbf{x}_s)$, where e is the energy per degree of freedom and s is the entropy per degree of freedom of the single glassy state. What happens, furthermore, is that the *typical* glassy states all end up being energetically and entropically degenerate: $f_f(\beta) = e(\beta) - T s(\beta)$. Each metastable state brings an (approximately equal) entropic contribution to s. The glassy minima are many, we can denote their number by $\mathcal{N}$, and all together yield a contribution to entropy equal to

$$s_c \equiv \lim_{N \to \infty} \frac{1}{N} \log \mathcal{N}.$$

In a rather viscous liquid, not too much above the glass temperature, we can imagine having a sufficient separation of time scales to maintain this description. Both slow and fast processes will be at equilibrium. In this case, the total entropy of the system is given by the sum of the contribution of the configurations of the single state s and the contribution of all metastable states s_c (which, for the record, glass investigators call *configurational* entropy):

$$s_{\text{tot}} = s + s_c.$$

When dropping below the glass temperature, at which the liquid remains locked out of equilibrium, in this description only one metastable state is selected: the s_c contribution vanishes. This can be seen experimentally from measurements of the specific heat as a function of temperature, as explained, for example, in ref. [96], but here we will not go into this any further. So, below the glass transition temperature,

$$s_{\text{tot}} = s.$$

Hence, s_c has also been called the entropy of hidden states, as well as configurational entropy, or glass complexity function.

Mathematical Appendices

14.A Derivation of Shannon Information and Entropy

We devote this appendix to some technical but very didactically instructive passages to understand the definition of Shannon's information function and entropy function by means of their direct derivation.

14.A.1 Construction of the Information Function

We want to define a quantity $I(P_n)$ that satisfies some reasonable requirements to be called *information*. It needs to be:

1. non-negative, $I(P_n) \geq 0$;
2. a decreasing function of the probability P_n of the event n;
3. null if the event is certain, $I(1) = 0$; and
4. additive if it is the information of two independent events, $I(P_1 P_2) = I(P_1) + I(P_2)$.

We expect I to become smaller and smaller as P_n increases because, if a very likely event occurs, we do not get much information about the process. In the limit in which n is a certain event, of probability 1, then, from its observation, we do not obtain any information; the process we are observing is not even stochastic. If the event is unlikely, instead, its occurrence provides more information about the phenomenon to which it pertains. The function

$$I(P_n) \equiv \log \frac{1}{P_n} \tag{14.97}$$

decreases as P_n increases. We can show that this is, indeed, the right choice that satisfies all the requirements listed above [84, 85]. One way of looking at it is to take two independent events i and j, whereby the probability of both i and j occurring is $u = P_i P_j$. Let us assume that the information of this joint event, $I(u)$, satisfies the additivity requirement 4:

$$I(P_i P_j) = I(P_i) + I(P_j).$$

If we differentiate first with respect to P_i,

$$P_j I'(P_i P_j) = I'(P_i),$$

and then with respect to P_j (or vice versa), we obtain the equation

$$I'(u) + u I''(u) = \frac{d[u I'(u)]}{du} = 0, \quad \text{with } u \equiv P_i P_j ,$$

from which

$$u \frac{dI}{du} = -K \quad \Longrightarrow \quad \int_{I(u)}^{I(1)} dI = -K \int_u^1 \frac{du'}{u'} ,$$

and, using requirement 3, $I(1) = 0$,

$$I(u) = -K \ln u \,,$$

where for requirements 1 and 2 one has $K > 0$. The formula (14.97) results in the correct definition of information, with a positive multiplicative constant.

14.A.2 Proof of Shannon's Entropy Formula

How does the formula (14.1) for entropy come about? Once we know the distribution $\{P\}$ of N events, we ask ourselves how we can measure the uncertainty of the outcome. In other words, we ask ourselves how much choice there is in the selection of the event that will occur. We try to construct a *choice measure*, also called uncertainty in information theory, which also corresponds to the rate of information production. A lot of information means a lot of choice, or even a lot of uncertainty about the final outcome, depending on the point of view. This measure of choice should reasonably be

(A) a continuous function of P_n,
(B) increasing with the number of events N in the case of a homogeneous distribution $P_n = 1/N, n = 1, \dots, N$, and
(C) additive in the choices, so that if a given choice is made up of two subsets of choices, the size of the first must be equal to the sum of the sizes of the second.

We now show that the choice measure that satisfies these properties is precisely the Shannon entropy (14.1).

Proof of Shannon's entropy formula.   Let us first start with a uniform distribution of events. Let us consider the entropy S, a measure of the possible choices of N equiprobable events:

$$S\left(\frac{1}{N}, \dots, \frac{1}{N}\right) \equiv f(N) \,. \tag{14.98}$$

By the condition (B), i.e., that the choice measure of a uniform distribution grows with the total number of possible events, $f(N)$ will be monotonically increasing in N.

Consider a set of V blocks of U equiprobable events, each one with uncertainty measure equal to $f(U)$. In total, they form a set of U^V equiprobable events. We want the measure to be additive in the choices (condition (C)), so

$$f(U^V) = V f(U) \,.$$

The same applies when changing the distribution of events into N blocks of M elements, so we can always write

$$f(M^N) = N f(M) \,.$$

By choosing N arbitrarily large, we can always find a V such that

$$U^V \le M^N \le U^{V+1} \quad \Longrightarrow \quad V \log U \le N \log M \le (V+1) \log U$$
$$\Longrightarrow \quad \frac{V}{N} \le \frac{\log M}{\log U} \le \frac{V}{N} + \frac{1}{N} \, .$$

By calling $\epsilon = 1/N$ (arbitrarily small), the expression above can be rewritten as

$$\left| \frac{V}{N} - \frac{\log M}{\log U} \right| < \epsilon \, . \tag{14.99}$$

Furthermore, from condition (B) of increasing monotonicity of f, we have $f(U^V) \le f(M^N) \le f(U^{V+1})$, i.e., $Vf(U) \le Nf(M) \le (V+1)f(U)$, from which we arrive at

$$\left| \frac{V}{N} - \frac{f(M)}{f(U)} \right| < \epsilon \, , \tag{14.100}$$

where again $\epsilon = 1/N$. Putting together (14.99) and (14.100) we have

$$\left| \frac{\log M}{\log U} - \frac{f(M)}{f(U)} \right| < 2\epsilon \, ,$$

from which it follows that $f(M) = K \log M$, with $K > 0$, which as seen in (14.6) is the entropy of M equiprobable events.

We now turn to a non-uniform distribution, constructed by grouping $\mathcal{N}$ equiprobable events into N subsets of variable size n_k, whose probability is defined as

$$P_k = \frac{n_k}{\mathcal{N}} , \quad \text{with } \mathcal{N} = \sum_{k=1}^{N} n_k \, . \tag{14.101}$$

Here the n_k are integers and the probabilities are, therefore, rational numbers. The measure of the total choice among all $\sum_k n_k$ possibilities is

$$K \log \sum_k n_k \, .$$

We can decompose the total choice into a sequence of two choices. First, the choice of groups of n_k events, each one with its probability P_k, and then, within each group, the choice with uniform probability among the n_k elements. The choice measure of groups distributed with (14.101) is precisely the entropy of the distribution (14.101), $S(\{P\})$; while the choice measure of n_k equiprobable events is $K \log n_k$, for each group labeled by k. Each group k, then, has a probability P_k of being drawn and the choice measure of all events within all groups is, therefore, equal to

$$K \sum_k P_k \log n_k$$

for this decomposition.

Putting all this together, the choice measure between all events $\mathcal{N} = \sum_k n_k$ is equal to the choice measure of the just defined decomposition in groups, i.e.,

$$K \log \sum_{k=1}^{N} n_k = S[P] + K \sum_{k=1}^{N} P_k \log n_k \, .$$

From this equality we have

$$S[P] = K \sum_{k=1}^{N} P_k \log \sum_{k=1}^{N} n_k - K \sum_{k=1}^{N} P_k \log n_k = -K \sum_{k=1}^{N} P_k \log P_k \,,$$

completing the proof. ∎

If the logarithm is the natural logarithm ln, then the positive constant K is the Boltzmann constant k_B in thermodynamics, or $1/\ln 2$ in information theory (number of bits $S_2[P] = S[P]/\ln 2$).

14.B Self-Delimiting Messages

The problem of message delimitation arises if we want to send a series of separate messages. If we imagine sending a stop signal to delimit the messages, we have a serious problem. In fact, at any instant, we can transmit

$$0, \quad 1, \quad \text{STOP}.$$

In this way we transmit $\log_2 (3)$ bits of information and not a single bit ($\log_2 (2) = 1$) as we assumed, so entropy increases.

The problem can be alleviated by introducing an alphabet of 2^K characters (for $K = 8$ we have the byte, for example). One of the 2^K characters is not used for transmission and is identified with the stop signal.[8] In this case, with each K bit sent, we transmit

$$\log_2(2^K - 1) \approx K - \frac{2^{-K}}{\ln 2} \tag{14.102}$$

bits of information. This is less than the information K we would have without STOP but for high K the inclusion of the stop signal in the language lengthens the message by very little.

A more elegant solution is to use self-delimiting messages, i.e., such that a stop signal is not needed to determine their end. The simplest (not optimal) method is to add a *header* to the message. In this header, we write in binary the number of bits in the message. At this point, we need to know when the header ends. The simplest solution is to send (i) as many zeros at the beginning of the message as there are binary digits for the length of the message, then (ii) the length of the message, and lastly (iii) the message itself.

Let us take a concrete example. Suppose that the message to be sent is

$$10101110001110100,$$

i.e., a message of 17 digits. The self-delimiting message that is sent is

$$00000 \ 10001 \ 10101110001110100,$$

---

[8] Note that nature has chosen this mechanism for the encoding of proteins in DNA. In this case, $K = 6$ if we consider the codon (three triplets of each of the four possible bases) to be a character.

where spaces are not transmitted but help the reader to separate the various parts of the message.

When we receive this message, if we count the number of zeros at the beginning, we realize that the length of the message is a number of 5 digits (the length of the message is a number that necessarily begins with 1.[9] Consequently the number of characters in the message is 10001 (17 in binary). The next 17 digits constitute the actual message.

In this way, instead of transmitting, for example, three messages of 17, 15, and 6 bits as

$$10101110001110100\text{STOP}101011100011101\text{STOP}10100,$$

we can transmit them without using a special symbol to separate the messages as

$$00000\ 10001\ 10101110001110100\quad 0000\ 1111\ 101011100011101\quad 000\ 101\ 10100,$$

where again the spaces are not transmitted but have the function of making them easier to read. Obviously, we still need to know when transmission begins, otherwise we are in trouble!

In general, if a message is originally N bits, its self-delimiting version that transmits the same information is $N + 2\log_2 N$ long. For large N the increase in bits required for communication is practically negligible. The method described is very simple and easy to remember. Using a more complicated method, the length can be reduced to approximately $N + \log_2 N$ (precisely to $N + \log_2 N + O(\log_2 \log_2 N)$) and one can prove that such a method is optimal. If, however, we miss a bit during transmission, the whole subsequent message remains corrupted. We save bits compared to an explicit STOP, but we increase the fragility of the communication.

---

[9] We disregard the trivial case of messages of zero characters.

References

[1] R. von Mises. *Mathematical Theory of Probability and Statistics*. Academic Press, 1964.

[2] P. Diaconis and B. Skyrms. *Ten Great Ideas About Chance* (Classics in Applied Mathematics). Princeton University Press, 2018.

[3] J. Bertrand. *Calcul des Probabilités*. Gauthier-Villars, 1889.

[4] J. M. Keynes. *A Treatise on Probability*. Cambridge University Press, 1921.

[5] È. Borel. *Traité du Calcul des Probabilités et de ses Applications*. Gauthier-Villars, 1934.

[6] B. de Finetti. Sul significato soggettivo della probabilità. *Fundamenta Mathematicae*, **17**, 298–329, 1931.

[7] D. H. Mellor. Cambridge philosophers I: F. P. Ramsey. *Philosophy*, **70**, 243–262, 1995.

[8] M. C. Galavotti. Pragmatism and the birth of subjective probability. *European Journal of Pragmatism and American Philosophy*, **XI-I**, 1509, 2019.

[9] D. Hilbert. Mathematical problems. *Bulletin of the American Mathematical Society*, **8**, 437–479, 1902.

[10] A. N. Kolmogorov. *Foundations of the Theory of Probability*. Chelsea, 1956.

[11] R. Von Mises. Grundlagen der Wahrscheinlichkeitsrechnung. *Mathematische Zeitschrift*, **5**, 52–99, 1919.

[12] M. van Lambalgen. Randomness and foundations of probability: von Mises' axiomatisation of random sequences. In T. Ferguson *et al.* (eds.), *Probability, Statistics and Game Theory, Papers in Honor of David Blackwell*. Institute for Mathematical Statistics, 1996.

[13] P. Billingsley. *Probability and Measure*. Wiley, 1995.

[14] P. R. Halmos. *Measure Theory*. Springer, 1974.

[15] G. B. Folland. *Real Analysis: Modern Techniques and Their Applications*. Wiley, 1984.

[16] W. Rudin. *Real and Complex Analysis*. McGraw-Hill, 1987.

[17] P. R. Halmos. *Naive Set Theory*. Van Nostrand, 1960.

[18] R. M. Solovay. A model of set-theory in which every set of reals is Lebesgue measurable. *Annals of Mathematics, Second Series*, **92**, 1–56, 1970.

[19] G. Vitali. *Opere sull'analisi reale e complessa. Carteggio* [Works on real and complex analysis. Correspondence]. Edizioni Cremonese, 1984.

[20] R. von Mises. Über Aufteilungs- und Besetzungswahrscheinlichkeiten. In P. Frank *et al.* (eds.), *Selected Papers of Richard von Mises*, vol. 2, pp. 313–334. American Mathematical Society, 1964.

[21] R. A. Fisher. Moments and product moments of sampling distributions. *Proceedings of the London Mathematical Society*, **30**, 199–238, 1929.

[22] S. J. Gould. *Bully for Brontosaurus: Reflections in Natural History*. W. W. Norton, 2008.

[23] E. Artin. *The Gamma Function*. Dover, 2015.

[24] C. M. Bender and S. A. Orszag. *Advanced Mathematical Methods for Scientists and Engineers*. McGraw-Hill, 1978.

[25] K. Knopp. *Theory and Application of Infinite Series*. Hafner, 1951.

[26] R. B. Dingle. *Asymptotic Expansions: Their Derivation and Interpretation*. Academic Press, 1973.

[27] R. B. Paris and D. Kaminski. *Asymptotics and Mellin–Barnes Integrals*. Cambridge University Press, 2001.

[28] W. Feller. *An Introduction to Probability Theory and Its Applications*. Wiley, 1970.

[29] P. Billingsley. *Ergodic Theory and Information*. Wiley, 1965.

[30] C. W. Gardiner. *Handbook of Stochastic Methods*. Springer, 1997.

[31] P. Lévy. Theorie des erreurs. La loi de Gauss et les lois exceptionelles. *Bulletin de la Société Mathématique de France*, **52**, 49–85, 1924.

[32] B. V. Gnedenko and A. N. Kolmogorov. *Limit Distributions for Sums of Independent Random Variables*. Addison-Wesley, 1954.

[33] M. Kanter. Stable densities under change of scale and total variation inequalities. *The Annals of Probability*, **3**, 697–707, 1975.

[34] F. W. J. Olver. Bessel functions of integer order. In M. Abramowitz and I.A. Stegun (eds.), *Handbook of Mathematical Functions with Formulas, Graphs and Mathematical Tables*. National Bureau of Standards, 1972.

[35] I. S. Gradshteyn and I. M. Ryzhik. *Table of Integrals, Series, and Products*. Academic Press, 1980.

[36] M. Reed and B. Simon. *Functional Analysis*. Academic Press, 1972.

[37] M. Loève. *Probability Theory*, vols. I and II. Springer, 1977–78.

[38] K. G. Wilson and J. Kogut. The renormalization group and the ϵ expansion. *Physics Reports*, **12**, 75–199, 1974.

[39] D. J. Amit and V. Martin-Mayor. *Field Theory, the Renormalization Group, and Critical Phenomena: Graphs to Computers*, 3rd edn. World Scientific, 2005.

[40] G. Benfatto and G. Gallavotti. *Renormalization Group*. Princeton University Press, 1995.

[41] D. E. Knuth. *The Art of Computer Programming*, vol. 2, *Seminumerical Algorithms*, 3rd edn. Addison-Wesley, 1997.

[42] W. H. Press, S. A. Teukolsky, W. A. Vetterling, and B. P. Flannery. *Numerical Recipes: The Art of Scientific Computing*. Cambridge University Press, 2007.

[43] SFB Project (formerly The pLab Project). Quasi-Monte Carlo Methods: Theory and Applications. Online at www.sfb-qmc.jku.at/.

[44] S. Geisser. On prior distributions for binary trials. *The American Statistician*, **38**, 244–247, 1984.

[45] J. M. Bernardo and A. F. M. Smith. *Bayesian Theory*. Wiley, 1994.

[46] J. B. S. Haldane. The precision of observed values of small frequencies. *Biometrika*, **35**, 297–300, 1948.

[47] H. Jeffreys. *Theory of Probability*, 2nd edn. Clarendon Press, 1948.

[48] W. Bialek, C. G. Callan, and S. P. Strong. Field theories for learning probability distributions. *Physical Review Letters*, **77**, 4693–4697, 1996.

[49] R. G. Miller. The jackknife – a review. *Biometrika*, **61**, 1–15, 1964.

[50] B. Efron. Bootstrap methods: Another look at the jackknife. *The Annals of Statistics*, **7**, 1–26, 1979.

[51] B. Efron. Computers and the theory of statistics: Thinking the unthinkable. *SIAM Review*, **21**, 460–480, 1979.

[52] H. Flyvbjerg. Error estimates on averages of correlated data. In J. Kertesz and I. Kondor (eds.), *Advances in Computer Simulation* (Lecture Notes in Physics, 501), pp. 88–103. Springer, 1998.

[53] P. Young. *Everything You Wanted to Know About Data Analysis and Fitting but Were Afraid to Ask*. Springer, 2015.

[54] V. Ambegaokar and M. Troyer. Estimating errors reliably in Monte Carlo simulations of the Ehrenfest model. *American Journal of Physics*, **78**, 150–157, 2010.

[55] R. Brown. A brief account of microscopical observations made in the months of June, July, and August, 1827, on the particles contained in the pollen of plants; and on the general existence of active molecules in organic and inorganic bodies. *The Edinburgh New Philosophical Journal*, **5**, 358–371, 1828.

[56] H. Risken. *The Fokker–Planck Equation: Methods of Solution and Applications*. Springer, 1996.

[57] K. Huang. *Statistical Mechanics*. Wiley, 1987.

[58] J. J. Sakurai, J. Napolitano, and S. Forte. *Meccanica Quantistica Moderna*. Zanichelli, 2014.

[59] R Kubo. The fluctuation–dissipation theorem. *Reports on Progress in Physics*, **29**, 255–284, 1966.

[60] N. G. van Kampen. *Stochastic Processes in Physics and Chemistry*. Elsevier, 1999.

[61] N. W. Ashcroft and N. D. Mermin. *Solid State Physics*. Saunders College, 1976.

[62] G. H. Hardy and J. E. Littlewood. Tauberian theorems concerning power series and Dirichlet's series whose coefficients are positive. *Proceedings of the London Mathematical Society*, **2**, 174–191, 1914.

[63] E. C. Titchmarsh. *The Theory of Functions*. Oxford University Press, 1939.

[64] H. Minc. *Nonnegative Matrices*. Wiley, 1988.

[65] A. Berman and R. J. Plemmons. *Nonnegative Matrices in the Mathematical Sciences* (Classics in Applied Mathematics, vol. 9). SIAM, 1994.

[66] R. B. Bapat and T. E. S. Raghavan. *Nonnegative Matrices and Applications* (Encyclopedia of Mathematics and Its Applications, vol. 64). Cambridge University Press, 1997.

[67] C. R. MacCluer. The many proofs and applications of Perron's theorem. *SIAM Review*, **42**, 487–498, 2000.

[68] M. R. F. Smyth. A spectral theoretic proof of Perron–Frobenius. *Mathematical Proceedings of the Royal Irish Academy*, **102A**, 29–35, 2002.

[69] O. Perron. Zur Theorie der Matrizen. *Mathematische Annalen*, **64**, 248–263, 1907.

[70] G. Frobenius. Über Matrizen aus nicht negativen Elementen. *Sitzung der physikalisch-mathematischen Classe, Preussischen Akademie der Wissenschaften*, **26**, 456–477, 1912.

[71] N. G. van Kampen. Remarks on non-Markov processes. *Brazilian Journal of Physics*, **28**, 90–96, 1972.

[72] E. Giné, G. R. Grimmett, and L. Saloff-Coste. *Lectures on Probability Theory and Statistics* (Lecture Notes in Mathematics, vol. 1665). Springer, 1997.

[73] A. Einstein. Kinetische Theorie des Wärmegleichgewichtes und des zweiten Hauptsatzes der Thermodynamik. *Annalen der Physik*, **9**, 417–433, 1902.

[74] A. Einstein. Eine Theorie der Grundlagen der Thermodynamik. *Annalen der Physik*, **11**, 170–197, 1903.

[75] N. Metropolis, A. W. Rosenbluth, M. N. Rosenbluth, A. H. Teller, and E. Teller. Equation of state calculations by fast computing machines. *Journal of Chemical Physics*, **21**, 1087–1092, 1953.

[76] L. M. Barone, E. Marinari, G. Organtini, and F. Ricci-Tersenghi. *Programmazione Scientifica. Linguaggio C, Algoritmi e Modelli nella Scienza*. Pearson, 2006.

[77] A. Crisanti. *YaC-Primer: Yet Another C-Primer*, vol. 1. Aracne, 2006.

[78] K. Yang, Y.-F. Chen, G. Roumpos, C. Colby, and J. Anderson. High performance Monte Carlo simulation of Ising model on TPU clusters. In *Proceedings of the International Conference for High Performance Computing, Networking, Storage and Analysis*, art. no. 83. Association for Computing Machinery, 2019.

[79] V. I. Manousiouthakis and M. D. Deem. Strict detailed balance is unnecessary in Monte Carlo simulation. *Journal of Chemical Physics*, **110**, 2753–2756, 1999.

[80] T. Kato. *Perturbation Theory for Linear Operators*. Springer, 1976.

[81] G. Udny Yule. A mathematical theory of evolution, based on the conclusions of Dr. J. C. Willis, F.R.S. *Philosophical Transactions of the Royal Society, London, Series B*, **213**, 21–87, 1924.

[82] S. Carnot. *Réflexions sur la Puissance Motrice du Feu et sur les Machines Propres à Développer cette Puissance*. Bachelier, 1824.

[83] R. Clausius. Über die bewegende Kraft der Wärme und die Gesetze, welche sich daraus für die Wärmelehre selbst ableiten lassen. *Annalen der Physik*, **79**, 368–397, 500–524, 1850.

[84] C. E. Shannon. A mathematical theory of communication. *The Bell System Technical Journal*, **27**, 379–423, 623–656, 1948.

[85] E. T. Jaynes. Information theory and statistical mechanics. *Physical Review*, **106**, 620–630, 1957.

[86] P. Charbonneau, E. Marinari, M. Mézard, G. Parisi, F. Ricci-Tersenghi, G. Sicuro, and F. Zamponi (eds.). *Spin Glass Theory and Far Beyond – Replica Symmetry Breaking after 40 Years*. World Scientific, 2023.

[87] G. Boffetta and A. Vulpiani. *Probabilità in Fisica, un'Introduzione*. Springer, 2012.

[88] D. J. C. MacKay. *Information Theory, Inference and Learning Algorithms*. Cambridge University Press, 2003.

[89] T. M. Cover. *Elements of Information Theory*. Wiley, 2006.

[90] A. N. Kolmogorov. Entropy per unit time as a metric invariant of automorphism. *Doklady of Russian Academy of Sciences*, **124**, 754–755, 1959.

[91] Y. G. Sinai. On the notion of entropy of a dynamical system. *Doklady of Russian Academy of Sciences*, **124**, 768–771, 1959.

[92] H. Goldstein. *Classical Mechanics*. Addison-Wesley, 1980.

[93] U. Marini Bettolo Marconi, A. Puglisi, L. Rondoni, and A. Vulpiani. Fluctuation–dissipation: Response theory in statistical physics. *Physics Reports*, **461**, 111–195, 2008.

[94] G. Parisi. *Statistical Field Theory*. Addison-Wesley, 1998.

[95] G. Parisi, P. Urbani, and F. Zamponi. *Theory of Simple Liquids*. Cambridge University Press, 2020.

[96] L. Leuzzi and T. A. Nieuwenhuizen. *Thermodynamics of the Glassy State*. Taylor & Francis, 2008.